# A Guide to the Automation Body of Knowledge

## 2nd Edition

# A Guide to the Automation Body of Knowledge

## 2nd Edition

Vernon L. Trevathan, Editor

**Notice**

The information presented in this publication is for the general education of the reader. Because neither the author nor the publisher have any control over the use of the information by the reader, both the author and the publisher disclaim any and all liability of any kind arising out of such use. The reader is expected to exercise sound professional judgment in using any of the information presented in a particular application.

Additionally, neither the author nor the publisher have investigated or considered the affect of any patents on the ability of the reader to use any of the information in a particular application. The reader is responsible for reviewing any possible patents that may affect any particular use of the information presented.

Any references to commercial products in the work are cited as examples only. Neither the author nor the publisher endorse any referenced commercial product. Any trademarks or tradenames referenced belong to the respective owner of the mark or name. Neither the author nor the publisher make any representation regarding the availability of any referenced commercial product at any time. The manufacturer's instructions on use of any commercial product must be followed at all times, even if in conflict with the information in this publication.


67 Alexander Drive
P.O. Box 12277
Research Triangle Park, NC 27709



Printed in the United States of America.
10 9 8 7 6 5 4 3 2

ISBN 1-55617-984-7
ISBN 978-1-55617-984-6



**Library of Congress Cataloging-in-Publication Data**
A Guide to the automation body of knowledge / Vernon L. Trevathan, editor. – 2nd ed.
p. cm.
ISBN 1-55617-984-7 (hardcover)
1. Automation–Handbooks, manuals, etc. I. Trevathan, Vernon L.
T59.5.G85 2006
629.8–dc22

2006014162

# Table of Contents

# Preface

## Preface to the Second Edition

In the few months of its existence, this "AutoBoK" is already making an impact by better defining the automation profession, both to automation practitioners and to academicians working to improve automation curricula, and by helping automation professionals prepare for the Certified Automation Professional (CAP) exam.

Now, circumstances have allowed us to make revisions and improvements to the book much sooner than we expected. A team working to provide technical expertise to a new self-study program for CAP, "ISA CAP Learning System" — Lee Lane, Nick Sands, and Joe Bingham — has identified several areas of improvement and has worked to make those changes happen. Their efforts have resulted in a much better book.

The topics on Analog Communications, Motion Control, and Electrical Installations have been strengthened to better cover the material in those topics; the topics on Digital Communications, Industrial Networks and Maintenance have had minor changes; and topics have been added on Continuous Emissions Monitoring Systems and on Custom Software to give more weight to those topics.

In addition, an appendix has been added on Control Equipment Structure. This appendix goes into some historical detail since an understanding of that history is useful to an understanding of how automation systems are structured today.

Prior to the 1970s when programmable control hardware for discrete applications began to see widespread use, discrete control was much more difficult and was thus limited in functionality. However, available analog controllers were very capable for continuous control; so most of the more complex control application work and most of the automation professionals worked on process applications.

Today, virtually all new control is performed in programmable devices which do a very good job for both discrete and continuous applications. However, the tremendous opportunities in manufacturing (discrete and motion) applications have caused this area to overtake the continuous applications and a major fraction of automation professionals today work outside the process industries.

Though a variety of control devices today can do a variety of discrete, motion, and process applications, many of the devices are best suited to specific areas. The design of these devices is influenced by the historical expectations of that manufacturing area as well as by the true requirements of the automation task.

April, 2006

## Preface to the First Edition

This book is intended to be read and studied both by automation professionals and by those who want to learn about automation—not just used for looking up facts and figures. However, it contains so much technical information you may also find it useful as a reference.

Because each topic was written by a highly respected expert, or experts, in that subject area, there is much more detailed information than you would expect in a typical overview. Even so, each topic is short enough to be read like a summary.

This book emerged from the work to develop ISA's Certified Automation Professional (CAP) program. However, its value is much broader than just as a helpful text for CAP.

The topics in this book represent THE scope of automation application, which makes it one of the most important books published by ISA. This is a unique book that will go a long way toward defining the scope of automation and helping to establish automation "engineering" as a profession.

The term "automation" includes all topics that have traditionally been identified using names such as *instrumentation, instruments and control, process control, process automation, control systems, automation and control, manufacturing control, manufacturing automation,* and *system integration*. Automation professionals are the practitioners responsible for the direction, design, and deployment of systems and equipment for manufacturing and control systems.

A number of organizations in recent years have developed a document that defines their knowledge base, frequently referred to as the *Body of Knowledge*. Some of those documents are bibliographies of the relevant literature, some are a critique of the literature, and some are an overview of the topics. This book is obviously the latter type. Because this book contains only a small fraction of all the information in automation knowledge, it might be titled "Overview," "Summary," or "Introduction." However, following the lead of the Project Management Institute with their *A Guide to Project Management Body of Knowledge,* which is now an ANSI standard, many organizations use the *Guide to* . . . nomenclature. We also have chosen to follow that usage.

This book is intended to serve as a technical summary of automation knowledge for those who need a comprehensive perspective on automation in their job, including:

- Automation professionals who need to understand the basics of an unfamiliar topic. They might, for example, need to determine if it is useful in the application on which they are currently working. Or, they may have been assigned to use that technology, and they need to begin to learn about it.
- Anyone who knows something about a topic, but needs to gain a better understanding of the range of information in the topic
- Academicians who need guidance in developing and improving curriculum or courses, and who wish to expand their own knowledge and that of their students
- Managers who need a better perspective of all aspects of automation, enabling them to better set direction and make staffing decisions
- Those who work in fields related to automation, and who need a comprehensive understanding of what automation is all about. For example, just as automation professionals need to learn much more about information technology (IT), people in IT working on systems related to manufacturing need to learn more about plant floor control and information systems.
- Students, novices, and others evaluating career decisions
- Those studying for ISA's Certified Automation Professional (CAP) exam

To be useful in all these ways, each of the 35 topics in the book needed to be understandable to those who know very little about that topic while, at the same time, useful to those knowledgeable and experienced in that subject. And, each topic needed to include real technical information—not just a newsy overview. Achieving all that was a real challenge in the short space that could be devoted to each topic.

Some users will find the material fully meets their educational or reference needs on a particular topic; others, who find they need more depth or more background, will find it useful to study some of the listed references. Those studying for the CAP exam, for example, may find that this material meets their needs in topics where they have some familiarity, but in topics where they know very little, they may also need to consult other sources to adequately understand the material.

Also, while this book attempts to cover all the topics in the scope of the CAP exam, many CAP exam questions will not be covered in this book, because CAP questions can be drawn from any book or referred paper in automation. Still, a good knowledge of the material in this book will be a big step towards preparing for the CAP exam.

The organization of the 35 topics in this book is somewhat arbitrary and evolved from more than a year of work trying to capture as many topics as possible in a logical category. The continuous numbering of the topics from 1 to 35 is to indicate the topics themselves are really the most important headings. The seven topic categories are for convenience only.

Deciding what not to include was as big a challenge as deciding what should be covered. Older technologies less used today have been skipped, as well as technologies that are very specialized. Techniques used less frequently—even ones very important in some applications— are often not covered simply because of space limitations. Applications used only by a particular industry are also not included. For example, automation professionals working in chemical and refining applications may feel that distillation column control is so basic that it should have been included. However, those outside those industries—a majority of today's automation professionals—may hardly know, or care, what a distillation column is.

You may feel your area of automation is slighted, or find that some topics you consider important in the scope of automation are not addressed. We welcome hearing from you about topics that you think should be included in future editions. If there are errors or topics that need further clarification, please let us know. Send your comments to info@isa.org.

The idea for this book first came from Ken Baker, who, as a member of the CAP Steering Team, realized the value of putting the entire scope of automation together into one book. Chip Lee, Director of Publications on the ISA staff, was also a strong proponent from the beginning and continued to pursue the idea even though some, including me, initially thought a useful coverage of such a broad scope in one reasonably sized book was not practical.

Lois Ferson on the ISA publications staff undertook the big task of identifying authors for each topic and convincing them to deliver a comprehensive technical document in fewer words than they probably imagined possible. Jim Strothman did the editing for format and style and to overcome our tendency as engineers to use too many unnecessary words. But it is the authors of the 35 topics who are most responsible for this book. Though many of them initially resisted the tough assignment of covering significant technical detail in such a small space, particularly with a tight time schedule, each did an outstanding job addressing their topic. Many of the authors have written books specifically on their topic, and a number of the authors are *the* recognized expert on their topic. About a third of the authors are ISA Fellows.

Vernon Trevathan
September, 2005

# I

# Basic Continuous Control

*Some might call this category "process control" or "instrumentation and control," because the material in these topics is closest to the historical scope of ISA and is consistent with ISA's original name,* Instrument Society of America. *But today, proportional, integral, derivative (PID) and other continuous control techniques are used in many applications outside the process industries. For example, they are used in areas such as automotive paint shop controls, motion controls, electrical equipment controls, building automation, ship stabilization controls, and many, many other areas.*

*PID is so pervasive that one could not be considered well educated as an automation professional if he or she does not know the basic concepts of PID and PID tuning. In fact, the CAP Steering Team decided early on that, while many people today work with manufacturing automation information technology that does not involve basic plant floor control, it did not make sense for them to be called automation professionals without knowing the basics of plant floor control—including continuous control.*

*Measurements are extremely important in any automation task. It is really true you can control well only those things that you can measure—and accuracy and reliability requirements are continuing to increase. Continuous instrumentation is required in many applications throughout automation, although here we call it process instrumentation because the particular type of transmitter packaging discussed is more widely used in process applications.*

*There are so many measurement principles and variations on those principles that this topic can only scratch the surface of all the available ones, but it hopefully covers the more popular types.*

*Analytical Instrumentation and Control Valves are more applicable to process applications, although these also are used in environmental and other applications in a variety of industries. The type of control system documentation discussed here is also more specific to process industries, but many outside processes can learn from the high degree of development of these documentation conventions. The control equipment topic in this category covers what has traditionally been called distributed control systems (DCSs), although that designation no longer has intrinsic meaning.*

*Control valves are critical components of a control loop in process and utility industries. It has often been demonstrated that, in nearly all process plants, control valve problems are a major cause of poor loop performance. A general knowledge of the impact of the control valve on loop performance is critical to process control.*

*It has become commonplace for automation professionals today to delegate the selection and specification of instrumentation and control valves and the tuning of controllers to technicians. However, performance in all of these areas may depend on advanced technical details that require the attention of an automation professional; there are difficult issues, including instrument selection, proper instrument installation, loop performance, advanced transmitter features, and valve dynamic performance. A knowledgeable automation professional could likely go into any process plant in the world and drastically improve the performance of the plant by tuning loops, redesigning the installation of an instrument for improved accuracy, or determining a needed dynamic performance improvement on a control valve—at minimal cost. More automation professionals need that knowledge.*

# 1 Process Instrumentation

*By Vernon L. Trevathan*

## Topic Highlights

*Pressure*
*Level*
*Flow*
*Temperature*
*Smart Instruments*

## 1.1 Introduction

Good control requires measurements that are accurate, reliable, responsive and maintainable. These factors are influenced by the choice of principle used for the measurement, the detailed specifications and features of the instrument selected and specified, how well the instrument and its installation is maintained, and particularly the installation details.

The vast majority of physical measurements in processes are of the big four: flow, pressure, level, and temperature. This topic will focus on the more popular methods for measuring these variables. Analytical measurements are covered in the next topic. While this topic is titled "process" instrumentation because the larger number of applications are in process and utility applications, much of this instrumentation is used in many other areas of automation application wherever continuous measurements are needed.

Much of the focus in this topic is on compact "transmitters"—devices that combine the sensor and communicating electronics in one package. Some temperature measurements are an exception to this, since those devices often separate the sensor and the communicating electronics. These transmitters sometimes take on different names: level transmitters become level gauges and flow transmitters become flowmeters.

### 1.1.1 Measurement Concepts

All continuous measurements share certain parameters of accuracy, repeatability, linearity, turndown and speed of response.

*Accuracy* is the ratio of the error to the full-scale output, generally expressed as a percentage of span.

*Repeatability* is how well an instrument gives the same output for the same input when the input is applied in the same way over a short time period. It is also often expressed as the error as a percent of span.

*Linearity* only applies to measurements that are supposed to be linear; then, it also is a percent of span of the deviation of the measurement versus actual value from a straight line.

*Response speed* is defined as the length of time required for the measured value to rise to within a certain percentage of its final value as a result of a step change in the actual value. A 98% response time, for example, while indicative of the time to get a good measurement, is much longer than a first order time constant. The first order time constant of the measurement is of most interest in the performance of a control loop.

Other response characteristics like hysteresis, dead band, and stiction are primarily related to mechanical equipment, such as control valves, and do not normally apply to electronic transmitters.

Measuring instruments can generally be adjusted for span and zero. Span error is how well the full scale output of the instrument matches a full span change in the actual variable, usually expressed as a percent of span. Zero error is the output of the instrument for a measurement that is at the low end of the span, usually expressed as a percent of span. A zero error causes a constant offset for any measurement.

The turndown ratio is the ratio of the maximum to minimum measurable value. For example, if the maximum flow that can be measured is 100 gallons per minute (gpm) and the turndown ratio is 3 (typical for an orifice), then the minimum flow that can be accurately measured will be 33 gpm. However, if the turndown ratio is 100 (possible for a Coriolis meter), then the minimum flow that can be accurately measured will be 1 gpm.

## 1.2 Pressure

Transmitters for measuring the pressure of a liquid or gas are very common in process and utility applications, since they are used both for actual pressure and also frequently used in the measurement of level and flow.

Pressure is generally measured in pounds per square inch or in inches of water column. The pressure measurement can be designed to measure the amount that the pressure is above atmospheric pressure (positive pressures only—a.k.a. "gauge"), the amount the pressure is above or below atmospheric pressure (positive and negative pressures—a.k.a. "compound range"), or it can be the amount that the pressure is above absolute zero pressure ("absolute"). At sea level, atmospheric pressure is 14.7 pounds, but it varies about 0.5 psi per 1,000 feet of elevation.

Pressure measurements can also be either simple pressures (i.e., a single input port) or differential pressure (i.e., two input ports). Differential pressure transmitters are critical—for example, when measuring a small differential pressure, say 20 inches of water in the presence of a high common pressure, say 1,000 pounds, it would not be possible to measure each pressure and then take the difference electronically. The inaccuracy of the transmitters and subtraction devices would make the resulting difference hopelessly inaccurate.

The ideal gas law says that pV/T is a constant where p is the pressure, V is the volume, and T is the absolute temperature. Obviously, then, the pressure is highly dependent on temperature and volume.

While a variety of pressure measurement methods are available, such as manometers, bourdon tubes and bellows, most pressure transmitters today, both single pressure and differential, measure pressure by sensing the deflection of a diaphragm. The sensing device for that deflection is a strain gauge or other technique and is often on a secondary diaphragm for temperature and shock protection. Figure 1-1 shows the internals of a differential pressure transmitter and the secondary diaphragm that is coupled by oil filled channels. The output of the sensor is then amplified for transmission. The sensor is analog whether the signal conditioning and transmission is digital or analog.

The diaphragm that contacts the process fluid must be of a material that will withstand the temperature and corrosive effects of the process. Since the diaphragms are thin, they have little tolerance for corrosion. Diaphragms are available in stainless steel, a variety of alloys, and ceramic.

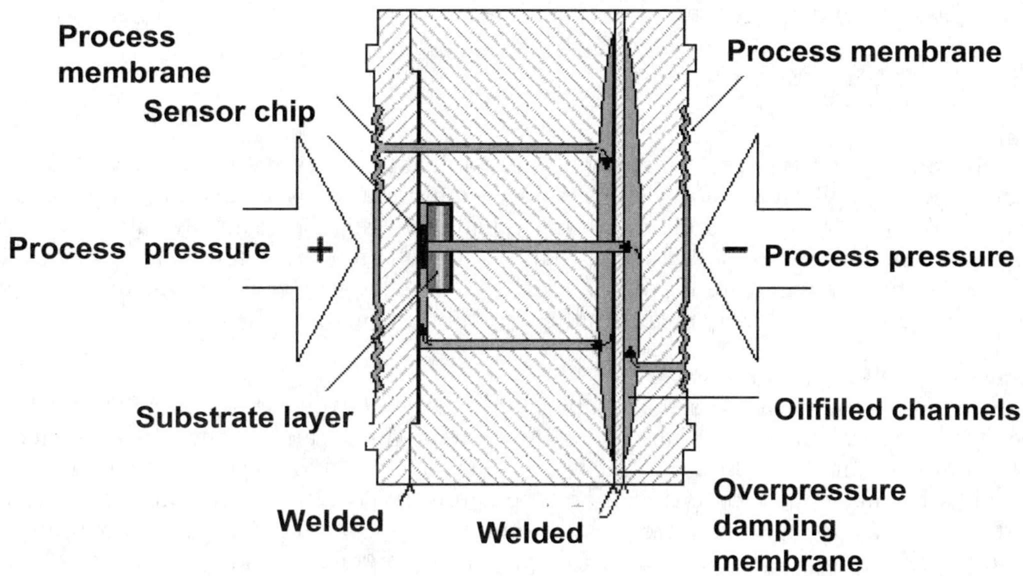


*Figure 1-1: Differential Pressure Diaphragm and Sensor Assembly (Courtesy: Endress + Hauser)*

Pressure transmitters may be connected to the process by a length of tubing or the diaphragm can be mounted flush to the process vessel using a pressure transmitter specially configured for that purpose. Some prefer the transmitter to be located for convenient maintenance access, which may mean that long tubing connections to the process piping or vessels are required. Others prefer for the transmitter to be close-coupled to the process piping or vessel to minimize leakages and tubing pluggauge and fill problems. That is, you can locate the transmitter for easy access so that, when it has a problem, it can be easily serviced. Or, you can locate it for reliability so it is less likely to need to be serviced.

Span and zero calibration is a major issue with analog pressure transmitters. Digital pressure transmitters tend to be much more accurate and stable than all analog transmitters. In addition, digital transmitters have a number of other functional advantages.

Various types of devices for developing pressures in the field have long been used by instrument technicians for calibration of span. These are used by first valving off the pressure transmitter from the process, and then connecting the transmitter to this portable pressure source. A known pressure measurement gauge can then be used to compare to the transmitter output. With analog instruments, this is the only way to change the span setting—from, say 0-100 in. $H_2O$ to 0-200 in. $H_2O$. Calibration of digital transmitters can be done entirely within the digital electronics and remotely via the communications wiring. In addition, digital transmitters today are likely to be more accurate than the pressure gauge that can be handled in the field. Because of this, field calibration is diminishing.

## 1.3 Level

Level measurements of liquids or solids are used extensively in all types of bulk manufacturing and storage facilities plus many utilities. The level measurement may be for accurate inventory, to determine the contents in a vessel where reactions are taking place, or just be to keep the tank from overflowing or from going empty. The location of the surface may be measured directly for solids or liquids. For liquids, level can be inferred from the pressure at the bottom of the tank. In difficult applications, the tank can be weighed. Solid level measurement is often inaccurate because the surface is an upward cone shape under the filling location or downward cone shape over the discharge location.

Even liquids may have turbulent surfaces from boiling or agitation which can cause inaccuracies in some types of level measurements.

### 1.3.1 Direct Level Measurement

**Float**

The most obvious measurement method is to use a float to determine a liquid level. This method is used in process applications, but possibly its most important use is in very large tanks with expensive contents. In those large tank applications it is called tank gauging. To achieve maximum accuracy the gauging system also utilizes vessel shape changes due to atmospheric temperatures and fill bloating and many other seemingly minor things. The float connects via a cable or tape to a measuring device outside the tank that precisely measures the length to the float.

**Ultrasonic and Radar (Microwave)**

These measurements work by sending a pulsed wave signal from the top of the tank that hits the surface of the material and reflects back to the instrument. The distance to the surface is then determined by the transmission time. Ultrasonic measurements have the advantages of no contact with the process and are suitable for various liquids and bulk products. Their disadvantages are that the process must not produce too much surface foam, and they are not suitable for high temperature, pressure or vacuum. Radar has the advantage of broad applicability on most liquids and measurement independent of pressure, temperature and vapor. Disadvantages are that the measurement may be lost due to heavy agitation of the liquid or the formation of foam. Radar instruments are now approaching the price of ultrasonic and are the fastest growing type of level measurement.

**Capacitance**

A metal probe is located vertically in the tank and electrically isolated from the tank. The probe and the walls of the tank form a capacitor that has a value that depends on the amount of material in the tank and the medium between the probe and the wall. When only vapor is present, the capacitance will be low. The capacitance will increase incrementally as the process material covers the probe. This method is suitable for liquids or solids, has no moving parts, and is suitable for highly corrosive media. The disadvantages are limited application for products with changing electrical properties and may be sensitive to coatings on the probe. Sensor selection is critical to the measurement, particularly if the sensed material is conductive.

**Radioactive**

A radioactive source—either point or strip—is placed on one side and outside the tank, and a radiation detector (Geiger counter), or series of detectors, is placed on the other side. The amount of radiation reaching the detector(s) is dependent on the amount of material in the tank. This type is expensive and requires stringent personnel safety requirements and licensing, so it is used only as a last resort. The measurement is very nonlinear unless a strip source and a series of detectors are used.

### 1.3.2 Inferring Level from Head Measurement

**Displacer**

A displacer is a vertical body that is heavier than the fluid being measured. When placed so it is partly submerged, an upward force is generated that is based on the difference between the weight of the displacer and the amount of liquid displaced. Since the displacer is often installed in a vertical pipe attached to the tank at both ends, it can see a very still liquid surface and is very accurate. A displacer is expensive to install and maintain.

**Bubbler**

In this type of measurement, a tube is placed in the tank from the top and connected to a source of air. A needle valve in the air stream is adjusted to allow a slow flow of air at maximum level, as determined by bubbles escaping the bottom of the tube, and also typically by a flow indicator. The pressure of the air stream downstream of the needle valve is measured and is equal to the head generated at the bottom of the tube. This method is very simple and is widely used in open vessels and sumps.

**Differential Pressure Transmitter**

Probably the most common method of determining level of a liquid is by measuring the pressure or head at some point in the tank below the zero level. Since this method is often used in closed tanks, it is necessary to also measure the pressure in the vapor space at the top of the tank and subtract that pressure. A differential pressure (dP) transmitter is ideal for this application.

Since there will be process fluid in the tubing connecting the dP cell to the bottom of the tank, this has to be taken into account in the calibration of the transmitter. It may be intended that the tubing connecting the dP cell to the top of the tank contains only gas which has little impact on the calibration, but often that leg will become filled with liquid from condensation or from an occasional high level in the tank. Alternately, if it is intended that that tubing be filled with liquid, the liquid may evaporate unless it is continually replenished with a purge flow. Either unintended situation will cause a significant error in the reading.

Transmitters are available that bolt flush to the bottom of the tank and thus eliminate that tubing connection; transmitters are also available that have a hydraulic filled tube between the dP cell diaphragm and a remote diaphragm. These remote diaphragms can be connected flush to the top and bottom of the tank, eliminating all tubing with process fluid. Figure 1-2 shows a differential pressure transmitter with diaphragm seals. Filling these systems requires utmost care to eliminate all air bubbles before being filled with the hydraulic fluid. In spite of their additional cost, the advantages of filled systems make them popular and some companies use them for all appropriate applications.

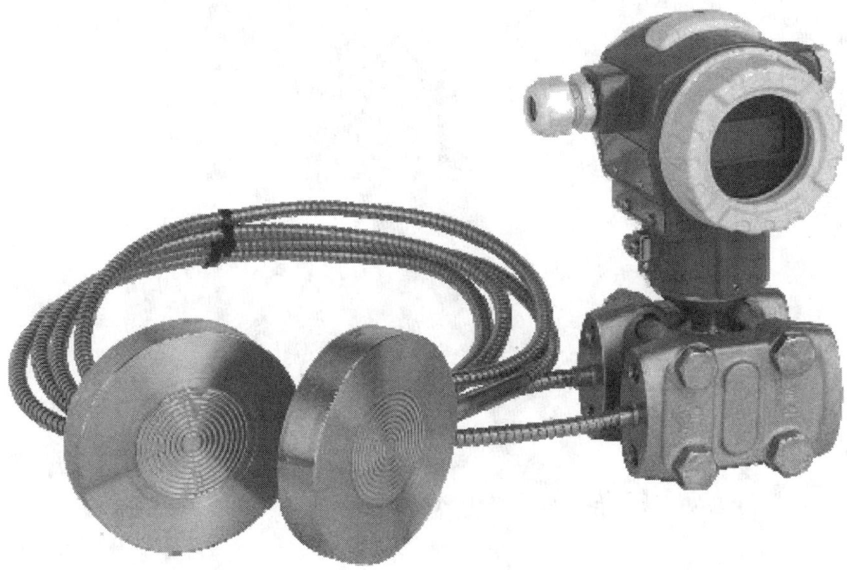

*Figure 1-2: dP Transmitter with Filled System Connecting to Remote Diaphragms (Courtesy: Endress+Hauser)*

Since the head or pressure of the material in the tank is a function of both level and density, changes in density will introduce errors into the level calculation.

### 1.3.3 Level Switches

Since high and low levels are so important in tanks, level switches are often used instead of a continuous measurement. Several types are available, such as a rotating paddle wheel for solids and a tuning fork for either liquids or solids.

In the paddle wheel type, the paddle is rotated by an electric motor through a clutch. When the paddle becomes covered with material, the paddle stalls and triggers a microswitch.

In the tuning fork type, the vibrating fork is driven to its resonant frequency in air by a piezoelectric crystal. When immersed in a liquid, the resonant frequency will shift approximately 10-20%. This shift in resonant frequency is picked up by a receiver crystal. Figure 1-3 shows a tuning fork switch. Tuning forks used in solids/particulates also vibrate at their resonant frequency, but detection is based on monitoring the decreased amplitude of fork motion when covered by solids.

These level switches are low cost and likely more accurate and reliable than a continuous level measurement, even if buildup occurs on the sensor.

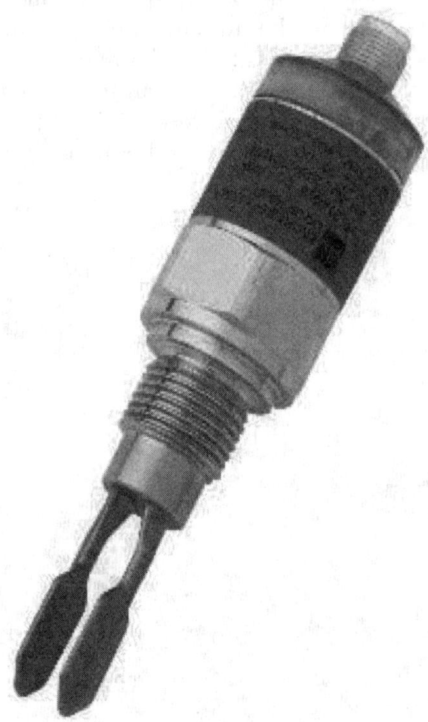

*Figure 1-3: Tuning Fork Level Switch (Courtesy: Endress+Hauser)*

## 1.4 Flow

This flow discussion will focus on measuring flow in closed pipes. Flow measurement in open channels is not discussed, though that is an important type of measurement in large utility streams.

Flow is laminar or turbulent, depending on the flow rate and viscosity. This can be predicted by calculating the Reynolds number, which is the ratio of inertial forces to viscous forces:

$$Re = 123.9\, pVD/u \tag{1-1}$$

where:

| | | |
|---|---|---|
| $Re$ | = | Reynolds number |
| $p$ | = | density in lbs./ft.$^3$ |
| $V$ | = | average velocity in ft/sec. |
| $D$ | = | pipe diameter in inches |
| $u$ | = | viscosity in centipoises |

Reynolds numbers less than 2000 indicate laminar flow and above 4000 indicate turbulent flow. However, some velocity meters require values above 20,000 to be absolutely certain that the flow is truly turbulent and that a good average velocity profile is established that can be measured from a single point on the flow profile. Most liquid flows are turbulent while highly viscous flows like polymers or very low flow rates are laminar.

Flow measurements can be of the average velocity, velocity at one point, volume of material flowing, or the mass of material. Velocity measurements in particular require that the flow stream velocity be relatively consistent across the diameter of the pipe. Less than fully turbulent flow creates lower velocities near the pipe wall.

Fittings, valves—anything else other than straight, open pipe upstream of the sensor—will cause velocity variations across the diameter of the pipe. Figure 1-4 illustrates the variations in velocity that can occur from pipe fittings. To achieve uniform flow, different types of flowmeters require straight pipe runs upstream and downstream of the measurement. These run requirements are expressed as a certain number of straight, open pipe diameters. For example, for a 6-inch pipe, 20 diameters would be 10 feet. There are no consistent recommendations even for a particular flowmeter type; it is best to follow the manufacturer's recommendations. Recommendations vary from 1 to 20, or even more, upstream diameters and a smaller number of downstream diameters.

Flow measurements can be grouped into four categories:

- Inferential methods
- Velocity methods
- Mass methods
- Volumetric methods

### 1.4.1 Inferential Methods

Placing an obstruction in the flow path causes the velocity to increase and the pressure to drop. The difference between this pressure and the pressure in the pipe can be used to measure the flow rate of most liquids, gases, and vapors, including steam. In turbulent flow, the differential pressure is proportional to the square of flow rate.

An orifice plate is the most common type of obstruction, and, in fact, differential pressure across an orifice is used more than any other type of flow measurement. The installed base of orifice meters is probably as great as all other flowmeters combined. The orifice plate is a metal disc with typically a round hole in it, placed between flanges in the pipe. Differential pressure can be measured at the pipe flanges directly upstream and downstream of the orifice or further upstream and downstream. The calculation formulas of differential pressure for a given orifice size and given location of the pressure taps are well developed, so no field calibration based on actual flow is needed (although the dP cell may have to be calibrated).

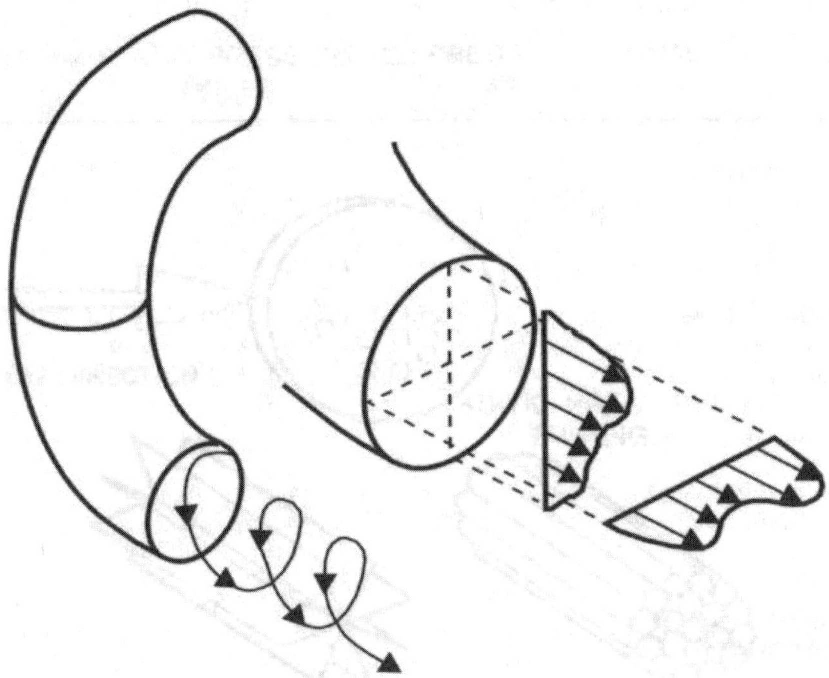

*Figure 1-4: First Fitting Causes Profile Distortion; Second Fitting Superimposes the Swirl*

Orifice flow measurements are relatively cheap to purchase but have relatively high installation costs. They have high operating costs because they create a fairly large unrecoverable pressure loss. Also, they have low turndown, in part due to the squared relationship. Orifices are suitable for high temperature and pressure, and are best for clean liquids, gases, and low velocity steam flows. They require long straight runs upstream and downstream. They are subject to a number of errors, such as flow velocity variations across the pipe and wear or buildup on the orifice plate. Because of these error sources, they are not generally very accurate even when highly accurate differential pressure transmitters are used.

Other types of obstructions include venturis and flow tubes which have less unrecoverable flow loss. A pitot tube is a device that can be inserted in large pipes or ducts to measure a differential pressure.

## 1.4.2 Velocity Methods

**Magnetic Flowmeters**

Magnetic flowmeters depend on the principle that motion between a conductor (the flowing fluid) and a magnetic field develops a voltage in the conductor that is proportional to the velocity of the fluid.

Coils outside the pipe generate a pulsed DC magnetic field. The material to be measured flows through the meter tube, which is lined with a non-conductive material such as Teflon, polyurethane, or rubber. Measuring electrodes protrude through the liner and contact the fluid and sense the generated voltage. Figure 1-6 shows the location of the coils and sensing electrodes.

The flowing fluid must be conductive, but there are very few other restrictions—most aqueous fluids are suitable. There are fewer Reynolds number limitations; the instrument is the full diameter of the pipe so there is no pressure loss; a wide range of sizes are available from a very small 1/8 inch to an enormous 10 feet in diameter; the flowing material can be liquids, slurries and suspended solids; and

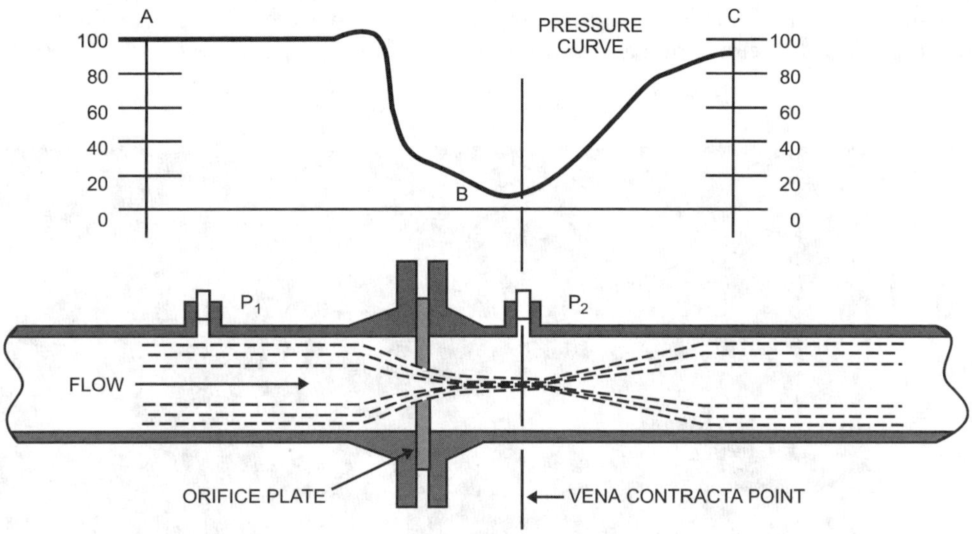


*Figure 1-5: Typical Pressure Profile of an Orifice*

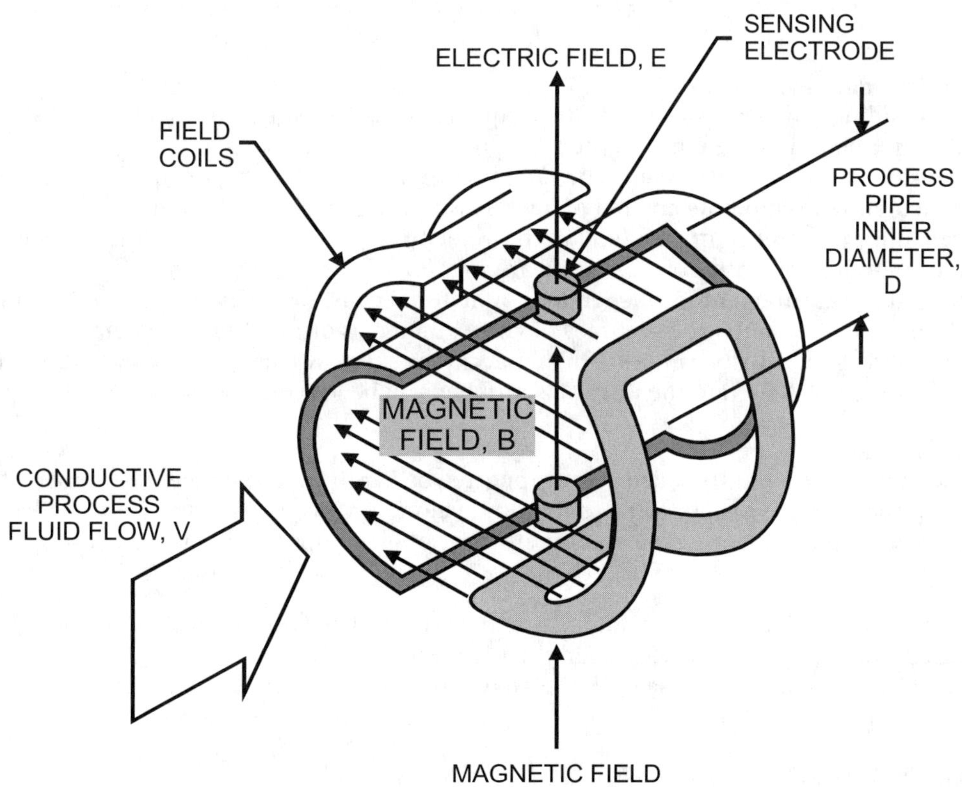


*Figure 1-6: Magnetic Flowmeter Principle of Operation*

there are minimum straight run requirements. Figure 1-7 shows two very large meters. These meters are widely used in utility as well as process applications and are particularly widely used in Europe. Calibration is factory determined and is rarely checked in the operating facility.

*Figure 1-7: Magnetic Flowmeter (Courtesy: Endress+Hauser)*

**Vortex Shedding Flowmeters**
Vortex shedding flowmeters measure the frequency of vortices shed from a blunt obstruction, called a "bluff body", placed in the pipe. As the flow divides to go around the bluff body, vortices are created on each side of the divided stream. The rate of vortex creation is proportional to the stream velocity. Since each vortex represents an area of low pressure, the presence-then-absence of low pressures is counted and the count is proportional to the velocity. Vortex flowmeters provide good measurement accuracy with liquids, gases, or steam and are tolerant of fouling. They have high accuracy at low flow rates and the measurement is independent of material characteristics. They require long runs of straight pipe. Even though the accuracy of vortex meters is often stated as a percent of flow rate rather than of full scale which does indicate higher accuracies, below a certain flow rate they cannot measure at all. At some low flow rate the Reynolds number will be low enough so that no vortices will be shed.

**Turbine Meters**
Turbine meters use a multi-bladed rotor supported by bearings in the pipe. The flowing fluid drives the rotor at a speed that is proportional to the fluid velocity. Movement of the rotor blades is sensed by a magnetic pickup outside the pipe and the number of blade tips passing the pickup is counted to get rotor speed.

These meters have high accuracy for a defined viscosity. They are suitable for very high and low temperatures and high pressures. However, they are sensitive to viscosity changes, and the rotor is easily damaged by overspeed. Because of the relatively high failure rate of their moving parts, they are not used as much as in the past.

**Ultrasonic Flowmeters**
Ultrasonic flowmeters send sound waves through the flowing stream. They can measure either the Doppler shift as ultrasonic waves are bounced off particles in the flow stream, or the time differential of ultrasonic waves with the flow stream compared to against the flow stream. Either method gives a

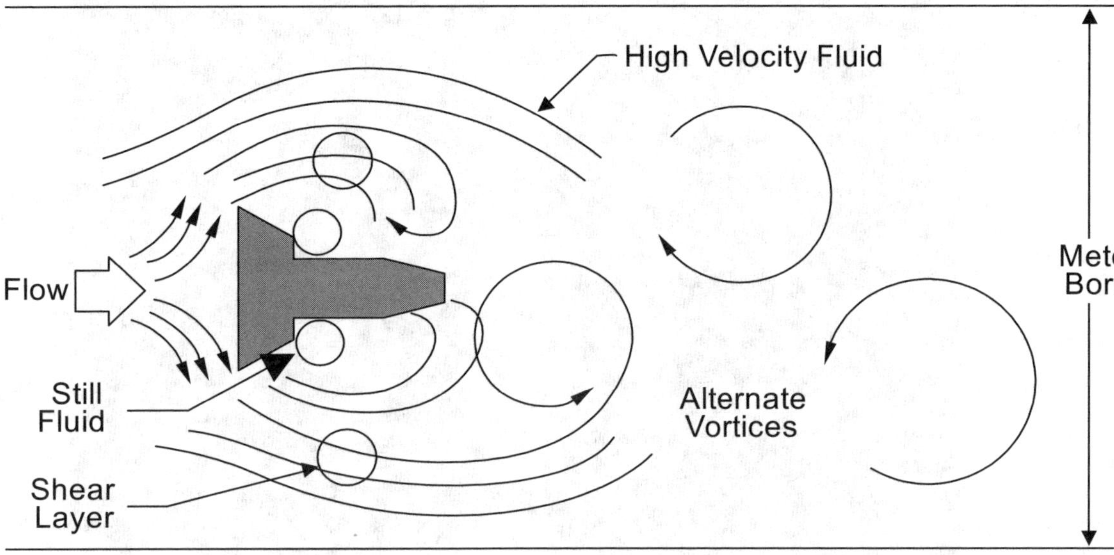


*Figure 1-8: Vortex Shedding Phenomenon*

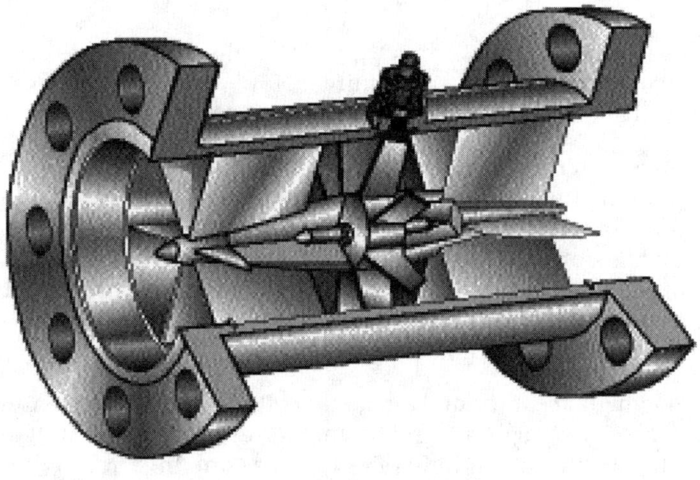

*Figure 1-9: Turbine Meter*

signal which is proportional to flow velocity. The Doppler method works with liquids with suspended solids, and the Transit time method works with liquids and gases. In both methods, the signal is proportional to flow velocity.

Ultrasonic meters are non-invasive but are relatively low accuracy. A clamp on meter is shown in Figure 1-10, which requires no connection to the pipe. Because these clamp-on meters are so easy to install, they can be used temporarily to verify the flowmeter that is permanently installed in the pipe. Since the same meter can do a variety of sizes, they are particularly cost effective in large sizes.

### 1.4.3 Mass Methods

Mass flowmeters measure actual mass flow. While it is possible to calculate mass flow from a velocity or inferential measurement and other variables like temperature for known fluids, only one meter

*Figure 1-10: Clamp-on Ultrasonic Flow Meter (Courtesy: Endress + Hauser)*

type commonly measures liquid mass directly, the Coriolis meter. This meter used to be applied only for when highly accurate, mass flow was required. Now with lower prices, a wider range of configurations and easier installation, it is being applied more routinely.

The heart of a Coriolis meter is a tube(s) that is vibrated at resonant frequency by magnetic drive coils. When fluid flows into the tube during the tube's upward movement, the fluid is forced to take on the vertical momentum of the vibrating tube. Therefore, as the tube moves upwards in the first half of the vibration cycle, the fluid entering the tube resists the motion of the tube and exerts a downward force. Fluid in the discharge end of the meter has momentum in the opposite direction, and the difference in forces causes the tube to twist. This tube twist is sensed as a phase difference by sensors located on each end of the tube arrangement, and twist is directly proportional to mass flow rate.

In addition to having high accuracy and a true mass flow measurement, Coriolis meters have no upstream and downstream straight run requirements, are independent of fluid properties, are low maintenance, and have a turndown ratio of as much as one hundred. While the meters originally were only available in a double U-shape, they are now available in a variety of configurations. Figure 1-11 shows a single, straight, full bore tube design. Coriolis meters are available in sizes up to 10 inches.

### 1.4.4 Positive Displacement Meters

This type of meter separates the flow stream into known volumes and by vanes, gears, pistons or diaphragms, and then counts the segmented volumes. They have good-to-excellent accuracy, can measure viscous liquids, and have no straight run requirements. However, they do have a non-recoverable pressure loss, and their moving parts subject to wear.

## 1.5 Temperature

Temperature is measured in Kelvin, Celsius, Fahrenheit, or Rankin. Unlike the other "big four" measurements, in many temperature applications the sensor is separate from the transmitter. While the

*Figure 1-11: Coriolis Mass Flowmeter with Single, Straight Tube (Courtesy: Endress+Hauser)*

transmitter or amplifier can be located in the housing of the sensor, it can also be remote—either in a field box containing a number of temperature transmitters, in the control room, or the output of the sensor can be connected directly to a temperature logger or directly to a DCS or PLC controller.

### 1.5.1 Thermocouples

The thermocouple is the most popular type of sensor. Thermocouples are based on the principle that two wires made of dissimilar materials connected at either end will generate a potential between the two ends that is a function of the materials and temperature difference between the two ends.

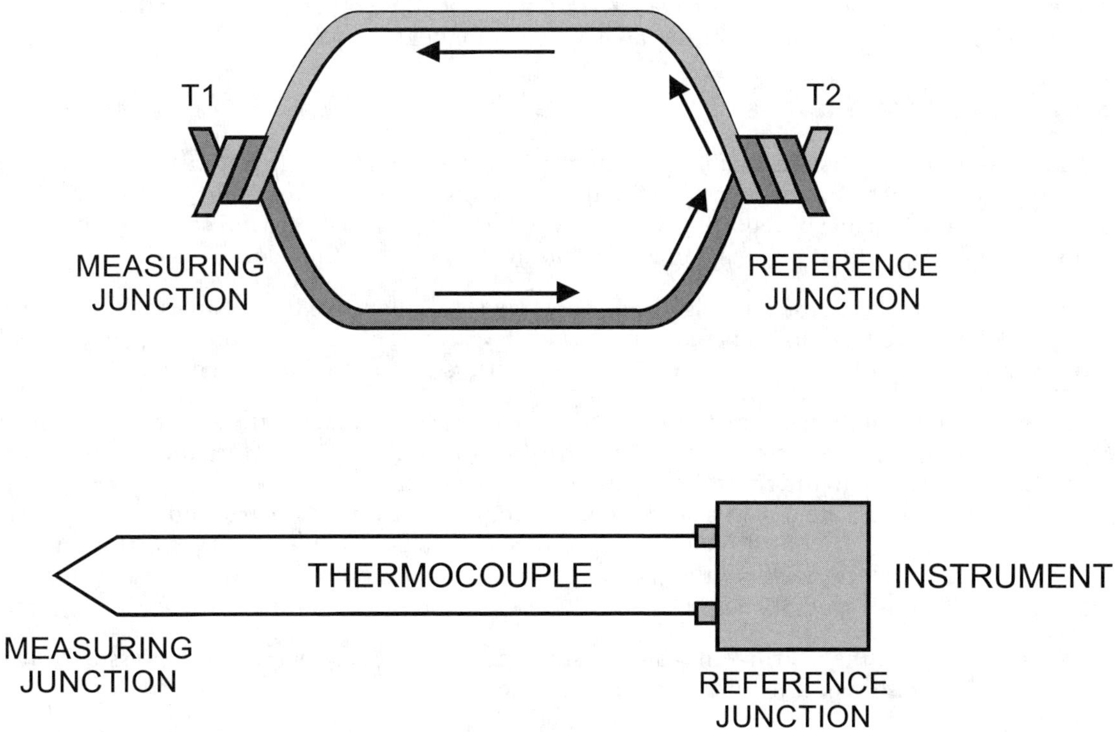


*Figure 1-12: Thermocouples*

A number of material choices are in common use. Base metal thermocouples are useful for measuring temperatures under 1000 degrees C. This class includes iron/constantan (Type J), Chromel/Alumel (Type K) and a number of others. Nobel metal thermocouples are useful to about 2000 degrees C. This class includes tungsten-rhenium alloy thermocouples and others.

The potential generated is in millivolts and is a nonlinear function of temperature. In practice, one end is placed near the material to be measured and the other end is connected to the instrument. Since the thermocouple materials are not typically good materials for transmission, wires with similar characteristics are used when the transmitting instrument is remote.

### 1.5.2 Resistive Temperature Detectors (RTDs)

RTDs are made of a metal wire or fiber or of semiconductor material that responds to temperature change by changing its resistance. Platinum, nickel, and tungsten and other metals are used that have high resistivity, good temperature coefficient of resistance, good ductile or tensile strength, and chemical inertness with packaging and insulation materials. When the material is a semiconductor, the sensor is called a thermistor.

The change in resistance can be determined using a bridge circuit. Since resistance changes in the connection wire due to ambient temperature changes can also affect the resistance reading, a third wire is used from another leg in the bridge to balance that change.

RTDs are generally more accurate than thermocouples, but are less rugged and cannot be used at as high temperatures.

All types of temperature measuring devices suffer from slow response, since it is necessary for the heat to conduct through the protective sheath, and through any installed well. Locating the well (or unprotected sensor) so that it sees as high a velocity of process material as possible helps reduce this lag, as does having the sensor contact the well. A bare thermocouple touching the sheath and/or well, however, generates a ground and requires an isolated amplifier.

## 1.6 Smart Instruments

The ISA *Automation, Systems, and Instrumentation Dictionary* defines a "smart" instrument as one that is microprocessor-based, may be programmed, has memory, can be communicated with from a remote location, and is capable of reporting faults and performing calculations and self-diagnostics. This does not specifically say that they have to communicate digitally, but they do.

For some automation professionals, common usage is to call transmitters "smart" that use the HART protocol and to call transmitters that communicate with some type of fieldbus as simply "fieldbus" transmitters. Regardless of the name, these transmitters give tremendous benefits:

- The calibration can be changed remotely by pushing a few buttons on a hand-held calibrator connected anywhere to the signal wiring or by entering the information from a computer connected to the control/asset management system. This makes it unnecessary, for example, to go to the field with a stack of equipment, pump up a pressure, and carefully turn screws until the span of a dP cell is changed to a new value. (Reading this "smart" information from transmitters is beyond the capability of past generation control systems, and the need to provide this capability has been part of the justification for some control system upgrades.)
- Many of the instruments can measure and report several variables: a pressure transmitter may also report temperature for example.
- The transmitter may be capable of reporting its specifications such as model number, materials of construction, calibration, tag number, and other items.

- Many types of transmitter failures can be detected by the transmitter itself and reported—ideally to an asset management system that will in turn, report the failure to a computerized maintenance management system (CMMS) so a repair work order can be issued.
- The transmitter can monitor its internal parameters: for example, a Coriolis meter might report its excitation current, frequency of the tubes, and other internal variables which can assist in troubleshooting and error detection.
- Some transmitters are even able to detect a change in the noise level of the signal and relate that to plugged process connection tubing and also call for repair.

## 1.7 References

Anderson, Norman A. *Instrumentation for Process Measurement and Control.* Third Edition. CRC Press, 1998.

Boyes, Walt, Editor. *Instrumentation Reference Book.* Third Edition. Butterworth-Heinemann, 2003.

Coggan, D.A., Editor. *Fundamentals of Industrial Control.* Second Edition. Practical Guides for Measurement and Control Series. ISA, 2005.

Goettsche, L. D., Editor. *Maintenance of Instruments and Systems.* Second Edition. Practical Guides for Measurement and Control Series. ISA, 2005.

Hashemian, H. M. *Sensor Performance and Reliability.* ISA, 2005.

Hughes, T. A. *Measurement and Control Basics.* Third Edition. ISA, 2002.

ISA. *The Automation, Systems, and Instrumentation Dictionary.* Fourth Edition. ISA, 2003.

Lipták, Béla G., Editor-in-Chief. *Instrument Engineer's Handbook: Process Measurement and Analysis, Volume I, Fourth Edition.* ISA & CRC Press, 2003.

### 1.7.1 Acknowledgment

The author would like to thank Endress+Hauser for permission to excerpt various materials from Endress+Hauser's employee training document, titled *Instrumentation Prerequisites.*

## About the Author

**Vernon L. Trevathan**, PE, PMP, has worked in process control and project management for 40 years, mostly with Monsanto Chemical Co. in senior engineering and manufacturing management positions. He also was Vice President of Benham Co. in charge of the St. Louis and St. Paul System Integration offices. An ISA Fellow, he currently consults and teaches automation project management and general control. Trevathan is chair of the ISA CAP Steering Team.

# 2 Analytical Instrumentation

*By James F. Tatera*

## Topic Highlights

*Sample Point Selection*
*Instrument Selection*
*Sample Conditioning Systems*
*Process Analytical System Installation*
*Maintenance*
*Utilization of Results*

## 2.1 Introduction

Process analytical instruments are a unique category of process control instruments. They are a special class of sensors that enable the control engineer to control and/or monitor process and product characteristics in significantly more complex and varying ways than traditional physical sensors—such as pressure, temperature and flow—allow.

Today's safety and environmental requirements, time sensitive production processes, inventory reduction efforts, cost reduction efforts, and process automation schemes have made process analysis a requirement for many processes' control strategies. Most process analyzers are providing real-time information to the control scheme that many years ago would have been the type of feedback the production process would have received from a plant's quality assurance laboratory. Most processes require faster feedback to control the process, rather than just being advised that their process was or wasn't in control at the time the lab sample was taken, and/or the sample was or wasn't in specification.

Different individuals have made several attempts to categorize the large variety of monitors typically called process analyzers. None of these classification schemes has been widely accepted; the result is there are a number of simultaneous categorization schemes in use. Most of these schemes are based on either the analytical technology being utilized by the monitor, the application to which the monitor is being applied, or the sample type. There are no hard and fast definitions for analyzer types. Consequently, most analytical instruments are classed under multiple and different groupings. Table 2-1 depicts a few of the analyzer type labels commonly used.

An example of how a single analyzer can be classed under many types would be a pH analyzer. This analyzer is designed to measure the pH (an electrochemical property of a solution—usually water-based). As such it can be used to report the pH of the solution and may be labeled as a pH analyzer or electrochemical analyzer (its analytical technology label). It may be used to monitor plant's water outfall and, as such, may be called an Environmental or Water Quality Analyzer (based on its application and sample type). It may be used to monitor the acid or base concentration of a process stream and, as

*Table 2-1: Examples of Analyzer Types/Classification*

| Compositional | Single Component |
|---|---|
| Physical Properties | Sulfur in Oil |
| Oxygen | Moisture |
| Inferential/Virtual | Auto Titrators |
| Chromatography | Spectroscopy |
| Safety | Environmental |
| Air Quality | Water Quality |
| BTU | LEL |

such, be labeled a single component concentration analyzer (based on its application and the desired result being reported). This is just an example and is only intended to assist you to understand you will probably come in contact with many process analyzers that will be labeled under multiple classifications. Please don't allow this to confuse or bother you.

There are too many process analyzer technologies to try to mention them all in this chapter, so only a few will be used as examples. Many books are published on the subject of process analysis. A few are listed in the reference summary at the end of this chapter. The balance of this chapter will be used to introduce some concepts and technical details important in the application of process analyzers.

## 2.2 Sample Point Selection

Which sample to analyze is usually an iterative process that is based on several factors and inputs. Some of these factors include regulatory requirements, product quality, process conditions, control strategies, economic justifications, and more. Usually, the final selection is a compromise that may not be optimum for any factor, but is the overall best of the options under consideration. Too often, mistakes are made on existing processes when selections are based on a simple guideline like the final product or the intermediate sample that has been routinely taken to the lab. True, you usually have relatively good data regarding the composition of that sample. But, is it the one that will help you best control your process to continuously make good product? Or, is it the one that will just tell you that you have or haven't made good product? Both are useful information, but usually the latter is more effectively accomplished in a laboratory environment.

When you consider all the costs of procurement, engineering, installation, and maintenance; you rarely save enough money to justify installing process analyzers to shut down or replace some of your lab analyzers. Large savings are usually achieved through improved process control, based on the analyzers input. If you can see a process moving out of control, and bring it back before it has gone out, you can avoid making any bad product, rework, waste, etc. If you have to rely on detecting final product that is already moving out of specification, you are more likely to make additional bad product.

Think of a distillation process. Typically, lab samples are taken from the top or bottom of columns and the results usually contain difficult-to-measure, very low concentrations of lights or heavies, as they have been distilled out. Often you can better achieve your goal by taking a sample from within the column at or near a major concentration break point. This sample can tell you that lights or heavies are moving in an undesired direction before a change has reached the column take-offs, and you can adjust the column operating parameters (temperature, pressure, and/or flow) in a way that returns the column operation to what you want. This sample, from within the distillation column, usually contains analytes in concentrations that are less difficult and more robust to analyze.

In order to accomplish this type of selection process, a multidisciplinary team is a good approach. In the distillation example mentioned above, you would probably want a process engineer, controls engineer, analyzer specialist, quality specialist, and possibly others on the team to help identify the best sampling point. If you have the luxury of an appropriate pilot plant, it is often the best place to start. You do not have to risk interfering with production as you possibly experiment with different sample points and control strategies. Pilot plants are also often allowed to intentionally make bad product and demonstrate undesirable operating conditions.

Another tool that is often used to help identify desirable sample points and control strategies is multiple Temporary Relocatable Process Analyzers (TURPAs). With an appropriate selection of these instruments, you can do a great job modeling the process and evaluating different control strategies. Once you have identified the sample point and desired measurement, you are ready to begin the instrument selection phase of the process.

## 2.3 Instrument Selection

Process analyzer selection is also typically best accomplished through a multidisciplinary team. This team's members often include the process analytical specialist, quality assurance and/or research lab analysts, process analyzer maintenance personnel, process engineers, instrument engineers, and possibly others. The list should include all the appropriate individuals who have a stake in the project. Of these job categories, the ones most often not contacted until the selection has been made—and the ones probably most important to its long-term success—are the maintenance personnel. Everyone relies on them having confidence in it and keeping it working over the long term.

At this point, you have identified the sample to be measured (concentration range to be monitored under normal and upset operating conditions, required measurement precision and accuracy, speed of analysis required to support the identified control strategy, etc.). You also should have identified other components/materials that could be present under normal and abnormal conditions. These other components/materials are used in the method selection process to ensure they will not interfere with the selected method/technology.

You are now identifying possible technology candidates and trying to select the best for your total situation. If you have a current lab analytical method for this or a similar sample, you should not ignore it, but more often than not, it is not selected as the best technology for the process analysis. It is often too slow, complex, fragile, expensive, or has other issues, to successfully pass the final cut. Lastly, if you have more than one good candidate, consider the sites' experience maintaining these technologies. Are they something with which maintenance has experience and training? Does the site have existing spare parts and compatible data communications systems? If not, what are the spare parts supply and maintenance support issues?

All of these items and more maintenance concerns should be brought up by the maintenance representative on your selection team and need to be taken seriously in the selection process. The greatest measurement in the world that cannot be maintained and kept performing in a manner consistent with your processes operations will not meet your long-term needs.

At this point, your selection team needs to review its options and select the analytical technology that you believe will best serve your needs over the long term. The analytical technology selected does not need to be the one that will yield the most accurate and precise measurement. It needs to be the one that will provide the measurement that you require in a timely, reliable and cost effective manner over the long term.

## 2.4 Sample Conditioning Systems

Sample conditioning sounds simple. You simply take the sample that the process provides and condition/modify it in ways that allow the selected analyzer to accept it. In spite of this relatively simple sounding mission, most process analyzer specialists attribute 50 to 80% of process analyzer failures to sample conditioning issues. Recall that the system has to deliver acceptable sample to the analyzer under a variety of normal, upset, start up, shut down, and other process conditions. Usually the sampling system not only has to consider the interface requirements of getting an acceptable sample from the process to the analyzer, but it usually has to dispose of that sample in a reliable and cost effective manner. Disposing the sample often involves returning it to the process and sometimes conditioning it to make it appropriate for the return journey.

What is considered an acceptable sample for the analyzer? Usually, this is defined as one that is "representative" of the process stream and compatible with the analyzer's sample introduction requirements. In this case "representative" can be a confusing term. It usually doesn't have to represent the process stream in a physical sense. Typical sample conditioning systems usually change the process sample's temperature, pressure and some other parameters in order to make the process sample compatible with the selected process analyzer. In many cases, conditioning goes beyond the simple temperature and pressure issues and includes things that actually change the composition of the sample—things like filters, demisters, bubblers, scrubbers, membrane separators, and more. These are things that actually change the sample composition. However, as long as the resulting sample is compatible with the analyzer and the resulting analysis is correlatable/representative of the process it is considered a good job of sample conditioning.

Some sampling systems also provide stream switching and/or auto calibration capabilities. Since the process/calibration samples often begin at different conditions and to reduce the possibility of cross contamination, most of these include multiple process sampling systems with stream selection capabilities (often double block and bleed valving) just before the analyzer.

Figure 2-1 depicts a typical boxed sample conditioning system. A total sampling system would normally also include the sample extraction approach (possibly a tee, probe, vaporizer, etc.,) transport lines, return lines, fast loop, slow loop, sample return, etc. In the figure you can see an assortment of filters, flow controllers, valves, etc. Figure 2-2 depicts an in situ analyzer installation that has required very little sample conditioning and is essentially installed in the process loop.

Most analyzers are designed to work on clean, dry, non-corrosive samples in a given temperature and pressure range. The sample system should accomplish the process sample's conversion to the analyzer's sample requirements in a timely, "representative," accurate, and usable form. A well-designed, operating, and maintained sampling system is necessary for the success of the process analyzer project. Sampling is a crucial art and science for successful process analysis projects. It is a topic that is too large to more than touch on in this chapter. For more information on sampling, refer to some of the references listed.

## 2.5 Process Analytical System Installation

The installation requirements of process analyzers vary dramatically. Figure 2-2 depicts a very environmentally hardened process viscometer installed directly in the process environment with very little sample or environmental conditioning. Figure 2-3 depicts process gas chromatographs (GC) installed in an environmentally conditioned shelter. These GC analyzers require a much more conditioned/controlled sample and installation environment.

Figure 2-4 shows the exterior of one of these environmentally controlled shelters. Note the heating and ventilation unit near the door on the left and the sample conditioning cabinets mounted on the shelter wall, under the canopy.

The next most important thing to measurement and sampling technologies, when installing a process analyzer—like real estate—is location. If the recommended environment for a given analyzer is not appropriate, the project is likely doomed to failure. Also, environmental conditioning can be very expensive and needs to be included in the project cost. In many cases the cost of a shelter and/or other environmental conditioning can easily exceed the costs of the instrument itself.

*Figure 2-1: Example of a Boxed Sampling System (Courtesy of Technical Automation Service Corp. [TASC])*

Some analyzers can operate with a moderate amount of process-induced vibration, sample condition variation, and ambient environmental fluctuations. Others require highly stable environments (almost a lab type environment). Some analyzers are suitable for installation in various hazardous process environments, while others may not be. This all needs to be taken into consideration in the earlier technology selection processes. To obtain the best choice for the situation, you need to consider sampling issues, such as how far you will have to transport a sample to install the analyzer in a suitable location.

Typical installation issues, for example, include the process sample area's hazardous rating, as compared to the process analyzers hazardous area certification. Are there a number of large pumps and compressors that may cause a lot of vibration and require special or distant analyzer installation techniques? Each analyzer comes with detailed installation requirements. These should be reviewed prior to purchasing. Generalized guidelines for various analyzer types are mentioned in some of the references cited at the end of this chapter.

*Figure 2-2: Process Viscometer Installation (Courtesy of Brookfield Engineering Labs Inc.)*

## 2.6 Maintenance

Maintenance is the backbone of any analyzer project. If you can't maintain it and keep it performing when the process wants to use its results, you shouldn't have installed it. No analyzer user company has an objective of buying analyzers. They buy them to save money and/or to keep their plants running.

A cadre of properly trained and dedicated craftsmen with access to appropriate maintenance resources is essential to keep process analyzers working properly. It is not uncommon for a complex process analyzer system to require 5 to 10% of its purchase price in annual maintenance. One Raman spectroscopy application actually required a $25,000 laser twice a year. The system only cost $150,000. The result was a 33% procurement cost to annual maintenance ratio. This is high, but not necessarily unacceptable, depending on the benefit the analyzer system is providing.

Plants using a number of maintenance approaches have successfully used analyzers. Most of these approaches have included a combination of predictive, preventive, and break-down maintenance. Issues like filter cleaning, utility gas cylinder replacement, mechanical valve and other moving part overhauls, and many others tend to lend themselves to predictive and/or preventive maintenance. Many process analyzers are complex enough to require microprocessor controllers, and many of these have excess capacity that vendors have put to work performing appropriate diagnostics to advise the maintenance men of failures and/or approaching failures.

*Figure 2-3: An Inside View of an Analyzer Shelter (Courtesy of ABB Process Analytics)*

Analyzer shelters have helped encourage frequent and good maintenance checks as well as appropriate repairs. Analyzer hardware is likely to receive better attention from your maintenance department if it is easily accessible and housed in a desirable work environment (like a heated and air conditioned shelter). Figure 2-5 depicts a moderately complex GC analyzer oven. It has multiple detectors, multiple separation columns and multiple valves to monitor, control and synchronize the GC separation and measurement.

Obviously this complex of a system will be more easily maintained in a well-lighted and environmentally conducive work environment. Complex analyzers like many GC systems and spectroscopy systems have typically demonstrated better performance when installed in environmentally stable areas. Figure 2-3 depicts one of these areas with three process GCs. The top section of the unit includes a microprocessor and the complex electronics required to control the analyzer functions and operation, communicate with the outside world, and sometimes to control a variety of sample system features. The middle section is the utility gas control section and controls the required flows of several application essential gases. The lower section is the guts, so to speak. It is the thermostatically controlled separation oven with an application specific assortment of separation columns, valves, and detectors (internally depicted in Figure 2-5).

Utilizing an acceptable (possibly not the best) analytical technology that your maintenance department is already familiar with can have many positive benefits. Existing spare parts may be readily available. Maintenance technicians may already be generally trained in the technology and, if the demonstrated applications have been highly successful, they may go into the start-up phase with a

*Figure 2-4: An Outside View of an Analyzer Shelter*
*(Courtesy of Technical Automation Service Corp. [TASC])*

very positive attitude. Your control system may already have connections to appropriate GC or other data highways.

Calibration is generally treated as a maintenance function. Calibrations are treated differently in different applications and facilities. Most regulatory and environmental applications require frequent calibrations (auto or manual). Many plants that are strongly into SPC and six sigma have come to realize they can actually induce some minor instability into a process by over calibrating their monitoring and control instruments. These organizations have primarily begun averaging multiple calibrations and using the average number for their calibration. They then conduct benchmark calibrations, or calibration checks, and as long as the check is with in the statistical guidelines of the original calibration series, they make no adjustments to the instrument. If the results are outside of the acceptable statistical range, they not only recalibrate but also go over the instrument/application to try to figure out what may have cause the change.

With the declining price of microprocessors, they are finding their way into more and more even simpler analytical instruments. With the excess computing capability that comes with many of these systems, an increasing number of vendors have been developing diagnostic and maintenance packages to aid in maintaining these analytical systems. They are typically called performance monitoring and/or diagnostics systems. They often monitor the status of the instrument, its availability to the process, keep a failure and maintenance history, contain software maintenance documentation/manuals, and provide appropriate alarms to the process and maintenance departments.

*Figure 2-5: Example of a Process GC Oven (Courtesy of ABB Process Analytics)*

Lastly, maintenance work assignments and priorities are especially tricky for process analytical instruments. Most are somewhat unique in complexity and other issues. Two process GC systems that look similar may require significantly different maintenance support because of process, sample, and/or application differences. Consequently it is usually best to develop your maintenance workload assignments based on actual maintenance histories when available, averaged out to eliminate individual maintenance worker variations, and time weighted to give more weight to recent history and consequently give more weight to the current (possibly improved or needing replacement) installation.

Maintenance priorities are quite complex and require a multidisciplinary team effort to determine. What the analyzer is doing for you at any given time can impact its priority. Some analyzers are primarily used during start up and shut down and have a higher priority as these operations approach. Others are absolutely required to run the plant (environmental and safety may be in this category) and, consequently, have an extremely high priority. Others can be prioritized based on the financial savings they provide for the company.

The multidisciplinary team must decide which analyzers justify immediate maintenance, including call-ins and/or vendor support. Some may only justify normal available workday maintenance activities. After you have gone through one of these priorities setting exercises, you will likely have a much better understanding of the value of your analytical installations to your operation. If you don't have adequate maintenance monitoring programs/activities in place, it can be very difficult to assess workloads and/or priorities. The first step in implementing these activities must be to collect the data that is necessary to appropriately make these types of decisions.

## 2.7 Utilization of Results

Process analytical results are used for many purposes. The following list covers some of the most prominent uses:

- Closed Loop Control
- Open Loop Control

- Process Monitoring
- Product Quality Monitoring
- Environmental Monitoring
- Safety Monitoring

With the large installed data communications base that exists in most modern plants, the process analysis results/outputs arc used by most major systems including process control systems, laboratory information management systems, plant information systems, maintenance systems, safety systems, enterprise systems, and regulatory reporting systems (like environmental reports to the EPA). How these results are used by the various groups have been alluded to in several of the previous sections and is more completely discussed in some of the cited references.

## 2.8 References

Carr-Brion, K. G. and J. R. P. Clarke. *Sampling Systems For Process Analysers.* Second Edition. Butterworth–Heinemann, 1996.

Houser, E. A. *Principles of Sample Handling and Sampling Systems Design For Process Analysis.* ISA, 1972.

Lipták, B. G., Editor-in-Chief. *Instrument Engineers' Handbook–Process Measurement and Analysis (Volume 1).* Fourth Edition. CRC Press and ISA, 2003.

Sherman, R. E., editor, and L. J. Rhodes. *Analytical Instrumentation.* Practical Guides for Measurement and Control Series. ISA, 1996.

Sherman, R. E. *Process Analyzer Sample-Conditioning System Technology.* Wiley Series in Chemical Engineering. John Wiley and Sons Inc., 2002.

## About the Author

**James F. (Jim) Tatera** is a Senior Process Analysis Consultant with Tatera & Associates Inc. For several years, he has provided process analytical consulting/contracting services to user, vendor and academic organizations. His 32-plus year career has included over 27 years of working with process analyzers for Dow Corning, including both U.S. and international assignments in analytical research, process engineering, project engineering, production management, and maintenance management. He is an ISA Fellow, is one of the original Certified Specialists in Analytical Technology (CSAT), is active in U.S. and IEC process analysis standards activities, and has received several awards for his work in the process analysis field. He is the ANSI USNC Technical Advisor to IEC SC65D (Analyzing Equipment in Industrial Process Measurement and Control) and has served as the ISA SP76 (Compositional Analyzers) committee chairman and member.

# 3 Continuous Control

*By Harold Wade*

## Topic Highlights

*Process Characteristics*
*Feedback Control*
*Controller Tuning*
*Advanced Regulatory Control*

## 3.1 Introduction

Continuous control refers to a form of automatic process control in which the information—both from sensing elements and to actuating devices—can have any value between minimum and maximum limits. This is in contrast to discrete control, where the information normally is in one of two states, such as on-off, open-closed, run-stop, etc.

Continuous control is organized into feedback control loops, as shown in Figure 3-1. In addition to a controlled process, each control loop consists of a sensing device that measures the value of a controlled variable, a controller that contains the control logic plus provisions for human interface, and an actuating device that manipulates the rate of addition or removal of mass or energy or some other property that can affect the controlled variable. (In emerging technology, the control logic may be located at either the sensing or the actuating device.)

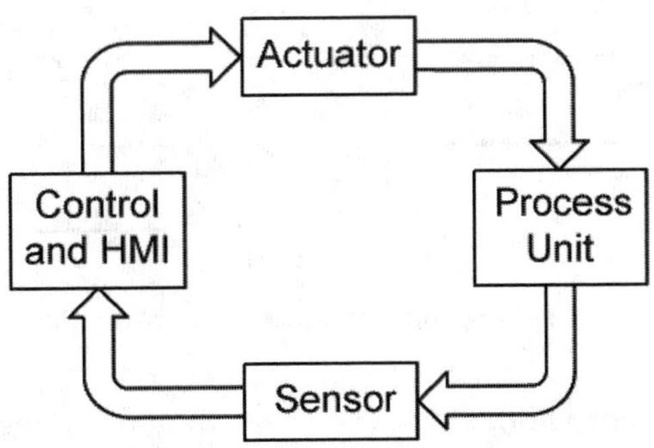


*Figure 3-1: Components and Information Flow in a Feedback Control Loop*

Continuous process control is used quite extensively in industries where the product is in a continuous, usually fluid, stream. Representative industries are petroleum refining, chemical and petrochemical, power generation, and municipal utilities. Continuous control can also be found in processes in which the final product is produced in batches, strips, slabs, or as a web in, for example, the pharmaceutical, pulp and paper, steel and textile industries. There are also applications for continuous control in the discrete industries—for instance, a temperature controller on an annealing furnace, or motion control in robotics.

Since the components in a feedback control loop may be physically separated by 50 meters (m) to 500 m or more, some form of signal communication must be employed. An early technology for signal transmission used in the process industries employed pneumatic signals, where 3 pounds per square inch (psi) [20.684 kiloPascals (kPa)] represented the minimum value and 15 psi (103.42 kPa) represented the maximum value. This technology has largely been replaced by electric signal transmission, utilizing a signal range of 4-to-20 milliamps (mA) direct current (DC). Electric signal transmission is the dominant technology used today, although this is being replaced by shared digital signals and by wireless communication.

The central device in a control loop, the controller, may be built as a stand-alone device or may exist as shared components in a digital system, such as a distributed control system (DCS) or programmable logic controller (PLC).

Process control systems, plus the related functions of measurements and alarms, are represented using special symbols on "piping and instrument diagrams" (P&IDs). A P&ID shows the outline of the process units and connecting piping as well as a standard symbolic representation of the instrumentation and control (I&C) systems. Figure 3-2 is an example of a small P&ID for a heat exchanger. In actual practice, however, a typical P&ID will encompass many process units, measurements and controls, and will be densely drawn on one or more large sheets of paper.

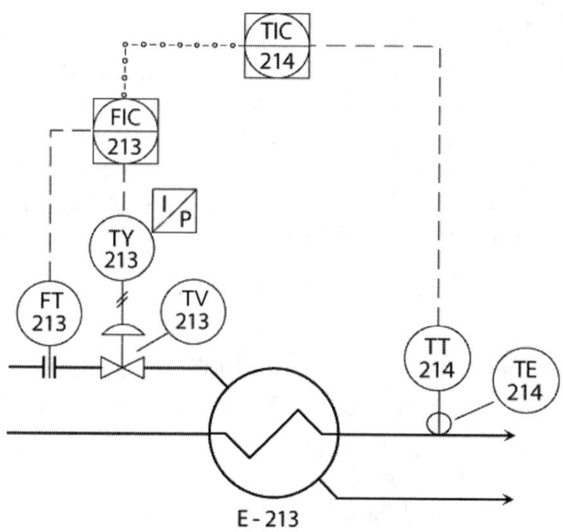


*Figure 3-2: Piping and Instrumentation Diagram (P&ID)*

The symbols used to represent various types of instrumentation devices, the type of communication between devices, and the nomenclature for device identification are defined by the standard ISA-5.1-1984 (R1992) -*Instrumentation Symbols and Identification.* Figure 3-3 shows the symbols used for continuous controllers and other general instrumentation, as defined in this standard.

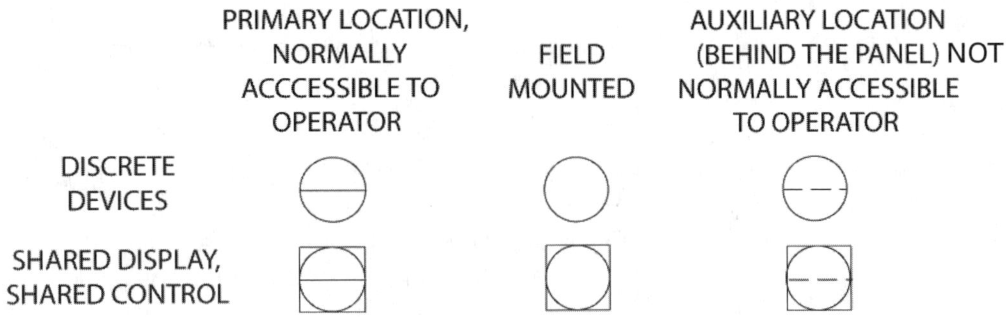


*Figure 3-3: Symbols for Continuous Controllers and Other General Instruments, from ISA-5.1-1984 (R1992) - Instrumentation Symbols and Identification*

## 3.2 Process Characteristics

In order to understand feedback control loops, one must understand the characteristics of the controlled process. Listed below are characteristics of almost all processes, regardless of the application or industry.

- Industrial processes are non-linear; that is, they will exhibit different responses at different operating points.
- Industrial processes are subject to random disturbances, due to fluctuations in feedstock, environmental effects, and changes or malfunctions of equipment.
- Most processes contain some amount of dead time; a control action will not produce an immediate feedback of its effect.
- Many processes are interacting; a change in one controller's output may affect other process variables besides the intended one.
- Most process measurements contain some amount of noise.
- Most processes are unique; processes using apparently identical equipment may have individual idiosyncrasies.

A typical response to a step change in signal to the actuating device is shown in Figure 3-4.

In addition, there are physical and environmental characteristics that must be considered when selecting equipment and installing control systems.

- The process may be toxic, requiring exceptional provisions to prevent release to the environment.
- The process may be highly corrosive, limiting the selection of materials for components that come in contact with the process.
- The process may be highly explosive, requiring special equipment housing or installation technology for electrical apparatus.

## 3.3 Feedback Control

The principle of feedback control is that, if a controlled variable deviates from its desired value (setpoint), corrective action moves a manipulated variable (the controller output) in a direction that

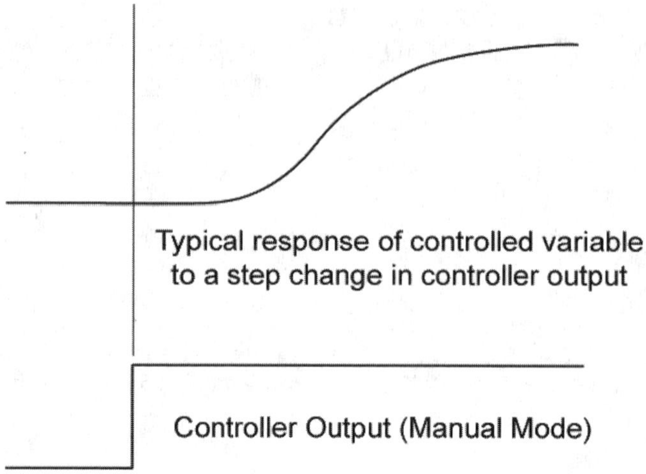


*Figure 3-4: Typical Response of Controlled Variable to Step-Change in Controller Output (Manual Mode)*

causes the controlled variable to return toward setpoint. Most feedback control loops in industrial processes utilize a proportional-integral-derivative (PID) control algorithm. There are several forms of the PID. There is no standardization for the names. The names "ideal," "interactive," and "parallel" are used here, although some vendors may use other names.

### 3.3.1 Ideal PID Algorithm

The most common form of PID algorithm is the *ideal* form (also called the "ISA" form). This is represented in mathematical terms by Equation 3-1, and in block diagram form by Figure 3-5:

$$m = K_C\left(e + \frac{1}{T_I}\int edt + T_D\frac{de}{dt}\right) \tag{3-1}$$

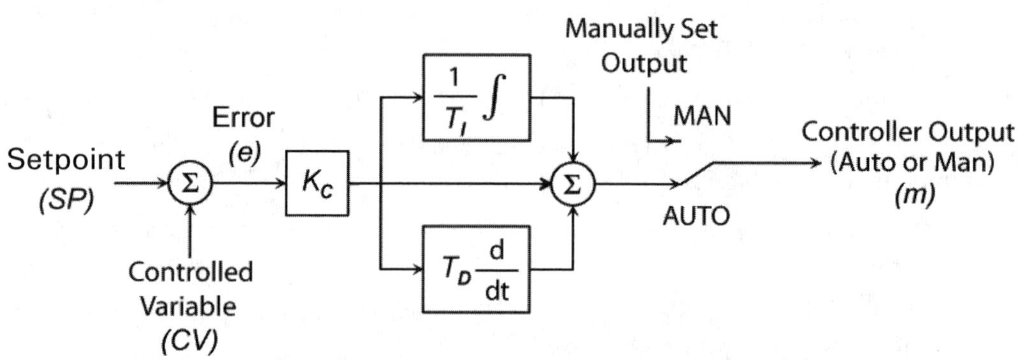


*Figure 3-5: Block Diagram of "Ideal" PID Algorithm,*
*also Showing Functional Form of Automatic – Manual Switch*

Here, $m$ represents the controller output; $e$ represents the error (difference between setpoint and controlled variable). Both $m$ and $e$ are in percent of span. The symbols $K_C$ (controller gain), $T_I$ (integral time) and $T_D$ (derivative time) represent tuning parameters that must be adjusted for each application.

The terms in the algorithm represent the proportional, integral, and derivative contributions to the output. The proportional mode is responsible for most of the correction. The integral mode assures that, in the long-term, there will be no deviation between setpoint and controlled variable. The derivative mode may be used for improved response of the control loop. In practice, the proportional and integral modes are almost always used; the derivative mode is often omitted, simply by setting $T_D = 0$.

There are other forms for the tuning parameters. For instance, controller gain may be expressed as proportional band (PB), defined as the amount of measurement change (in percent of measurement span) required to cause 100% change in the controller output. The conversion between controller gain and proportional band is shown by Equation 3-2:

$$K_C = \frac{100}{PB} \qquad PB = \frac{100}{K_C} \tag{3-2}$$

The integral mode tuning parameter may be expressed in reciprocal form, called *reset rate*. Whereas $T_I$ is normally expressed in "minutes per repeat," reset rate is expressed in "repeats per minute." The derivative mode tuning parameter, $T_D$, is always in time units, usually in minutes. (Traditionally, the time units for tuning parameters has been "minutes." Today, however, many vendors are expressing the time units in "seconds.")

### 3.3.2 Interactive PID Algorithm

The *interactive* form, depicted by Figure 3-6, was the predominant form for analog controllers and is used by some vendors today. Other vendors provide a choice of the ideal or interactive form. There is essentially no technological advantage to either form; however, the required tuning parameters differ if the derivative mode is used.

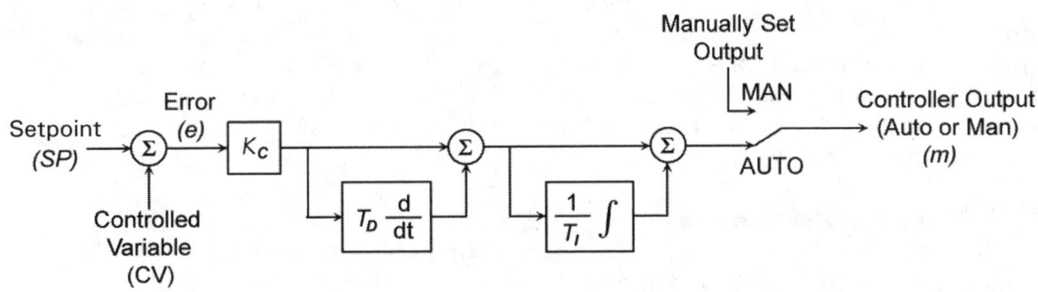


*Figure 3-6: Block Diagram of Interactive PID Algorithm*

### 3.3.3 Parallel PID Algorithm

The *parallel* form, shown in Figure 3-7, uses independent gains on each mode. This form has traditionally been used in the power generation industry and in such applications as robotics, flight control, motion control, etc. Other than power generation, it is rarely found in the continuous process industries. With compatible tuning, the ideal, interactive, and parallel forms of PID produce identical performance; hence no technological advantage can be claimed for any form. The tuning procedure for the parallel form differs decidedly from that of the other two forms.

### 3.3.4 Time Proportioning Control

Time proportioning refers to a form of control in which the PID controller output consists of a series of periodic pulses whose duration is varied to relate to the normal continuous output. For example, if the fixed cycle base is 10 seconds, a controller output of 30% will produce an "on" pulse of 3 seconds and an "off" pulse of 7 seconds. An output of 75% will produce an "on" pulse of 7.5 seconds and an "off"

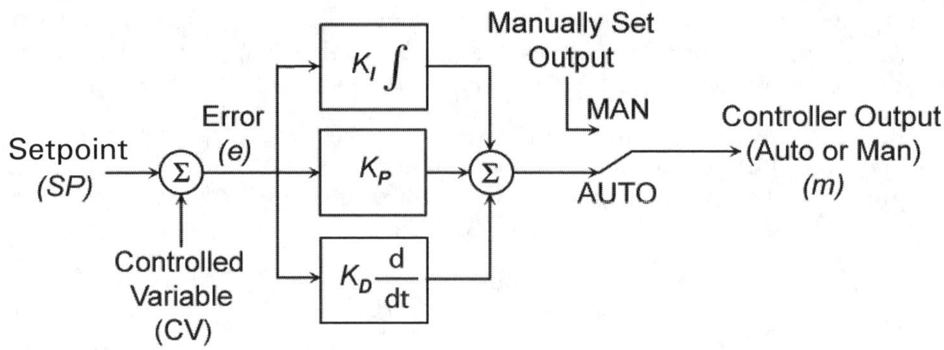


*Figure 3-7: Block Diagram of Parallel Algorithm (Showing Independent Gains on Each Mode)*

pulse of 2.5 seconds. This type of control is usually applied where the cost of an on-off final actuating device is considerably less than the cost of a modulating device. In a typical application, the "on" pulses apply heating or cooling by turning on a resistance heating element, an SCR (silicon controlled rectifier) or a solenoid valve. The mass of the process unit, say, a plastics extruder barrel, acts as a filter to remove the low-frequency harmonics and apply an even amount of heating or cooling to the process.

### 3.3.5 Manual-Automatic Switching

It is desirable to provide a means for process operator intervention in a control loop in event of abnormal circumstances, such as a sensor failure or a major process upset. Figures 3-5, 3-6, and 3-7 show a manual-automatic switch that permits switching between manual and automatic modes. In the manual mode, the operator can set the signal to the controller output. However, when the switch is returned to the automatic position, the automatic controller output must match the operator's manual setting or else there will be a "bump" in the controller output. (The term *bumpless transfer* is frequently used.) With older technology, it was the operator's responsibility to prevent bumping the process. With current technology, bumpless transfer is built into most control systems; some vendors refer to this as initializing the control algorithm.

### 3.3.6 Direct and Reverse Acting

For safety and environmental reasons, most final actuators such as valves will close in the event of loss of signal or power to the actuator. There are instances, however, when the valve should open in the event of signal or power failure. Once the failure mode of the valve is determined, the action of the controller must be selected. Controllers may be either *direct acting* (DA) or *reverse acting* (RA). If a controller is direct acting, an increase in the controlled variable will cause the controller output to increase. If the controller is reverse acting, an increase in the controlled variable will cause the output to decrease. Since most control valves are fail-closed, then the majority of the controllers are set to be reverse acting. The setting—DA or RA—is normally made at the time the control loop is commissioned. With some DCSs, the DA/RA selection can be made without considering the failure mode of the valve; then a separate selection is made as to whether to reverse the analog output signal. This permits the human-machine interface (HMI) to display all valve positions in a consistent manner, 0% for closed and 100% for open.

### 3.3.7 Activation for Proportional and Derivative Modes

Regardless of algorithm form, there are certain configuration options that every vendor offers. One configuration option is the DA/RA setting. Other configuration options pertain to the actuating signal for the proportional and derivative modes. Note that, in any of the forms of algorithms, if the derivative mode is being used ($T_D \neq 0$), a setpoint change will induce an undesirable spike on the controller output. A configuration option permits the user to make the derivative mode sensitive only to changes in the controlled variable, not to the setpoint. This choice is called "derivative-on-error" or "derivative-on-measurement."

Even with derivative-on-measurement, on a setpoint change, the proportional mode will cause a step change in controller output. This, too, may be undesirable. Therefore, a similar configuration option permits the user to select "proportional-on-measurement" or "proportional-on-error." Figure 3-8 shows both proportional and derivative modes sensitive to measurement changes alone. This leaves only the integral mode on error, where it must remain, since it is responsible for assuring the long-term equality of setpoint and controlled variable. In event of a disturbance, there is no difference in the responses of derivative-on-measurement, proportional-on-measurement, and the configuration with all modes on error.

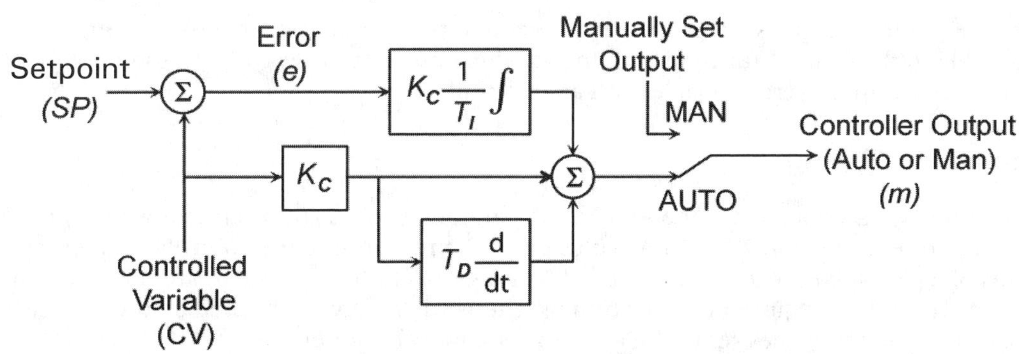


*Figure 3-8: Block Diagram of Ideal PID Algorithm with Both Proportional and Derivative Modes on Measurement*

### 3.3.8 Discrete Forms of PID

The algorithm forms presented above, using calculus symbols, are applicable to analog controllers that operate continuously. However, control algorithms implemented in a digital system are processed at discrete sample instants (for instance, one-second intervals), rather than continuously. Therefore, a modification must be made to show how a digital system approximates the continuous forms of the algorithm presented above. Digital processing of the PID algorithm also presents an alternative that was not present in analog systems. At each sample instant the PID algorithm can calculate either a new position for the controller output or the increment by which the output should change. These forms are called the *position* and the *velocity* forms, respectively. Assuming that the controller is in the automatic mode, the following equations are processed at each sample instant for the position algorithm. The subscript "*n*" refers to the nth processing instant, "*n-1*" to the previous processing instant, and so on.

Compute the error: $e_n = SP_n - CV_n$ (3-3)

Increment sum of errors: $S_n = S_{n-1} + e_n$

Compute controller output: $m_n = K_C\left[e_n + \frac{\Delta T}{T_I}S_n + \frac{T_D}{\Delta T}(e_n - e_{n-1})\right]$

Save $S_n$ and $e_n$ values for the subsequent processing time.

The velocity mode or incremental algorithm is similar. It computes the amount by which the controller output should be changed at the $n^{th}$ sample instant.

Compute change in controller output:

$$\Delta m_n = K_C\left[(e_n - e_{n-1}) + \frac{\Delta T}{T_I}e_n + \frac{T_D}{\Delta T}(e_n - 2e_{n-1} + e_{n-2})\right] \tag{3-4}$$

Add the incremental output to the previous value of controller output, to create new value of output:

$$m_n = m_{n-1} + \Delta m_n \tag{3-5}$$

Save $m_n$, $e_{n-1}$, and $e_{n-2}$ values for the subsequent processing time.

From a user point of view, there is no advantage of one form over the other. Vendors, however, may prefer a particular form due to the ease of incorporation of features of their system, such as tuning and bumpless transfer.

The configuration options—DA/RA, derivative- and proportional-on-measurement, or error—are also applicable to the discrete forms of PID. In fact, there are more user configuration options offered with digital systems than were available with analog controllers.

## 3.4 Controller Tuning

In the previous section, it was mentioned the parameters $K_C$, $T_I$ (or their equivalents, proportional band and reset rate), and $T_D$ must be adjusted so the response of the controller matches the requirements of a particular process. This is called "tuning the controller." There are no hard and fast rules as to the performance requirements for tuning. These are largely established by the particular process application and by the desires of the operator or controller tuner.

### 3.4.1 Acceptable Criteria for Loop Performance

One widely used response criterion is the loop should exhibit a quarter-amplitude decay following a setpoint change. See Figure 3-9. For many applications, however, this is too oscillatory. A smooth response to a setpoint change with minimum overshoot is more desirable. A response to setpoint change that provides minimum overshoot is considered less aggressive tuning than quarter-amplitude decay. The penalty for less aggressive tuning is that a disturbance will cause a greater deviation from setpoint or a longer time to return to setpoint. The controller tuner must decide the acceptable criterion for loop performance before actual tuning.

Controller tuning techniques may be divided into two broad categories: those that require testing of the process, either with the controller in automatic or manual, and those that are less formal, often called *trial-and-error* tuning.

### 3.4.2 Tuning from Open Loop Tests

The open-loop process testing method uses only the manually-set output of the controller. A typical response to a step change in output was shown in Figure 3-4. It is often possible to approximate the response with a simplified process model containing only three parameters—the process gain ($K_p$), the dead time in the process ($T_d$), and the process time constant ($\tau_p$). Figure 3-10 shows the response of a first-order-plus-dead-time (FOPDT) model that approximates the true process response.

Figure 3-10 also shows the parameter values, $K_p$, $T_d$ and $\tau_p$. There are a number of published correlations for obtaining controller tuning parameters from these process parameters. The best known is based upon the Ziegler-Nichols reaction curve method. Correlations for P-only, PI, and PID controllers are given here in Table 3-1.

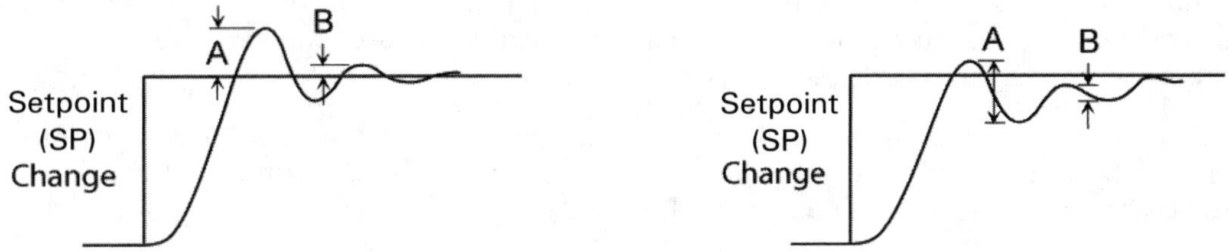


$$Decay\ Ratio = \frac{Second\ deviation\ from\ SP}{First\ deviation\ from\ SP} = \frac{B}{A}$$

a. Decay Ratio Definition in Normal Situations

$$Decay\ Ratio = \frac{Second\ (peak - valley)}{First\ (peak - valley)} = \frac{B}{A}$$

b. Decay Ratio Definition in Abnormal Situations (Also can be used to determine decay ratio following a disturbance)

*Figure 3-9: Decay Ratio Definitions*

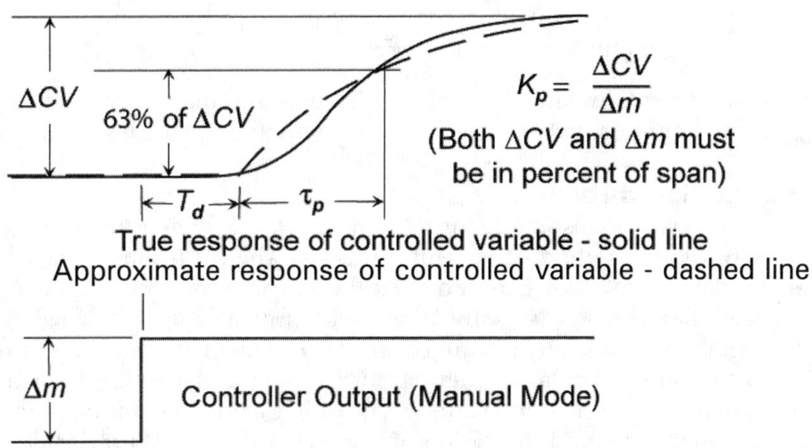


*Figure 3-10: Approximating the Open-Loop (Controller in Manual) Response to Step Change in Output with a Simplified Process Model*

*Table 3-1: Controller Tuning Parameters Based Upon Open Loop Test Data*

| | **P-Only** | **PI** | **PID** |
|---|---|---|---|
| $K_C$ | $\frac{\tau_p}{K_p T_d}$ | $\frac{0.9\tau_p}{K_p T_d}$ | $\frac{1.2\tau_p}{K_p T_d}$ |
| $T_I$ | --- | $3.33T_d$ | $2.0T_d$ |
| $T_D$ | --- | --- | $0.5T_d$ |

Another tuning technique that uses the same open-loop process test data is called "lambda tuning." The objective of this technique is for the setpoint response to be an exponential rise with a specified time constant, $\lambda$. This technique is applicable whenever it is desired to have a very smooth setpoint response, at the expense of degraded response to disturbances.

There are other elaborations of the open-loop test method, including multiple valve movements in both directions, numerical regression methods for obtaining the process parameters, etc. Despite its simplicity, the open-loop method suffers from the following problems:

- It may not be possible to interrupt normal process operations to make the test.
- If there is noise on the measurement, it may not be possible to get good data, unless the controlled variable change is at least five times the amplitude of the noise. For many processes, that may be too much disturbance.
- The technique is very sensitive to parameter estimation error, particularly if the ratio of $T_d/\tau_p$ is small.
- The method does not take into consideration the effects of valve stiction.
- The actual process response may be difficult to approximate with an FOPDT model.
- A disturbance to the process during the test will severely deteriorate the quality of the data.
- For very slow processes, the complete results of the test may require one or more working shifts.
- The data is valid only at one operating point. If the process is nonlinear, additional tests at other operating points may be required.

Despite these problems, under relatively ideal conditions—minimal process noise, minimal disturbances during the test, minimal valve stiction, etc.—the method provides acceptable results.

### 3.4.3 Tuning from Closed Loop Tests

Another technique is based on testing the process in the closed-loop. (Ziegler-Nichols referred to this as the "ultimate sensitivity" method.) To perform this test, the controller is placed in the automatic mode, integral and derivative actions are removed (or a proportional-only controller is used), a low controller gain is set, then the process is disturbed—either by a setpoint change or a forced disturbance—and the oscillating characteristics are observed. The objective is to repeat this procedure with increased gain until sustained oscillation (neither increasing nor decreasing) is achieved. At that point, two pieces of data may be obtained: the value of controller gain (called the "ultimate gain," $K_{CU}$) that produced sustained oscillation, and the period of the oscillation, $P_U$. With this data, one can enter Table 3-2 and calculate tuning parameters for a P-only, PI, or PID controller.

*Table 3-2: Controller Tuning Parameters Based Upon Closed Loop Test Data*

| | P-Only | PI | PID |
|---|---|---|---|
| $K_C$ | $0.5K_{CU}$ | $0.45K_{CU}$ | $0.6K_{CU}$ |
| $T_I$ | --- | $0.83P_U$ | $0.5P_U$ |
| $T_D$ | --- | --- | $0.125P_U$ |

There are also problems with the closed-loop method.

- It may not be possible to subject the process to a sustained oscillation.

- Even if that were possible, it is difficult to predict or to control the magnitude of the oscillation.
- Multiple tests may be required, resulting in long periods of interruption to normal operation.

Despite these problems, there are certain advantages to the closed-loop method.

- Minimal uncertainty in the data. (Frequency, or its inverse, period, can be measured quite accurately.)
- The method inherently includes the effect of a sticking valve.
- Moderate disturbances during the testing can be tolerated.
- No *a priori* assumption as to the form of the process model is required.

A modification of the closed-loop method, called the *relay method,* attempts to exploit the advantages while circumventing most of the problems. The relay method utilizes the establishment of maximum and minimum limits for the controller output. For instance, if controller output normally is 55%, the maximum output can be set at 60% and the minimum at 50%. While this does not establish hard limits for excursion of the controlled variable, persons familiar with process will feel comfortable with these settings or will reduce the difference between the limits.

The process is then tested by a setpoint change or a forced disturbance, using an on-off controller. If an on-off controller is not available, then a P-only controller with a maximum value of controller gain can be substituted. The controlled variable will oscillate above and below setpoint, with the controller output at either the maximum or minimum value, as shown in Figure 3-11.

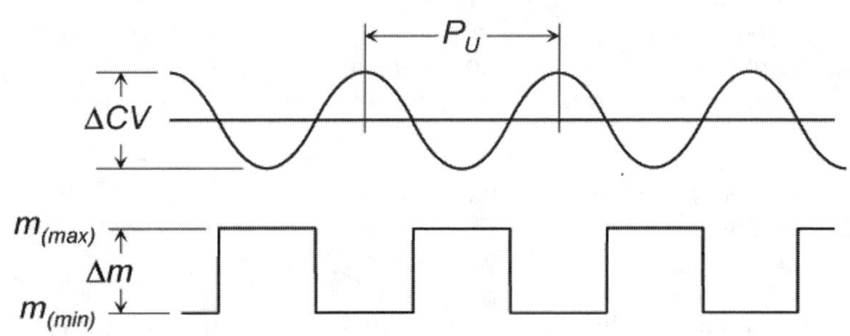


*Figure 3-11: On-Off Controller Output and Oscillating Controlled Variable in a Relay Test*

If the period of time when the controller output is at the maximum setting exceeds the time at the minimum, then both the maximum and limits should be shifted upward by a small but identical amount. After one or more adjustments, the output square wave should be approximately symmetrical. At that condition, the period of oscillation, $P_U$, is the same as would have been obtained by the previously described closed-loop test. Furthermore, the ultimate gain can be determined from a ratio of the controller output and CV amplitudes:

$$K_{CU} = \frac{4}{\pi}\frac{\Delta m}{\Delta CV} \tag{3-6}$$

Thus the data required to enter Table 3-2 and calculate tuning parameters has been obtained in a much more controlled manner than the unbounded closed loop test.

While the relay method is a viable technique for manual testing, it can also be easily automated. For this reason, it is the basis for some vendors' self-tuning techniques.

### 3.4.4 Trial-and-Error Tuning

Despite these tools for formal process testing for determination of tuning parameters, many loops are tuned by trial-and-error. That is, an unsatisfactory loop closed-loop behavior is observed, and an estimate (often merely a guess) is made as to which parameter(s) should be changed and by how much. Good results often depend upon the person's experience. Various methods of visual pattern recognition have been described but, in general, such tuning techniques remain more of an art than a science.

A recently published technique, Wade's *Basic and Advanced Regulatory Control: System Design and Application* (see 3.6.1) called *improving as-found tuning*, or "intelligent trial-and-error tuning," attempts to place controller tuning on a more methodological basis. The premise of this technique, which is applicable only to PI controllers, is that a well-tuned controller exhibiting a slight oscillation (oscillations that are decaying rapidly) will have a predictable relation between the integral time and period of oscillation. The following relation has been found to provide acceptable results:

$$1.5 \le \frac{P}{T_I} \le 2.0 \qquad (3\text{-}7)$$

Further insight into this technique can be gained by noting that the phase shift through a PI controller, from error to controller output, depends very strongly on the ratio $P/T_I$ and only slightly on the decay ratio. For a control loop with a quarter-amplitude decay, the limits above are equivalent to specifying a phase shift of approximately $15^o$.

If a control system engineer or instrumentation technician is called upon to correct the errant behavior of a control loop, then (assuming that it is a tuning problem and not some external problem) the "as-found" behavior is caused by the "as-found" tuning parameter settings. The behavior can be characterized by the decay ratio *(DR)* and the period *(P)* of oscillation. The as-found data set—$K_C$, $T_I$, *DR*, *P*—represents a quanta of knowledge about the process. If either an open-loop or closed-loop test were made in an attempt to determine tuning parameters, then the existing knowledge about the process would be sacrificed.

From Equation 3-7, upper and lower limits for an acceptable period can be established.

$$1.5\,T_I \le P \le 2.0\ T_I \qquad (3\text{-}8)$$

If the as-found period *P* meets this criteria, the implication is the integral time is acceptable. Hence, adjustments should be made to the controller gain $K_C$ until the desired decay ratio is obtained. If the period is outside this limit, then the present period can be used in the inverted relation to determine a range of acceptable new values for $T_I$:

$$0.5\ P \le T_I \le 0.67\,P \qquad (3\text{-}9)$$

Wade's *Basic and Advanced Regulatory Control: System Design and Application* (see 3.6.1) and "Trial and error: an organized procedure" (see 3.6.2) contain more information, including a flow chart, describing this technique.

### 3.4.5 Self-Tuning

Although self-tuning, auto-tuning and adaptive-tuning have slightly different connotations, they will be discussed collectively here.

There are two different circumstances where some form of self-tuning would be desirable:

1. If a process is highly nonlinear and also experiences a wide range of operating points, then a technique that automatically adjusts the tuning parameters for different operating conditions would be highly beneficial.
2. If a new process unit with many control loops is to be commissioned, it would be beneficial if the controllers could determine their own best tuning parameters.

There are different technologies that address these situations.

For initial tuning, there are commercial systems that in essence automate the open-loop test procedure. On command, the controller will revert to the manual mode, test the process, characterize the response by a simple process model, then determine appropriate tuning parameters. Most commercial systems that follow this procedure display the computed parameters and await confirmation before entering the parameters into the controller. An automation of the relay tuning method described previously falls into this category.

The simplest technique addressing the nonlinearity problem is called *scheduled tuning.* If the nonlinearity of a process can be related to a key parameter such as process throughput, then a measure of that parameter can be used as an index to a lookup table (schedule) for appropriate tuning parameters. The key parameter may be divided into regions, with suitable tuning parameters listed for each region. Note that this technique depends upon the correct tabulation of tuning parameters for each region. There is nothing in the technique that evaluates the loop performance and automatically adjusts the parameters based upon the evaluation.

There are also systems that attempt to recognize features of the response to normal disturbances to the loop. From these features, heuristic rules are used to calculate new tuning parameters. These may be displayed for confirmation, or they may be entered into the algorithm "on the fly." Used in this manner, the system tries to adapt the controller to the random environment of disturbances and setpoint changes as they occur.

There are also "third party" packages, typically running in a notebook computer, that access data from the process, such as by transferring data from the DCS data highway. The data is then analyzed and advisory messages are presented that suggest tuning parameters and provide an indication of the "health" of control loop components, especially the valve.

## 3.5 Advanced Regulatory Control

If the process disturbances are few and not severe, feedback controllers will maintain the average value of the controlled variable at setpoint. But in the presence of frequent or severe disturbances, feedback controllers permit significant variability in the control loop. This is because a feedback controller *must* experience a deviation from setpoint in order to change its output. This variability may result in an economic loss. For instance a process may operate at a safe margin away from a target value to prevent encroaching on the limit and producing off-spec product. Reducing the margin of safety will produce some economic benefit, such as reduced energy consumption, reduced raw material usage or increased production. Reducing the variability cannot be done by feedback controller tuning alone. It may be accomplished by the use of more advanced control loops such as ratio, cascade, feedforward, decoupling, and selector control.

### 3.5.1 Ratio Control

Often, when two or more ingredients are blended or mixed, the flow rate of one of the ingredients paces the production rate. The flow rates for the other ingredients are controlled to maintain a specified ratio to the pacing ingredient. Figure 3-12 shows a ratio control loop. Ratio control systems are found in batch processing, fuel oil blending, combustion processes where the air flow may be ratioed to the fuel flow, and many other applications. The pacing stream is often called the "wild" flow, since it may or may not be provided with an independent flow rate controller—only a measurement of the wild flow stream is utilized in ratio control.

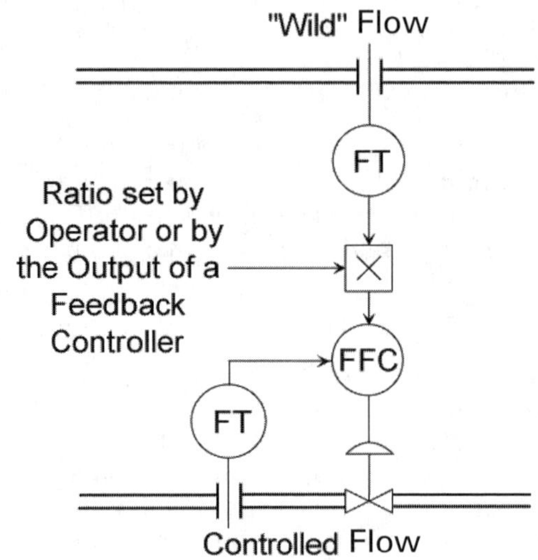


*Figure 3-12: Ratio Control Strategy*

The specified ratio may be manually set, automatically set from a batch recipe, or adjusted by the output of a feedback controller. An example of the latter is a process heater that uses a stack oxygen controller to adjust the air-to-fuel ratio. When the required ratio is automatically set by a higher-level feedback controller, the ratio control strategy is merely one form of feedforward control.

### 3.5.2 Cascade Control

Cascade control refers to control schemes that have an inner control loop nested within an outer loop. The feedback controller in the outer loop is called the "primary" controller. Its output sets the setpoint for the inner loop controller, called the "secondary." The secondary control loop must be significantly faster than the primary loop. Figure 3-13 depicts an example of cascade control applied to a heat exchanger. In this example a process fluid is heated with a hot oil stream. A temperature controller on the heat exchanger output sets the setpoint of the hot oil flow controller.

If the temperature controller directly manipulated the valve, there would still be a valid feedback control loop. Any disturbance to the loop, such as a change in the process stream flow rate or a change in hot oil supply pressure, would require a new position of the control valve. Therefore, a deviation of temperature from setpoint would be required to move the valve.

With the secondary loop installed as shown in Figure 3-13, a change in hot oil supply pressure will result in a change in hot oil flow. This will be rapidly detected by the flow controller which will then make a compensating adjustment to the valve. The momentary variation in hot oil flow will cause minimal, if any, disturbance to the temperature control loop.

In the general situation, all disturbances within the secondary loop—a sticking valve, adverse valve characteristics, or (in the example) variations in supply pressure—are confined to the secondary loop and have minimal effect on the primary controlled variable. A disturbance that directly affects the primary loop, such as a change in process flow rate in the example, will require a deviation at the primary controller for its correction regardless of the presence or absence of a secondary controller.

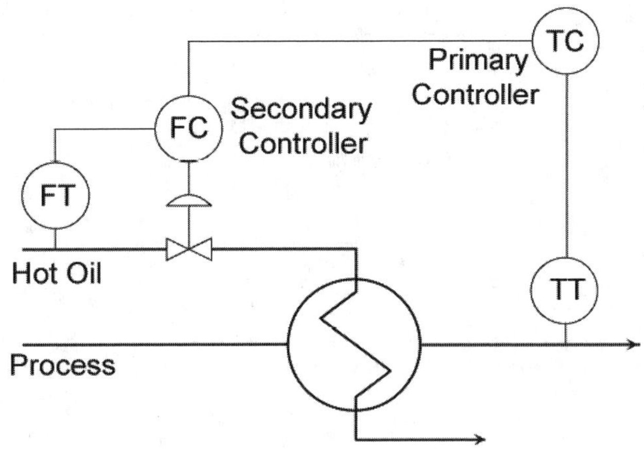


*Figure 3-13: Example of Cascade Control Strategy*

When examining a process control system for possible improvements, consider whether intermediate control loops can be closed to encompass certain of the disturbances. If so, the effect of these disturbances will be removed from the primary controller.

### 3.5.3 Feedforward Control

Feedforward control is defined as the manipulation of the final control element—valve position or setpoint of a lower-level flow controller—using a measure of a disturbance rather than the output of a feedback controller. In essence, feedforward control is open loop control. Feedforward control requires a process model in order to know how much and when correction should be made for a given disturbance. If the process model were perfect, feedforward control alone could be used. In actuality, the process model is never perfect; therefore, feedforward and feedback control are usually combined.

The example in the previous section employed cascade control to overcome the effect of disturbances caused by variations in hot oil supply pressure. It was noted, however, that variations in process flow rate would still cause a disturbance to the primary controller. If the process and hot oil flow rates varied in a proportionate amount, there would be only minimal effect on the process outlet temperature. Thus a ratio between the hot oil and process flow rates should be maintained. While this would eliminate most of the variability at the temperature controller, there may be other circumstances, such as heat exchanger tube scaling, that would necessitate a long-term shift in the required ratio. This can be implemented by letting the feedback temperature controller set the required ratio as shown in Figure 3-14.

Ratio control, noted earlier as an example of feedforward-feedback control, corrects for the steady-state effects on the controlled variable. Suppose that there is also a difference in dynamic effects of the hot oil and process streams on the outlet temperature. In order to synchronize the effects at the outlet temperature, *dynamic compensation* may be required in the feedforward controller.

To take a more general view of feedforward, consider the generic process shown within the dotted lines in Figure 3-15. This process is subject to two influences (inputs)—a disturbance and a control effort. The control effort may be the signal to a valve or to a lower level flow controller. In this latter case, the flow controller can be considered as a part of the process. Transfer functions *A(s)* and *B(s)* are mathematical abstractions of the dynamic effect of each of the inputs on the controlled variable. A feedforward controller *C(s)*, a feedback controller, and the junction combining feedback and feedforward are also shown in Figure 3-15.

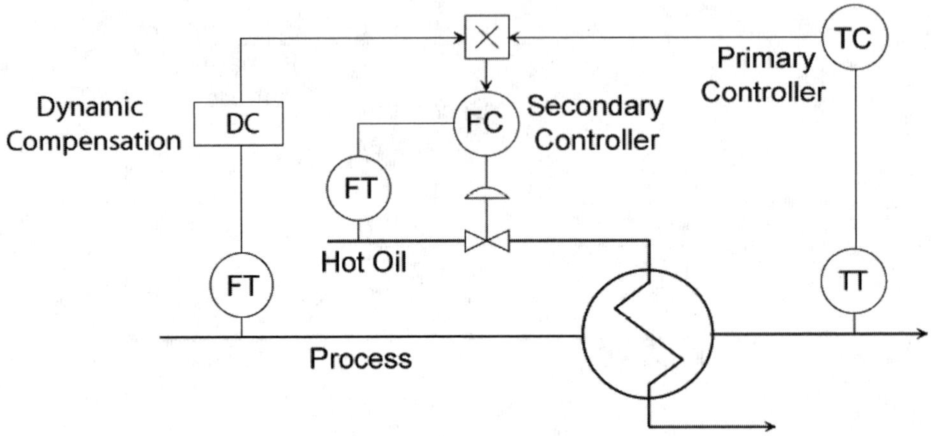


*Figure 3-14: Combined Feedback-Feedforward Control with Dynamic Compensation*

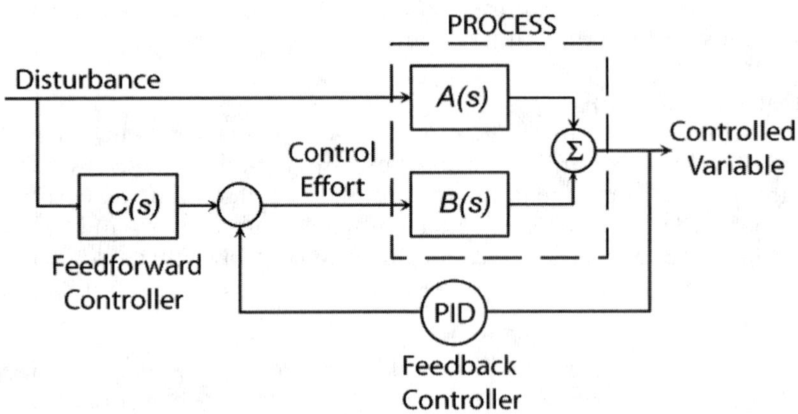


*Figure 3-15: Generic Feedback-Feedforward Control Structure*

There are two paths of influence from the disturbance to the controlled variable. If the disturbance is to have no effect on the controlled variable (that is the objective of feedforward control), these two paths must be mirror images that cancel out each other. Thus the feedforward controller must be the ratio of the two process dynamic effects, with an appropriate sign adjustment. The correct sign will be obvious in any practical situation. That is:

$$C(s) = -\frac{A(s)}{B(s)} \tag{3-10}$$

If both A(s) and B(s) have been approximated as FOPDT models (see Section 3.4.2), then *C(s)* is comprised of, at most, a steady-state gain, a lead-lag and a dead-time function. These functions are contained in every vendor's function block library. The dynamic compensation can often be simpler than this. For instance, if the dead times through *A(s)* and *B(s)* are identical, then no dead-time term is required in the dynamic compensation.

Now consider combining feedback and feedforward control. Figure 3-15 shows a junction for combining these two forms of control but does not indicate how they are combined. In general, feedback and feedforward can be combined by adding or by multiplying the signals. A multiplicative combination is essentially the same as ratio control. In situations where a ratio must be maintained between disturbance and control effort, multiplicative combination of feedback and feedforward will provide a relatively constant process gain for the feedback controller. If the feedback and feedforward were combined additively, variations in process gain seen by the feedback controller would require frequent retuning. In other situations, it is better to combine feedback and feedforward additively, a control application often called "feedback trim."

Regardless of the method of combining feedback and feedforward, the dynamic compensation terms should be only in the feedforward path, not the feedback path. It would be erroneous for the dynamic compensation terms to follow the combining junction in Figure 3-15.

Feedforward control is one of the most powerful control techniques for minimizing variability in a control loop. It is often overlooked due to lack familiarity with the technique.

### 3.5.4 Decoupling Control

Frequently in industrial processes, a manipulated variable—a signal to a valve or to a lower-level flow controller—will affect more than one controlled variable. If each controlled variable is paired with a particular manipulated variable through a feedback controller, interaction between the control loops will lead to undesirable variability.

One way of coping with the problem is to pair the controlled and manipulated variables so as to reduce the interaction between the control loops. A technique for pairing the variables, called *relative gain analysis*, is described in most texts on process control, as well as in both books referenced in 3.6.1. If, after applying this technique, the residual interaction is too great, the control loops should be modified for the purpose of decoupling. With decoupled control loops, each feedback controller output affects only one controlled variable.

Figure 3-16 shows a generic process with two controlled inputs—a signal to valves or setpoints to lower-level flow controllers—and two controlled variables. The functions $P_{11}$, $P_{12}$, $P_{21}$ and $P_{22}$ represent dynamic influences of inputs on the controlled variables. With no decoupling, there will be interaction between the control loops. However, decoupling elements can be installed so that the output of PID#1 has no effect on CV#2, and PID#2 output has no effect on CV#1.

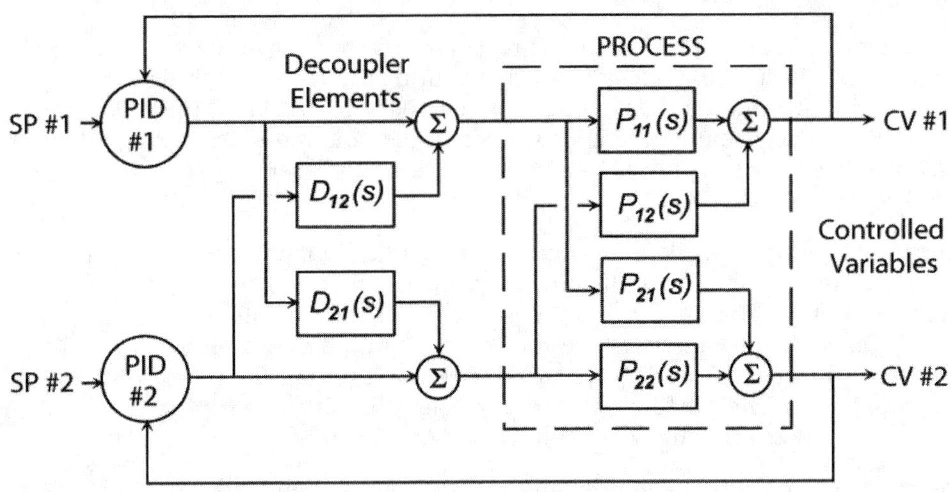


*Figure 3-16: Multiple-Input, Multiple-Output (2x2) Process with Decoupled Feedback Control Loops*

Using an approach similar to feedforward control, note that there are two paths of influence from the output of PID#1 to CV#2. One path is through the process element $P_{21}(s)$. The other is through the decoupling element $D_{21}(s)$ and the process element $P_{22}(s)$. For the output of PID#1 to have no effect on CV#2 these paths must be mirror images that cancel out each other. Therefore, the decoupling element must be

$$D_{21}(s) = -\frac{P_{21}(s)}{P_{22}(s)} \tag{3-11}$$

In a practical application, the appropriate sign will be obvious. In a similar fashion, the other decoupling element is given by

$$D_{12}(s) = -\frac{P_{12}(s)}{P_{11}(s)} \tag{3-12}$$

If the process elements are approximated with FOPDT models as in Section 3.4.2, the decoupling elements are, at most, comprised of gain, lead-lag and dead-time functions, all of which are available from most vendors' function block library.

The decoupling technique described here can be called "forward decoupling." Inverted decoupling, an alternative described in Wade's "Inverted Decoupling, A Neglected Technique" (see 3.6.2), has certain advantages as well as possible disadvantages.

If one variable is of greater priority than the other, partial decoupling should be considered. Suppose that CV#1 in Figure 3-16 is a high-valued product and CV#2 is a low-valued product. Variability in CV#1 should be minimized, whereas variability in CV#2 can be tolerated. Therefore the decoupling element $D_{12}(s)$ can be implemented and $D_{21}(s)$ omitted.

### 3.5.5 Selector (Override) Control

Selector control, also called "override" control, differs from the other techniques because it does not have as its objective the reduction of variability in a control loop. It does have an economic consequence, however, because the most economical operating point for many processes is near the point of encroachment on a process, equipment, or safety limit. Unless a control system is present that prevents such encroachment, the tendency will be to operate well away from the limit, at a less-than-optimum operating point. Selector control permits operating closer to the limit.

As an example, Figure 3-17 illustrates a process heater. In normal operation, an outlet temperature controller controls the firing rate of the heater. During this time, a critical tube temperature is below its limit. Should, however, the tube temperature encroach on the limit, the tube temperature controller will override the normal outlet temperature controller and reduce the firing rate of the heater. The low-signal selector in the controller outputs provides for the selection of the controller that is demanding the lower firing rate.

If ordinary PI or PID controllers are used for this application, one or the other of the controlled variables will be at its setpoint, with the other variable less that its setpoint. The integral action of the non-selected controller will cause it to wind up—that is, its output will climb to 100%. In normal operation, this will be the tube temperature controller. Should the tube temperature rise above its setpoint, its output must unwind from 100% to a value that is less than the other controller's output before there is any effect on heater firing. Depending upon the controller tuning, there may be a considerable amount of time when the tube temperature is above its limit.

When the tube temperature controller overrides the normal outlet temperature controller and reduces heater firing, there will be a drop in heater outlet temperature. This will cause the outlet temperature

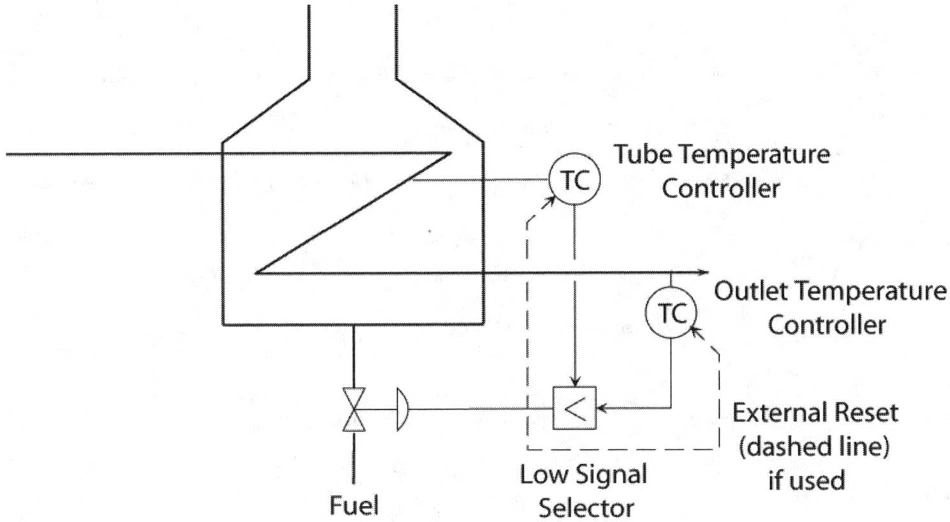


*Figure 3-17: Application Example of the Use of Selector (Override) Control*

controller to wind up. Once the tube temperature is reduced, returning to normal outlet temperature control is as awkward as was the switch to tube temperature control.

These problems arise because ordinary PID controllers were used in the application. Most vendors have PID algorithms with alternative functions to circumvent these problems. Two techniques will be briefly described.

Some vendors formulate their PID algorithm with "external reset." The integral action is achieved by feeding the output of the controller back to a positive feedback loop that utilizes a unity-gain first-order lag. With the controller output connected to the external feedback port, the response of a controller with this formulation is identical to that of an ordinary PID controller. Different behavior occurs when the external reset feedback is taken from the output of the selector, as shown in Figure 3-17. The non-selected controller will not wind up. Instead, its output will be equal to the selected controller's output plus a value representing its own gain times error. As the non-selected controlled variable (for instance, tube temperature) approaches its limit, the controller outputs become more nearly equal, but with the non-selected controller's output being higher. When the non-selected controller's process variable reaches the limit, the controller outputs will be equal. Should the non-selected controller's process variable continue to rise, its output will become the lower of the two—hence it will be selected for control. Since there is no requirement for the controller to unwind, the switch-over will be immediate.

Other systems do not use the external feedback. The non-selected controller is identified from the selector switch. As long as it remains the non-selected controller, it is continually initialized so that its output equals the other controller output plus the value of its own gain times error. This behavior is essentially the same as external feedback.

There are many other examples of selector control in industrial processes. On a pipeline, for instance, a variable speed compressor may be operated at the lower speed demanded by suction and discharge pressure controllers. For distillation control, reboiler heat may be set by the lower of the demands of a composition controller and a controller of differential pressure across one section of a tower, indicative of tower flooding.

## 3.6 References


### 3.6.1 Books

*The Automation, Systems, and Instrumentation Dictionary,* Fourth Edition. ISA, 2003.

Shinskey, F.G. *Process Control Systems: Application Design and Tuning.* 4th Ed. McGraw-Hill, 1996.

Wade, H.L. *Basic and Advanced Regulatory Control: System Design and Application.* ISA, 2004.

### 3.6.2 Articles

Wade, H.L. "Inverted Decoupling, A Neglected Technique." *ISA Transactions,* Vol. 36, No. 1. 1997.

Wade, H.L. "Trial and error: An organized procedure." *InTech,* May 2005.

### 3.6.3 Standards

ISA-5.1-1984 (R1992), *Instrumentation Symbols and Identification.*


## About the Author

**Harold Wade, Ph.D.,** is president of Wade Associates, Inc., a Houston-based consulting firm specializing in control systems engineering, instrumentation, and process control training. He has more than 40 years of instrumentation and control industry experience with Honeywell, Foxboro, and Biles and Associates. A Senior Member of ISA and a licensed professional engineer, he is the author of *Basic and Advanced Regulatory Control: System Design and Application,* 2nd Ed., published by ISA in 2004. He started teaching for ISA in 1987. He was a 2002 inductee into *Control Magazine's* "Automation Hall of Fame."

# 4 Control Valves

*By James Reed*

## Topic Highlights

*Valve Types*
*Standards and Codes*
*Valve Selection*
*Operation*
*Actuators and Accessories*

## 4.1 Introduction

A control valve is a power-actuated device that modifies the fluid flow rate in a process control system. The valve is connected to an actuator mechanism that is capable of changing the position of the valve's closure member in response to a signal from the controlling system.

A control valve is used to control properties such as upstream pressure, downstream pressure, flow rate, or liquid level. The control valve responds to a controller that measures and compares one of these properties to a setpoint. The controller's signal to the control valve will vary to maintain the property at the setpoint. Functionally, a control valve can be described as a controlled variable orifice.

The control valve's actuator positions the closure member, in the valve body, in a position consistent with the controller's signal. The closure member, often called a plug or a disc, creates a flow area in conjunction with a seat or a guiding cage that modifies the flow rate.

## 4.2 Valve Types

Control valves have two basic styles: linear and rotary motion. The valve stem of linear motion valves moves linearly up and down. The valve shaft of rotary motion valves rotates without any linear motion. A globe valve is a typical linear motion valve; ball valves and butterfly valves are both rotary motion valves. Linear motion valves are commonly used for more severe duty, with the rotary motion valves generally used in moderate to light duty service in sizes above 2-inch. For the same line size, rotary valves are smaller and lighter than linear motion valves and the rotary motion valves are more economical in cost, particularly as the line sizes increase.

A ball valve, used as a control valve, will usually have design modifications to improve its performance. Instead of a full spherical ball, it will typically have a 1/3 ball segment. This reduces the amount of seal contact and, therefore, reduces friction, allowing for more precise positioning. The leading edge of the ball segment may have a V-shaped groove to improve the control characteristic. A ball valve's trim material is generally 300 series stainless steel.

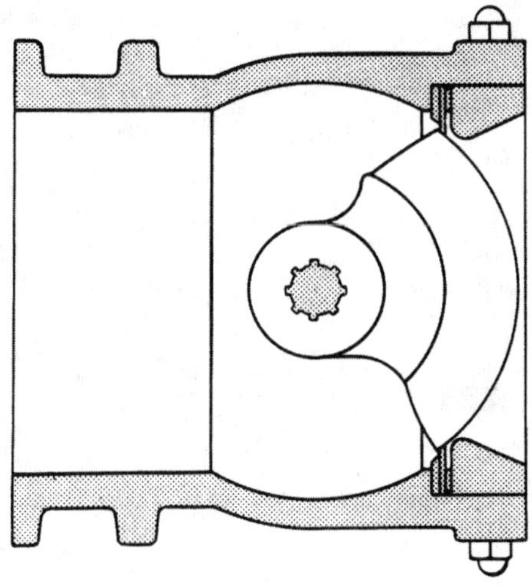

*Figure 4-1: Segmental Ball Valve Cross Section (Courtesy: Masoneilan/Dresser)*

A butterfly valve used as a control valve may have a somewhat S-shaped disc to reduce the flow-induced torque on the disc, allowing for more precise positioning. A butterfly valve's trim material may be bronze, ductile iron, or 300 series stainless steel.

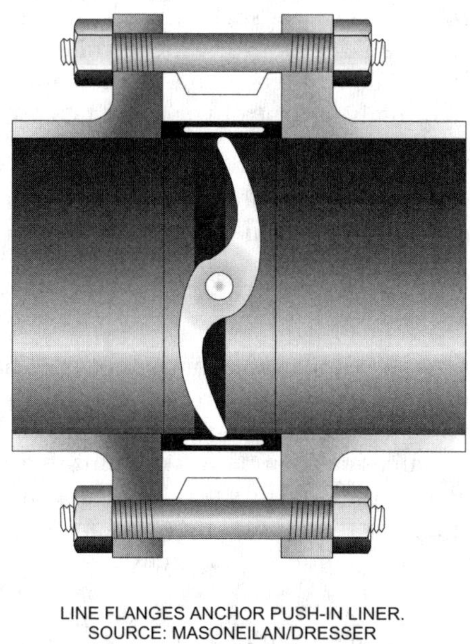


*Figure 4-2: Butterfly Valve (Courtesy: Masoneilan/Dresser)*

Another rotary control valve is the eccentric disc with the closure member shaped similar to a mushroom and is attached slightly offset to the shaft. The style provides good control along with tight shutoff, as the offset supplies leverage to flex the disc face into the seat. This valve's advantage is tight shutoff without the elastomeric seat seals that are used in ball and butterfly valves. Eccentric disc valves' trim material is generally 300 series stainless steel, which may be clad with stellite hardfacing.

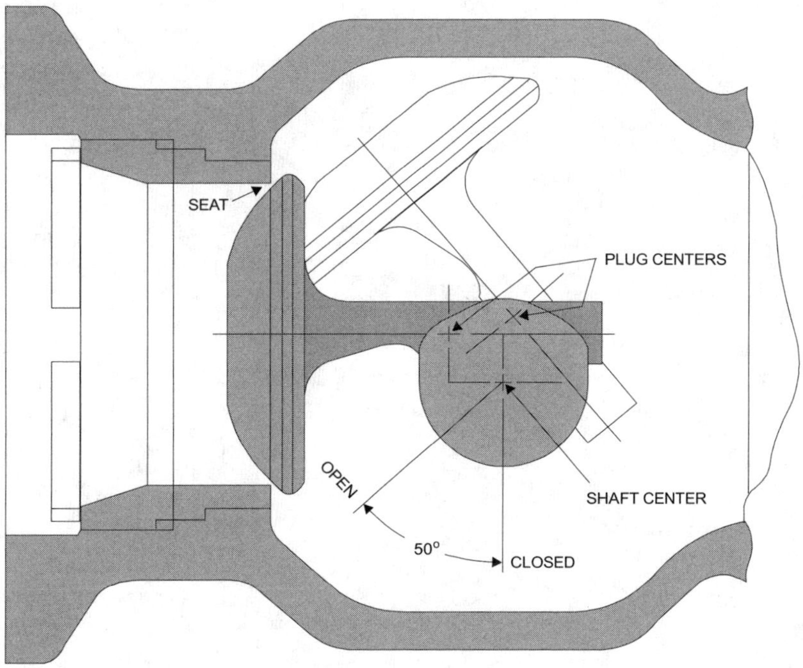


*Figure 4-3: Eccentric Control Valve*

Linear motion control valves have two common styles: post-guided and cage-guided.

Post-guided valves have the moving closure member guided by a bushing in the valve's bonnet. The closure member is usually unbalanced, and the fluid pressure drop acting on the closure member can create significant forces. Post-guided trims are well suited for slurries and fluids with entrained solids. The post-guiding area is not in the active flow stream, reducing the chance of solids entering the guiding joint. Post-guided valve trim materials are usually either 400 or 300 series or 17-4PH stainless steel.

Cage-guided valves have a cylindrical cage between the body and bonnet. Below the cage is a seat ring. The cage/seat ring stack is sealed with resilient gaskets on both ends. The cage guides the closure member, also known as the plug. The plug is often pressure balanced with a dynamic seal between the plug and cage. The balanced plug will have passageways through the plug to eliminate the pressure differential across the plug and the resulting pressure induced force. The trim materials for cage guided valves are often either 400 or 300 series or 17-4PH stainless steel sometimes nitrided or stellite hard faced.

The valve closure member is connected to the actuator with either a stem or a shaft. Stem is the term used for linear motion valves, and shaft is used for rotary motion. The stem/shaft connects from inside the valve body to the actuator outside. The fluid pressure is sealed with a packing box; a cylindrical

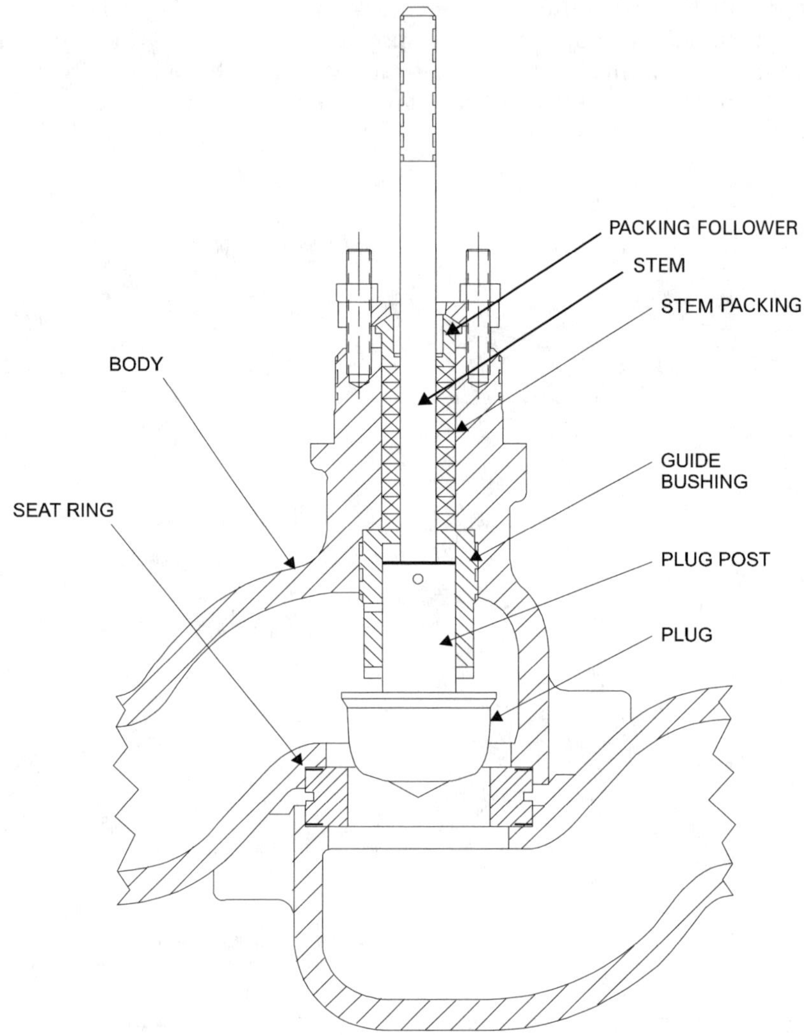


*Figure 4-4: Post-Guided Control Valve*

chamber with a guiding bushing; generally a packing spring or other means of applying a resilient compressive force to the packing; several rings of packing; and a packing follower. The packing rings need to be compressed to effect a tight seal. The fluid pressure will usually provide sufficient compression force on the packing rings. The packing rings usually lose some of their volume, reducing the stacked height of the packing. The packing spring serves to accommodate the volume change of the packing rings as well as providing a sufficient initial force as the valve is being pressurized to ensure a tight seal.

## 4.3 Standards and Codes

A control valve is a pressure vessel and must meet an industry code such as ASME/ANSI B16.34 - 1996 - *Valves Flanged, Threaded, and Welding End*. This code provides calculations for minimum wall thicknesses, along with temperature/pressure ratings for 40 different body materials. It also provides rules for marking and hydrostatic testing. Valves have pressure ratings indicating their allowed working pressures at various temperatures. The pressure ratings are Classes 150, 300, 400, 600, 900, 1500,

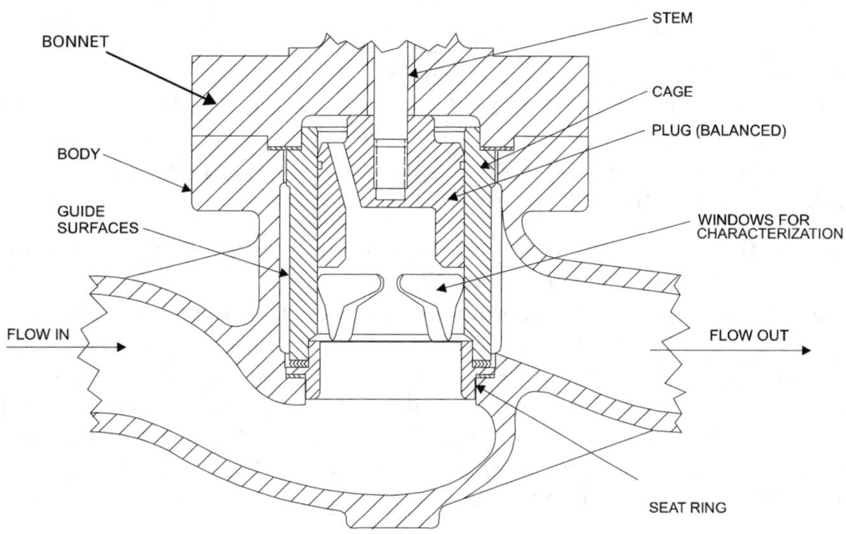


*Figure 4-5: Cage-Guided Control Valve*

2500, and 4500. The pressure class does not directly relate to a specific pressure/temperature relationship but is a generic designation for the pressure rating class. The class number matches the allowable pressure at an intermediate temperature around 600°F to 800°F. The valve's body and bonnet materials are, most commonly, carbon steel, chrome moly steel, and stainless steel. Chrome moly steel is used for elevated temperature applications such as power generation, and stainless steel is used for corrosive applications such as petrochemical.

There is a broad range of valve end connections used to connect a control valve to a pipeline: flanged, butt weld, socket weld, threaded end, and flangeless. The flangeless connection is installed between two pipeline flanges using long studs to clamp the valve between the flanges.

## 4.4 Valve Selection

Choosing the appropriate control valve for a particular application is a multistep process. The initial and most important consideration is determining the required flow capacity by calculating the valve sizing coefficient, $C_V$. The basic formula for incompressible fluids is

$$C_V = Q\sqrt{\frac{G_f}{\Delta P}}$$

where

Q is the flow rate

$G_f$ is the fluid's specific gravity

$\Delta P$ is the pressure drop across the valve

ANSI/ISA-75.01.01-2002 (60534-2-1 Mod) - *Flow Equations for Sizing Control Valves* fully describes the variations of the basic formula for liquid, vapor and two phase flows with consideration for cavitation, flashing, and sonic velocity—all affecting the required $C_V$. Most control valve manufacturers have flow sizing programs available for use on their Web sites. Manufacturers rate their valves to the maximum and minimum controllable $C_V$ to enable the proper selection of valve for the particular application.

Control valves create a pressure drop in the flow stream to modify the flow rate. A reduced flow area is developed by the position of the closure member relative to the seat or cage port. The flowing fluid loses pressure and accelerates as it passes through the reduced flow area. The point of highest fluid velocity and lowest pressure is called the vena contracta. Downstream of the vena contracta, the fluid pressure partially recovers. The degree of pressure recovery is a major element in choosing the appropriate control valve for a particular situation.

Figure 4-6 shows the pressure drops and pressure recovery as fluid passes through a restriction. In the example shown, the lowest pressure is at the vena contracta ($P_{VC}$) with the pressure recovering about 35% to the outlet pressure. The vena contracta pressure becomes very significant for liquid flow when the fluid vapor pressure is greater than the vena contracta pressure. If the liquid vapor pressure shown as Vapor Pressure 1 is higher than the vena contracta pressure and below the outlet pressure, then liquid vaporizing at the vena contracta will recondense to a liquid, resulting in cavitation in the region between Vapor Pressure 1 and the vena contracta. If the vapor pressure shown as Vapor Pressure 2 is higher than the outlet pressure, then the liquid will vaporize or "flash" leaving the valve as a liquid/vapor mixture. Each of these phenomena can cause operational problems.

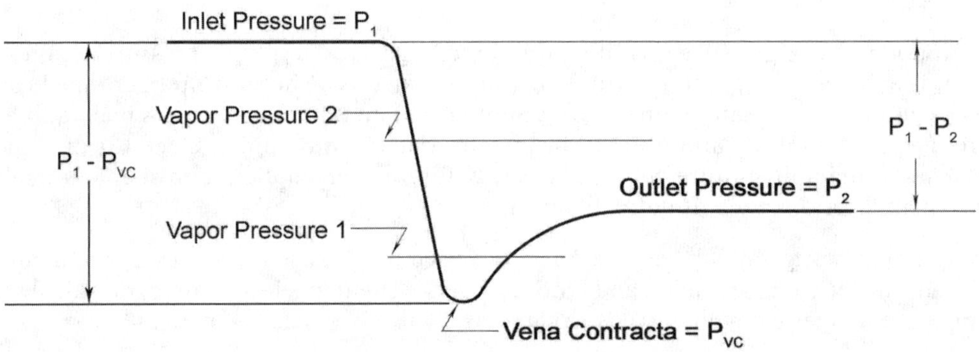


*Figure 4-6: Pressure Drops and Recovery as Fluid Passes Through Restriction*

Figure 4-6 shows the flow profile for a typical linear motion globe valve. A rotary motion valve, such as a ball or butterfly valve, with the same inlet and outlet pressures will require the pressure drop between the inlet pressure and the vena contracta to be about 75% larger, for the same overall pressure drop. This is due to its straight-through, streamlined flow path and higher pressure recovery characteristic. This phenomenon makes these valves more susceptible to cavitation.

The vena contracta pressure is also significant for gas and vapor flow. The gas or vapor flow rate cannot exceed the sonic velocity. The valve's maximum flow velocity occurs at the vena contracta and, once the flow velocity becomes sonic, the flow is choked and further reductions in the downstream pressure will not increase the flow rate through the valve.

Cavitation occurs in liquid flow when the fluid pressure drops below the liquid's vapor pressure and the vapor pressure is below the outlet pressure. In this region, the liquid will flash to a vapor and then suddenly collapse back to a liquid. This sudden collapse creates very high localized shock waves that will cause surface damage when it occurs at, or adjacent to, components of the valve.

Flashing occurs in liquid flow when the fluid pressure drops below the liquid's vapor pressure and the vapor pressure is above the outlet pressure. The liquid will continue to flash downstream of the valve as the fluid pressure decreases. The flowing mixture of liquid and vapor may cause erosion of the valve and pipe surfaces and create a higher flow noise level than for liquid flow.

Cavitation and flashing flow with entrained vapor may limit the flow rate due to the increased volume of the liquid/vapor.

## 4.5 Operation

Effective control valve operation requires a rigorous application process. The operating conditions must be known, as most misapplications come from having incorrect or incomplete information. The application information may include identification of the flowing fluid, along with its viscosity and density, inlet and outlet pressures, fluid temperature, and range of flow rates. Not all applications require each of these data elements, but the more severe service requires a comprehensive specification. The primary element is the calculation of the $C_V$ value.

The next step is usually choosing between linear or rotary motion valves. If there is a wide range in the flow rate, the minimum $C_V$ should be calculated. Rangeability is the ratio of maximum to minimum controllable $C_V$ values. Each valve style has its rangeability limits. Fluid flow noise is also a consideration, since excessive flow noise, above the permissible noise levels of OSHA regulations, can harm a person's hearing permanently and can shorten valve life. Aerodynamic noise from vapor flow can be predicted using the standard ISA-75.17-1989 - *Control Valve Aerodynamic Noise Prediction*. The $C_V$ calculation process will indicate if cavitation or flashing is likely and also if the flow rate will be choked from cavitation, flashing, or sonic velocity.

Control valves are seldom fully open or closed and generally are moving in response to the control signal. However, when the valves are closed, the allowable seat leakage is important, affecting the choice of trim style, seat material, and actuator size.

A control valve's seat tightness is specified by Leakage Classes as defined in the ANSI/FCI 70-2 *Control Valve Leakage* standard. Six leakage classes are specified: Classes I, II, III IV, V, and VI. Classes I through VI progressively require increasingly stringent sealing performance. As examples, Class II allows 0.5% of the valve's rated flow capacity (that is .5% of the valve's rated maximum $C_V$). Class VI, normally applied to resilient seated valves, allows 0.45 ml of liquid leakage per minute at 50 psi for a 2" valve.

The required $C_V$ value can be calculated using ANSI/ISA-75.01.01-2002 (60534-2-1 Mod) - *Flow Equations for Sizing Control Valves*. Aerodynamic noise from vapor flow can be predicted using ISA-75.17-1989 - *Control Valve Aerodynamic Noise Prediction.*

The relationship between the valve's trim travel and the trim's flow coefficient, $C_V$, is the "Inherent Characteristic" determined by flow testing with a constant differential pressure applied across the valve throughout the valve travel. The two more common characteristics are Equal Percentage and Linear.

The linear characteristic $C_V$ value is linear to the valve travel. The equal percentage characteristic is defined as an equal percentage increase in $C_V$ for each equal increment of valve travel. The equal percentage $C_V$ value increases slowly at the lower end of the valve travel and then increases more rapidly as the trim opens. Inherent valve flow characteristics are chosen with respect to operating conditions such that the resulting installed characteristic is approximately linear. An equal percentage characteristic has two important advantages: it offers more precise control at the lower end of travel and, in applications where the pressure drop across the valve significantly decreases as the valve opens, the resulting installed characteristic approximates a linear characteristic, providing effective control.

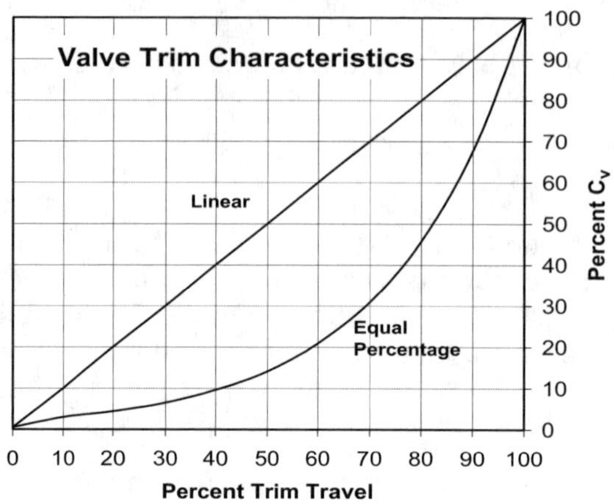


*Figure 4-7: Valve Trim Characteristics*

## 4.6 Actuators and Accessories

A control valve actuator will precisely position the closure member in response to the control signal. There are three major groups of actuators: pneumatic, electric, and hydraulic.

The most common is the pneumatic actuator, mainly in a spring/diaphragm style but also in a piston style, with or without a spring. The spring in a pneumatic actuator opposes the force of the pneumatic pressure—usually air—against the diaphragm or piston. The pneumatic pressure is increased to create motion by compressing the spring, or the pressure is reduced to allow motion in the opposite direction by partially uncompressing the spring. The spring/diaphragm actuator can be used for rotary motion valves by converting the linear motion to rotary with lever linkage. The spring/diaphragm actuator can be configured so the spring can open the valve (direct acting) or so the spring can close the valve (reverse acting). The reverse acting mode will fail closed when actuating pressure is lost. The direct acting mode will fail open.

Pneumatic actuators' main advantage is lower cost than the other styles, together with low friction and low dead band.

The hydraulic actuator is very similar to the pneumatic actuator, except its force is provided by pressurized liquid. A hydraulic actuator's style is usually piston. An advantage of a hydraulic actuator is the actuating fluid is incompressible, resulting in a more stable device that is less affected by rapidly changing closure member forces than the pneumatic actuator. The hydraulic actuator's other advantage is its construction allows higher actuating pressures.

The electric actuator uses an electric motor with a gear train to rotate a threaded stem nut, for linear valves, or a shaft for rotary valves.

Actuators have three major accessories: positioners, limit switches, and pressure transducers.

A positioner measures the valve's stem/shaft position, comparing it to the signal. The positioner will adjust the force applied by the actuator—whether pneumatic, hydraulic, or electric—to maintain the position of the valve closure member relative to the signal. The positioner's signal may be pneumatic, electric (variable current), or a digital electronic input. A positioner can be considered a valve controller. A digital "smart" positioner can communicate back to the controller, sending the positioner's actual position, supply pressure, actuator pressure, and a performance diagnosis of itself and the valve/actuator system.

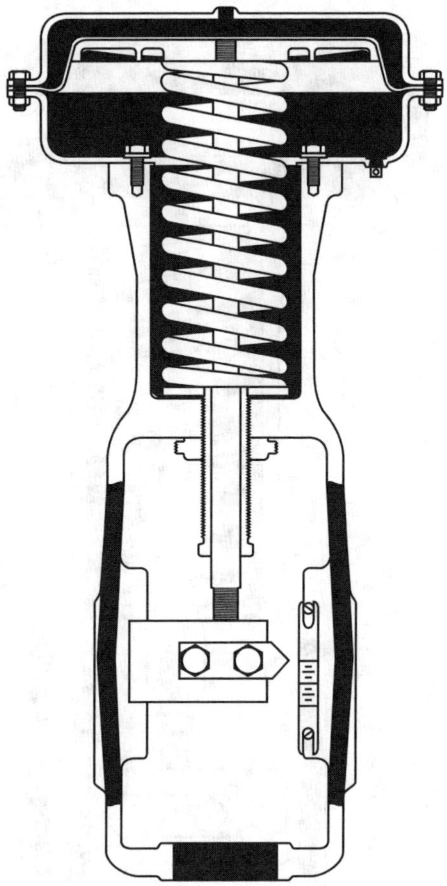

SERIES 127 DIRECT ACTING

*Figure 4-8: Direct Acting Spring/Diaphragm Actuator (Courtesy: Norriseal)*

The limit switches monitor the stem/shaft position, opening or closing the switch's contacts when the stem/shaft is in a particular position, usually fully open or closed.

Pressure transducers are usually used to convert an electric input signal to pneumatic or hydraulic pressure.

Precision positioning is a major goal of a control valve actuator, with friction being a major cause for losses of precision. Friction may also create significant deadband. Deadband is the range the signal may be varied, upon reversal of direction, without stem/shaft motion. If the static friction is much higher than the dynamic friction, the pneumatic actuator will increase pressure to overcome the high static friction and then may surge past its intended position, as the lower dynamic friction cannot contain the excess pressure. After this happens, the positioner reverses direction and tries to achieve the intended position again, overshooting going into a hunting cycle. High friction can also cause hysteresis between the increasing signal position and the position of the same decreasing signal. Some valve applications require very accurate control. Achieving this requires an accurate positioner and low friction on all moving parts.

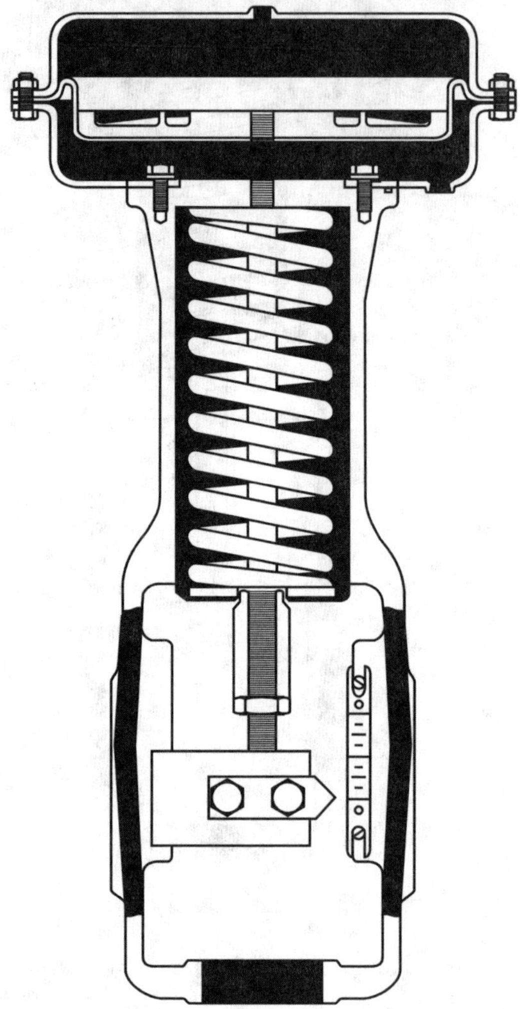

SERIES 137 REVERSE ACTING

*Figure 4-9: Reverse Acting Spring/Diaphragm Actuator (Courtesy: Norriseal)*

## 4.7 References

Borden, Guy, Jr., ed., and P. G. Friedmann, style ed. *Control Valves.* Practical Guides for Measurement and Control Series. ISA, 1998.

Hutchinson, J. W. *ISA Handbook of Control Valves.* Second Edition. ISA, 1976.

**Standards**

ANSI/FCI-70-2-2003. *Control Valve Seat Leakage.*

ANSI/ISA-75.01.01-2002 (60534-2-1 Mod.). *Flow Equations for Sizing Control Valves.*

ASME/ANSI-B16.34-1996. *Valves Flanged, Threaded and Welding End.*

ISA-75.17-1989. *Control Valve Aerodynamic Noise Prediction.*

## About the Author

**James Reed**, now retired, was Vice President of Engineering for Norriseal in Houston. He is a graduate of Pennsylvania State University. He has been responsible for product design/development and technical personnel management for three control valve companies: Norriseal, Masoneilan, and Copes Vulcan. He has been an active member of seven ISA SP75 standards subcommittees for more than 20 years and is currently chairman of two.

# 5 Analog Communications

*By Richard Caro and Lawrence (Larry) M. Thompson*

## Topic Highlights

*Pneumatic Signals*
*Current Signals*
*Suppression and Elevation of Zero*
*Other Types of Signals*
*Analog-to-Digital and Digital-to-Analog Conversion*

## 5.1 Introduction

The earliest process control instruments were mechanical devices in which the sensor was directly coupled to the control mechanism, which in turn was directly coupled to the control valve. Usually, a dial indicator was provided so the technician could read the process variable value. These devices are still being used today and are called "self-actuating controllers" or often just "regulators." These mechanical controllers often take advantage of a physical property of some fluid to operate the final control element. For example, a fluid-filled system can take advantage of the thermal expansion of the fluid to both sense temperature and operate a control valve. Likewise, process pressure changes can be channeled mechanically or through filled systems to operate a control valve. Such controllers are proportional controllers, with some gain adjustment available through mechanical linkages or some other mechanical advantage. We now know they can exhibit some offset error.

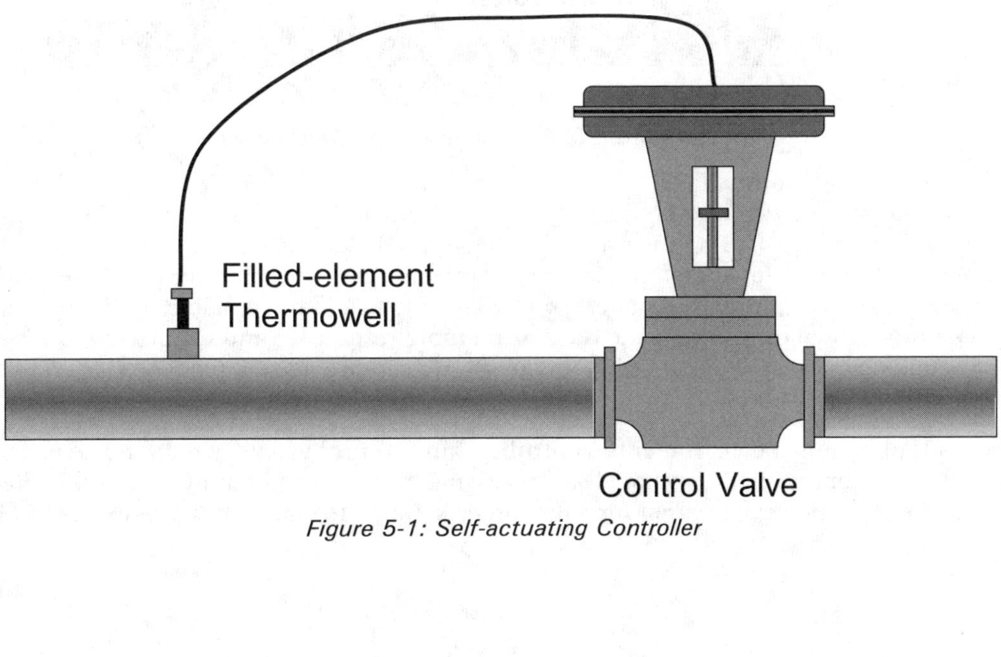


*Figure 5-1: Self-actuating Controller*

## 5.2 Pneumatic Signals

Although self-actuating controllers are usually low-cost devices, the industry soon recognized it would be easier and safer for the process operator to monitor and control processes if there was an indication of the process variable in a more convenient and protected place. Therefore, a need was established to communicate the process variable from the sensor that remained in the field to a remote operator panel. The mechanism created for this communications was air pressure over the range 3-to-15 pounds per square inch gauge (psig). This is called pneumatic transmission. European standardization in international units was 20-to-100 kilopascal (kPa), which is very close to the same pressures as 3-to-15 psi. The value of using 3 psi (or 20 kPa) rather than zero is to detect failure of the instrument air supply. The measurement 15 psi (or 100 kPa) is selected for 100 percent, because it is well below nominal pressures of the air supply for diagnostic purposes.

However, the operator still had to go to the field to change the set point of the controller. The solution was to build the controller into the display unit mounted at the operator panel using pneumatic computing relays. Organizations could also more easily service the panel-mounted controller than a controller in the field.

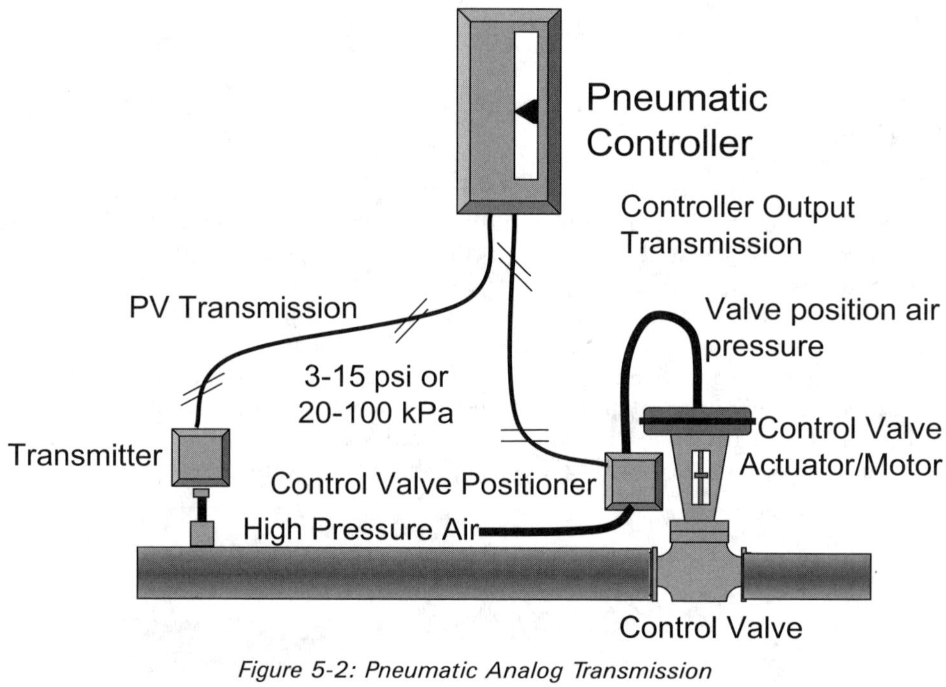


*Figure 5-2: Pneumatic Analog Transmission*

The controller output was also in the 3-to-15 psi air pressure range and piped to a control valve that was, by necessity, mounted on the process piping in the field. The control valve was operated by a pneumatic actuator or force motor using higher pressure air for operation. Once the pneumatic controller was created, innovative suppliers soon were able to add integral and derivative control to the original proportional control to make the control more responsive and to correct for offset error. Additionally, engineers created pneumatic valve positioners to provide simple feedback control of control valve position.

Thousands of pneumatic instruments, controllers, and control valves remain in use more than 50 years after the commercialization of electronic signal transmission—and well into the digital signal transmission age. However, except for a few processes for the manufacture of extremely hazardous

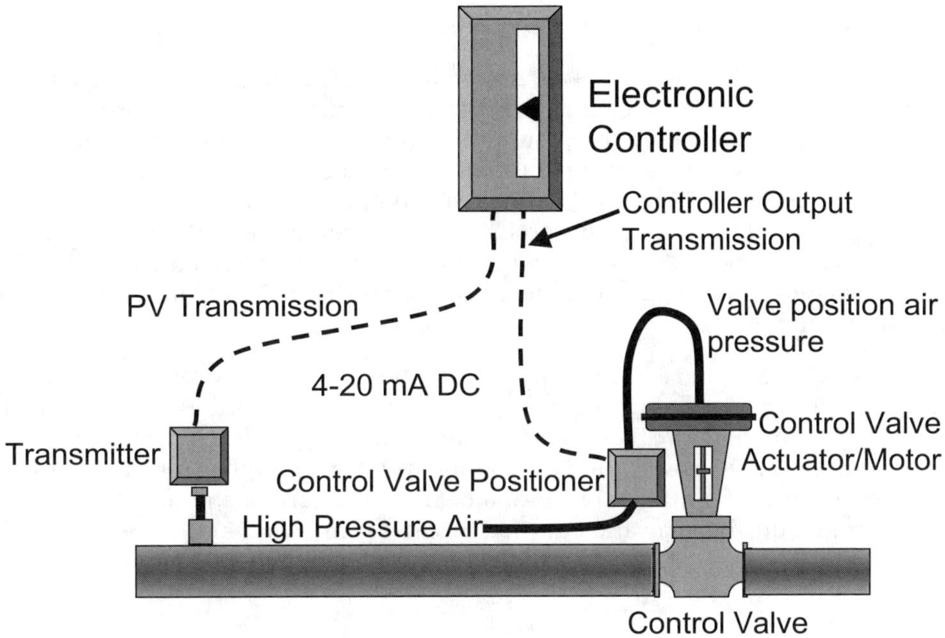


*Figure 5-3: Electronic Analog Transmission*

gases and liquids such as ether, there is no growth in pneumatic instrumentation and signal transmission. Many pneumatic process control systems are being modernized to electronic signal transmission or directly to digital data transmission and control. Although pneumatic data transmission and control proved to be highly reliable, it is relatively expensive to interconnect sensors, controllers, and final control elements with leak-free tubing. Frequent maintenance is required to repair tubing and also to clean instruments of entrained oil from air compressors and silica gel from air driers.

## 5.3 Current Signals

The replacement for pneumatic signal transmission was decided in the 1960s to be a small analog direct current (DC) signal, which could be used over considerable distances on small gauge wiring without amplification. While the majority of supplier companies agreed that the range of 4-to-20 milliamp (mA) was probably the best, one supplier persisted in its demand for 10-to-50 mA, since its equipment was not sensitive enough to transmit and receive at the lower range. The first ANSI/ISA S50.1-1975 standard (now ANSI/ISA-50.00.01-1975 [R2002] – Compatibility of Analog Signals for Electronic Industrial Process Instruments) was for 4–20 mA DC, with an alternative at 10-50 mA. Eventually, that one supplier changed technologies and accepted 4–20 mA DC analog signal communications. The alternative for 10-50 mA was removed for the 1982 edition of this standard.

There are different explanations for the selection of 4 mA for the low end of the transmission range—it provides a live zero to prove that the field instrument is operating and it is also required to provide the minimal electrical power necessary to energize the field instrument.

One reason for selecting a current-based signal is that sufficient electrical power to energize the sensor can be delivered over the same pair of wires as the signal. Another reason for using a current signal is that, within overall resistance limits, current does not vary with the length of the wire as would a voltage-based signal. Using two wires for both the signal and power reduces the cost of installation. Some field instruments require too much electrical energy to be powered from the signal transmission line and are said to be "self-powered," meaning they are powered from another source than the 4-20 mA transmission line.

Because a 250-ohm resistor is typically used at the input of the controller to generate a 1–5V signal, this low impedance resistor reduces noise effects.

Although the transmitted signal is a 4–20 mA analog current, the control valve is most often operated by high-pressure pneumatic air, because it is the most economic and responsive technology to move the position of the control valve. This requires that the 4–20 mA output from the controller be used to modulate the high-pressure air driving the control valve actuator. These devices may be simply a converter from 4–20 mA to 3–15 psi (or 20–100 kPa), commonly called an I/P converter. The output of the I/P then goes to a pneumatic valve positioner. However, more often the conversion actually takes place in an electronic control valve positioner that uses feedback from the control valve itself and modulates the high-pressure pneumatic air to achieve the position required by the controller based on its 4–20 mA output.

The 4–20 mA signal is achieved by the field transmitter or the controller acting as a current regulator or variable resistor in the circuit. The two-wire loop passing from the DC power source through the field transmitter can, therefore, have a maximum total resistance such that the total voltage drop cannot exceed that of the DC power source—nominally 24V. One of the voltage drops is the resistor placed across the analog controller or the analog input I/O card of a multiplexer in a digital control system. Other instruments may also be wired in series connection in the same two-wire current loop as long as the total loop resistance does not exceed approximately 800 ohms.

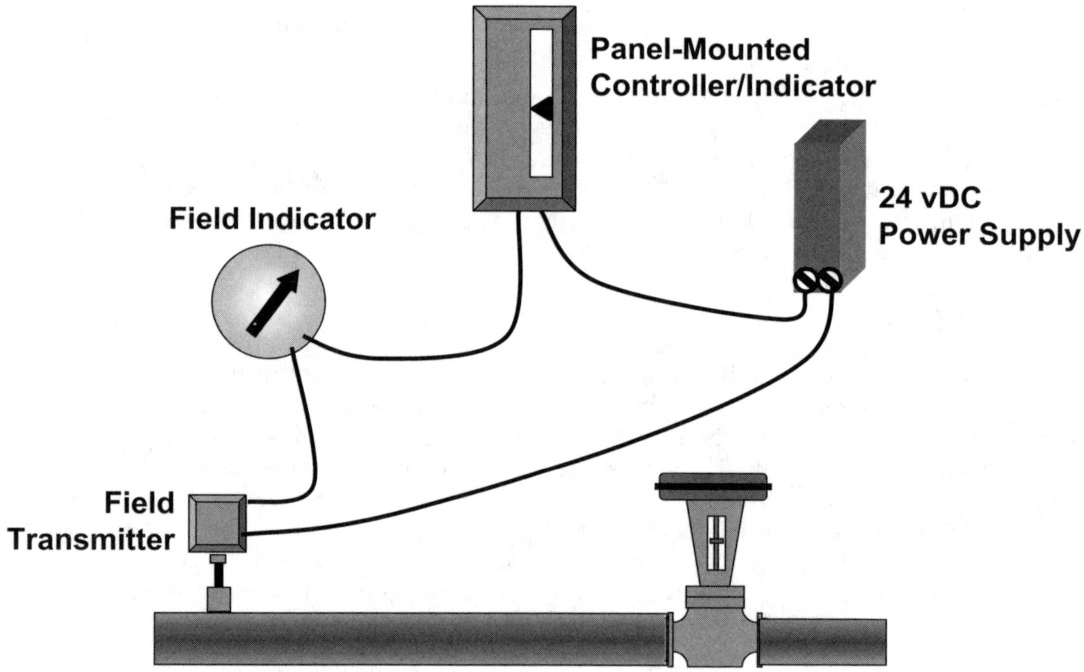


*Figure 5-4: Loop-powered Devices*

Even 20 years after the work began to develop a digital data transmission standard, 4–20 mA DC still dominates the process control market for both new and revamped installations because it now serves as the primary signal transmission method for Highway Addressable Remote Transducer (HART) protocol. HART uses its one 4–20 mA analog transmission channel for the primary variable, usually the process variable measurement value, and transmits all other data on its digital signal channels.

Because it is a continuous signal, analog electronic signal transmission remains the fastest way to transmit a measured variable to a controller. This is especially true when the measurement mechanism itself continuously modulates the output current as in force-motor driven devices. However, even in more modern field transmitters that use inherent digital transducers, delays to produce the analog signal are very small compared to process dynamics and are virtually continuous themselves.

Continuous measurement, transmission, and analog electronic controllers are not affected by the signal aliasing errors that can occur in sampled data digital transmission and control.

The design of process manufacturing plants is usually documented on process and instrumentation diagrams (P&IDs), which attempt to show the points at which the PV is measured, where control valves are located, and the interconnection of instruments and the control system. While the documentation symbols for the instruments and control valves are covered elsewhere in this book, the P&ID graphic representations for the instrumentation connections are covered in ISA-5.1-1984 (R1992) – Instrumentation Symbols and Identification.

## 5.4 Suppression and Elevation of Zero

Although not pertaining to the analog signal directly, what zero actually means is a characteristic of the measurement side. In this case 0% is always 4 mA; however, what does 0% represent? If it represents the true zero, such as 0 psi, 0 gpm flow, 0 gage pressure, or 0 fill level, the measurement is referred to as zero based and 0% = 0 (measurement value) = 4 mA.

Alternatively, when using pressure to determine level in an open tank, the measurement sensor is usually some distance above the tank bottom. It is best not to uncover the sensor, so some distance above the sensor is called the 0% of fill value, meaning there is a positive pressure on the transmitter that is calibrated as 0%. As an example, say the empty level is 10 inches above the sensor. Then, even though the transmitter is actually measuring 10 inches of water, the 4 mA (0%) point is set there. This is zero suppression, because the true zero pressure has been suppressed below the instrument zero.

If, however, you have a pressurized tank and are using a differential pressure transmitter to measure the tank level and it is a condensing fluid, use a wet leg to the low side (or use remote seals, which have mostly replaced the wet leg). Figure 5-5 is a simple diagram illustrating this scenario.

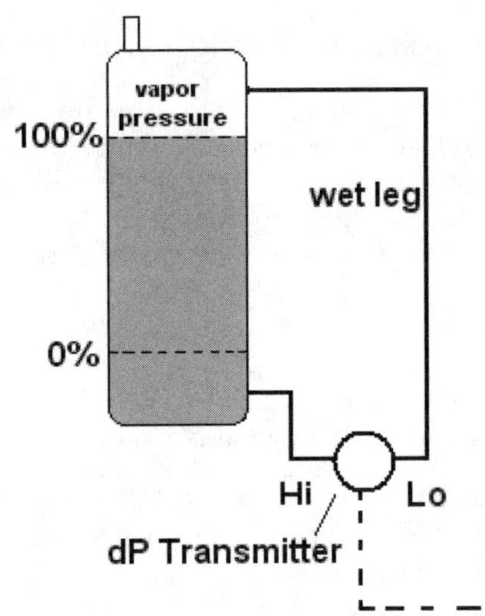


*Figure 5-5: Wet Leg*

In this case, the pressure on the wet leg connected to the low side of the transmitter (when the take is in the lower level portion, which is perhaps all the time) will be greater than the pressure on the high side. When this occurs, it is called a negative pressure. In this case, setting 0% (4 mA) of the instrument to represent the negative pressure as zero results in elevating the true zero.

It is easy to keep straight. If you set the instrument zero to a positive value, you are performing zero suppression; to a negative value, zero elevation; to zero, zero based.

## 5.5 Other Signals

### 5.5.1 Motion Control Signals

There are other communications standards that use analog signals, particularly those that input into PLC analog I/O. Motion controls typically use either a ±5V or a ±10V signal to indicate both direction and velocity, and depending upon the number of axis can determine position and other physical metrics.

### 5.5.2 Voice Channel

With the advent of communication links into the control network hierarchy came, at least for a while, the voice channel. It is still widely used in SCADA. The voice channel is the telephone electrical signal representation (analog) of a voice for transmission over a pair of wires.

Assuming the worst-case scenario, the voice channel was designed to pass intelligence, not fidelity. It starts at 300 Hz (voice frequencies below 300 Hz do not materially contribute to signal intelligence), and the upper frequencies were limited to 3000 Hz (frequencies above that add to the quality of a signal but not the intelligence). As a result, the pass band (span) of a voice channel is 3 KHz.

Because of the lower limit of 300 Hz, signals with extended stays at one value or another (digital) cannot pass in raw form over this channel, which is probably the most common electrical circuit in the world. This is why a modem (modulator-demodulator) is required for digital operation over a voice channel, but that is the subject of a different section. Electrical signals found on this channel vary immensely. Voice conversation is generally measured in dbm (db relative to a milliwatt), but the typical voice battery when off the hook is about 4–6V. Ringing is 90 VDC interrupted 20 times a second (typically). Other than a design impedance of 600 ohms (500 in some cases) and the pass band, signaling requirements for a voice channel in terms of amplitude are determined by the operational unit.

## 5.6 Analog-to-Digital and Digital-to-Analog Conversion

Modern electronic controllers are digital. Although some transmitters and the majority of systems controllers communicate in digital format, the measurements found in industry operate in a continuous world: the analog world of measurement. For a digital device to communicate with and control analog devices, analog-to-digital (A/D) and digital-to-analog (D/A) conversions are required. There are different methods for performing either type of conversion, and this section will outline some of the more prevalent techniques.

### 5.6.1 Binary Codes

Analog-digital conversions involve coding. Different converters output (A/D) and input (D/A) different codes. To properly understand analog-digital conversions, you must understand the coding.

The binary number system can be used in its natural format. If the binary number system uses binary 0 to represent the least positive voltage and a binary 1 to represent the most positive voltage, then the coding system is called *natural binary.*

### 5.6.1.1 Natural Binary

Natural binary is also called *unipolar,* because it is used to represent voltages (such as currents) of only one polarity (e.g., 0 to +5V). The binary number system previously discussed (values 0 to 15 represented by 0 to F Hex) would be natural binary if it were used to represent 0 to some positive value. Binary numbers are used in their fractional form in many industrial settings. Table 5-1 illustrates a four-bit fractional code.

*Table 5-1: Natural Binary Values*

| MSB | B2 | B1 | LSB | Decimal | Fractional | |
|---|---|---|---|---|---|---|
| 1 | 1 | 1 | 1 | 0.9375 | 15/16 | |
| 1 | 1 | 1 | 0 | 0.8750 | 14/16 | 7/8 |
| 1 | 1 | 0 | 1 | 0.8125 | 13/16 | |
| 1 | 1 | 0 | 0 | 0.7500 | 12/16 | 3/4 |
| 1 | 0 | 1 | 1 | 0.6875 | 11/16 | |
| 1 | 0 | 1 | 0 | 0.6250 | 10/16 | 5/8 |
| 1 | 0 | 0 | 1 | 0.5625 | 9/16 | |
| 1 | 0 | 0 | 0 | 0.5000 | 8/16 | 1/2 |
| 0 | 1 | 1 | 1 | 0.4375 | 7/16 | |
| 0 | 1 | 1 | 0 | 0.3750 | 6/16 | 3/8 |
| 0 | 1 | 0 | 1 | 0.3125 | 5/16 | |
| 0 | 1 | 0 | 0 | 0.2500 | 4/16 | 1/4 |
| 0 | 0 | 1 | 1 | 0.1875 | 3/16 | |
| 0 | 0 | 1 | 0 | 0.1250 | 2/16 | 1/8 |
| 0 | 0 | 0 | 1 | 0.0625 | 1/16 | |
| 0 | 0 | 0 | 0 | 0.0000 | 0 | |

Note that if a four-bit number is used to represent 0 to 1 volt, or 0 to 100 percent of full scale, there is an *error inherent* in the representation. This is 1/16 or 0.0625. There is always one least significant bit (LSB) error in the binary representation of a range if 0 is chosen to be the binary zero value and the scale corresponds to the binary fractions. At any part of the input range of the A/D conversion process there is a constant value between digital codes (in this case ±0.0625V). This is the least amount of error (with the number of bits used) for the system. In real systems this error can be determined as shown in equation 5-1.

$$\text{Quantization Error: } Q = \frac{\text{Full scale}}{\text{No. of bits}} \qquad (5\text{-}1)$$

In this equation, $Q$ = quantization error or quantization noise, which is the uncertainty of the measurement due to the conversion process. The only way to reduce this error is by increasing the number of bits used. In industrial systems the following bit numbers can be found in Table 5-2.

*Table 5-2: Common Conversion Word Sizes*

| Bits in ConversionWord | ± (Error) |
|---|---|
| 8 | 0.00391 |
| 10 | 0.00097 |
| 12 | 0.00024 |
| 14 | 0.00006 |
| 16 | 0.000015 |

Generally full scale is standardized at either 0 to +5 volts or 0 to +10 volts for unipolar converters.

### 5.6.1.2 Bipolar Codes

If you wish to represent ± values, you will use a bipolar coding. The standard bipolar values are ±2.5V, ±5.0V, and ±10.0V. To represent these values, straight or natural binary can be used by allowing the "all zeros" state to represent the most negative value, and the "all ones" state to represent the most positive value. Table 5-3 illustrates a four-bit natural binary bipolar coding.

*Table 5-3: Bipolar Coding using Natural Binary*

| Natural Binary | Fraction |
|---|---|
| 1 1 1 1 | + 7/8 |
| 1 1 1 0 | + 6/8 |
| 1 1 0 1 | + 5/8 |
| 1 1 0 0 | + 4/8 |
| 1 0 1 1 | + 3/8 |
| 1 0 1 0 | + 2/8 |
| 1 0 0 1 | + 1/8 |
| 1 0 0 0 | + 0/8 |
| 0 1 1 1 | - 1/8 |
| 0 1 1 0 | - 2/8 |
| 0 1 0 1 | - 3/8 |
| 0 1 0 0 | - 4/8 |
| 0 0 1 1 | - 5/8 |
| 0 0 1 0 | - 6/8 |
| 0 0 0 1 | - 7/8 |
| 0 0 0 0 | - 8/8 |

When natural binary is used to represent bipolar values, the halfway value, 1000, represents the value 0. So 1000 (binary) is the "offset" from binary 0000. This is why natural binary is called offset binary when it is used to represent bipolar values.

Although there are many other codes to represent bipolar values, the two's complement is the most common in computer-driven systems. Two's complement coding is illustrated in Table 5-4.

*Table 5-4: Two's Complement*

| Natural Binary | Twos Complement | Decimal |
|---|---|---|
| 1 1 1 1 | 0 1 1 1 | + 7/8 |
| 1 1 1 0 | 0 1 1 0 | + 6/8 |
| 1 1 0 1 | 0 1 0 1 | + 5/8 |
| 1 1 0 0 | 0 1 0 0 | + 4/8 |
| 1 0 1 1 | 0 0 1 1 | + 3/8 |
| 1 0 1 0 | 0 0 1 0 | + 2/8 |
| 1 0 0 1 | 0 0 0 1 | + 1/8 |
| 1 0 0 0 | 0 0 0 0 | 0/8 |
| 0 1 1 1 | 1 1 1 1 | - 1/8 |
| 0 1 1 0 | 1 1 1 0 | - 2/8 |
| 0 1 0 1 | 1 1 0 1 | - 3/8 |
| 0 1 0 0 | 1 1 0 0 | - 4/8 |
| 0 0 1 1 | 1 0 1 1 | - 5/8 |
| 0 0 1 0 | 1 0 1 0 | - 6/8 |
| 0 0 0 1 | 1 0 0 1 | - 7/8 |
| 0 0 0 0 | 1 0 0 0 | - 8/8 |

Several items concerning two's complements coding should be noted. Zero value is represented by 0 binary. If you add a positive number and the same negative number (for example, +2/8 added to –2/8), the result is 0 with a carry. Actually, two's complement is the offset (natural) binary system with the most significant bit inverted (complemented). Most binary computers perform arithmetic operations using two's complement, so its use in these systems is understandable.

### 5.6.2 Digital-to-Analog Conversion

Digital-to-analog conversions are discussed first for a number of reasons, the primary one being that most successive approximation analog-to-digital converters use a digital-to-analog converter (either internally or externally) as a reference.

Many different techniques have been used for converting digital values to either voltage or current values. Almost all contemporary converters are of the parallel type, meaning that they convert the entire number of bits simultaneously to the voltage or current value.

#### 5.6.2.1 Weighted Resistor Networks

One of the more popular methods used by discrete circuitry or hybrid integrated circuit converters is one using a weighted resistor network. This is illustrated in Figure 5-6.

In the figure the switches, are either on or off, and the current-limiting resistors have a binary weight. This of course is a simplified diagram. The switches are put in the ON state by a positive voltage (representing the one state), and OFF by 0 volts (representing the zero state). If the natural binary number "1 0 0 0," with "1" the MSB, is applied to this circuit, the switch with R will be on; the others are off. The output voltage will be VREF/2, because R is equal to R. This means that the voltage at the output will be what you would expect with a four-bit system using natural binary coding as binary "1000" = ½ full scale.

Each bit has a resistor that is twice the value of the preceding resistor. Because the currents are summed, if more than one resistor is ON, you only have to add the current values to obtain the output.

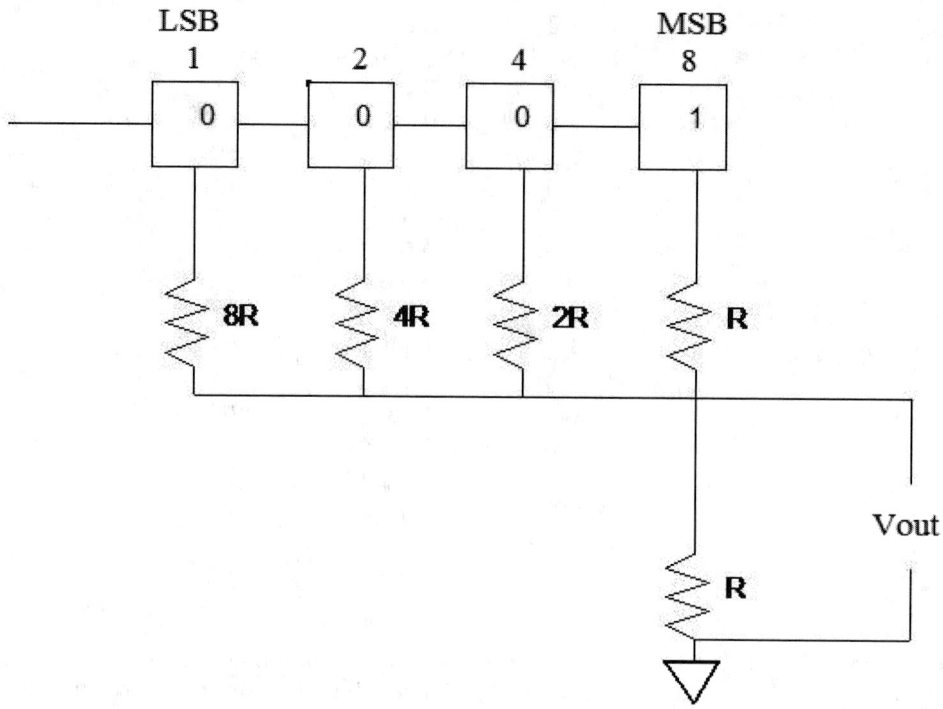


*Figure 5-6: Weighted Resistor Network*

**5.6.2.2 R-2R Networks**

One of the more common methods of D/A conversion uses a resistive network comprised of only two values. It is called the R-2R ladder method. It is used primarily with the successive approximation type A-to-D converters and is quite suitable for integrated circuit construction, as the range of resistance values required are just two.

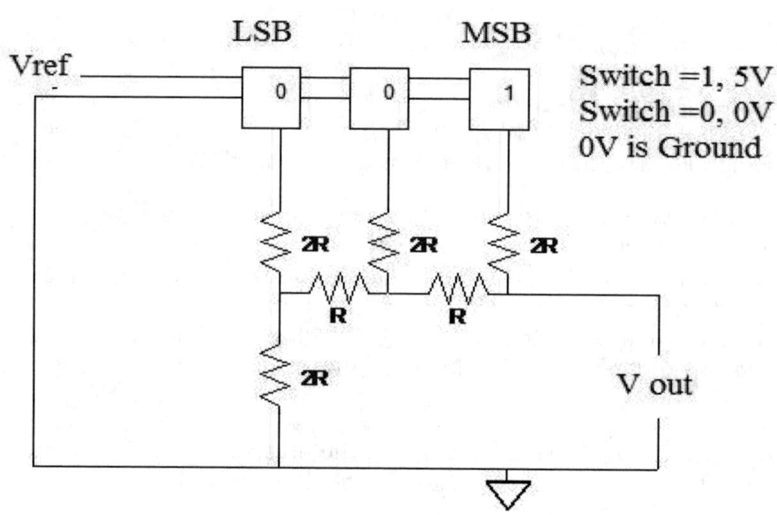


*Figure 5-7: R-2R Ladder*

Figure 5-7 illustrates a R-2R ladder. Only a three-bit ladder is shown for simplicity. Determining the operation is really just an exercise in Ohm's law. If two resistors of the same value are in parallel, the equivalent resistance is one-half the value. In this case, if you have 2R || 2R, then the equivalent is R.

#### 5.6.2.3 Other Considerations

Regardless of which method is used, the output will not be continuous but rather a series of levels, switching each time a new binary value is placed into the D-to-A unit. To smooth out the switching transients, the output signal is averaged over time (integrated). This may be accomplished by using a low-pass filter, or an integrating device. Although the integrated output is an approximation of the output, the more bits used, the closer the approximation to the binary representation. The lower the rate of change from one level to another, in other words, the lower the frequency of the signal, the more accurately the output represents the binary value.

### 5.6.3 Analog-to-Digital Conversion

The three types of conversions discussed in this text are integrating, successive approximation, and parallel conversions.

#### 5.6.3.1 Integrating

One common method for converting low frequency signals, which includes most industrial process variables, is the integrating type. There are two different techniques presently used. These are:

- dual slope
- voltage to frequency

#### 5.6.3.2 Successive Approximation

Successive approximation is one of the most widely used techniques for analog-to-digital conversion. Compared to the integrating methods it is quite complex and has the disadvantage of losing some code combinations if the design is not carefully considered. However, it is very fast. Modern integrated circuit designs can approach 500,000 conversions per second in a relatively inexpensive package. The basic block diagram is illustrated in Figure 5-8. Note there is a D/A used as a reference. The successive approximation converter uses the principle of binary division to make the least amount of decisions necessary to locate a random number in its range.

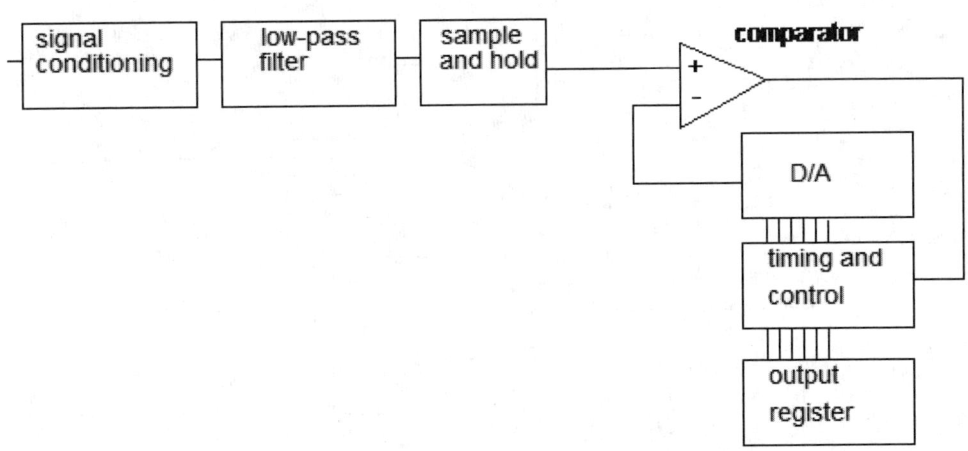


*Figure 5-8: Successive Approximation Block Diagram*

The operation is as follows:

1. The signal is input through conditioning circuits, some form of attenuator and amplitude limiter. There is also a low-pass filter, which limits the upper frequency of the input signal and is crucial to proper operation of the converter.

2. The signal is then gated through to one leg of a comparator; the other leg of the comparator is connected to the D/A.

3. At the start of conversion the D/A is fed by a timing and control register with the binary ½ full-scale value placed in it by the control circuit. Therefore, if the input voltage is above the reference voltage (above ½ full scale), the comparator output will be a "1" state. If the input voltage had been below the reference, the output of the comparator would have been a "0" state.

4. The input signal is again tested against this new value, with the same operation and each succeeding bit treated the same way. In this way successive tests approximate the input signal level closer and closer to its true value.

The sample clock (the frequency at which the sample and hold circuit is switched) must be at least twice the highest frequency expected. It could be many times more, but at the very least it must be twice as much. This is to ensure at least two samples per waveform. You may correctly conclude that the bit rate is the number of bits times the sample rate.

#### 5.6.3.3 Flash (Parallel) Conversion

Figure 5-9 illustrates parallel conversion, also known as "flash" conversion. This is the fastest method of A/D conversion. The speed is limited only by the settling time of the comparators and the gate propagation time of the decoder logic. A precision reference is divided down between each of the comparators. The number of comparators required is one less then two raised to the number of bits (i.e., an eight-bit flash converter will require 255 comparators). Because this is a rather large number of comparators even for modern integrated circuitry, methods of combining to four-bit converters (each requiring 15 comparators) are used at the penalty of slightly slower operation. Technology has been used to reduce the price of flash converters where they are now competitive against successive approximation types, and still have the very high speed of conversion.

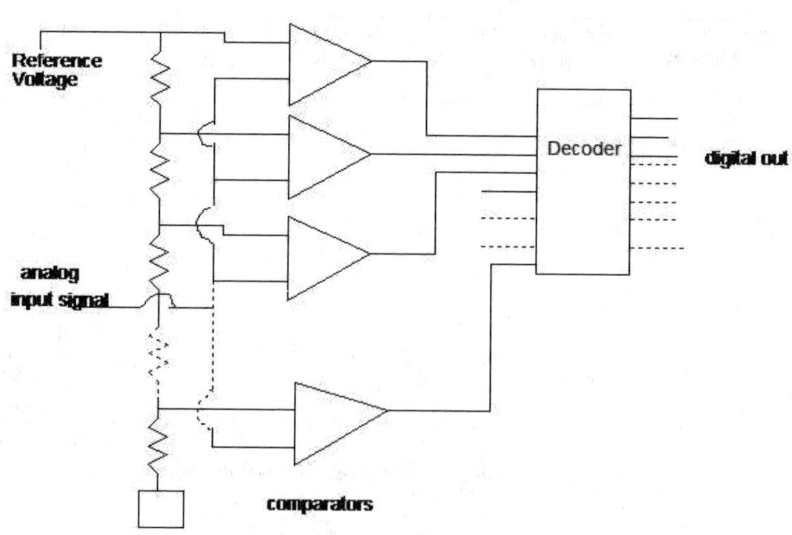


*Figure 5-9: Flash A/D Conversion*

## 5.7 References

ANSI/ISA-50.1-1982 (R2002) – *Compatibility of Analog Signals for Electronic Industrial Process Instruments.*

ANSI/ISA-5.1-1984 (R1992) – *Instrumentation Symbols and Identification.*

## About the Authors

**Richard "Dick" Caro** is CEO of CMC Associates, a business strategy and professional services firm in Acton, Mass. Before working at CMC, he was Vice President of ARC Advisory Group in Dedham, Mass. He is the Chairman of ISA SP50 and formerly of the IEC (International Electrotechnical Committee) Fieldbus Standards Committees. Before joining ARC, he held the position of Senior Manager with Arthur D. Little, Inc., in Cambridge, Mass. He was founder of Autech Data Systems and was Director of Marketing at ModComp. In the 1970s, Foxboro employed him in both development and marketing positions. He holds a BS and an MS in chemical engineering and an MBA.

**Lawrence (Larry) M. Thompson** is General Manager and Owner of Electronic Systems: development and training company (EsdatCo), a training and development company in Waco, Texas. An adjunct instructor for ISA for more than 20 years, Larry has served in a multitude of positions ranging from technician through test engineering supervisor in various aspects of automation and communications. A 20-year veteran of the USAF, primarily in Electronic Encryption Equipment, his career has been focused on automation and computing. Thompson holds a BAAS in technology from Tarleton State University and is the author of several ISA books, including *Industrial Data Communications* and *Basic Electricity and Electronics for Control, Fundamentals and Applications.*

# 6 Control System Documentation

*By Fred Meier*

## Topic Highlights

*Reasons for Documentation*
*Types of Documentation*
*Process Flow Diagram (PFD)*
*Piping and Instrument Diagrams (P&IDs)*
   *Loop Numbering*
*Instrument Lists*
*Specification Forms*
*Logic Diagrams*
*Location Plans (Instrument Location Drawings)*
*Installation Details*
*Loop Diagrams*
*Standards and Regulations*
   *Mandatory Standards*
   *Consensus Standards*
*Operating Instructions*

## 6.1 Reasons for Documentation

The documentation used to define modern control systems has evolved over the past 50 years. Its purpose is to impart, efficiently and clearly, to a knowledgeable viewer enough information so that the result is an operating plant producing the desired product. The documents described in this chapter form a typical set for use in the design of a continuous process plant. Some of the documents are also used in other process types. The typical set is not necessarily a standard set. Some designs may not include all of the described documents, and some designs include documents not described.

All of the illustrations and much of the description used in this section were published in 2004 by ISA in *Instrumentation and Control System Documentation* by Frederick A. Meier and Clifford A. Meier. That book includes many more illustrations and much more explanation.

ISA is the abbreviation for The Instrumentation, Systems, and Automation Society. For this reason, this section uses the term "automation and control" (A&C), rather than "instrument and control" (I&C) used in the Meiers' book to describe the engineers and designers developing the control system documentation.

## 6.2 Types of Documentation

Descriptions and typical sketches are included for the following:

- Process Flow Diagrams (PFD)
- Piping and Instrument Diagrams (P&ID)
- Loop Numbering
- Instrument Lists
- Specification Forms
- Logic Diagrams
- Location Plans (Instrument Location Drawings)
- Installation Details
- Loop Diagrams
- Standards and Regulations
- Operating Instructions

Figure 6-1, or the timeline, illustrates a possible sequence for document development. Information from one document is used to develop succeeding documents.

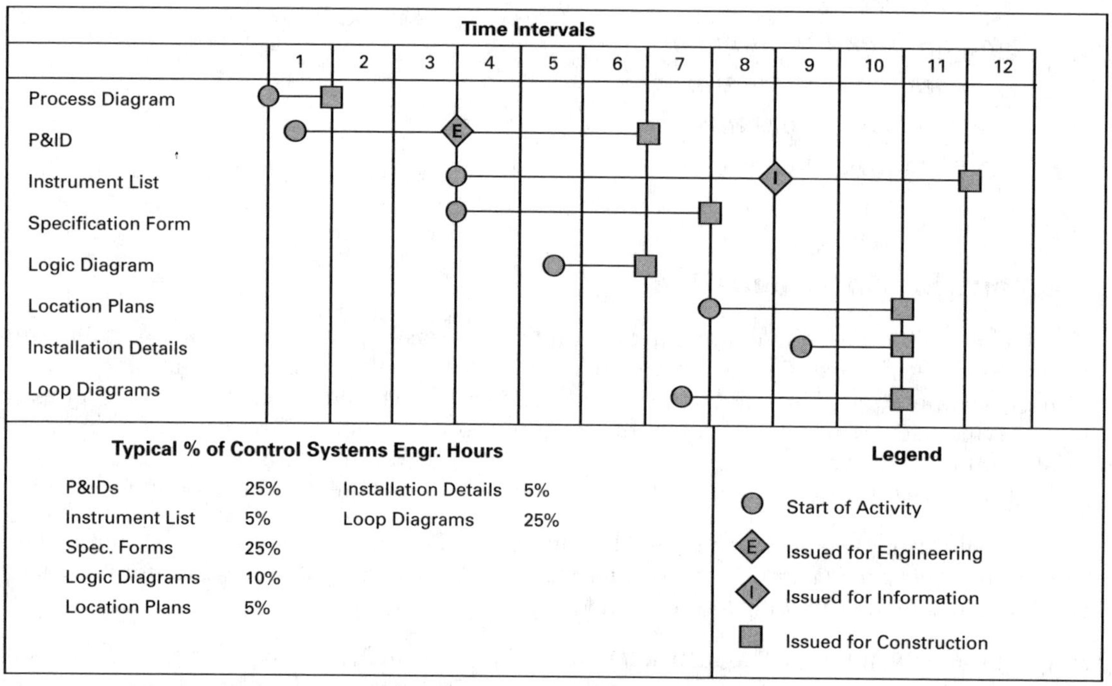


*Figure 6-1: Instrument Drawing Schedule*

The time intervals vary. The intervals might be days, weeks, or months, but the sequence remains the same. The documents listed are not all developed or used solely by a typical automation and control group (A&C). However, the A&C group contributes to, and uses, the information contained in them during plant design.

## 6.3 Process Flow Diagram (PFD)

A Process Flow Diagram defines a process schematically. PFDs are most valuable for continuous process chemical plants. The PFD shows what and how much of each product a plant might make; descriptions and quantities of the raw materials necessary; by-products produced; critical process conditions—pressures, temperatures, and flows; necessary equipment; and major process piping.

Figure 6-2 shows a simple PFD of a knockout drum, which separates the liquid from a wet gas stream. Process engineers frequently produce PFDs. Some PFDs include basic, important—or high-cost—A&C components. There is no ISA standard for PFDs, but ISA-5.1-1984 (R1992) *Instrument Symbols and Identification* and ISA-5.3-1983 *Graphic Symbols for Distributed Control/Shared Display Instrumentation, Logic, and Computer Systems* contain symbols that can be used to show A&C components.

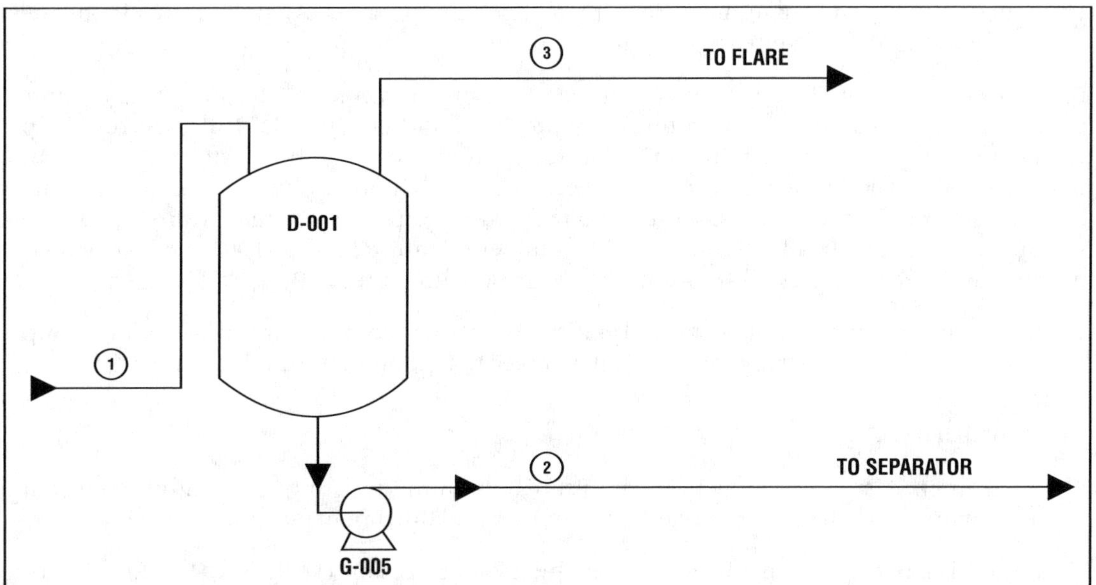


| STREAM NUMBER | FLOW | DESCRIPTION | TEMP | PRESSURE | SP GRAVITY |
|---|---|---|---|---|---|
| 1 | 10,000#/Hr | WET GAS | 90° - 180° F | 20 psi | - |
| 2 | 1,000#/Hr | DEGASSED MATERIAL | 70° - 170° F | 50 psi | 0.9 AT 60°F |
| 3 | 9,000#/Hr | LIGHT ENDS TO FLARE | 80° - 140° F | 4 psi | - |

| ISA COURSE FG15 |
|---|
| PROCESS FLOW DIAGRAM |
| PLANT 001 KNOCKOUT DRUM 0-001 |
| DRG #PFD-1 |

*Figure 6-2: Process Flow Diagram*

Batch process plants may configure their equipment in various ways, as raw materials and process parameters change. Many different products are often produced in the same plant. A control recipe, or formula, is developed for each product. A PFD may be developed for each recipe.

## 6.4 Piping and Instrument Diagrams (P&ID)

The acronym P&ID is widely understood within the process industries to mean the principal document used to define the process—the equipment, piping, and all A&C components. ISA's *Automation, Systems, and Instrumentation Dictionary* definition for P&ID tells us what they do. P&IDs "show the interconnection of process equipment and instrumentation used to control the process."[1]

P&IDs are developed in steps by members of the various design disciplines as a project proceeds. Information placed on a P&ID by one discipline is used by other disciplines as the basis for their design.

The P&ID shown in Figure 6-3 has been developed from the PFD in Figure 6-2. The P&ID includes the control system definition using symbols from ISA-5.1 and 5.3. There are two electronic loops which are part of the shared display/distributed control system (DCS): FRC-100, a flow loop with control and recording capability, and LIC-100, a level loop with control and indicating capability. There is one field-mounted pneumatic loop, PIC-100, with control and indication capability. There are several switches and lights on a local (field) mounted panel, including hand operated switches and lights HS and ZL-400, HS and HL-401, and HS and HL-402. There are other control system components shown, in addition to the above. The P&ID now includes more piping and mechanical equipment details.

### 6.4.1 Loop Numbering

Letter designations and tag numbers identify all A&C components. All devices in a loop have the same tag number but different letter designations.

Figure 6-4 consists of LT-100, a field mounted electronic transmitter; LI-100, a field mounted electronic indicator; LIC-100, a level controller which is part of the DCS; LY-100, a current-to-pneumatic (I/P) converter; and LV-100, a pneumatic butterfly control valve. ISA-5.1 states that loop numbers may be parallel, requiring a new number sequence for each process variable, or serial, using a single numeric sequence for all process variables. Figure 6-3: P&ID uses a parallel numbering system. There is a flow loop FRC-100, a level loop LIC-100, and temperature loop TI-100. The level gauges, pressure gauges, and thermometers all are numbered starting with 1: LG-1, PI-1, TI-1.

Figure 6-5 shows how tag marks may also identify the loop location or service. Other numbering systems are used that tie instruments to a P&ID, a piece of equipment or a location.

## 6.5 Instrument Lists

The Instrument List (or instrument index) is an alphanumeric listing of all tag-marked components. Each tag mark will reference the relevant drawings and documents for that device.

Figure 6-6 is a partial listing which includes the level devices on D-001—K.O. drum: LG-1, level gauge; LT-100, level transmitter; and LI-100, level indicator (all from Figure 6-3: P&ID). In addition, the list shows other instruments on other P&IDs, not included. Figure 6-6 has six columns—for P&ID, Spec Form, Req. #, Location Plan, Installation Detail, and Piping Drawing.

The Instrument List is developed by the A&C group. There is no ISA standard defining an Instrument List. With the advent of computer-aided design techniques, the Instrument List may contain a large number of columns for various uses during project design, construction, and operation.

## 6.6 Specification Forms

The A&C group defines the tag-marked devices so suppliers may quote and supply the correct device. A Specification Form (or data sheet) is filled out for each device.

1. *The Automation, Systems, and Instrumentation Dictionary,* 4th edition (ISA, 2003), pg. 273.

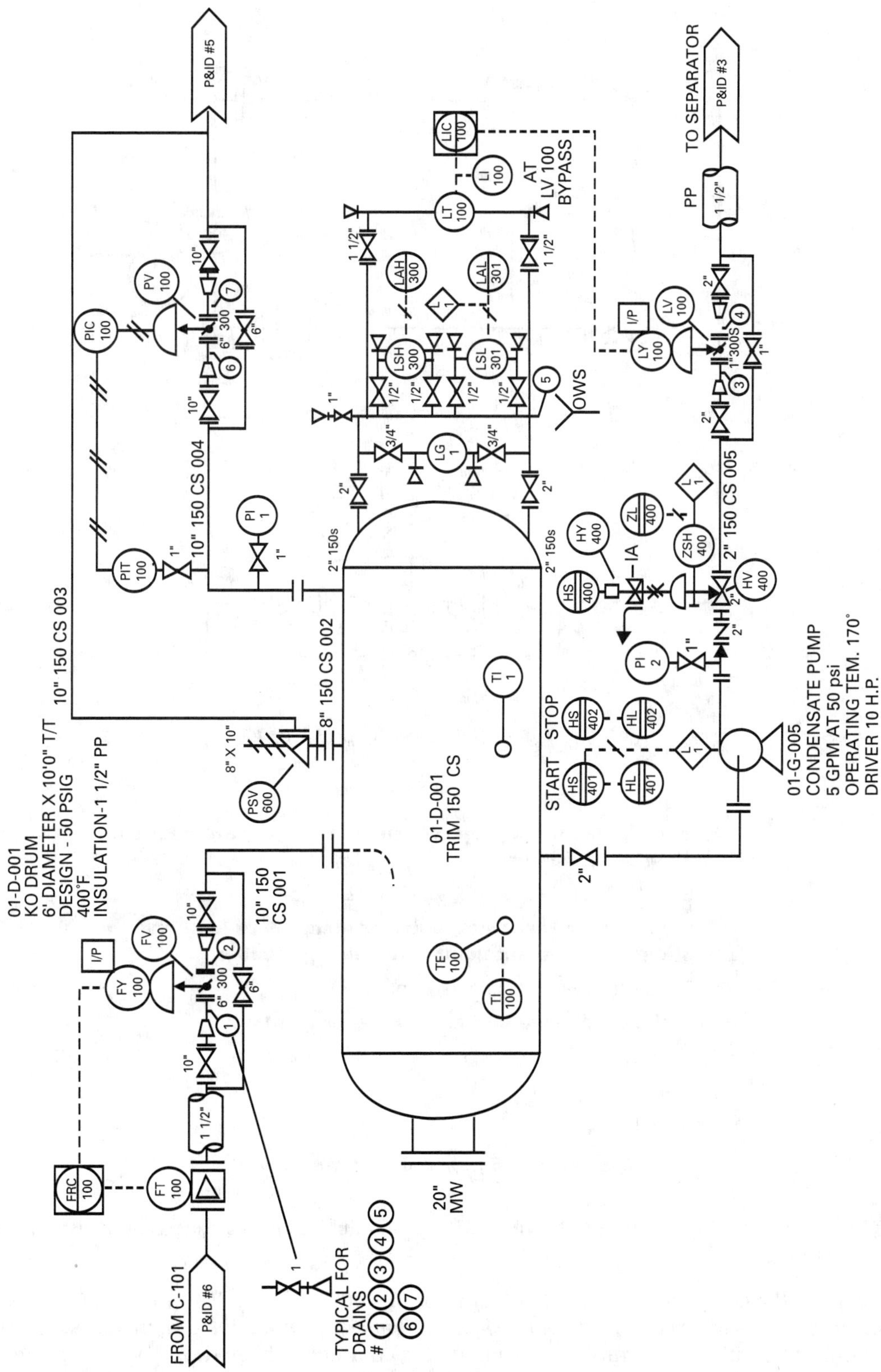


*Figure 6-3: P&ID*

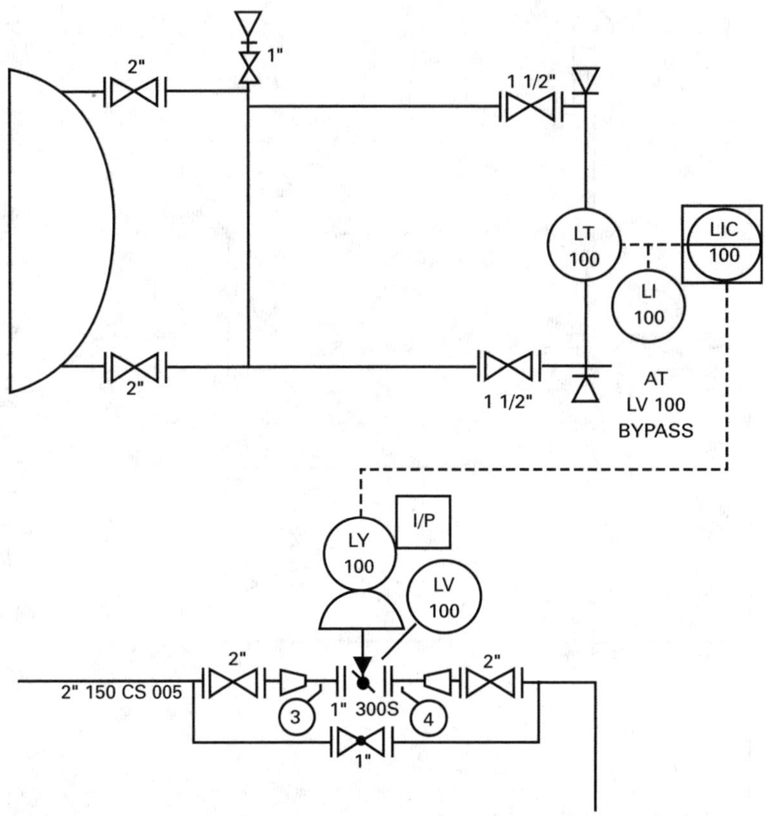


*Figure 6-4: Level Loop LIC-100*

- **Use Basic Number if project is small and there are no area, unit, or plant numbers:**
  - Basic Number FT-2 or FT-02 or FT-002
- **If project has a few areas, units, or plants (9 or less), use the first digit of the plant number as the tag number:**
  - FT-102 (1 = area, unit, or plant number)
- **If project is divided into area, units, or plants:**
  - 1-FT002
  - 01-FT002
  - 001-FT002

*Figure 6-5: Instrument Numbering*

Let's look at LT-100 from Figure 6-3. The P&ID symbol defines it as an electronic displacement-type level transmitter.

Figure 6-7 is the completed Specification Form for LT-100. This form is from ISA-20-1981, *Specification Forms for Process Measurement of Control Instruments, Primary Elements and Control Valves*. There are many variations of Specification Forms. Most engineering contractors have developed a set, some control component suppliers have their set, and ISA has another newer set in technical report, ISA-

| Tag # | Desc. | P&ID # | Spec Form # | REQ # | Location Plan # | Install. Detail | Piping Drawing |
|---|---|---|---|---|---|---|---|
| LG-1 | D-001-K.O. Drum | 1 | L-1 | L-1 | – | – | ISO-010 |
| LG-2 | D-001 Distil. Column | 2 | L-1 | L-1 | – | – | ISO-015 |
| LG-3 | C-002 Stripper | 3 | L-1 | L-1 | – | – | ISO-016 |
| LT-100 | D-001 K.O. Drum | 1 | L-100 | T-1 | LP-1 | ID-001 | ISO-010 |
| LI-100 | D-001 K.O. Drum | 1 | I-100 | I-1 | LP-1 | ID-002 | – |
| LT-101 | C-001- Distil. Column | 2 | L-100 | T-1 | LP-4 | ID-001 | ISO-015 |
| LT-102 | C-002 Stripper | 3 | L-100 | T-1 | LP-5 | ID-001 | ISO-016 |

*Figure 6-6: A Typical Instrument List*

TR20.00.01-2001, *Specification Forms for Process Management & Control - Part 1: General Considerations*. The purpose of all of the forms is to aid the A&C group to organize the information needed to fully and accurately define control components so they may be quoted on, and supplied, by vendors. Specification Forms are filled out by the A&C group. Their development is a significant part of the group's effort.

## 6.7 Logic Diagrams

Continuous process control is shown clearly on P&IDs. Different presentations are needed for on/off control. Logic Diagrams are one form of these presentations. ISA's set of symbols are defined in ISA-5.2-1976(R1992) - *Binary Logic Diagrams for Process Operations.*

ISA symbols AND, OR, NOT and MEMORY (FLIP-FLOP) with an explanation of their meaning are shown in Figures 6-8 and 6-9. Other sets of symbols and other methods may be used to document on/off control. Some examples: text descriptions, a written description of the on/off system; ladder diagrams; or electrical elementaries.

Some designers develop a Functional Specification or Operation Description to document the entire system. These documents usually include a description of the on-off control of the process.

A motor start circuit is shown in Figure 6-10 in ISA logic form and also by an elementary diagram.

## 6.8 Location Plans (Instrument Location Drawings)

There is no ISA standard that defines a Location Plan or an Instrument Location Drawing. Location Plans show the location and elevation of control components on plan drawings of a plant.

Figure 6-11 shows one approach for a Location Plan. It shows the approximate location and elevation of the tag-marked devices included on the P&ID, Figure 6-3, air supplies for the devices, and interconnection tubing needed to complete the pneumatic loop. Other approaches to Location Plans might include conduit and cabling information and fitting and junction box information. Location Plans are developed by the A&C or electrical groups. They are used during construction and by maintenance personnel after the plant is built to locate the various devices.

| LEVEL INSTRUMENTS (DISPLACER OR FLOAT) | | | | SHEET | | OF |
|---|---|---|---|---|---|---|
| NO | BY | DATE | REVISION | SPEC. NO. 321 | | REV. 0 |
| 0 | FAM | 12/15/2003 | | CONTRACT 1234 | | DATE 1/3/2003 |
| | | | | REQ. - P.O. J-6 J-12 | | |
| | | | | BY FAM | CHK'D CHK CAM | APPR. LF |

| | | | | | | |
|---|---|---|---|---|---|---|
| | 1 | Tag Number | | LT-100 | | |
| | 2 | Service | | K.O. DRUM | | |
| | 3 | Line Number / Vessel Number | | 01-D-001 | | |
| BODY/CAGE | 4 | Body or Cage Material | | C.S. | | |
| | | Rating | | 300 psi | | |
| | 5 | Conn Size & Location Upper | | 1 1/2" TOP | | |
| | | Type | | 300 psi FLG | | |
| | 6 | Conn Size & Location Lower | | 1 1/2" BTM | | |
| | | Type | | 300 psi FLG | | |
| | 7 | Case Mounting | | SIDE | | |
| | | Type | | | | |
| | 8 | Rotatable Head | | NOT REQ | | |
| | 9 | | | | | |
| | 10 | Orientation | | LEFT HAND | | |
| | 11 | Cooling Extension | | NOT REQ | | |
| | 12 | | | | | |
| DISPLACER OR FLOAT | 13 | Dimensions | | 48" | | |
| | 14 | Insertion Depth | | | | |
| | 15 | Displacer Extension | | | | |
| | 16 | Disp. or Float Material | | 304 S.S. | | |
| | 17 | Displacer Spring/Tube Mtl. | | MFG. STD. | | |
| | 18 | | | | | |
| | 19 | | | | | |
| XMTR/CONT. | 20 | Function | | TRANSMITTER | | |
| | 21 | Output | | 4-20 mAdc | | |
| | 22 | Control Modes | | | | |
| | 23 | Differential | | | | |
| | 24 | Output Action: Level Rise | | INCREASE | | |
| | 25 | Mounting | | INTEGRAL | | |
| | 26 | Enclosure Class | | NEMA 8 | | |
| | 27 | Elec. Power or Air Supply | | 24Vdc from shared | | |
| | 28 | | | display | | |
| SERVICE | 29 | Upper Liquid | | WET GAS | | |
| | 30 | Lower Liquid | | DEGASSED MTL. | | |
| | 31 | Sp. Gr.: Upper | Sp. Gr.: Lower | | .9 @ 60 F | |
| | 32 | Press. Max. | Normal | 50 PSI | 4 PSI | |
| | 33 | Temp. Max. | Normal | 400 F | 90-150 F | |
| | 34 | | | | | |
| | 35 | | | | | |
| OPTIONS | 36 | Airset | Supply Gage | | | |
| | 37 | Gage Glass Connections | | | | |
| | 38 | Gage Glass Model No. | | | | |
| | 39 | Contact: No. | Contact: Form | | | |
| | 40 | Contact Rating | | | | |
| | 41 | Action of Contacts | | | | |
| | 42 | | | | | |
| | 43 | | | | | |
| | 44 | | | | | |
| | 45 | | | | | |
| | 46 | Manufacturer | | LATER | | |
| | 47 | Model Number | | LATER | | |
| | 48 | | | | | |

NOTES:

 ISA FORM S20.26

*Figure 6-7: Level Instrument - Specification Form*

## 6.9 Installation Details

Installation Details define the requirements to correctly install the tag-marked devices. The Installation Details show process connections, pneumatic tubing, or conduit connections, insulation and winteriz-

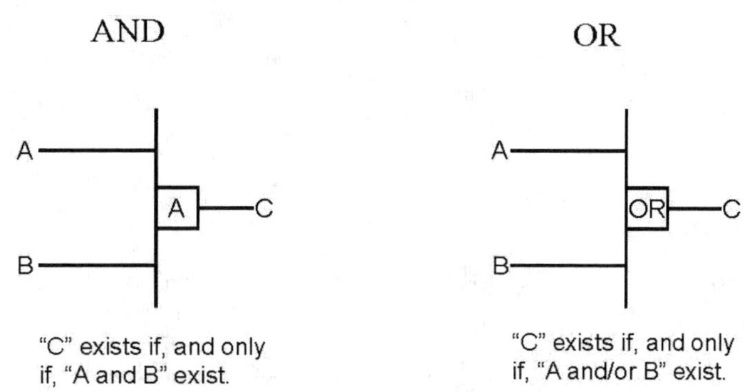


*Figure 6-8: Binary Logic Symbols - AND & OR*

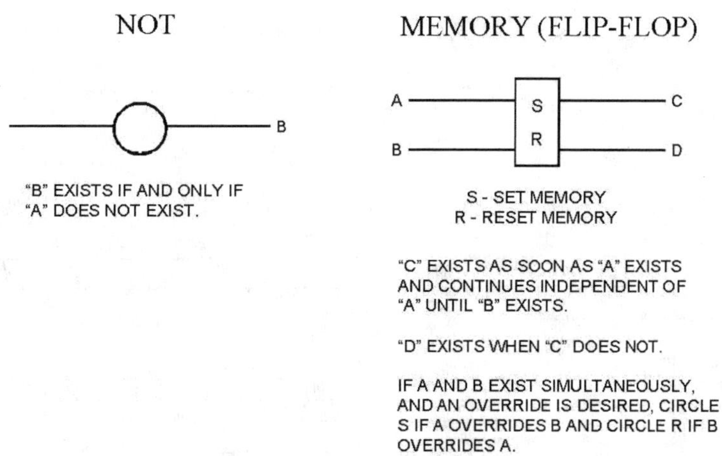


*Figure 6-9: Binary Logic Symbols - NOT & MEMORY (FLIP-FLOP)*

ing requirements, and support methods. There is no ISA Standard that defines Installation Details. However, libraries of Installation Details have been developed and maintained by engineering contractors, A&C device suppliers, some plant owners, installation contractors, and some individual designers. They all have the same aim—successful installation. They may differ in details as to how to achieve it, however.

Figure 6-12 shows one approach. This drawing includes a material list to aid in procuring installation materials and assisting installation personnel.

Installation Details may by developed by the A&C group during the design phase. However, they are sometimes developed by the installer during construction or by an equipment supplier for the project.

## 6.10 Loop Diagrams

ISA's *Automation, Systems, and Instrumentation Dictionary* defines a Loop Diagram as "a schematic representation of a complete hydraulic, electric, magnetic or pneumatic circuit."[1] The circuit is called a loop. For a typical loop see Figure 6-4. ISA-5.4-1991, *Instrument Loop Diagrams* presents six typical loop dia-

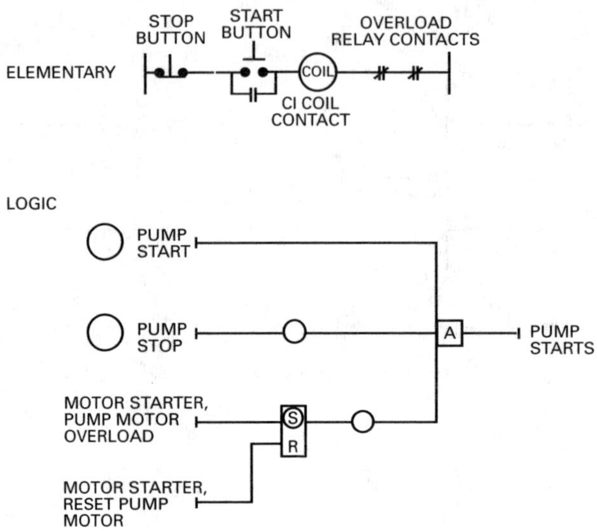


*Figure 6-10: Motor Start Logic*

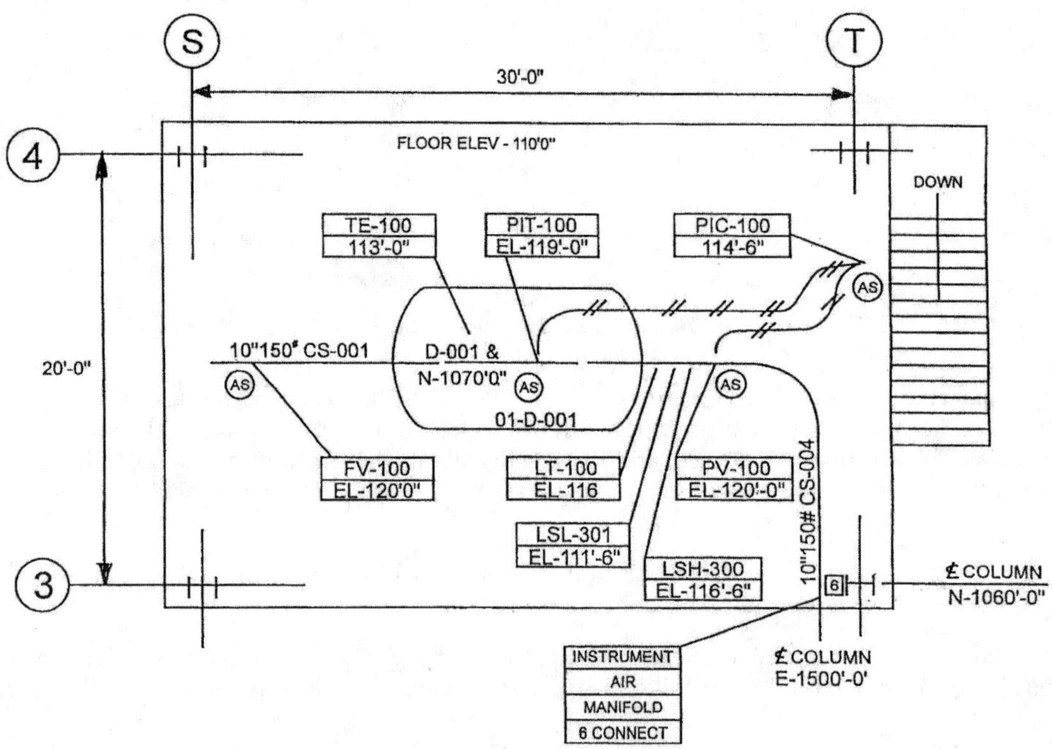


*Figure 6-11: Location Plan, Approach A*

1. Ibid., pg. 299.

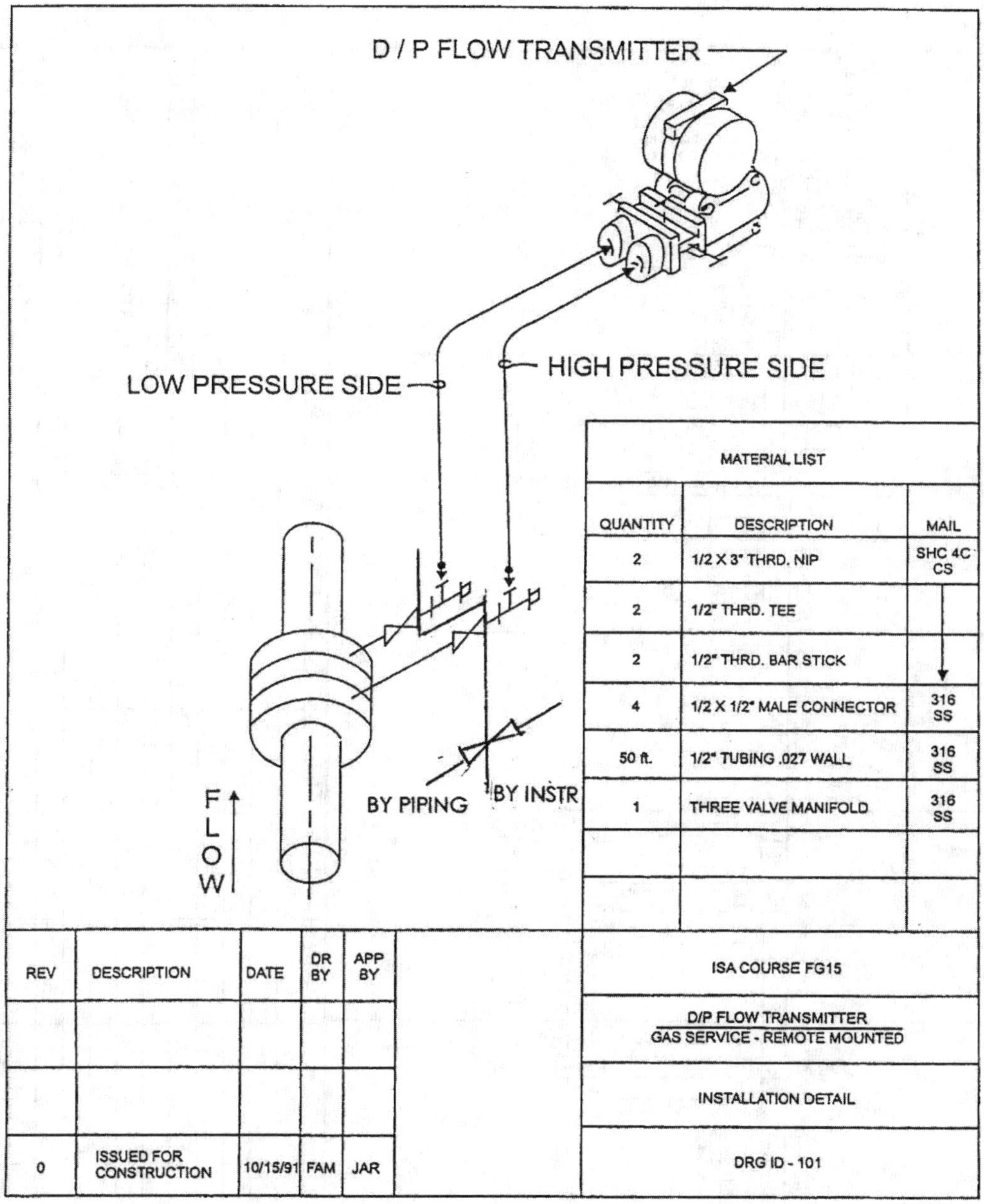


*Figure 6-12: Installation Detail, Type 2 - Flow Transmitter*

grams, two each for pneumatic, electronic, and distributed control (DCS). One of each type shows the minimum items required, and the other shows additional optional items.

Figure 6-13 is a Loop Diagram for electronic flow loop FIC-301. Loop Diagrams are not always included in a design package. Some plant owners do not believe they are worth their cost, which is significant. Loop Diagrams are sometimes produced by the principal project A&C supplier, the installation contractor, or by the plant owner's operations, maintenance, or engineering personnel. Sometimes Loop Diagrams are produced on an "as needed" basis after the plant is running.

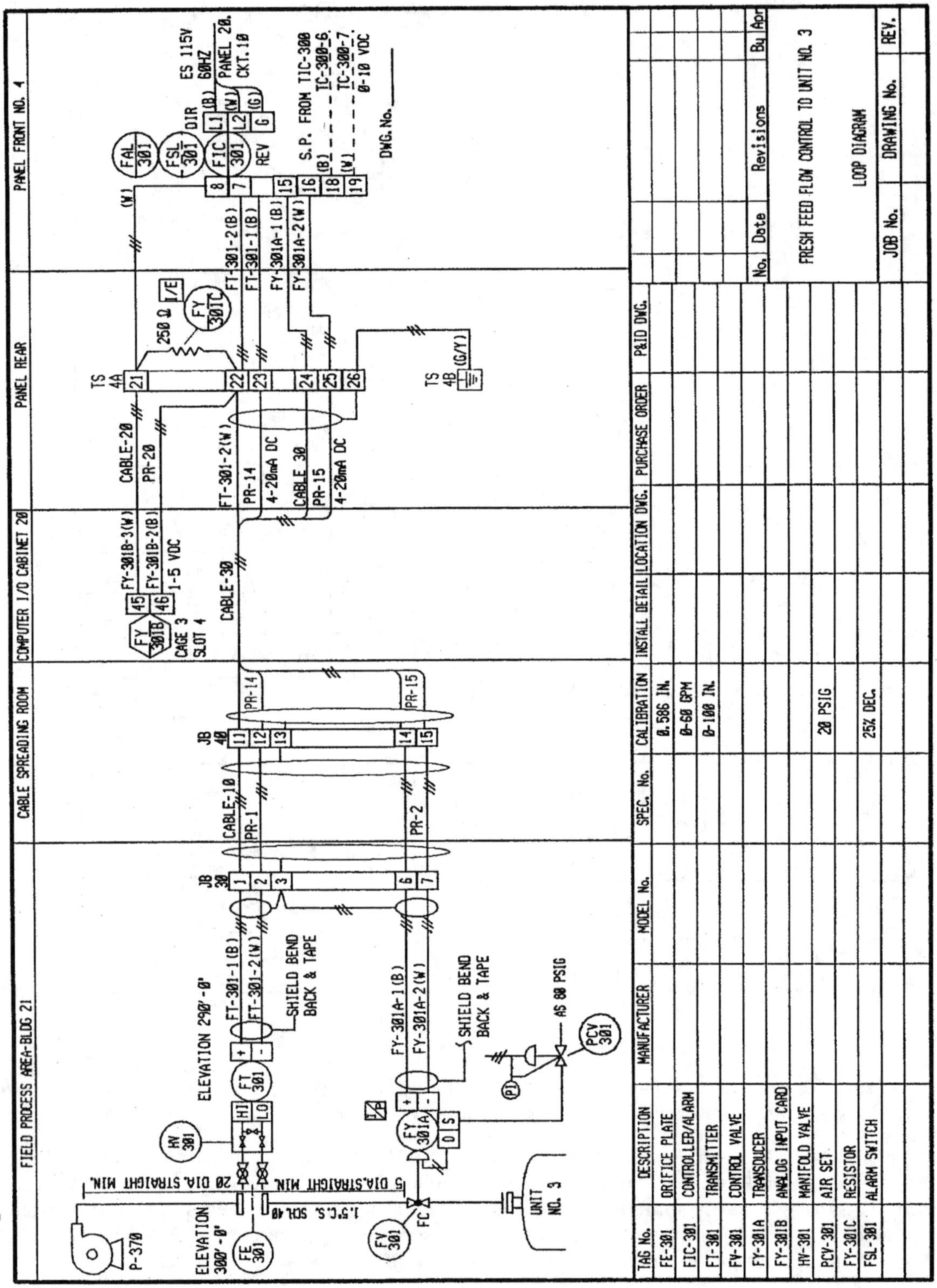


*Figure 6-13: Loop Diagram, Electronic Control, Minimum Required Items Plus Optional Items*

## 6.11 Standards and Regulations

### 6.11.1 Mandatory Standards

Federal, state, and local laws establish mandatory requirements: codes, laws, regulations, requirements, etc. The Food and Drug Administration issues Good Manufacturing Practices. The National Fire Protection Association (NFPA) issues Standard 70, the *National Electric Code* (NEC). The United States government manages about 50,000 mandatory standards. The Occupational Safety and Health Administration (OSHA) issues many regulations including government document 29 CFR 1910.119, *Process Safety Management of Highly Hazardous Chemicals* (PSM). There are three paragraphs in the PSM which list documents required if certain hazardous materials are handled. Some of these documents require input from the plant A&C group.

### 6.11.2 Consensus Standards

Consensus Standards include recommended practices, standards, and other documents developed by professional societies and industry organizations. The standards developed by ISA are the ones used most often by A&C personnel. Relevant ISA standards include: ISA-5.1-1984-(R1992), *Instrumentation Symbols and Identification*, which defines symbols for A&C devices; ISA-5.2-1976-(R1992), *Binary Logic Diagrams for Process Operations*, which provides additional symbols used on Logic Diagrams; and ISA-5.3-1983, *Graphic Symbols for Distributed Control/ Shared Display Instrumentation, Logic and Computer Systems*, which contains symbols useful for DCS definition. The key elements of ISA-5.3 are now included in ISA-5.1, and ISA-5.3 will be withdrawn in the future.

ISA-5.4 *Instrument Loop Diagrams* includes additional symbols and six typical instrument Loop Diagrams. ISA-5.5 *Graphic Symbols for Process Displays* establishes a set of symbols used in process display. Other ISA standards of interest include ISA-20-1981, *Specification Forms for Process Measurement and Control Instruments, Primary Elements and Control Valves*. ISA TR20.00.01-2001, *Specification Forms for Process Control and Instrument, Part 1: General Considerations*, updates ISA-20. ANSI/ISA-84.00.01-2004 - *Functional Safety: Safety Instrumented Systems for the Process Industry Sector*, defines the requirements for safe systems. ANSI/ISA-88.01-1995, *Batch Control Part I Models and Terminology*, shows the relationships involved between the models and the terminology.

In addition to ISA, other organizations develop documents to guide professionals. These organizations include American Petroleum Institute, American Society of Mechanical Engineers, National Electrical Manufacturers Association, Process Industry Practice, and Technical Association of the Pulp and Paper Industry.

## 6.12 Operating Instructions

Operating Instructions are necessary to operate a complex plant. They range from a few pages describing how to operate one part of a plant to a complete set of books covering the operation of all parts of a facility. They might be included in a functional specification or an operating description. There is no ISA standard to aid in developing Operating Instructions. They might be prepared by a group of project, process, electrical and A&C personnel during plant design. Some owners prefer plant operations personnel to prepare these documents. The Operating Instructions guide plant operators and other personnel during normal and abnormal plant operation, including start-up, shutdown, and emergency operation of the plant.

OSHA requires operating procedures for all installations handling hazardous chemicals. Their requirements are defined in government document 29 CFR 1910.119(d) *Process Safety Information*, (f ) *Operating Procedures* and (l) *Management of Change*. For many types of food processing and drug manufacturing, the Food and Drug Administration issues Good Manufacturing Practices.

## About the Author

**Fred Meier**'s career in engineering and engineering management spans 50 years. He has been an active member of ISA for more than 40 years. He has earned an ME from Stevens Institute of Technology and an MBA from Rutgers University and has held Professional Engineer licenses in the United States and in Canada. Fred and his son, Clifford, are authors of *Instrumentation and Control System Documentation* published by ISA in 2004. He and his wife Jean live in Chapel Hill, NC.

# 7 Control Equipment

*By Douglas C. White*

## Topic Highlights

*Input/Output (I/O)*
- *Pneumatic Component Interface – 3-15 psi*
- *Standard Electronic Analog I/O – 4-20 mA*
- *Discrete I/O*
- *Serial Communication*
- *HART*
- *Digital Buses*
- *Specialized I/O*
- *Wireless*

*Control Network*

*Control Modules*
- *Database*
- *Configuration*
- *Redundancy*
- *Backup*
- *Online Version Upgrade*

*Human Machine Interface (HMI)*
- *Keyboard*
- *Standard Displays*
- *Alarms*
- *Sequence of Events*
- *Historian/Trend Package*

*HMI – System Workstation*

*Application Servers*
- *Remote Accessibility*
- *Connectivity*

*Other Control Systems*
- *Emergency Shutdown Systems (ESD)*
- *Programmable Logic Controllers*
- *HMI-SCADA Systems*

*Future DCS Evolution*

## 7.1 Introduction and Overview

Sensors, actuators, and control algorithms were discussed in previous sections. Integrating the hardware and software of these three elements to produce a control loop is the subject of this section.

The earliest process plants were characterized by numerous local measurement indicators and many local control valves. Operators would walk around the plant and make manual adjustments to valves, based on their experience and readings on the indicators. Perhaps they might carry a clipboard or equivalent on which to record a few key measurement values and some overall coordinating instructions. The operator was the controller, as well as the connection between the sensor and the actuator.

The earliest direct integration between sensors and controllers is generally believed to be the governor on a steam engine, developed by James Watt in the 1780s, which regulated steam pressure via mechanical action. In the 1870s, William Fisher developed a regulator that adjusted steam flow to a pump based on its outlet pressure. Around 1900, pneumatic controllers appeared that combined a pneumatic sensor that detected flow, level, or pressure; calculated a required change in output; and sent a pneumatic signal to a diaphragm valve. These controllers were mounted locally and, again, the operator would walk through the plant and make individual adjustments to the controller settings.

Pneumatic transmitters appeared in the 1930s and permitted remote display of variable response and remote adjustment of controller setpoints. This led to the first centralized control rooms where coordinated control action on many different controllers could be made in one location. Typically, large circular charts were used to display responses, and the control rooms would have walls filled with these charts. Electronic single loop analog controllers appeared in the 1940s and 1950s. They eventually were adopted in place of pneumatic controllers because of improved control response, increased accuracy, reduced size, and reduced maintenance costs.

In the late 1950s through the 1960s there were several installations of control systems using centralized computer systems to execute all plant control algorithms directly. These systems were called direct digital control (DDC). However, cost and reliability issues limited widespread adoption of this technology. During the 1970s, distributed control systems (DCSs), based on commercial microprocessors, were introduced and widely adopted. These systems converted all data to digital form, executed multiple controllers in a single electronic component, used cathode ray tubes (CRTs) for the operator interface and keyboards for control, and connected all components together with a single digital data network. The "distributed" in DCS implies that various control software tasks are performed in different physical devices with an overall coordinating and scheduling software program. Development in the 1990s of lower-cost personal computers (PCs) with their associated servers, standard operating systems such as Microsoft Windows, external communication standards, and standard digital bus protocols permitted a new generation of DCSs to be introduced.

The elements of a modern DCS are shown in the Figure 7-1.

Major components of a typical system include input/output connectivity and processing, control modules, operator stations, system workstations, application servers, and a process control bus. Each is discussed further in the following sections.

## 7.2 Input/Output (I/O)

The first step in control is to convert the sensed measurement into a reading that can be evaluated by the control algorithms. It has been common in the past to bring all the I/O in the plant to marshalling panels located in a control building, from which connections are made to the controllers. This facilitates maintaining and upgrading equipment. There are many types of equipment that have to be connected to a modern DCS, with different specific electronic and physical requirements for each interface. Each input/output type discussed below requires its own specialized I/O interface card that will convert the signal to the digital value used in the DCS.

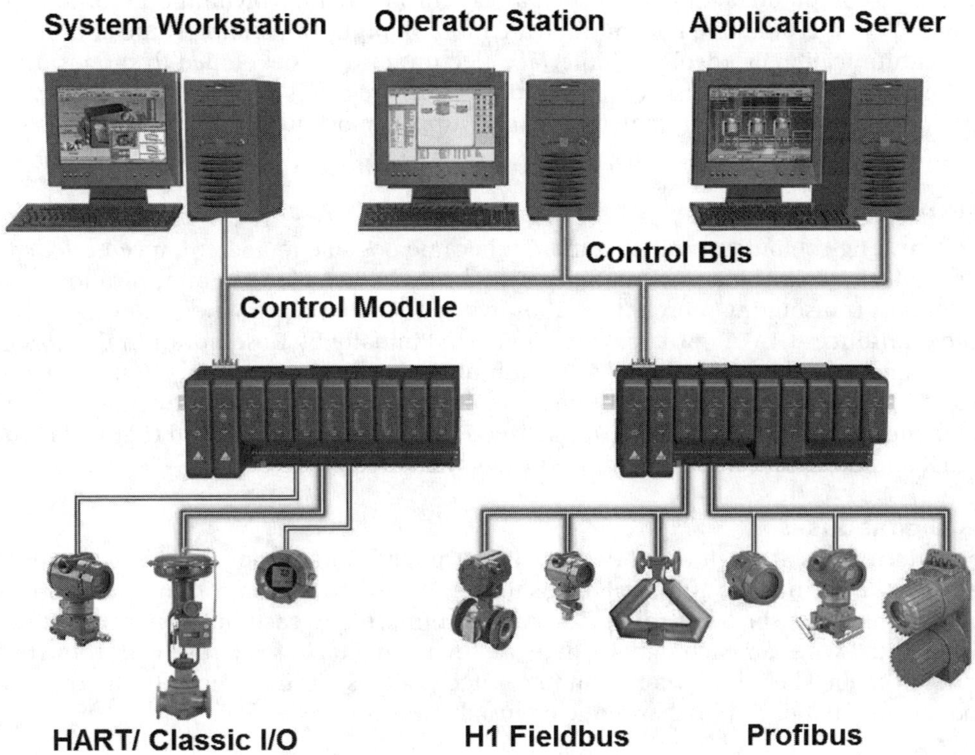


*Figure 7-1: Elements of a Modern DCS*

### 7.2.1 Pneumatic Component Interface – 3-15 psi

With the development of pneumatic controllers and transmitters, it became apparent that a standard input and output range was required so equipment from multiple manufacturers could be used in the same plant. The 3-15 pounds per square inch (psi) range was adopted with 3 psi representing 0% of span and 15 psi representing 100%. There are many pneumatic controllers still operating, and interfacing them to a modern DCS typically requires a pressure/current (P/I) converter on the input and an I/P converter on the output from the I/O card to the controller with the actual interface card a standard electronic analog type.

### 7.2.2 Standard Electronic Analog I/O – 4-20 mA

When electronic instrumentation was first introduced, there were also many different standards for the electronic I/O signals. Eventually the 4-20 milliampere (mA) analog direct current (DC) signal was adopted as standard and is still the most widely used input/output format. Each signal used for control has its own wire, which is brought back to the marshalling panel and then connected to the controller. The most common analog I/O cards used are 16-bit converters, meaning the maximum resolution is about 0.1%, i.e., 4 significant digits.

### 7.2.3 Discrete I/O

There are a number of devices, such as limit switches and motors, whose state is binary, i.e., off or on. These require separate I/O processing than the analog I/O and, most commonly, separate I/O cards.

### 7.2.4 Serial Communication

For relatively complicated equipment such as gas chromatographs, vibration monitors, turbine controls, and PLCs, it is desirable to communicate more than just an analog value and/or a digital signal. Serial communication protocols and interface electronics were developed to support this communication. Modbus is one of the most common protocols. I/O registers within the device support predefined read/write commands. Recent implementations support block data transfer as well as single value transmission.

### 7.2.5 HART

With continuing computer miniaturization, it became possible to add enhanced calculation capability to field devices ("smart" instrumentation). This created a need for a communication protocol that could support transmitting more information without adding more wires. The Highway Addressable Remote Transducer (HART) protocol was developed initially by Rosemount in the 1980s and turned over to an independent body, the HART Communication Foundation, in 1993. The HART protocol retains the 4-20 mA signal for measurement and transmits other diagnostic information on the same physical line via a digital protocol. A specialized HART I/O card is required that will read both the analog signal and diagnostic information from the same wire.

### 7.2.6 Digital Buses

Since modern DCSs are digital and new field instrumentation supports digital communication, there was a natural demand for a fully digital bus to connect them. Such a communication protocol reduces wiring requirements, since several devices can be connected to each bus segment; it is not necessary to have individual wires for each signal. There are several digital buses in use with some of the more popular ones described below. It is common to connect the bus wiring directly to the controller, bypassing the marshalling panel and further reducing installation costs.

#### Fieldbus

FOUNDATION fieldbus is a digital communication protocol supporting interconnection between sensors, actuators, and controllers. It provides power to the field devices, supports distribution of computational functions, and acts as a local network. For example, it is possible, with FOUNDATION fieldbus, to digitally connect a smart transmitter and a smart valve and execute the PID controller algorithm connecting them locally in the valve electronics—with the increased reliability that results from such an architecture. The FOUNDATION fieldbus standard is administered by an independent body, the Fieldbus Foundation.

#### Profibus

Profibus, or PROcess FieldBUS, was originally a German standard and is now a European standard. There are several variations. Profibus DP or *Decentralized Peripherals* is focused on factory automation, while Profibus PA targets the process industries. It is administered by the Profibus International organization.

#### AS-i

AS-i, or Actuator Sensor interface, provides a low-cost digital network that transmits information that can be encoded in a few digital bits. It is popular for discrete devices supporting on/off indicators such as motor starters, level switches, on/off valves, and solenoids.

#### DeviceNet

DeviceNet is a digital communication protocol that can support bi-directional messages up to eight bytes. It is commonly used for variable speed drives, solenoid valve manifolds, discrete valve controls, and some motor starters.

#### Ethernet

Some field devices, such as sophisticated on-line analyzers, are now supporting direct Ethernet connectivity using standard TCP/IP protocols.

### 7.2.7 Specialized I/O

There are many specialized measurements with equally special interface requirements.

#### Thermocouples

Thermocouples may be cold junction compensated or uncompensated. Each of these input formats requires different calculations to convert the signal to a reading. Multiplexers are often used for multiple thermocouple readings to reduce I/O wiring.

#### Pulse Counts

For turbine meters and some other devices it is necessary to accumulate the number of pulses transmitted since a predefined start time.

### 7.2.8 Wireless

Wireless communication is becoming more popular, particularly for non-critical readings. Standards in this area are rapidly evolving but should stabilize in the near future.

## 7.3 Control Network

The control bus network is the backbone of the DCS and supports communication among the various components. Data and status information from the I/O components is transferred to and from the controller. Similarly, data and status information from the I/O and controllers goes to and from the HMI. Transit priorities are enforced—data and communication concerning control is the highest priority with process information and configuration changes lower priority. Typically, the bus is redundant and supports high speed transfer. With early DCSs, each vendor had their own proprietary bus protocol that defined data source and destination addressing, data packet length, and the speed of data transmission. Today, Ethernet networks with TCP/IP base protocol and IP addressing have become the most common choice. The International Organization for Standardization – Open System Interconnection (ISO-OSI) model is the defining standard for the overall control networks implementation and is followed by most vendors. It consists of seven layers of implementation references from the physical layer through the application layer.

## 7.4 Control Modules

The control modules in a typical DCS are connected to the I/O cards by a high speed bus that brings the raw data to the module. The module contains a microprocessor that executes, in real time, the following actions:

- Process variable processing
  - Scan execution frequency control
  - Process variable status checking
  - Process variable engineering units conversion—this may involve flow element conversion, mass flow compensation, square root extraction, linearization, filtering, and/or totalization
  - Comparison of variable against alarm limits, and alarming if outside limits
  - Signal characterization
  - Calculated variable generation—combining one or more variables via addition, subtraction, multiplication, division, integration, accumulation, high/low select and dynamic compensation, such as lead/lag and dead time
  - "Bad" input propagation through the calculations
- Control algorithms
  - Mode change—manual, auto, remote, computer

  - Algorithm initialization on first execution of new mode
  - Control loop execution at defined frequency—PID, ratio, override, cascade, feedforward
  - Advanced control function execution—Model Predictive Control, Fuzzy Logic Control, Adaptive Control
  - Windup protection
  - Loop performance monitoring—mode, I/O condition, variability

- Controlled variable output processing

  - Output clamping and rate of change limit implementation

- Discrete input variable processing

  - Change of state detection
  - Set/reset flip flops
  - Logic calculations via Boolean functions—and, or, not, nand, nor
  - Comparison logic—equal to, greater than, less than, not equal to

- Discrete output processing

  - Execute pulse or latching outputs
  - Manual execution

- Sequential control

  - Execute step
    - Execute algorithm after preset delay
    - Execute algorithm after counter reaches preset value
    - Execute algorithm when comparison logic is true
    - Execute algorithm when Boolean logic is true
  - Hold step if condition exists
  - Restart step if condition exists
  - Skip one or more steps if condition exists
  - Recycle to previous step if condition exists

- Historian/Trending

  - Store analog and digital values and status for later retrieval and trending
  - Store alarm information for later retrieval
  - Store operator entered information for later retrieval

- Diagnostics

  - Execution of performance diagnostics on the functions above
  - Field instrument diagnostic information capture

- System access control enforcement

The microprocessors used for real time execution have limited processor capacity and memory, though these limits are continually being raised with the ongoing developments in the computer industry. As a result, the number of I/O points, the number of control loops, and the number of calculations processed in a single module are limited. Multiple modules are used to handle the complete requirements for a process. Control modules are usually located in a climate controlled environment.

### 7.4.1 Database

Each module contains a database that stores the current information scanned and calculated as well as configuration and tuning information.

### 7.4.2 Configuration

Configuration of the control module is normally performed off-line in the engineering workstation with initial configuration and updates downloaded to the control module for execution. Configuration updates can be downloaded with only momentary interruption of process control. Today, this configuration is usually done in a GUI based system with drop-and-drag icons, dialog windows and fill-in-the-blank forms, with no programming required. A typical screen for configuration is shown in Figure 7-2, where the boxes represent I/O or PID control blocks. Lines represent data transfer between control blocks.

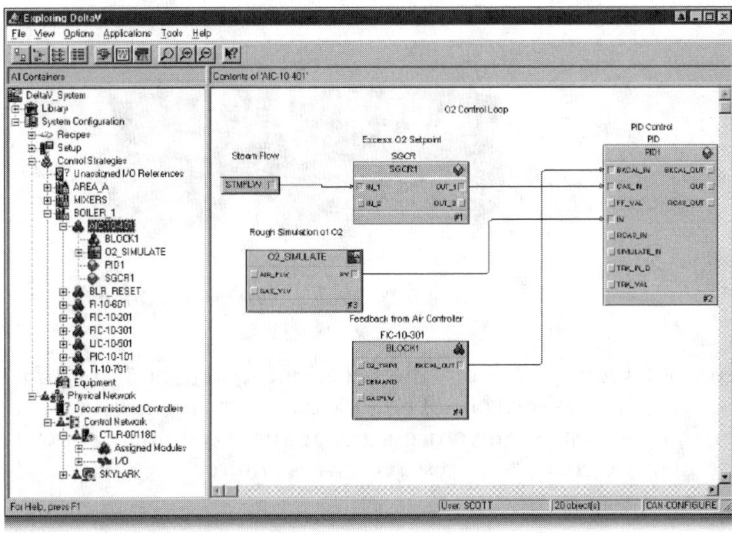


*Figure 7-2: Typical Configuration Screen*

Generally there will be predefined templates for standard I/O and control functions, which combine icons, connections, and alarming functions. Global search and replace is supported. Prior to downloading to the control modules, the updated configuration is checked for validity and any errors identified.

### 7.4.3 Redundancy

For critical control functions, redundant control modules are often used. These support automatic switching from the primary to the backup controller upon detection of a failure.

### 7.4.4 Backup

Current database and configuration information is normally generated without taking a control module offline and stored for backup purposes.

### 7.4.5 Online Version Upgrade

Hardware and software for DCSs continue to evolve, and it is desirable to be able to install new releases without taking the control module offline. If redundant modules are available, this can be done by running on the primary module, switching the hardware for the secondary module, and/or downloading the new release of software, switching to the secondary module for control (after appropriate initialization), and repeating the update for the primary module.

## 7.5 Human Machine Interface (HMI)

### 7.5.1 Operator Station

There are usually two different user interfaces for the DCS—one for the operator running the process, and a second one for system support used for configuration, system diagnostics, and maintenance. In a small application these two interfaces may be physically resident in the same workstation hardware. For systems of moderate or larger size, they will be physically separate. The operator interface is covered in this section and the system interface in the next section. A typical operator station is shown in Figure 7-3.

*Figure 7-3: Typical Operator Station*

The number of consoles required is set by the size of the system and the complexity of the control application. The consoles access the control module database via the control bus to display information about the current and past state of the process and are used to initiate control actions such as setpoint changes to loops and mode changes. Access security is enforced by the consoles through individual login and privilege assignment.

### 7.5.2 Keyboard

Standard computer keyboards and mice are the most common operator console interface, supplemented occasionally with dedicated key pads that include common sets of keystrokes preprogrammed into individual keys.

### 7.5.3 Standard Displays

The Graphical User Interface (GUI) consoles are equipped with standard display types commonly used by the operator.

#### Faceplates

Faceplate displays show dynamic and status parameters about a single control loop and permit an operator to change control mode and selected parameter values for the loop.

#### Custom Graphic Displays

These displays present graphic representations of the plant with real time data displays, regularly refreshed, superimposed on the graphics at a point in the display corresponding to their approximate location in the process. A standard display is shown in Figure 7-4, with a faceplate display superimposed.

Displays can be grouped and linked via hierarchies and paging to permit closer examination of the data from a specific section of a plant or an overview of the whole plant operation.

### 7.5.4 Alarms

Alarms generated will cause a visible display on the operator console such as a blinking red tag identifier and often an audible indication. Operators acknowledge active alarms and take appropriate action.

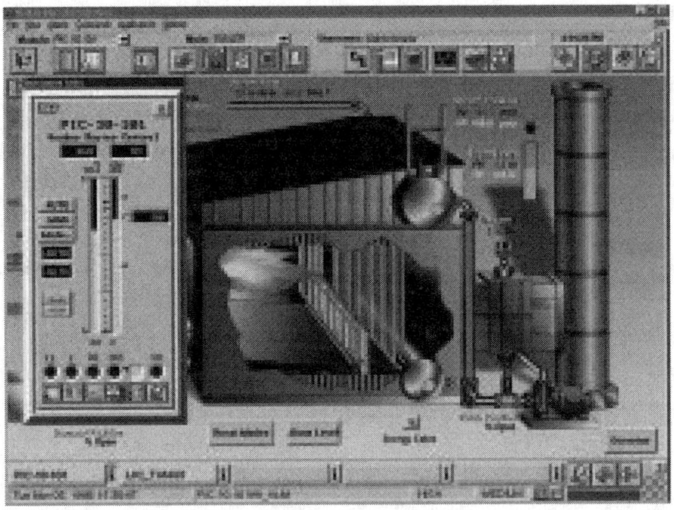

*Figure 7-4: Standard Display with Faceplate Display Superimposed*

Alarms are time stamped and stored in an alarm history system retrievable for analysis and review. Different operator stations may have responsibility for acknowledgement of different alarms. Alarm "floods" occur when a plant has a major upset, and the number of alarms can actually distract the operator and consume excessive system resources. Responding to these "floods" and providing useful information to the operator in real emergency situations is an area of active system development.

### 7.5.5 Sequence of Events

Other events such as operator logins, setpoint changes, mode changes, system parameter changes, status point changes, and automation equipment error messages are captured, time stamped, and stored in a sequence of events system—again retrievable for analysis and review. If sequence of events recording is included on specific process equipment, such as a compressor, it may be integrated with this system.

### 7.5.6 Historian/Trend Package

A historical data collection package is used to support trending, logging, and reporting. The trend package shows real time and historical data on the trend display. Preconfigured trends are normally provided along with the capabilities for user defined trends. A typical trend display is shown in Figure 7-5.

## 7.6 HMI—System Workstation

The system workstation supports the following functionality, which has been discussed previously:

- System and control configuration
- Database generation, edit and backup
- System access management
- Diagnostics access
- Area/plant/equipment group definition and assignment

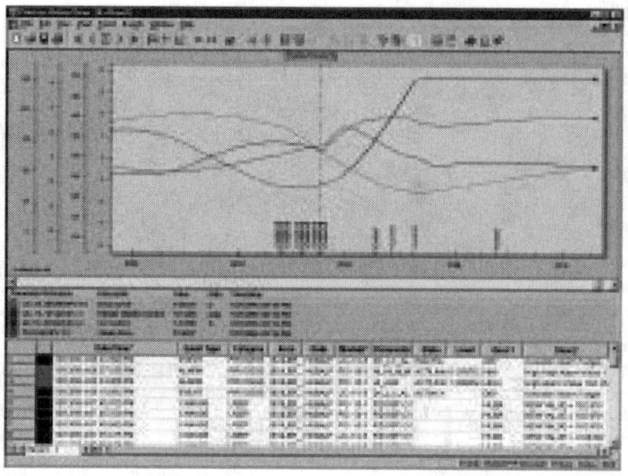

*Figure 7-5: Typical Trend Display*

Other functionality includes:

**Graphic Building**
A standard utility is provided to generate and modify user defined graphics. This uses preconfigured graphic elements, including typical ISA symbols and user fill-in tables. New graphics can be added and graphics deleted without interrupting control functionality.

**Simulation/Emulation**
It is desirable to test and debug configuration changes and graphics prior to downloading to the control module. Simulation/emulation capabilities permit this to be performed in the system workstation using the current actual plant configuration.

**Audit Trail/Change Control**
It is common to require an audit trail or record of configuration and parameter changes in the system, along with documentation of the individual authorizing the changes.

## 7.7 Application Servers

Application servers are used to host additional software applications that are computationally intensive, complicated, or transaction-oriented, such as batch execution and management, production management, operator training, online process/energy optimization, etc.

### 7.7.1 Remote Accessibility

It is desirable for users to be able to access information from the DCS remotely. Application servers can act as a secure remote terminal server, providing access for multiple users simultaneously and controlling privileges and area access.

### 7.7.2 Connectivity

The application server is also used to host communication software for external data transfer. There are several standards that are commonly used for this transfer.

**OPC**
The Object Linking and Execution (OLE) standard was initially developed by Microsoft to facilitate application interoperability. The standard was extended to become OLE for Process Control (OPC), which is a specialized standard for the data transfer demands of real time application client and serv-

ers. It is widely supported in the automation industry and permits communication among software programs from many different sources.

**XML/SOAP/Web Services**

Providing a single Web-based user interface to multiple applications is the goal of many software applications. The eXtensible Markup Language (XML) is a standard message/document protocol to permit communication between applications and the user interface. The Simple Object Access Protocol (SOAP) uses XML and Remote Procedure Calls (RPC) to permit programs in different programming languages to communicate. Web services use SOAP and XML-RPC to provide the user interface to multiple applications and permit web access to the information. This framework is used in Microsoft's .NET software architecture.

## 7.8 Other Control Systems

### 7.8.1 Emergency Shutdown Systems (ESD)

Emergency shutdown systems are specialized control systems installed in plants with the objective to automatically shutdown a plant and bring it to a safe state in the event of a major emergency. Typically they will have separate valves and transmitters to those used for normal control. In place of a control module is a logic solver that is programmed to detect specified unsafe conditions. If these conditions occur, the ESD is activated to shutdown the equipment in question or an entire plant.

### 7.8.2 Programmable Logic Controllers

Comparing PLCs with DCSs, PLCs typically cost less initially per I/O point but have, in general, less functionality and less redundancy. PLCs are the common choice for systems that are predominately discrete I/O with relatively fixed logic, and also for machine control and motion control where very high speed scanning is required. DCSs are most often chosen for continuous, semi-continuous and process batch applications, and applications where the analog I/O count is high. Other differences include separate databases for the I/O, the control, and the human-machine interface (HMI) for PLCs, while DCSs have a common database for all of these functions. Configuration of PLCs is predominately done with ladder logic, while DCSs have automated fill-in-the-blanks configuration editing and high-level language support. Many PLCs require the system be taken offline for control logic modifications while most DCSs can be updated online. Advanced automation applications' support, like batch and advanced control, is typically greater in a DCS, as compared with a PLC.

### 7.8.3 HMI-SCADA Systems

The development of cheaper personal computers (PCs) led to development of lower-cost systems, based on these PCs, that could be used to monitor equipment conditions; concentrate data, perhaps from geographically distinct areas and display it to operators and managers. These systems are often called HMI-SCADA systems, where SCADA stands for supervisory control and data acquisition. Typically control functionality is limited in these systems.

## 7.9 Future DCS Evolution

New functionality is continually added to DCSs with the ongoing evolution of computation and communication capabilities. Several trends are evident. One is that central control rooms are being installed physically remote from the actual plant, in some cases hundreds of miles distant, with responsibility for many plants simultaneously. This increases the demand for diagnostic information on both the instrumentation and other process equipment to better diagnose and predict process problems so that corrective action can be taken before they occur. A second related trend is the increased requirement for "sensor to boardroom" integration that imposes ever increasing communication bandwidth demands. Good, real-time corporate decisions depend on good, real-time information about the state of the plant. Secure integration of wireless field devices and terminals into the control system is an active area of current development.

## 7.10 References

For further information on the evolution of control see:

Feeley, J., et al. "100 Years of Process Automation." *Control Magazine*. (Vol. XII, No. 12), December 1999 (Special Issue).

### Standards

ANSI/ISA-50 Series, Parts 2-6 - *Fieldbus Standard for Use in Industrial Control Systems.*

IEC 61158 Series, Parts 1-6 - *Digital Data Communications for Measurement and Control – Fieldbus for Use in Industrial Control Systems.*

## About the Author

**Douglas C. "Doug" White** is Vice President, APC Services, for the Process Systems and Solutions Division of Emerson Process Management. Previously, he held senior management and technical positions with MDC Technology, Profitpoint Solutions, Aspen Technology, and Setpoint. In these positions, he has been responsible for developing and implementing state-of-the-art advanced automation and optimization systems in process plants around the world and has published more than 50 technical papers on these subjects. He started his career with Caltex Petroleum Corporation with positions at their Australian refinery and central engineering groups. He has a BChE from the University of Florida, an MS from California Institute of Technology, and an MA and PhD from Princeton University, all in chemical engineering.

# II

# Basic Discrete, Sequencing and Manufacturing Control

Discrete control *used to be considered the opposite of* process control, *but now the merging of industry needs and the merging of technologies has made the distinction between the two types of control less clear. In fact, it would have made sense to combine the continuous and discrete categories of this book into a single Basic Control category. However the literature is still fairly separate, and this separate categorization still seemed best.*

*Some of the sensors included in the topic on* Discrete Input and Output Devices and General Manufacturing Measurements *are continuous, which further confuses the distinction. As with process instrumentation, this topic only scratches the surface of the available types. The topics on* Motor and Drive Control *and on* Motion Control *are not "discrete" either, but they fit with PLCs and manufacturing control. Variable Speed Drives are also used in process control.*

*Motion control is an unknown area to many who work in other areas of automation. However, the tremendous increase in automated equipment is rapidly increasing the range of applications of motion controllers.*

# 8 Discrete Input & Output Devices and General Manufacturing Measurements

*By Kenneth C. Crater*

## Topic Highlights

*Actuation Technologies and Their Control*
*Sensing Technologies and Interfacing Techniques*
*Remote and Networked I/O*

## 8.1 Introduction

Historically, the practice of discrete control evolved quite differently from continuous process control, with very little overlap in technologies and applied principles. Early discrete manufacturing machines were wholly mechanical contrivances, harnessing waterpower through ingenious combinations of drive shafts, cams, and levers to accomplish ever more complex automation goals.

The advent of electrical motors caused little change at first, with mechanical control still the norm. Highly automated assembly and fabrication systems were often based on a central camshaft driven by a motor, with the entire sequence of the system defined by a single turn of the camshaft. All motions were driven by cams on this shaft, with lobes triggering activity at the appropriate interval.

### 8.1.1 Separating the Control Function

The first real separation of control and actuation came with the increasing use of electromechanical relays, especially in the automotive industry. This ultimately led to complex control configurations of these relays arranged to create latches and combinatorial functions, creating motion through the use of electrically driven hydraulic or pneumatic valves (solenoid-actuated valves or, simply *solenoid valves*).

In this new world of separate control and actuation, some functionality was actually lost. The actuation technologies were crude and simple: pneumatic and hydraulic cylinders which, when the related solenoid valve was triggered, would dumbly extend to the limit of their travel as defined by a mechanical stop. Speed of travel of these actuators could be controlled to some degree by installing a flow valve or pressure regulator in the hydraulic or pneumatic supply line, and these could be adjusted manually.

But the exacting control of acceleration and deceleration through the entirety of the stroke, which had been possible by skilled practitioners of cam technology, had been lost and would not be regained until new actuation technologies were introduced.

Beginning in some industries as early as the 1980s, additional sophistication of actuation and control became increasingly necessary. Compelled by the need for further automation spurred by increasing labor costs, as well as by increasing demands among consumers for broad product choices and customization, automation practitioners were required to produce more flexible machines. Often these machines were required to be soft-configurable—automatically adjustable under electronic control to manufacture slightly different products, or products of different sizes.

This trend led to the use of more flexible actuation technologies, including *servo motors, stepping motors,* and servo-controlled hydraulics. With corresponding advances in control, these technologies regained the precise control over motion, adding the benefit of automatic adjustment under software control.

### 8.1.2 Evolution in Sensing Technologies

A simultaneous trend dramatically affected the sensing technologies used in discrete control. In early systems, *limit switches* were often employed to sense the end of travel of mechanical motions driven by simple pneumatic cylinders. Often these limit switches were incorporated directly into a control scheme to trigger the next function in sequence.

Flexible machinery, however, brought with it the requirement for more flexible sensing. In many cases it was no longer adequate to sense when an actuator reached a single position; rather, it became necessary to sense multiple positions, or to continuously sense the absolute position of an actuator at all times. Sensing devices such as *encoders, linear potentiometers,* and *magnetostrictive transducers* came into more widespread use to accomplish this goal.

Sensing technology has also been driven by a parallel trend—that toward the imposition of ever-greater quality requirements, with the corresponding need to measure and control all relevant parameters relating to the manufacture of a product. It is now common for discrete manufacturing equipment to measure temperatures and hydraulic or pneumatic pressures present at the time of a product's manufacture, as well as to take critical dimensional measurements of a product either in-process or upon completion. This data is then used either in a control scheme to meet qualitative goals, or is stored as proof of product quality.

Therefore, where previously *digital inputs* (those capable of sensing only "on" or "off" states) were employed in discrete control, now *analog inputs* are increasing in widespread use to sense the many variables of interest to the control engineer.

The rest of this topic will look at the technologies and devices mentioned above in greater detail.

## 8.2 Actuation Technologies and Their Control

Actuation technologies in common use today include the following:

- Fluid power devices
    - Pneumatic cylinders, rotary actuators, and air motors
    - Hydraulic cylinders and motors
- Electrical motors
    - DC motors
    - AC synchronous motors and variable frequency drives
    - Servo motors
    - Stepping motors
    - There are also a number of specialty actuators in use, such as *piezoconstrictive devices* and *voicecoil actuators,* which will be described in passing.

The choice of an actuation technology is typically dictated by the characteristics of each technology and how those characteristics match the needs and circumstances of an application. For example:

Pneumatic cylinders can create linear motion with reasonably high degrees of force in a small space. The amount of force achievable is determined by the air pressure applied multiplied by the area of the cylinder's piston, as shown in Equation 8-1: Force Obtained from a Cylinder Actuator.

$$F = PA \tag{8-1}$$

where:

F = Force in pounds

P = Applied pressure in PSI

A = Area of cylinder in square inches

With adequate air supply, high actuation speeds are also possible, making pneumatic actuation suitable for many small, high-speed automation systems, and indeed the technology is in widespread use because of its advantages.

In applications requiring great levels of force, however, two limitations of pneumatic actuation become prominent. The first is the inherent compressibility of the actuation medium, air. This implies that there will be at least a small amount of resilience present in a pneumatic system.

The second limitation is the pressure available for actuation. For reasons of safety and practicality, air pressure as distributed in typical "shop air" systems is limited to roughly 80 PSI (5.5 bar). For a given cylinder diameter, this puts an upper limit on the amount of force achievable.

For higher force applications, hydraulic systems are often employed with pressures up to 5000 PSI (345 bar). Disadvantages include cost, since localized hydraulic pumps are often required, and the inevitable leakage of hydraulic fluid, which precludes use of the technology in clean environments.

As the cost of related control technology declines, increasing use is being made of electrical actuation technologies, including servo motors and stepping motors. With appropriate gearing, these devices can generate great amounts of force, although the force is typically in the form of rotary motion. To obtain linear motion, which is often the requirement in automated systems, a *lead screw* or belt drive is frequently employed. Drive and control costs, as well as the cost of conversion of rotary to linear motion, can be a substantial impediment to using electrical motor actuation, but the cleanliness and extent of control that are possible are often overriding factors.

### 8.2.1 Control of Fluid Power Actuation

The most common method of interfacing fluid power (hydraulic or pneumatic) actuators to a control system is through the use of *solenoid valves*. These are valves which are actuated by means of an electric solenoid—application of electrical power to a coil of wire generates a magnetic field, which acts upon a metallic actuator that switches the valve.

### 8.2.2 Configurations of Solenoid Valves

Pneumatic valves for control purposes are often categorized as two-way, three-way or four-way valves. A two-way valve possesses two ports to which pneumatic connections may be made. When actuated, the valve creates a connection between the two ports (assuming "normally closed" design of the valve) allowing flow to take place. Such a valve might be used in an air motor, for example, where air flow is applied to make the motor turn, and removed to make it come to a stop.

Three-way valves are outfitted with three ports. One of these (the "load" port) is switched between a connection to the "exhaust" port and, when actuated, the "supply" port. Such valves are often used to control *single-acting* pneumatic cylinders. These are cylinders driven by compressed air in only one direction, and are outfitted with a spring that will return them to a rest position. A source of compressed air is connected to the supply port, and the cylinder is connected to the load port. Upon actuation, these two ports are interconnected, allowing compressed air to flow to the cylinder, which overcomes its spring and extends. Upon deactivation, however, the pressure behind the cylinder's piston must be relieved or the cylinder will fail to retract. This is accomplished by leaving the exhaust port of the valve open to atmosphere. When the valve returns to its deactivated position, the load port (connected to the pressurized cylinder) is connected back to the exhaust port, allowing the pressure to release to atmosphere.

Four-way valves are typically used to control *double-acting* cylinders—those having pneumatic connections on either side of the piston. Two valve connections are provided for exhaust and supply, and two more for the two ends of the cylinder under control. In one position, the valve connects one end of the cylinder to the supply pressure and the other to exhaust. When the valve is switched, these connections are reversed.

Hydraulic valves work similarly, except that instead of exhausting compressed air to atmosphere, depressurized hydraulic fluid is returned via plumbing to a reservoir tank.

Miniature solenoid valves intended for small automated machines can be quite small—as little as 0.5 cu. in. (8 cc)—and are often mounted on air manifolds that provide a common air supply to the valves without individual plumbing. Some suppliers provide valve assemblies mounted on printed circuit boards, simplifying both pneumatic and electrical connection of the devices.

*Figure 8-1: Valve Manifolds Integrated with Electronic Circuit Boards (Courtesy: Clippard Instrument Laboratory, Inc.)*

Where greater air or hydraulic flows are required than are possible with small low-power valves, two options exist. The first is obvious: use a larger valve, switched by a higher-power solenoid, often requiring additional electrical interfacing to provide the necessary current. An alternative is to use a piloted valve—a valve which is actuated by pneumatic or hydraulic pressure—and use a low-power solenoid valve to supply the required actuation pressure.

The electrical interface to a solenoid valve depends on its voltage and current requirements. Although AC-powered valves (e.g., 120 VAC) were formerly the norm, the near universality of electronic control has resulted in 24 VDC valves predominating, with lower voltage valves common in some forms of equipment. In any case, solenoid valves are typically controlled through digital outputs of appropriate rating on a programmable logic controller (PLC) or other control system (see 8.2.6: Output Interfacing Techniques).

### 8.2.3 Proportionate Control Valves

In fluid power applications requiring more exacting control of motion characteristics, proportionate control valves are sometimes employed. These valves are similar in principle to standard solenoid valves, as described above, but are constructed and controlled such that intermediate states between the valve's two extreme positions are possible. The valve can be controlled to slowly meter air or hydraulic fluid to a cylinder, thus allowing more gradual control of the motion, all under electronic control. When used in conjunction with position sensing in a feedback loop, exacting control of position is also possible.

Proportionate control valves are sometimes provided with their own control electronics, and typically require an analog signal (e.g., 0 to ±10 VDC) as a control input.

### 8.2.4 Electrical Actuation Control

Several forms of electrical actuation are commonly applied in industrial automation today including, of course, AC and DC motors of various sorts and, increasingly, linear motors.

A common motor technology used in undemanding applications is the AC induction motor. These are commonly controlled through the use of *motor starters*, subsystems which typically include a switching mechanism and some form of overload protection. Often, motor starters are equipped with circuitry—typically resistors or inductors—which limit the initial starting current drawn by the motor. Once the motor has begun running, an additional switch is triggered which bridges the limiter, taking it out of the circuit.

Because of the amount of energy used in AC motors, they have been the focus of much innovation and product development in recent years, the details of which would fill many volumes. The common AC induction motor may still be the workhorse of industry, but it has given way in many applications to more advanced technologies.

One of the strongest trends has been toward the more widespread use of variable frequency drives (VFDs) which, instead of driving a motor with a constant 60 Hz (or 50 Hz) AC waveform, use electronic circuitry to vary the frequency of the power being supplied to the motor, thus varying its speed. In applications where the required output from the motor varies greatly and is often less than full output, the VFD can result in considerable savings of energy.

### 8.2.5 DC Motor Technologies

For many applications in discrete manufacturing, however, much smaller motors are often required, and these are often DC motors. With declining cost in drive technologies and computing power, stepping motors and servo motors are becoming increasingly widespread in use. Motors are covered in more detail in Topic 10, *Motor and Drive Control*, in this book.

The advantages offered by these technologies include the ability to position actuators quickly and accurately under program control, making them ideal for robotics and other applications in which flexibility is paramount. Even the need for linear motion—long a drawback of motors whose motion is inherently rotary—has been addressed through the development of linear motors. These are often patterned after stepping motors, but with the stator stretched out in linear form.

Other advanced and highly specialized forms of electrical actuation are employed in specific industries, such as microelectronics, where unusually demanding requirements are present. Such technologies include piezoconstrictive devices, which make use of microscopic dimensional changes in certain materials when an electrical current is applied, and voicecoil actuators, which make use of the same principle that moves the cone of a loudspeaker.

### 8.2.6 Output Interfacing Techniques

Virtually all of the actuation devices mentioned in this section will ultimately need to be made subject to electronic control, typically by a programmable controller or other computer-based system. In the case of programmable controllers, several types of output devices are commonly used for this function, typically in the form of output modules to be inserted into the programmable controller system. These output devices will be briefly discussed here.

### 8.2.7 DC Outputs

Although some industries historically have favored the use of AC-powered devices, low voltage DC devices are typically preferred for new applications, with 24 VDC being a commonly used control voltage. The reasons for this choice include safety, the relaxed wiring standards that are often possible with low-voltage devices, and the ease and low cost of interfacing such devices to electronic systems. Programmable controller DC output modules containing 16, or even 32 outputs are commonly available, greatly increasing the density of control possible.

The two most common forms of DC outputs are *open-collector transistor* and *field effect transistor (FET)* outputs. The application of these two forms is similar, although the terminology used may differ, so they will be treated identically for most of this discussion.

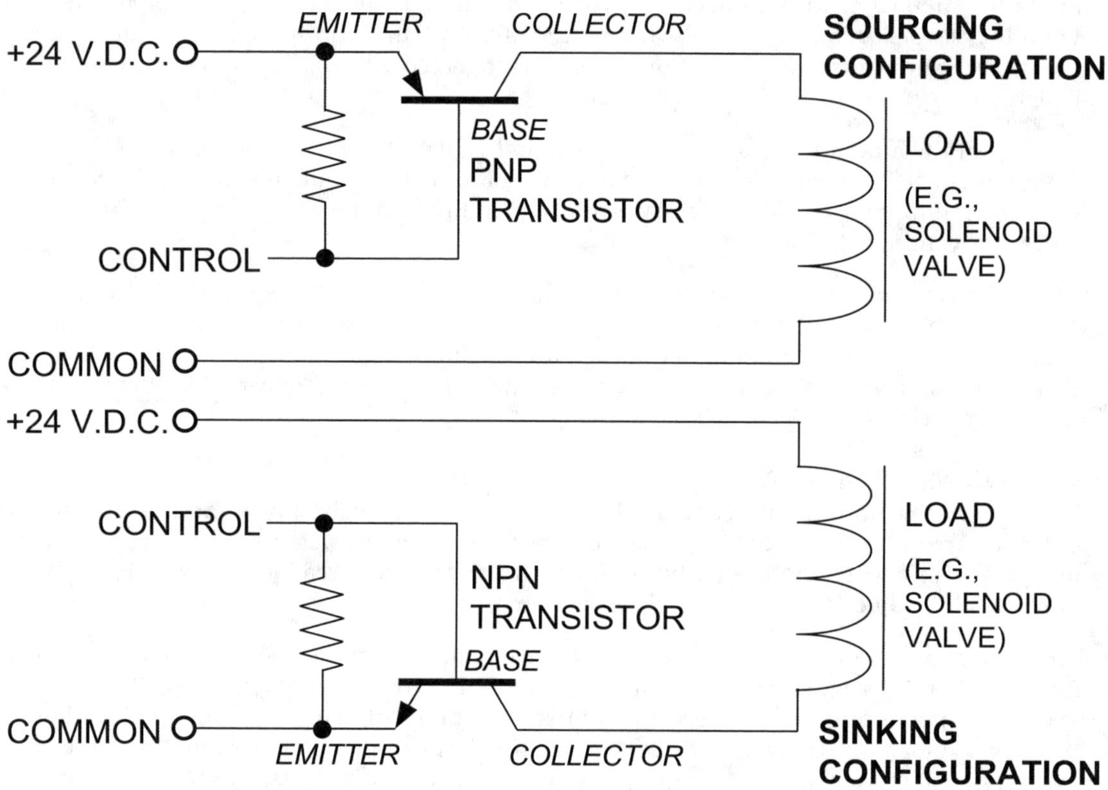


*Figure 8-2: Open-Collector Outputs*

Open-collector outputs employ transistor devices internally as switches, which are used to switch power to the external load (e.g., solenoid valve, etc.). Transistors are supplied with three terminals, named the *base, collector,* and *emitter* (in an FET, they are called the *gate, source,* and *drain*). In an open-collector output, the base and emitter are connected to internal circuitry in the output module. The base is the control terminal, often connected to microprocessor circuitry comprising the controller's logic, and the emitter acts as a common terminal, typically connected together with emitters of the other output channels on a given module. The collector terminal is brought out for external connection, hence the term *open-collector.*

### 8.2.8 Using Open-Collector Outputs

Externally, the load being controlled is connected between the collector and one pole of a power supply, the other pole of the power supply being connected to a *common* terminal on the output module. When a signal is sent from the microprocessor to turn on the transistor, the transistor allows current to flow from the collector to the emitter, essentially acting like a switch which has been turned on. This completes the circuit, sending power to the external load device.

There are two forms of open-collector output modules available—*sourcing* and *sinking*. In the case of sourcing outputs, the common terminal is connected to the positive pole of the DC power supply; the transistors, when switched, will *source* this positive voltage to the load device. In such a case, the other side of each load device is connected to the negative pole of the power supply.

With sinking outputs, the common terminal is connected to the negative pole of the power supply. The transistors, when switched on, will *sink* one terminal of the load device to this negative voltage (often referred to as ground, or common). Assuming the other side of the external load has a fixed connection to the positive pole of the power supply, it will then turn on.

### 8.2.9 DC Output Precautions

There is a general preference toward the use of sourcing outputs, due to the fact the negative terminal of the power supply is sometimes connected to electrical ground and, therefore, to the frame of an automated system. In the sourcing configuration, the loads have a fixed connection to the negative terminal, and, therefore, an accidental short circuit between a wire leading to a load device and the external metalwork will not cause the unintentional energization of an actuator. With sinking configurations, however, one side of each load device is always connected to a positive voltage source, such that an inadvertent short circuit leading from the other connection on the load device to ground would cause the device to turn on.

In addition, it must be realized that transistors are not perfect switches. When energized, there is a small offset voltage across the transistor, which generates heat equal to the product of current and voltage. This heat determines a maximum current rating for the output, which must be carefully obeyed.

Similarly, transistor outputs have a maximum voltage rating. Usually, the voltage of the DC power supply and control devices used is selected based on this rating. Even so, however, the rating can be inadvertently violated. Inductive devices, such as solenoid coils, motors, or even electromechanical relays, will generate a high voltage when current to the device is interrupted, such as when they are turned off. This voltage, called *back EMF,* can be high enough to destroy a transistor output unless precautionary measures are taken.

Often, a protection diode will be connected in reverse polarity across an inductive load, providing a harmless discharge path for this back-EMF and protecting the output transistor. Some output modules have integrated protection diodes, making this precaution unnecessary.

Another potential source of failure—either of an output transistor or the control system it is part of—is *electrostatic discharge (ESD)* and inducted *electrical noise*. Many processes, particularly high-speed processes involving plastic films, can generate high levels of static electricity, as can human beings walking across a wool carpet in rubber-soled shoes. When the eventual discharge path for this static electricity includes the exposed output or input circuitry of a control system, damage can result. In such environments, voltage limiting devices such as *metal-oxide varistors (MOVs)* may be used to provide some degree of protection, as can shielding of exposed wiring.

Another source of spurious and potentially-damaging high voltage is electrical noise that can be coupled from adjacent wiring. It is common practice to physically separate low-voltage and high-voltage wiring to avoid or limit the amount of inductive or capacitive coupling to sensitive circuits.

An almost universal protection technique used to avoid damage to or malfunction of microprocessor-based control systems is the use of optical isolation between any external connections (inputs or outputs) and the internal microprocessor circuitry. Microprocessors are highly susceptible to even minor fluctuations in voltage, which can cause them to reset or execute randomly, and optoisolation provides an important protective shield between the electrically noisy industrial environment and this sensitive circuitry.

### 8.2.10 AC Outputs

Electronic control systems employ two different technologies for controlling AC load devices: *triacs* and electromechanical relays. It is assumed that electromechanical relays are well understood—the control system energizes the relay's coil, which electromagnetically switches a mechanical switch, which in turn may be used to control an external load device. The drawbacks of relays are also well known. Over time, the contacts may become pitted and worn, particularly when used to control reactive loads, and will often require replacement well in advance of their solid-state counterparts.

Triacs are semiconductor switches that, when triggered, can pass current in both directions, making them well suited to the control of AC load devices. Some specific characteristics of triacs deserve attention, however, as they can affect the success and reliability of an installation.

When the control signal is removed from a triac, it will remain turned on until such time as the current through it falls to near zero. At this time, the voltage across it must also be near zero or the rapid rise in voltage across the device could spuriously trigger it into conduction again. Unfortunately, when driving an inductive load (such as a solenoid valve, motor, relay, or just about any other device you would typically drive with a triac), the load creates a phase shift between the voltage and current waveforms, such that the triac is unable to turn off. To allow for the triac's proper operation, a *snubber network* composed of a serially connected resistor and capacitor is typically connected across the triac's main terminals, controlling the rate of voltage rise across the device and allowing it to turn off.

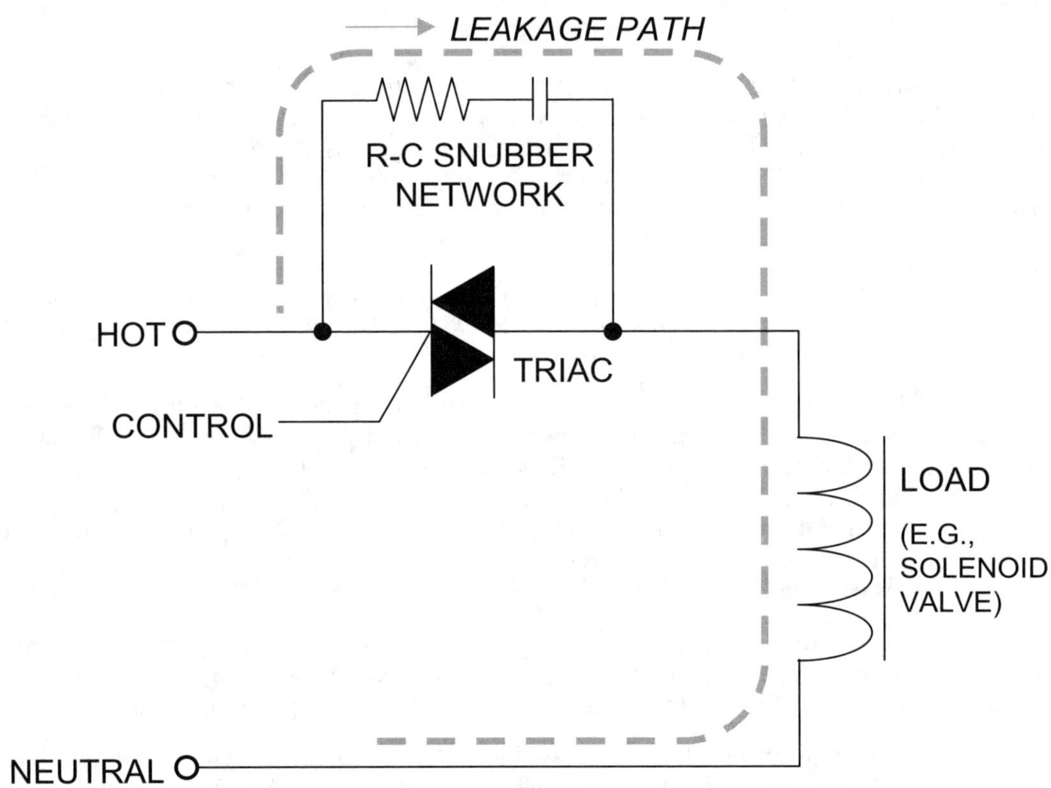


*Figure 8-3: Triacs, Snubbers, and the Effects of Leakage*

An unfortunate side-effect of this snubber network is that it will always leak a small amount of current. Although small, this current will sometimes be enough to keep a very-low-power device, such as a small solenoid valve, actuated even when the triac output has been turned off. This also means that a strategy of connecting a triac output to a controller's AC input for signaling purposes will often fail, because the low current requirement of the input will often be met by the leakage current provided by the snubber network, whether the triac output is turned on or not.

Even with this shortcoming, however, triacs and related semiconductor devices are popular and successful choices for AC control, when carefully applied with an understanding of their limitations.

## 8.3 Sensing Technologies and Interfacing Techniques

Sensing devices have increased dramatically in sophistication as the impetus for their use has evolved over the years. Simple limit switches originally served to detect the completion of a mechanical motion prior to initiation of the next motion, thus preventing damage and allowing a machine to operate at a speed determined solely by its mechanical capabilities.

*Figure 8-4: Limit Switches (Courtesy: Omron Electronics LLC)*

With the advent of flexible machinery, new sensing technologies such as encoders were employed to allow a single mechanism to accommodate different sizes or configurations of workpieces. Then, as quality issues began to predominate in industrial automation, still more sophisticated technologies such as machine vision systems were deployed to automate high-speed in-process inspection.

This progression represents a gradual supplanting of human involvement in the manufacturing process. Indeed, the earliest "sensors" were the eyes and ears of the operator, who then operated the machine through manual control switches. Although there are certainly a number of areas where electronic sensors exceeded human capabilities—in speed, exactitude, or consistency—some of the more basic judgments made by the human counterpart have taken decades to evolve technology with comparable performance. The simple question, "is the part upside down?" may not be so simple for a sensor to answer.

### 8.3.1 Limit Switches

In its original meaning, a *limit switch* is a switch placed to sense when a mechanism has reached the end, or limit, of its travel. Although some automated systems are designed to rely on the passage of time alone to infer that a motion has been completed, it is generally considered a safer technique to sense the actual completion of a motion prior to commencing the next part of a sequence of operation.

This is especially true if mechanical damage or danger to an operator could occur in the event of an undetected jam.

The simplest form of limit switch is a mechanical switch designed to be mounted to a machine component. Often, these are equipped with actuators that allow some degree of overtravel without damage to the switch. Depending on the environmental and regulatory requirements of an application, limit switches may be plastic-encased devices with exposed electrical terminals, or fully metal-enclosed devices sealed against moisture.

Driven in large part by reliability concerns, there is a strong trend to the use of non-contact electronic sensors in place of conventional mechanical switches. *Hall-effect sensors*, for example, sense the proximity of a magnetic field, and *proximity switches* are available to sense a variety of materials including non-metallics. *Photoelectric sensors* send out a beam of light—typically infrared—and sense its reflection off an approaching object. The light beam is often modulated so that ambient light can be effectively filtered from the reflected signal.

Each type of sensor has its applications:

- Some suppliers equip the piston inside a pneumatic or hydraulic cylinder with a magnetic element, allowing Hall-effect sensors to be mounted on the outside of the cylinder to sense the piston at various points in its stroke.
- Proximity sensors can be applied to discriminate between various materials, or to sense objects in environments where optical sensing might be unreliable due to dust, oil or other contaminants.
- Photoelectric sensors can be employed to sense across distances or, using fiber optic technology, in tight quarters where other types of sensing would not be practical.

### 8.3.2 Interfacing Concerns with Limit Switches

Careful attention must be paid to the electrical characteristics of a limit switch to determine the proper way to interface it to an electronic control system. This is true even with mechanical switches—the low voltages used in some control systems may not be sufficient to break through the layer of oxidation or contamination buildup on mechanical switch contacts, rendering them unreliable. For this reason, gold-plated contacts are frequently used on mechanical switches intended for interfacing to a control system.

Although some sensing devices—notably certain photoelectric sensors—are equipped with internal relays for interfacing, most devices instead have solid-state outputs with limitations that must be obeyed. Two-wire and three-wire devices are commonly available. Two-wire devices rely on a low level of current constantly passing through the device to power its internal electronics. For this reason, the electronic system to which it is connected must allow for this "leakage" current without falsely triggering. When the sensor enters its active state, it draws sufficient additional current to bring the voltage across the sensor down to a few volts—again, the control system must reliably see this state as an *on* condition, even in the presence of this *offset voltage.*

Interfacing three-wire sensors is less exacting, since the three wires leading to the sensor allow for power, a common connection, and a separate output connection from the sensor to the control system. The output is often of the *open-collector* variety and may be sinking or sourcing. Current and voltage ratings must be respected, of course, but the additional concerns of leakage and offset voltage tend not to be significant factors.

### 8.3.3 Position-Sensing Technologies

There is now frequently a need to obtain position information beyond the simple end-of-stroke indication provided by limit switches. Examples include:

- Feedback confirmation of mechanical positioning (e.g., for robotic arms, X-Y positioning tables, etc.)
- Triggering secondary events at specific points in a primary motion.
- Velocity control feedback, through differentiation of successive position indications.

The most common application for position sensing of this sort is in connection with servo motors (see Topic 11), and the technologies employed for this purpose are incremental and absolute encoders and resolvers.

Two other position sensing technologies are worthy of note, although their application is somewhat more specialized. *Potentiometric sensing* makes use of a linear or rotary *potentiometer,* which consists of an electrically resistive element on which an electrical contact rides. Often, the extreme ends of the resistive element are connected across a precise fixed voltage, at which point the voltage sensed on the moving contact is analogous to its position along the element. This approach provides an analog voltage directly proportionate to position (subject to the linearity of the resistive element). One drawback to be aware of, however, is the relatively limited life of the resistive element due to wear from the moving contact.

Finally, *magnetostrictive sensing* is sometimes applied for linear position sensing of long hydraulic cylinders, for example. This technique uses a long waveguide, down which an electromagnetic reference pulse is induced. A permanent magnet, typically connected to a moving mechanical element (e.g., a piston in a hydraulic cylinder) rides on this waveguide and, when the reference pulse reaches the magnetic field of this magnet a strain pulse is generated which travels back down the waveguide. Precise timing measurement between the induction of the reference pulse and the receipt of the strain pulse provides the position of the magnet, and therefore the piston. Magnetostrictive sensing allows the sensing elements to be completely sealed, making it an appropriate technology for many industrial environments.

### 8.3.4 Sensing Other Physical Variables

Automation of discrete manufacturing processes today often carries with it the need for sensing a very broad range of physical variables. Often, this is to create a permanent record of manufacturing conditions relating to a given workpiece. Other times, the measurement is used directly as feedback for the process being controlled. Topic 1 in this book discusses pressure and temperature measurements.

### 8.3.5 Tracking Product Using Electronic Identification

In many cases, automated systems must be able to identify a specific workpiece being acted upon. This is true where customization of individual products to a customer's specification is required, and also where regulatory authorities require tracking and recording of manufacturing data.

One approach to product tracking is the use of barcode technology, through the application of a unique barcode or Data Matrix code to a product or product carrier. Fixed-mount barcode readers are then employed to read and transmit the barcode information to a control system, typically via an RS-232 serial interface.

Another tracking technology that has some advantages for industrial application is RFID *(Radio Frequency IDentification)*. This approach consists of a "tag" capable of transmitting a unique identification number back to a nearby transceiver. *Passive tags* consist simply of an etched antenna and an integrated circuit, and derive sufficient power from the RF energy received by the antenna to power the IC to send a response. *Active tags* contain an internal power source (and are therefore slightly larger) and often have the capability to store additional information received from the transceiver. In environments where a ubiquitous plant network is not feasible, this allows a workpiece equipped with such a tag to accumulate data about its manufacture as it progresses through stages of production.

### 8.3.6 Machine Vision

Machine vision is a sensing technology which is seeing increasing interest and application due to its potential to extend the reach of automation into operations which formerly required human intervention. Applications for machine vision today include automated inspection for a wide range of characteristics such as placement, dimensional measurements, surface defects and color grading, as well as providing visual feedback for robotic positioning and component orientation.

Available systems span a broad range of cost and capability, from relatively simple and inexpensive systems capable of performing some pattern matching and basic orientation, to *smart cameras* incorporating integrated processors allowing some more advanced functionality, to larger and more complex systems which may be custom programmed for a highly sophisticated tasks. The latter often require the specialized skills of a system integrator experienced in vision system design for successful application.

The electrical interface of a machine vision system involves two aspects. First, a trigger input is typically provided, which may be a signal from a sensor detecting the presence of a workpiece, or a command from a control system. Often, vision systems are provided with high-speed trigger inputs to accommodate machines with high production rates and fast-moving product.

The output from a vision system often takes one of two forms—a simple "good/bad" signal, or more complex data which may include a captured image for subsequent processing or storage. Systems often incorporate discrete outputs which may be programmed to signal, for example, when a stored pattern is matched, and often have additional communications capabilities (serial or Ethernet) to communicate higher-level information to a control system.

## 8.4 Remote and Networked I/O

A discussion of input/output technologies used in automation control would not be complete without mention of the dramatic shift in I/O architecture taking place today, driven by the prevalence of networking technologies in the plant environment. Where formerly all sensors, valves, and other control devices would be wired back to a single programmable controller, increasingly a more distributed form of connectivity is being applied.

The first step in this progression was the use of "device-level" networks such as Modbus, Profibus and DeviceNet, which provided a means of reducing wiring costs by combining devices on a common serial network. Widespread use of these bus technologies encouraged the development of functional subsystems—for example, motion controllers and temperature controllers—such that time-critical information did not have to be relayed back to a central controller, and low level tasks could be offloaded to the local processing capability of the subsystem. These subsystems would then take their high-level commands from, and provide their high-level results to, the central controller across the bus.

With the near-universal acceptance of Ethernet for plant data communications, variants of each of the device-level serial networking protocols have been developed that may be used across Ethernet. Although the initial impetus for this migration was cost savings—to achieve commonality of equipment and wiring with that already used in the plant—the result has been another rethinking of I/O architecture.

With sensing and actuation no longer tied physically to a proximate control system, there is once more a blurring of actuation and control, such that smart sensors and smart actuators are coming into broad use. Localized processing—in some cases, localized control—has changed and will continue to change the face of sensing and actuation in discrete manufacturing. The functions mentioned in this section will increasingly appear in aggregated form as subsystems on a network, interchanging high-level information with peer and supervisory systems, a most interesting trend to watch.

For more information on networking technology, refer to Topic 22, *Digital Communications*, and Topic 23, *Industrial Networks*.

## 8.5 References

*The Automation, Systems, and Instrumentation Dictionary.* Fourth Edition. ISA, 2003.

Coggan, D.A., Editor. *Fundamentals of Industrial Control.* Second Edition. Practical Guides for Measurement and Control Series. ISA, 2005.

Hughes, Thomas A. *Programmable Controllers.* Fourth Edition. ISA, 2004.

IEC 61131 – *Programmable Controllers.*

IEC 61131-1 Ed. 2.0 en:2003. *Part 1: General information.*

IEC 61131-2 Ed. 2.0 en:2003. *Part 2: Equipment requirements and tests.*

IEC 61131-3 Ed. 2.0 en:2003. *Part 3: Programming languages.*

IEC/TR 61131-4 Ed. 2.0 en:2004. *Part 4: User guidelines.*

IEC 61131-5 Ed. 1.0 en:2000. *Part 5: Communications.*

IEC 61131-7 Ed. 1.0 b:2000. *Part 7: Fuzzy control programming.*

IEC/TR 61131-8 Ed. 2.0 en:2003. *Part 8: Guidelines for the application and implementation of programming languages.*

Liptak, Bela G., Editor-in-Chief. *Instrument Engineer's Handbook: Volume 1 - Process Measurement and Analysis.* Fourth Edition. CRC Press and ISA, 2003.

## About the Author

**Kenneth C. Crater** is Chairman of Control Technology Corp., a Hopkinton, Mass. manufacturer of web-enabled automation controllers. He is also founder and President of Modbus-IDA, a trade association for the Modbus protocol. Active in the past with ISA, he served as a Director of ISA Services Inc. and was a co-founder and first President of the Industrial Computing Society in partnership with ISA. He holds six patents in the field of industrial automation control.

# 9 Discrete and Sequencing Control

*By Bob Kretschmann and Jim Christensen*

## Topic Highlights

*Discrete/Sequential Control Concepts and Hardware Systems*
*Basic Functional Structure of a Programmable Controller System*
*User's Control Objectives and Application Requirements*
*Selecting a PLC System*
*Software, Programs, and Programming Languages*

## 9.1 Introduction

Discrete control is concerned with finite numbers of stable states—devices such as on/off valves, pumps and manifolds. The interest of the user is in an orderly transition from one state to the other, and whether a situation is normal or abnormal.

A related aspect of control in common industrial processes is known as sequential control. With sequential control, a process moves through a succession of distinct states, usually carried out by discrete control functions, but also occasionally by manipulating process outputs and monitoring process inputs.

To introduce this section for new engineers reading this, here are basic definitions, taken from ISA's *Automation, Systems, and Instrumentation Dictionary* (2003):

- **Discrete control** – On/Off control. One of the two output values is equal to zero.
- **Sequential control** – A class of industrial process control functions in which the objective of the control systems is to sequence the process units through a series of discrete states (as distinct from continuous control).

## 9.2 Discrete/Sequential Control Concepts and Hardware Systems

One of the principal means used by industry to achieve or execute discrete and sequential control methodologies is the programmable (logic) controller (PLC).

This technology has become so important that the International Electrotechnical Committee (IEC) created and maintains the standard IEC 61131, comprised of seven parts that describe, define, and provide application guidance for PLCs.

In this standard, a PLC is defined as:

*Digitally operating electronic system, designed for use in an industrial environment, which uses a programmable memory for the internal storage of user-oriented instructions for implementing specific functions such as logic, sequencing, timing, counting and arithmetic, to control, through digital or analog inputs and outputs, various types of machines or processes. Both the PLC and its associated peripherals are designed so that they can be easily integrated into an industrial control system and easily used in all their intended functions.*

Here is a basic definition, taken from ISA's *Automation, Systems, and Instrumentation Dictionary* (2003):

> ***PLC – Programmable Logic Controller –*** *1. A controller, usually with multiple inputs and outputs, that contains an alterable program (ANSI/ISA-5.1-1984[R1992]).*

### 9.2.1 Programmable Logic Controllers

Programmable logic controllers (PLCs) are high-speed control computers. Typical features of a PLC are:

- a high-speed CPU
- flexible I/O systems with specialized cards for handling motion control
- a separate human machine interface

PLCs were traditionally programmed using relay ladder logic. Ladder logic is a powerful language for handling electrical motor control, but though it replaces traditional hardwired control, it can be considered a low-level programming language. Adding more advanced elements to traditional ladder logic can give the programmer the power to design complex control applications.

The PLC offers a very flexible programming environment. However, it still does not offer all the built-in functionality of a distributed control system (DCS) in terms of sharing process information through tags and tracking data quality. PLCs offer redundancy features for most critical components (power supply, CPU, I/O). However, exploiting these features typically requires more effort from programmers than a DCS requires. PLCs are now used for medium- and large-size control of analog process variables.

## 9.3 Basic Functional Structure of a Programmable Controller System

The general structure with main functional components in a programmable controller system is illustrated in the Figures 9-1, 9-2 and 9-3. These functions communicate with each other and with the signals of the machine/process to be controlled.

The operation of most PLCs consists of a repeated cycle of four major steps.

1. All inputs from appropriate interfaces are scanned, providing a consistent "image" of control input data. These represent field sensors such as switches, proximities, temperatures, data from other PLCs, etc.
2. A single "scan" is made of the user's control program, calculating or deriving a new "image" of control output data. At the same time all program variables, timers, counters, etc., are updated.
3. The new "image" of control output data is transferred to the appropriate interfaces for transfer to the target control devices. These represent field actuators such as relays, motor drive commands, displays, data to other PLCs, etc.
4. Finally, "housekeeping" tasks are performed on a time available basis. These may include communication with operator stations, some supervisory communication, diagnostics routines, etc.

After completion of the "housekeeping" the cycle repeats.

There are variations on this method, but, in general, this represents the overall process.

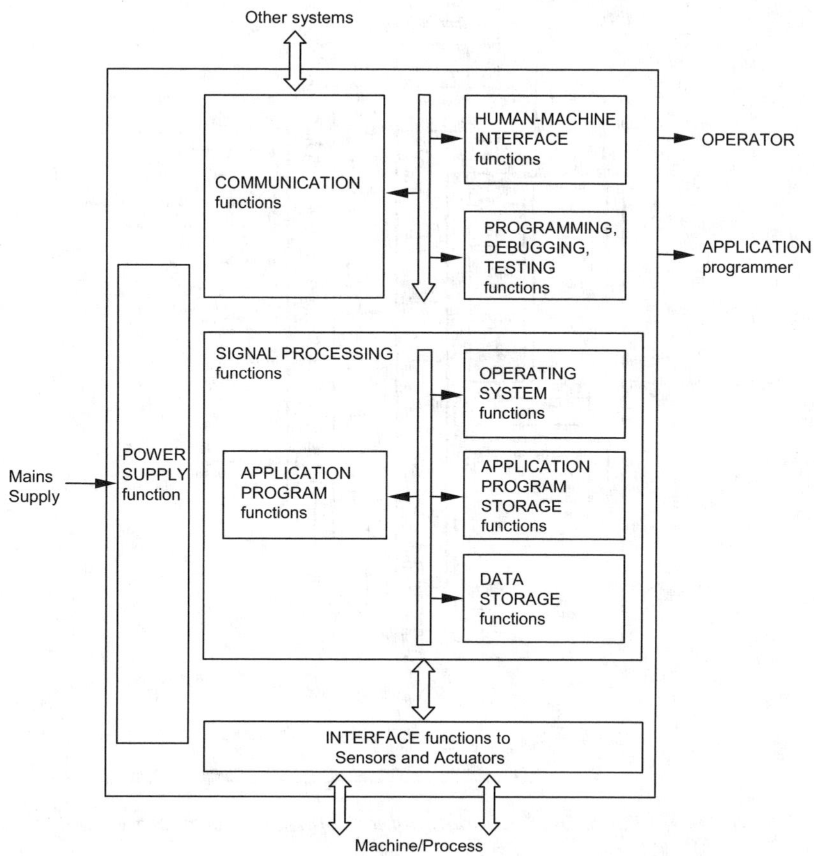


*Figure 9-1: Basic Functional Structure of a PLC System*

Some programmable controller systems with separate I/O and/or communications processors provide for overlapping the scanning of the user program with the scanning of the inputs (Step 1) and outputs (Step 3) and communication functions (Step 4). In these cases, special programming mechanisms may be needed to achieve concurrency and synchronization between the program and I/O scans, and between the program and communications processing.

A further feature of some PLCs is the incorporation of a multi-tasking operating system. As in general computing, the PLC operating system serves to coordinate the multitude of hardware and software resources and capabilities of the PLC. The incorporation of a multi-tasking operating system serves to allow the PLC to essentially execute multiple instances of the basic operation cycle described above, rather than a single instance, at the same time. This affords much increased flexibility and capability over a single-tasking operating system PLC.

In PLCs incorporating multi-tasking operating systems, mechanisms may be needed to achieve appropriate levels of task coordination and synchronization between the multiple instances of the basic operating cycle. As an example, one task's results may affect another task's decisions—especially with regard to I/O.

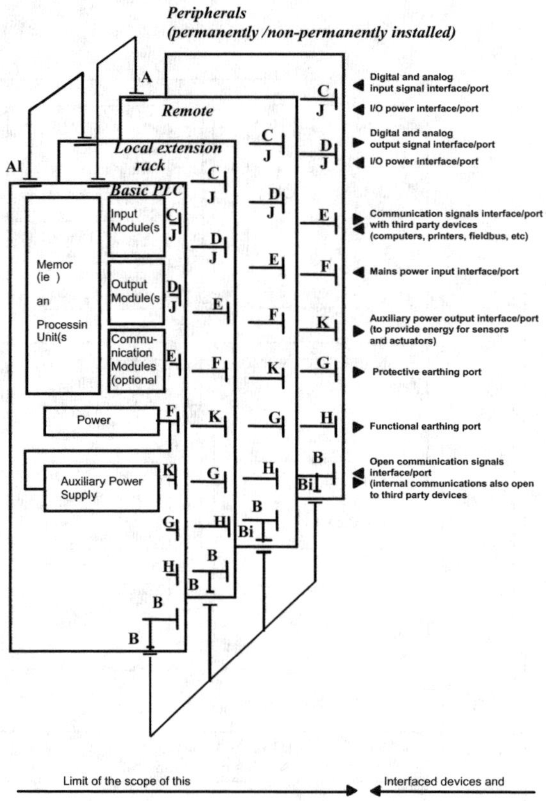


*Figure 9-2: Programmable Controller Hardware Model*

| AI | Communication interface/port for local I/O |
|---|---|
| Ar | Communication interface/port for remote I/O station |
| Be | Open communication interface/port also open to third party devices (e.g., personal computer used for programming instead of a PADT) |
| Bi | Internal communication interface/port for peripherals |
| C | Interface/port for digital and analog input signals |
| D | Interface/port for digital and analog output signals |
| E | Serial or parallel communication interfaces/ports for data communication with third party devices |
| F | Mains power interface/port. Devices with F ports have requirements on keeping downstream devices intelligent during power up, power down and power interruptions. |
| G | Port for protective earthing |
| H | Port for functional earthing |
| J | I/O power interface/port used to power sensors and actuators |
| K | Auxiliary power output interface/port |

*Figure 9-3: Interface/Port Descriptions for Figure 9-2*

## 9.4 User's Control Objectives and Application Requirements

With clear objectives and application requirements in mind, discussions with the vendor about controllers become immeasurably easier. This design information is essential and should include a general description of the process or equipment to be controlled. This will enable the user and the vendor or manufacturer to establish general equipment requirements and considerations.

Major topics that should be considered:

- process to be controlled
- user control objectives
- general control and operation requirements
- plant and personnel protection considerations
- general installation and environmental considerations
- expansion and integration requirements
- hardware configuration and regulations
- system availability requirements
- equipment breakdown implications
- spare parts requirements
- redundancy
- applicable local/national/international standards and regulations
- performance requirements
- interface to other systems
- maintenance
- cable wiring, routing and termination
- company regulations
- documentation: content, format
- regulatory/certification/approval requirements
- delivery and equipment installation schedule
- engineering responsibility: hardware, software, documentation, protective measures, testing, commissioning
- other considerations, as required

### 9.4.1 User System Description

This subclause refers to information to be presented to the vendor by the user as shown by the dashed lines in Figure 9-4.

The user must provide the vendor information about the existing system engineering, including current information about any third-party engineer. The vendor should be apprised of plant engineering, as well—both production engineering and maintenance engineering.

This system description should succinctly define the user's objective and give the vendor clear and relevant information concerning operations, monitoring, and configuration of any hardware and software. Additionally, corresponding documentation, such as diagrams, drawings, and signal (I/O) sequence and timing requirements should be used as support.

### 9.4.2 User System Characteristics

User system characteristics include such matters as:

- continuous or batch processing
- loop (PID) control
- distributed control
- changing of process (recipe supporting)
- downloading

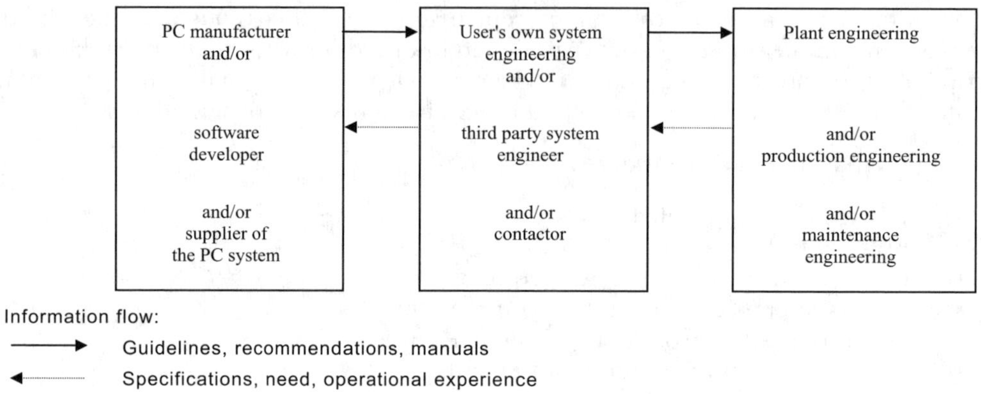


*Figure 9-4 : System Description Information Flow*

- autonomy of local stations
- system availability requirements
- total system response time
- redundancy
- multitasking
- alarm handling
- trending
- operator interface
- operator prompt
- remote supervision
- manual over-ride
- protection/safety considerations
- interlocks
- dynamic characteristics
- non-linearities
- authorization
- normal shutdown
- automatic restart
- data communication
- peripherals
- networks (LAN, WAN)

### 9.4.3 Control System Parameters

Important elements of the control system should be defined when process control is required. This can include the specific operation of the plant and equipment, the specific allowable process variables and values, the limits for actions and reactions, etc.

Other topics, that should also be considered, include:

- I/O listing
- sensors: types, signal level, power requirements
- signal conditioning
- control outputs: types, power requirements
- data transfer/access
- local/remote display

- logging and archiving
- electrical interference rejection criteria
- system availability
- I/O redundancy: single or voting (e.g. two out of three)

### 9.4.4. Alarms

The user's system description should include alarm requirements. The organization concept, priority, alarm operation and display should be defined.

Other major topics to be considered include:

- alarm sensing methods
- first fault identification
- fault discrimination
- dedicated alarm display
- alarm at man-machine interface
- alarm acknowledgment
- alarm logging

### 9.4.5 Human-Machine Interface

Human-machine interface (HMI) requirements include considerations for operator intervention, access control, and security arrangements.

Other major topics include:

- multistation display
- dedicated display areas
- dynamic graphics
- access control, authorization and passwords
- key locks
- software locks
- keyboard, track ball, mouse, touch screen, etc.
- ergonomics

### 9.4.6 Interlocks, Sequencing

Interlocks are used to arrange the control of machines or devices so their operation is interdependent, in order to assure they are coordinated properly (ISA-RP55.1-1975 [R1983]).

Interlocks may be:

- A physical device, equipment, or software routine that prevents an operation from beginning or changing function until some condition or set of conditions is fulfilled;

- A device, such as a switch, that prevents a piece of equipment from operating when a hazard exits;

- A device used to prove the physical state of a required condition and to furnish that proof to the primary control circuit;

- A device or group of devices (hardware or software) that are arranged so to sense a limit or off-limit condition or an improper sequence of events. To avoid an undesirable condition, they then shut down the offending or related piece of equipment or prevent it from proceeding in an improper sequence. (ANSI/ISA-77.44.01 &.02-1995).

In a similar fashion, sequencing is the ordering of overall process events, such that one control process properly completes its assigned task before handing off control to the next task process in the overall event path.

Major considerations for the interlocks and sequencing include:

- physical requirements of interlock
- types of systems to be interlocked
- systems and system communication
- requirements for data network.

### 9.4.7 System Outage

Topics relating to system outage include the following:

- power supply configuration
- system back-up
- diagnostics
- failure mode
- failure display levels: system, module, or card
- restart: cold, hot, or warm
- protection of personnel and equipment.

## 9.5 Selecting a PLC System

### 9.5.1 Main processing unit

The user should refer to IEC 61131-1 and IEC 61131-2 for relevant details, as well as Table 9-1 for a listing of selection considerations.

### 9.5.2 Human-Machine Interface

The human-machine interface (HMI) should be carefully specified and selected because it will be the operator's window to the plant control system. It may also be the access facility for programming and fault diagnosis for the PC system. Selection criteria for HMI are listed in Table 9-2.

*Table 9-1: Main Processing Unit—Selection Criteria*

| Criteria | Comments and considerations |
|---|---|
| User program memory | The organization and size of user application program memory |
| Memory back-up | Power back-up for volatile memory |
| System hardware configuration | - Racks<br>- Cables<br>- Bus expanders<br>- Power supplies<br>- Number of I/O modules per type<br>- Memory allocation per I/O type<br>- etc. |
| Programming languages supported | - Languages supported by the MPU<br>- Conformance to the P.C. language standard<br>- Any differences in objects, instructions, semantic and syntactic rules should be noted |
| Scan time | The calculation of scan time which includes:<br>- Scan<br>- Memory utilization<br>- Transfer<br>- Program execution<br>- User's program diagnostics |
| I/O memory processing | i.e., use of I/O image registers periodically refreshed, "get/put" type instructions, interrupt and event-driven programs, etc. and their effects on system response times, including restart (cold, warm, hot restart) |

*Table 9-2: Human-Machine Interface—Selection Criteria*

| Criteria | Comments and considerations |
|---|---|
| Types | 1) Integrated in PLC - uses PLC-MPU and memory, may restrict display features and PC operating features<br>2) Intelligent - permits extensive displays and operating features, which may include operator support |
| Display | - Brightness<br>- Contrast<br>- Screen size<br>- Definition<br>- Resolution<br>- Color purity<br>- Number of characters<br>- Labels<br>- Touch screen<br>- Mouse<br>- Rollerball<br>- Display refresh time<br>- Formats<br>- Windows<br>- Menus<br>- Native language support |
| Keyboard | - Tactile feel<br>- Ergonomics<br>- Mouse, rollerball |
| Access control | - Operator and/or programmer<br>- Keylock or software protected levels |
| Alarms | - Separate display<br>- Screen windows<br>- Active data points<br>- Alarm management |

## 9.6 Software, Programs and Programming Languages

Software is a general term that designates a set of computer programs designed to carry out specific tasks.

A program is a series of actions that a computer takes to achieve a certain result. Activities such as "locate," "read," "interpret," and "execute the instruction" are repeated in sequence for each instruction throughout the program.

A programming language is a set of representations, conventions, and rules used to convey information and instructions to the computer. The language of the digital electronic circuits inside the computer is comprised of the binary codes that represent the numbers, letters, symbols, and commands used by humans to give instructions to the computer. The language the human user uses must be converted into digital codes that the machine understands. This digital code is called machine language.

To facilitate communication between humans and machines, a high-level language is required. This is a general-purpose programming language such as C/C++, Java, FORTRAN, BASIC, Pascal, and ADA. These high-level languages are organized in a way that is directly related to the way humans solve problems. An intermediate language, called Assembly language, is also used to bridge the high level language with the machine language in the program compiler, although nowadays this is transparent to the user.

Many years ago, a programmer might have used assembly language to write a program using instruction abbreviations called mnemonics. The program was then converted into a machine-language program using an Assembler. When a facility purchases the digital control system, the computer is usually preprogrammed to communicate in a high-level language. Some systems are also equipped with configurable prewritten programs for specific applications. These systems are referred to as a configurable digital system. They can be easily configured by engineers without specialized computer backgrounds. The global capability of these configurable digital systems includes all conventional control system functions plus many capabilities not easily implemented using analog hardware, including feed-forward, dead-time compensation, and multi-variable control.

Traditionally, control systems have been implemented using a wide variety of languages from Assembler to general-purpose language such as BASIC, FORTRAN, or C. Over the intervening years, in an effort to simplify and enhance the control engineers experience and effectiveness, programming languages specific to the needs of the industrial process control application space have been developed.

These languages are:

- *Relay ladder logic*. This programming language was initially a software rendition of the hardwired logic control that was implemented using electromechanical relays. The roots of this language lie in discrete machine control. Ladder logic has evolved extensively from its origins to incorporate many sophisticated control elements, including mathematical expressions (integer and real), controllers (PID), and Table manipulation. The relay ladder logic language is traditionally used by programmable logic controllers (PLC).
- *Function block*. This language is a software representation of basic analog process control elements. Function block components include: scaling, alarming, PID, and many others. The function block language is used in single-loop controllers and distributed control systems (DCS).

The increasing demands of complete plant integration and the high level of automation now required by industrial operations have revealed many deficiencies in the conventional relay ladder and function block languages. Those shortcomings can be summarized as follows:

- Language implementations vary between different systems.
- It is difficult to reuse standard software elements (no object orientation).
- Complex data addressing using real address instead of symbolic variable names.
- It is difficult to program sequence operation.

### 9.6.1 IEC 61131 Features and Programming Languages

Because of the difficulties traditional programming languages pose for process control, the International Electrotechnical Commission (IEC) commissioned a standard to help standardize control system programming: IEC 61131, Part 3 (usually referred to as IEC 1131). It covers all important aspects of control systems programming.

The standard allows facilities to implement modern software architectures, for example: structured functional blocks, the definition of reusable software elements, and symbolic variable names.

This part of IEC 61131 specifies the syntax and semantics of a unified suite of programming languages for PLCs. These consist of two textual languages,

- Instruction List (IL), and
- Structured Text (ST)

and two graphical languages,

- Ladder Diagram (LD)
- Function Block Diagram (FBD)

Sequential Function Chart (SFC) elements are defined for structuring the internal organization of programmable controller programs and function blocks. Also, configuration elements are defined which support the installation of programmable controller programs into programmable controller systems.

The selection of software and hardware are complementary procedures and should be concurrently evaluated during system selection procedures. These PLC programming languages, defined within IEC 61131-3, are generally available for PLC hardware.

### 9.6.2 Sequential Function Charts (SFCs)

Sequential function chart (SFC) is a graphical language for depicting the sequential behavior of a control system (see Figure 9-5). It is used to define control sequences that are time and event-driven. SFC is an extremely effective graphical language for expressing both the high-level sequential parts of a control program and for programming low-level sequences, for example, to program an interface to a device.

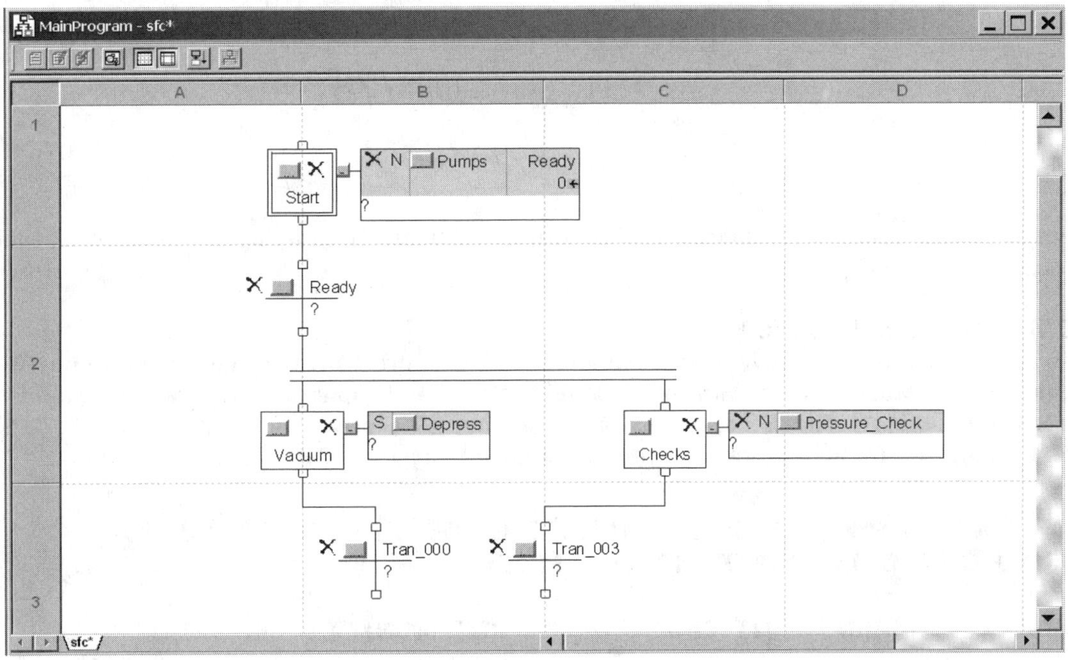


*Figure 9-5: Sequential Function Chart*

The SFC elements are used to structure the internal organization to perform sequential control functions in order to partition a set of steps and transitions that are connected by directed links. Only function blocks and programs can be structured using these.

A transition is the condition where the control passes from one step to another, along the corresponding directed link. Each transition step has an associated transition condition which is the result of evaluation by a single Boolean expression. Associated with each step will be zero or more actions.

For further information and additional details on best target applications (user guidance), as well as key features/capabilities, see reference IEC 61131-3, 2.6.

### 9.6.3 Instruction List

Instruction list (IL) is a low-level "assembler-like" language that is based on similar instruction list languages found in a wide range of today's PLCs (see Figure 9-6). An Instruction List program is composed of a series of relatively simple instructions. Precise syntax is required, although blank lines may be inserted between instructions. Each line must begin on a new line and contain an operator, along with optional modifiers, and, if necessary to the operation, one or more operands that must be separated by commas.

```
      LD    T1
    JMPC    Reset
      LD    Temp_1
      ST    Max_Temp
 Set: LD    0
      ST    D_V76
```

*Figure 9-6: Instruction List*

For additional details on best target applications (use guidance), as well as key features/capabilities, see reference IEC 61131-3, 3.2.

### 9.6.4 Structured Text (ST)

Structured text is a high-level textual language that encourages structured programming. It has a language structure (syntax) that strongly resembles Pascal and supports a wide range of standard functions and operators. The IEC 61131-3 standard defines ST's formal syntax (see Figure 9-7). Structured text is based on traditional software programming language look and feel.

```
INT_CMD := MANUAL_CMD & MANUAL_MODE OR AUTO_CMD & NOT
AUTO_CMD_CHECK & NOT MANUAL_MODE

CMD_ON_TMR(IN := INT_CMD, PT := T_CMD_MIN)

ALRM_XY(S1 := CMD_ON_TMR.Q & NOT PERM, R := OK)

ALRM := ALRM_XY.Q1
```

*Figure 9-7: Structured Text*

For further information and additional details on best target applications (user guidance), as well as key features/capabilities, see reference IEC 61131-3, 3.3.

### 9.6.5 Ladder Diagram (LD)

Ladder diagram is a graphical language that is based on traditional relay ladder diagrams. Diagrams represent power flow. This language was the first commonly used language to program traditional

PLCs (see Figure 9-8). However, the IEC ladder diagram language also allows user-defined function blocks and functions to be interconnected so they can be used in a hierarchical design.

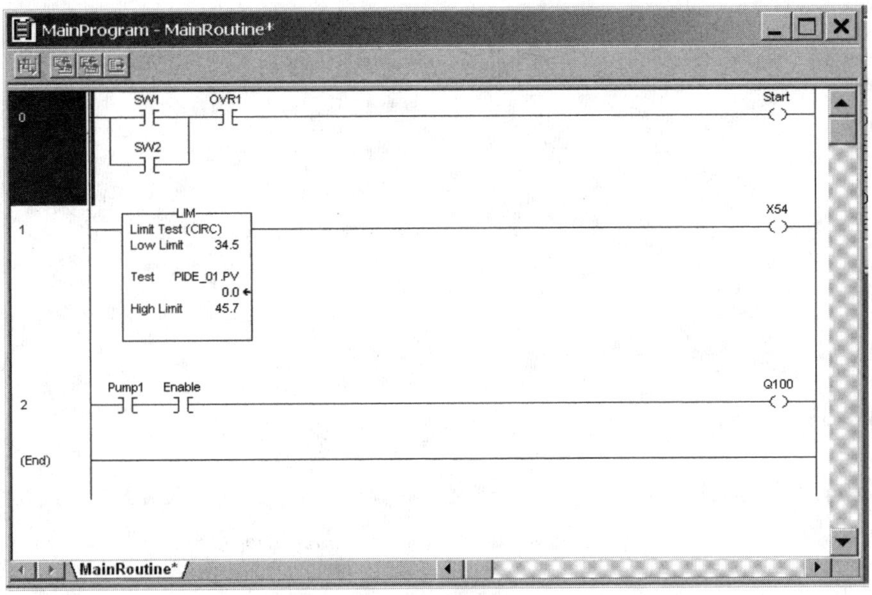


*Figure 9-8: Ladder Diagram*

For further information and additional details on best target applications (user guidance), as well as key features/capabilities, see reference IEC 61131-3, 4.2.

### 9.6.6 Function Block Diagram (FBD)

Function block diagram is a graphical language for depicting signal and data flows through function blocks that is, reusable software elements (see Figure 9-9). FBD is very useful for expressing the interconnection of control system algorithms and logic. It is based on the look and feel of the traditional logic diagram. In it, the diagrams represent data flow. The function block diagram is a network in which the nodes are function block instances, graphically represented functions (procedures), variables, literals, and labels.

For further information and additional details on best target applications (user guidance), as well as key features/capabilities, see reference IEC 61131-3, 4.3.1.

### 9.6.7 Compatibility of 61131 Languages

IEC 61131-3 takes into account the different courses of evolution of programmable controllers in North America, Europe, and Japan, and the wide variety of applications of programmable controllers in modern industry.

Figure 9-10, 9-11, and 9-12 below show the application of the LD, ST, and FBD languages to implement a simple command execution and monitoring function. In general, a desired functionality can be programmed in any one of the IEC languages. Hence, languages can be chosen depending on their suitability for each particular application. And, this demonstrates the mutual compatibility of the languages.

Programming discrete control devices for continuous control functionality is covered under an earlier topic in the book.

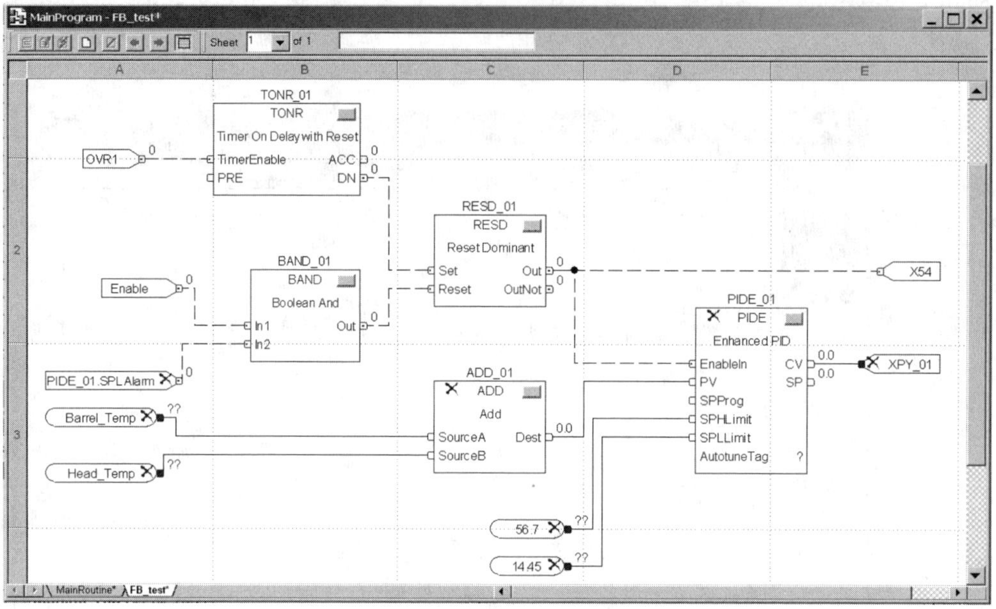


*Figure 9-9: Function Block Diagram*

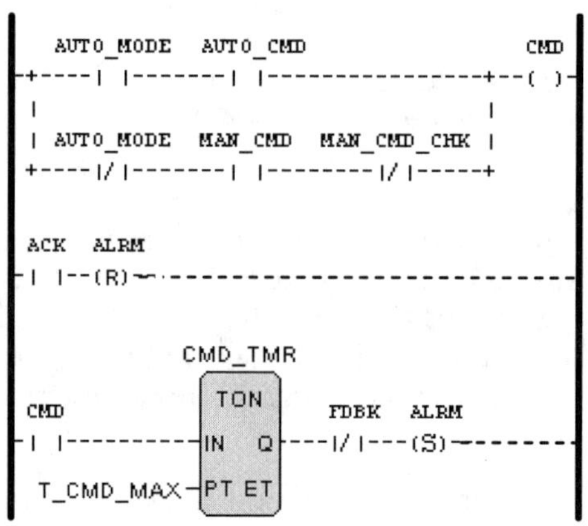


*Figure 9-10: Application Using Ladder Diagram Language*

**CMD := AUTO_CMD & AUTO_MODE OR MAN_CMD & NOT MAN_CMD_CHECK & NOT AUTO_MODE**
**CMD_TMR(IN := CMD, PT := T_CMD_MAX)**
**ALRM_FF(S1 := CMD_TMR.Q & NOT FBDK, R := ACK)**
**ALRM := ALRM_FF.Q1**

*Figure 9-11: Application Using Structured Text Language*

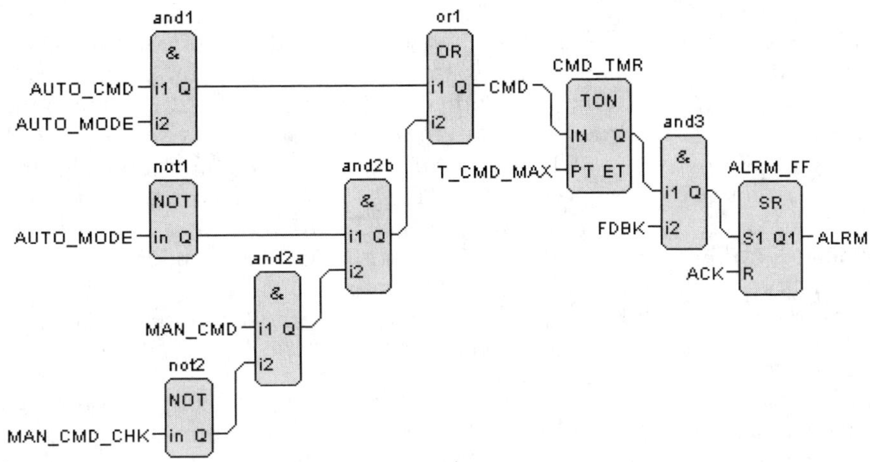


*Figure 9-12: Application Using Function Block Diagram Language*

### 9.6.8 System Software

Major topics to be considered include:

- programming structure/programming language
- PADT
- access to control programs: access methods, authorization
- documentation
- computer aided engineering tools
- on-line/off-line configuration.

## 9.7 References

*The Automation, Systems, and Instrumentation Dictionary.* Fourth Edition. ISA, 2003.

Coggan, D.A., Editor. *Fundamentals of Industrial Control.* Second Edition. Practical Guides for Measurement and Control Series. ISA, 2005.

Hughes, Thomas A. *Programmable Controllers.* Fourth Edition. ISA, 2004.

IEC 61131 – *Programmable Controllers.*

IEC 61131-1 Ed. 2.0 en:2003. *Part 1: General information.*
IEC 61131-2 Ed. 2.0 en:2003. *Part 2: Equipment requirements and tests.*
IEC 61131-3 Ed. 2.0 en:2003. *Part 3: Programming languages.*
IEC/TR 61131-4 Ed. 2.0 en:2004. *Part 4: User guidelines.*
IEC 61131-5 Ed. 1.0 en:2000. *Part 5: Communications.*
IEC 61131-7 Ed. 1.0 b:2000. *Part 7: Fuzzy control programming.*
IEC/TR 61131-8 Ed. 2.0 en:2003. *Part 8: Guidelines for the application and implementation of programming languages.*

John, Karl-Heinz and Michael Tiegelkamp. *IEC 61131-3: Programming Industrial Automation Systems – Concepts and Programming Languages, Requirements for Programming Systems, Aids to Decision-Making Tools.* Springer, 2001.

Lewis, R.W. *Programming Industrial Control Systems Using IEC 1131-3.* Revised Edition. IEEE, 1998.

## About the Authors

**Robert J. Kretschmann**'s 25-years at Rockwell Automation include industrial automation experience in product development, advanced research, and internal consulting. He has five additional years of experience in telecommunications and other work for the U.S. Air Force. Standards activities include: IEC TC65 Industrial-Process Measurement and Control Advisory Group, Technical Advisor; IEC USNC Technical Advisor for SC17B, Low Voltage Switchgear and Controlgear; IEC USNC Co-Technical Advisor for SC65B, Devices; NEMA THSC 17 Working Group – Control Circuit Device; CANENA THC 01IS; IEC Convenor TC65/SC65B/WG7 Programmable Logic Controllers; NEMA 1-IS Programmable Controllers Sub-Committee 22, Chairman; IEEE Standards Coordinating Committee 22, Power Quality – 1994-1997; and NEMA 1-IS Industrial Proximity & Presence Sensors Sub-Committee 15. A member of both ISA and IEEE, Kretschmann holds multiple U.S. patents. He frequently lectures, writes, and edits books and papers.

**James Christensen** retired from Rockwell Automation in April 2005 after a 23-year career. At Rockwell, he led development of the IEC 61131-3 Standard for Programming Languages for Programmable Controllers, which has become the global standard for this segment of the automation and control market. He also led development of the IEC 61499 standard, the successor to IEC 61131-3 for the next generation of industrial automation and control. His Function Block Development Kit (FBDK) software is the first, internationally used IEC 61499-compliant software tool kit. Before joining Rockwell, he worked two and one-half years with the Electronic Controls Division of Texas Instruments and three years with Strider Systems, Inc. He also was employed for eight years on the faculty of the University of Oklahoma, culminating in a tenured position as Associate Professor of Chemical Engineering; one year as a Ford Foundation Postdoctoral Fellow at Thayer School of Engineering of Dartmouth College; and two years as a research assistant at the Solar Energy Laboratory of the University of Wisconsin. Christensen is currently President of Holobloc, Inc., an Ohio-based corporation providing software tools, consulting and training for the application of IEC Standard 61499. His achievements in pioneering applications of object-oriented programming earned him the Rockwell International Engineer of the Year and Lynde Bradley Innovation Awards in 1991. He has written over 30 papers and book chapters, as well as delivered invited lectures and keynote addresses. He is a member of ISA and IEEE.

# 10 Motor and Drive Control

*By Dave Polka*

## Topic Highlights

*DC Motors and Their Principles of Operation*
*DC Motor Types*
*AC Motors and Their Principles of Operation*
*AC Motor Types*
*Choosing the Right Motor*
*Variable Speed Drives (Electronic DC)*
*Variable Speed Drives (Electronic AC)*
*Automation and the Use of VFDs*

## 10.1 Introduction

In order to truly understand modern day automation, it is vital that a study of motor and electronic drive principles be undertaken. The drive is the device that controls the motor. The two interact to provide the torque, speed, and horsepower (HP) necessary to operate the application.

## 10.2 DC Motors and Their Principles of Operation

There are two basic circuits in any direct current (DC) motor: the *armature* (device that rotates) and the *field* (stationary part with windings). The two components magnetically interact with one another to produce rotation of the armature. Both the armature and the field are two separate circuits and are physically next to each other, in order to promote magnetic interaction.

The armature ($I_A$) has an integral part, called a "commutator" (see Figure 10-1). The commutator acts as an electrical switch, always changing polarity of the magnetic flux to ensure there is a "repelling" force taking place. The armature rotates as a result of the "repelling" motion created by the magnetic flux of the armature, in opposition to the magnetic flux created by the field winding ($I_F$).

The physical connection of voltage to the armature is done through "brushes." Brushes are made of a carbon material that is in constant contact with the armature's commutator plates. The brushes are typically spring loaded to provide constant pressure of the brush to the commutator plates.

### 10.2.1 Control of Speed

The speed of a DC motor is a direct result of armature voltage applied. The field receives voltage from a separate power supply, sometimes referred to as a "field exciter." This exciter provides power to the field which, in turn, generates current and magnetic flux. In a normal condition, the field is kept at

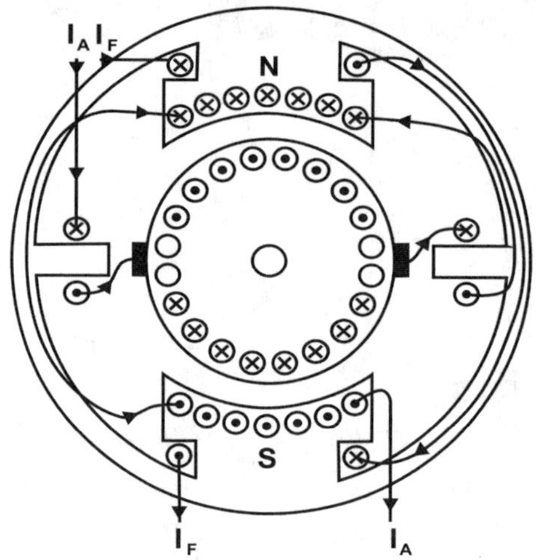


*Figure 10-1: Armature and Field Connections (Courtesy of ABB Inc.)*

maximum strength, allowing the field winding to develop maximum current and flux (known as the "armature range"). The only way to control the speed is through change in armature voltage.

### 10.2.2 Control of Torque

Under certain conditions, motor torque remains constant, when operating below base speed. However, when operating in the field weakening range, torque drops off inversely as $1/\text{Speed}^2$. If field flux is held constant, as well as the design constant of the motor, then torque is proportional to the armature current. The more load the motor sees, the more current that is consumed by the armature.

### 10.2.3 Enclosure Types and Cooling Methods

In most cases, to allow the motor to develop full torque at less than 50% speed, an additional blower is required for motor cooling. The enclosures most commonly found in standard industrial applications are:

- DPFG (Drip-proof - Fully Guarded) — This type of enclosure is self-ventilated and has no external means of cooling. Most DPFG designs can generate 100% rated torque down to 50% of base speed (see Figure 10-2).
- TENV (Totally Enclosed Non-Ventilated) — This type of enclosure has no external cooling, but uses an internal fan to circulate the air within the motor. This type of motor is capable of delivering 100% torque down to 10 or 5% of base speed.
- TEFC (Totally Enclosed Fan Cooled) — This type of enclosure has an externally mounted fan on the commutator end shaft. Air flow is a direct result of the speed of the motor, which also means that this type of enclosure is not suitable for low speed applications.

### 10.2.4 Protection and Ratings

- ***Ambient Temperature*** - Typical recommendations are for the motor ambient conditions not to exceed 40°C (104°F). Motors continuously used in higher temperatures will need a lower temperature rise class of insulation. Major insulation temperature classes refer to mechanical and dielectric strength, and are: A (lowest grade), B, F, and H (highest grade).

*Figure 10-2: DPFG Motor (Courtesy of ABB Inc.)*

- ***Over-Temperature Conditions*** - Placing the motor into overload conditions is one cause of over-temperature. High temperature inside the motor causes expansion stress in the wire insulation, resulting in cracks which, in turn, can cause contamination and eventual wire failure.
- ***Nameplate Ratings*** - Typical DC motor nameplate ratings are: Frame, HP, Amps/Field & Armature, Base/Max Speed. Additional ratings include: enclosure type, thermostat type, ambient rating, catalog and serial number, and tachometer, duty ratings.

## 10.3 DC Motor Types

### 10.3.1 Series Wound

A series wound DC motor has the armature and field windings connected in a series circuit. Starting torque developed can be as high as 500% of the full load rating. The high starting torque is a result of the field winding being operated below the saturation point. An increase in load will cause a corresponding increase in both armature and field winding current, which means both armature and field winding flux increase together. Torque in a DC motor increases as the square of the current value. Compared to a shunt wound motor, a series wound motor will generate a larger torque increase for a given increase in current.

### 10.3.2 Parallel (Shunt) Wound

Shunt wound DC motors have the armature and field windings connected in parallel (see Figure 10-3).

This type of motor requires two power supplies—one for the armature and one for the field winding. The starting torque developed can be 250-300% of the full load torque rating, for a short period of time. Speed regulation (speed fluctuation due to load) is acceptable in many cases, between 5-10% of maximum speed, when operated from a DC drive.

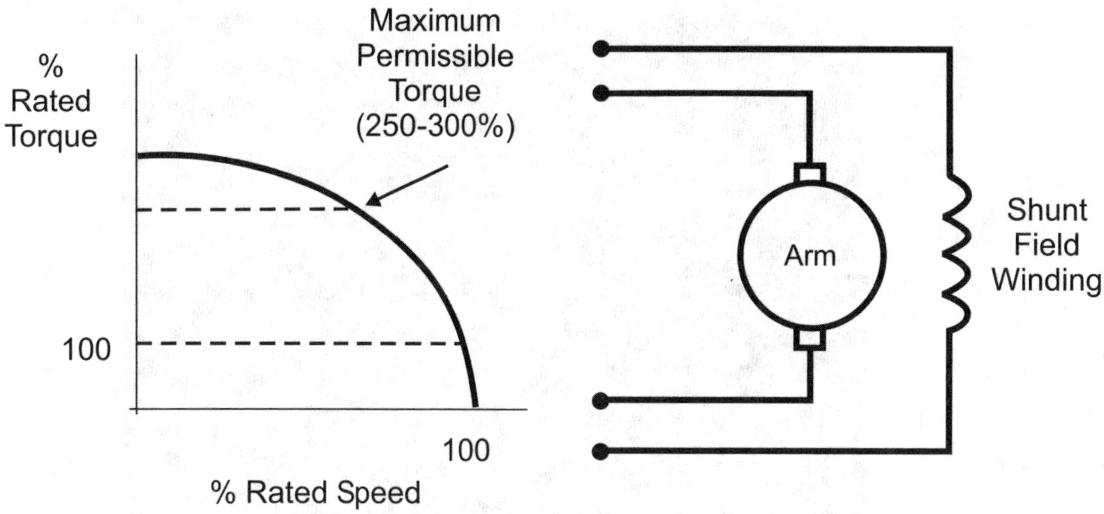


*Figure 10-3: Shunt Wound DC Motor and Curve*

### 10.3.3 Compound Wound

Compound wound DC motors are basically a combination of shunt and series wound configurations. This type of motor offers the high starting torque of a series wound motor and constant speed regulation (speed stability) under a given load. The torque and speed characteristics are the result of placing a portion of the field winding circuit, in series, with the armature circuit. When a load is applied, there is a corresponding increase in current through the series winding, which also increases the field flux, increasing torque output.

### 10.3.4 Permanent Magnet

Permanent magnet motors are built with a standard armature and brushes, but have permanent magnets in place of the shunt field winding. The speed characteristic is close to that of a shunt wound DC motor. This type of motor is simple to install, with only the two armature connections needed and also simple to reverse—simply reverse the connections to the armature. Though this type of motor has very good starting torque capability, the speed regulation is slightly less than that of a compound wound motor. Peak torque is limited to about 150%.

## 10.4 AC Motors and Their Principles of Operation

All alternating current (AC) motors can be classified into single-phase and polyphase motors (poly, meaning *many* phase or *3-phase*). For industrial applications, 3-phase induction motors are mainly used, due to efficiency. A more powerful motor can be built into a smaller frame, compared to a single-phase motor.

The main parts in an AC induction motor are the rotor (rotating element) and the stator (stationary element that generates the magnetic flux). The rotor consists of copper or aluminum bars, connected together at the ends with end rings. The rotor is filled with many individual discs of steel, called "laminations." The stator consists of cores that are also constructed with laminations. These laminations are coated with insulating varnish and then welded together to form the core (see Figure 10-4).

The revolving field set up by the stator currents cut the squirrel-cage conducting aluminum bars of the rotor. This causes voltage in these bars, with a corresponding current flow, which sets up north and south poles in the rotor. Torque (turning of the rotor) is produced due to the attraction and repulsion between these poles and the poles of the revolving stator field.

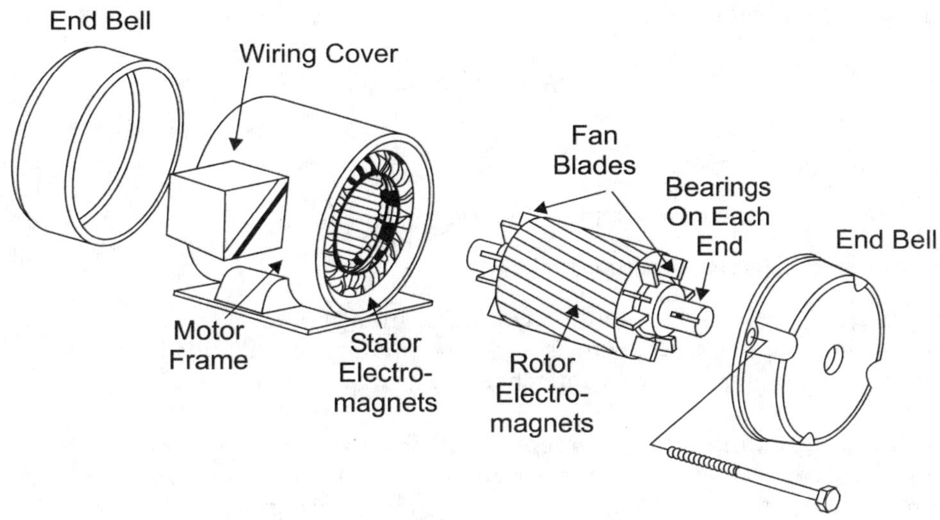


*Figure 10-4: Induction Motor Construction*

Each magnetic pole pair in Figure 10-5 is wound in such a way that allows the stator magnetic field to "rotate." A simple 2-pole stator shown in the figure has three coils in each pole group. (A 2-pole motor would have 2 poles x 3 phases = 6 physical poles.) Each coil in a pole group is connected to one phase of the 3-phase power source. With 3-phase power, each phase current reaches a maximum value at different time intervals. This is shown by maximum and minimum values in the lower part of Figure 10-5.

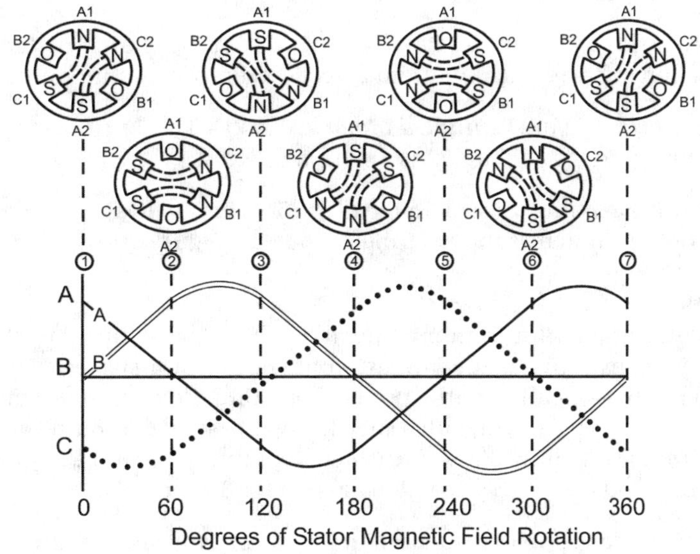


*Figure 10-5: Basic Two Pole Stator*

### 10.4.1 Control of Speed

The speed of a squirrel-cage motor depends on the frequency and number of poles for which the motor is wound (Equation 10-1).

$$N = \frac{120 \times F}{P} - \text{Slip} \qquad (10\text{-}1)$$

N = Shaft Speed (RPM)

F = frequency of the power supply (Hertz)

P = number of Stator poles (pole pairs)

Squirrel-cage motors are built with the slip ranging from about 3% to 20%. The actual "slip" speed is referred to as Base Speed, which is the speed of the motor at rated voltage, rated frequency, and rated load. Motor direction is reversed by interchanging any two motor input leads.

### 10.4.2 Control of Torque and Horsepower (HP)

HP takes into account the "speed" at which the shaft rotates (Equation 10-2). By rearranging the equation, a corresponding value for torque can also be determined.

$$HP = \frac{T \times N}{5252} \qquad (10\text{-}2)$$

T = Torque in lb-ft.

N = Speed in RPM

A higher number of poles in a motor means a larger amount of torque is developed, with a corresponding lower base speed. With a lower number of poles, the opposite would be true.

### 10.4.3 Enclosure Types and Cooling

The more common types of AC motor enclosures are:

- ***Open Drip-Proof Motor (ODP)*** - This enclosure type is constructed so that drops of liquid or solids falling on the machine from a vertical direction cannot enter the machine.
- ***Totally-Enclosed Non-Ventilated Motor (TENV)*** - This enclosure type is totally-enclosed and is not equipped for cooling from external devices.
- ***Totally-Enclosed Fan-Cooled Motor (TEFC)*** - This enclosure type has a shaft-mounted fan to blow cooling air across the external frame (see Figure 10-6).

### 10.4.4 Protection

To adequately protect the motor from prolonged overload conditions, motor overloads are installed, typically in the same enclosure as the 3-phase contactor (motor starter). These overloads (OLs) operate as "heater elements"—heating to the point of opening the circuit and mechanically disconnecting power. Overloads can be purchased with a specific time designed into the element. A Class 10 overload indicates the overload will allow 600% inrush current for 10 seconds, before opening the circuit. Class 20 would allow 600% for 20 seconds, Class 30 for 30 seconds.

An insulation system is a group of insulating materials in association with conductors and the supporting structure of a motor. Insulation systems are divided into classes according to the thermal rating of the system: Class A – (temperature up to 105° C), B – (temperature up to 130°C), F – (temperature up to 155°C), and H – (temperature up to 180°C).

*Figure 10-6: Totally Enclosed Fan Cooled (TEFC) AC Motor (Courtesy of ABB Motors)*

AC motors also include a voltage insulation system in the stator windings. These classes are designated by Class B, F, and H, for example. National Electrical Manufacturers Association (NEMA) standard MG1, Part 31 indicates the motor insulation voltage classes relative to use on AC drives.

### 10.4.5 Ratings

Because of the variety of torque requirements, NEMA has established different "designs" to cover almost every application. These designs take into consideration starting current and slip, as well as torque (see Figure 10-7).

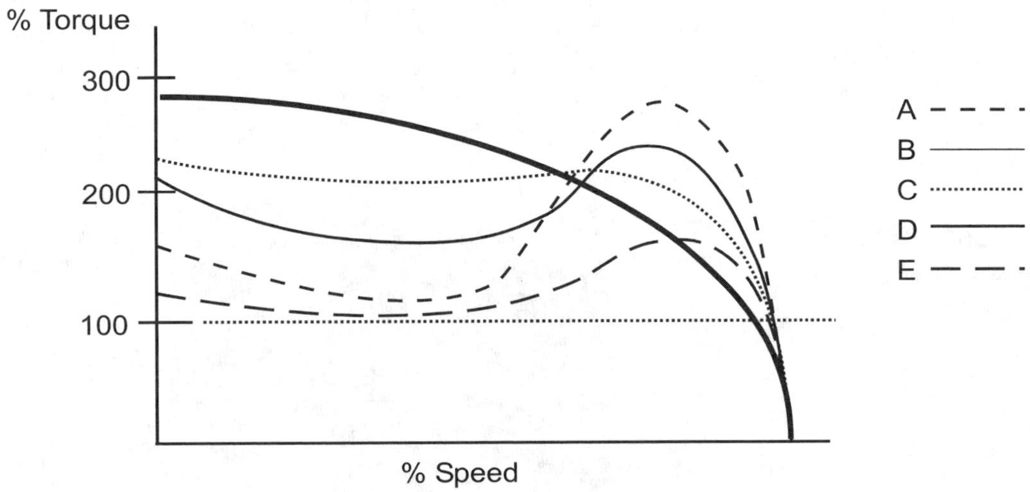


*Figure 10-7: Comparison of NEMA Designs, Across the Line (Speed/Torque Characteristics)*

- ***NEMA Design A*** – This type of motor has a high breakdown torque characteristic, compared to NEMA Design B motors. These motors are normally designed for specific use, with a slip characteristic usually less than 5%.
- ***NEMA Design B*** – This type of motor is designed for general purpose use, and accounts for the largest share of induction motors sold. The typical slip for a Design B motor is 3-to-5% or less.

- ***NEMA Design C*** – This type of motor has a high starting torque, with a relatively normal starting current and low slip. The type of load applied to a Design C is one where breakaway loads are high upon start.
- ***NEMA Design D*** – This type of motor has high starting torque, high slip, but also low full load speed. Because of its high slip (5-to-13%), the speed can easily fluctuate due to changes in load.
- ***NEMA Design E*** – This type of motor is known for high efficiency and is used mainly where the starting torque requirements are low (e.g., centrifugal fans and pumps).

It should be noted that in the motor world, there are two rating designations—NEMA and IEC. NEMA frame motors are in widespread use throughout U.S. industry. IEC is the acronym for the International Electrotechnical Commission. Though NEMA and IEC standards use different terms, they are essentially similar in ratings and, in many cases, are interchangeable. NEMA standards are probably more conservative, which allows for interpretations in design. IEC standards are more specific and categorized. NEMA MG-1, Part 31 standards indicate that motors operated on drives of 600V or less, should be capable of withstanding peak voltage of 1600V. Motors with a 1200V or 1000V insulation strength should not be applied to AC drives, unless additional precautions are taken.

## 10.5 AC Motor Types

### 10.5.1 Standard AC Induction

AC Motors can be divided into two major categories: asynchronous and synchronous. The induction motor is probably the most common type of asynchronous motor (meaning speed is dependent on slip). All standard motors include a small rectangular slot, cut lengthwise in the shaft, called a "keyway" or "keyseat." This slot includes a tapered cut rectangular piece of steel, called a "key," which is pressure fit into a coupler for a mechanical connection (see Figure 10-8).

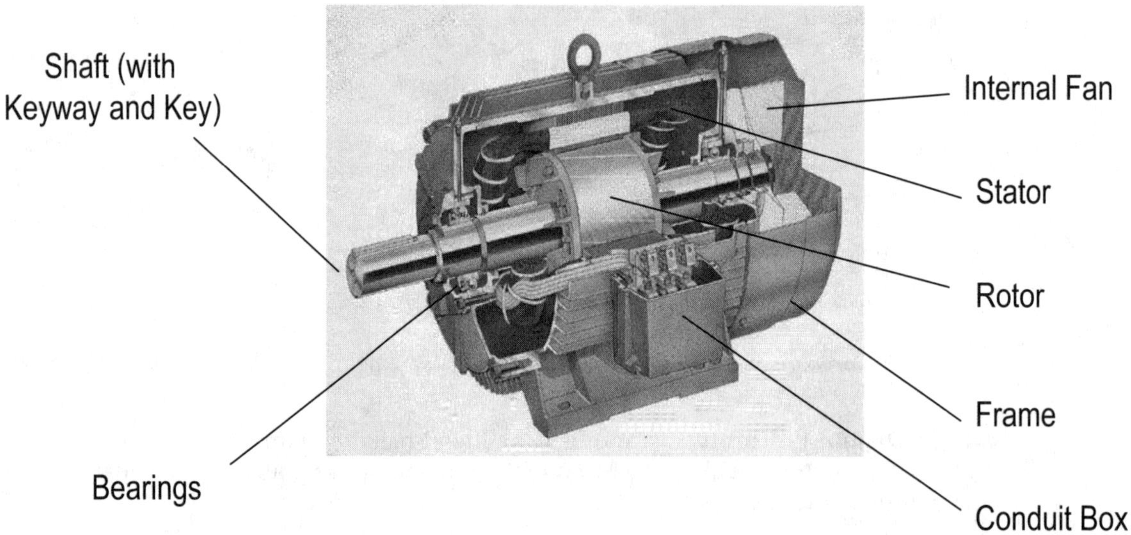


*Figure 10-8: AC Induction Motor Construction (Courtesy of ABB Motors)*

### 10.5.2 Wound Rotor

The "wound rotor" motor has controllable speed and torque characteristics. Different values of resistance are inserted into the rotor circuit to obtain various performance. They are normally started with a secondary resistance connected to the rotor circuit. The resistance is reduced to allow the motor to increase in speed. This type of motor can develop substantial torque and, at the same time, limit the amount of locked rotor current.

### 10.5.3 Synchronous

The two types of synchronous motors are: non-excited and DC-excited. Without complex electronic control, this motor type is inherently a fixed-speed motor. The synchronous motor could be considered a 3-phase alternator, only operated backwards. DC is applied directly to the rotor to produce a rotating electromagnetic field, which interacts with the separately powered stator windings to produce rotation. In reality, synchronous motors have little to no starting torque. An external device must be used for the initial start of the motor.

### 10.5.4 Multiple Pole

Multiple pole motors could be considered "multiple speed" motors. Most of the multiple pole motors are "dual speed." Essentially, the conduit box would contain two sets of wiring configurations—one for low-speed and one for high-speed windings. The windings would be engaged by electrical contacts or a two-position switch.

### 10.5.5 Specialty Motors (Stepper and Vector)

A "stepper" motor is one in which electrical pulses are converted into mechanical movements. A stepper motor rotates in fixed increments whenever it is "pulsed on." The size of the step, or step angle, is determined by the motor construction or by the type of controller connected. (Note: Step angle is determined in fractions of $360^o$). For example, the step resolution of 90° would be four steps per rev (revolution). Due to their exactness of rotation, Stepper motors are used, "open-loop," in control systems where position is critical. They are 2-phase, brushless motors that can deliver high torque at zero speed, with no drifting of the shaft position. The direction of rotation is reversed by reversing the direction of pulses from the controller.

A "vector" motor is a specific type that would be applied to an AC vector or flux vector drive. Principles of operation are identical to the standard AC induction motor. Vector control means the requirement of full torque at zero speed. The vector motor is specially designed to operate at low slip and be able to handle the heat generated by providing full torque at zero speed.

Principles of operation lie with analyzing voltage and flux vectors. The rotor is divided into $360^o$ of rotation, which is one complete rotation. A vector would be the direction and amount of a certain quantity in the motor circuit—in this case, rotor flux or stator flux (see Figure 10-9).

Physical torque developed is a by-product of the magnitude of the stator and rotor flux vectors. Stator flux is a function of the input voltage to the motor. (The voltage vectors are indicated by $U_1$ to $U_6$ in Figure 10-9.) We could consider the dashed curve pair the torque span developed in the motor. A vector or flux vector drive would control the amount of stator and rotor flux generated. In most cases, the vector motor must have provisions for the mounting of a feedback device (encoder or resolver) on the shaft end. The feedback device sends information back to the drive control, indicating exact rotor position.

## 10.6 Choosing the Right Motor

### 10.6.1 Application Related

Both torque and speed need to be assessed for any speeds during the operation cycle. The DC motor has a higher "starting torque" capability operated from a drive. Technology advancements have

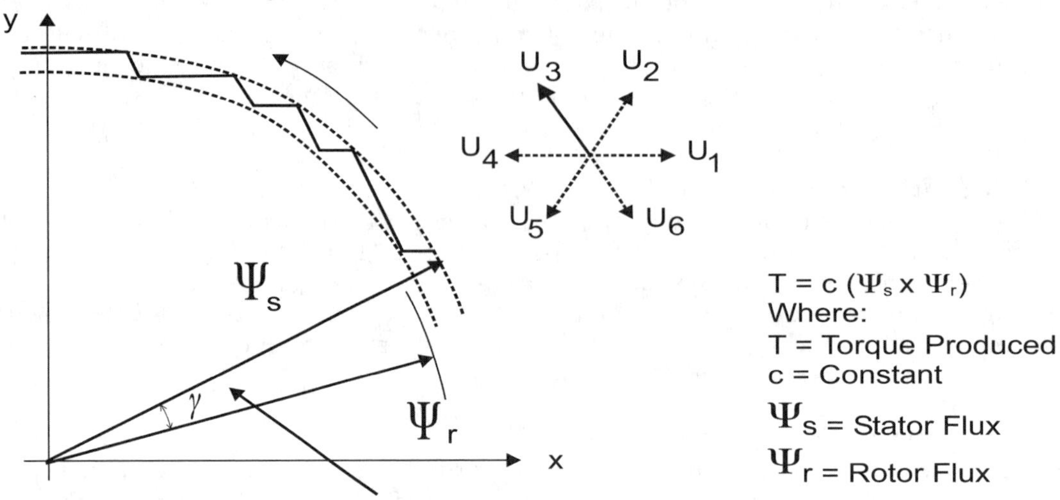


*Figure 10-9: Vector Motor Relationships – Stator and Rotor Flux (Courtesy of ABB Inc.)*

brought the AC motor starting torque close to that of the DC motor. Typical NEMA-D, AC motor classifications create over 300% starting torque across the line. Additional considerations relate to the environment, heat, humidity, accessibility, duty cycle, and overload requirements.

### 10.6.2 AC versus DC

There are no fundamental performance limitations that would prevent a flux vector variable frequency drive (VFD) from being used in any application where DC drives are used. In areas such as high speed operation, the inherent capability of AC motors exceeds the capability of DC motors. Inverter duty motors have speed range capabilities that are equal to or above the capabilities of DC motors. DC motors usually require cooling air forced through the interior of the motor in order to operate over wide speed ranges. Totally enclosed AC motors are also available with wide speed range capabilities.

Although DC motors are usually significantly more expensive than AC motors, the motor-drive package price for a VFD is often comparable to the price of a DC drive package. If spare motors are required, the package price tends to favor the VFD. Since AC motors are more reliable in a variety of situations, and have a longer average life, the DC drive alternative may require a spare motor while the AC drive may not. AC motors are available with a wide range of optional electrical and mechanical configurations and accessories. DC motors are generally less flexible and the optional features are generally more expensive.

DC motors are typically operated from a DC drive, which has reduced efficiency at lower speeds. Since DC motors tend to be less efficient than AC motors, they generally require more elaborate cooling arrangements. Most AC motors are supplied in totally enclosed housings that are cooled by blowing air over the exterior surface.

The motor is the controlling element of a DC drive system, while the electronic controller is the controlling element of an AC drive system. The purchase cost of a DC drive, in low horsepower sizes, may be less than that of its corresponding AC drive of the same horsepower. However, the cost of the DC motor may be twice that of the comparable AC motor. Technology advancements in VFD design have brought the purchase price gap closer to DC. DC motor brushes and commutators must be maintained and replaced after periods of operation. AC motors are typically less maintenance intensive, and are more "off-the-shelf" compared to comparable horsepower DC motors.

# 10.7 Variable Speed Drives (Electronic DC)

## 10.7.1 Principles of Operation

This type of drive converts fixed voltage and frequency AC to an adjustable voltage DC. A DC drive can operate a shunt wound DC motor or a permanent magnet motor. Most DC drives use silicon controlled rectifiers (SCRs) to convert AC to DC (see Figure 10-10).

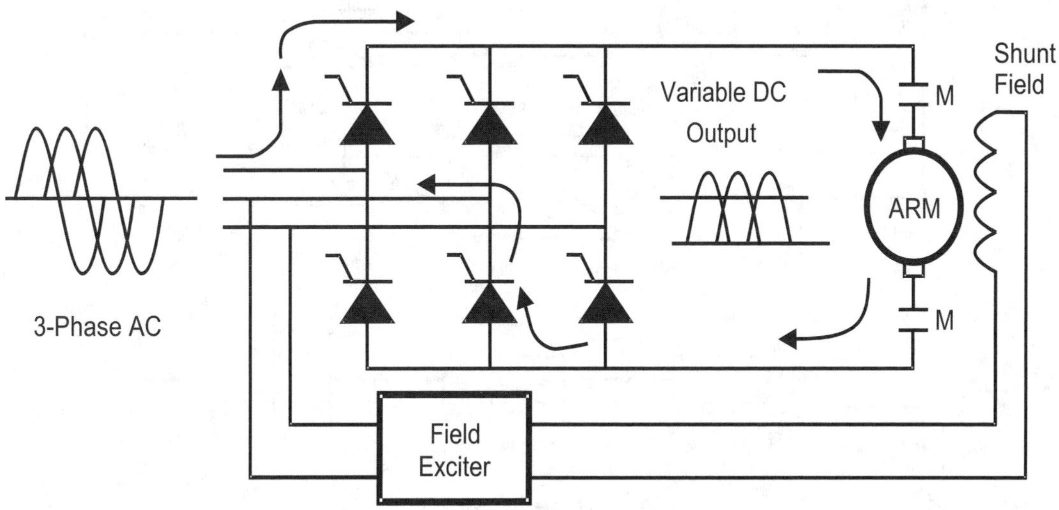


*Figure 10-10: SCR Full Wave Bridge Rectification*

SCRs provide output voltage when a small voltage is applied to the gate circuit. Output voltage depends on when the SCR is "gated on," causing output for the remainder of the cycle. When the SCR goes through zero, it automatically shuts off until it is gated "on" again. 3-phase DC drives, use six SCRs for full-wave bridge rectification. Insulated gate bipolar transistors (IGBTs) are now replacing SCRs in power conversion. IGBTs also use an extremely low voltage to gate "on" the device.

When the speed controller circuit calls for voltage to be produced, the M contactor (main contactor) is closed and the SCRs conduct. In one instant of time, voltage from the line enters the drive through one phase, is conducted through the SCR, and into the armature. Voltage flows through the armature and back into the SCR bridge and returns through the power line through another phase. At the time this cycle is about complete, another phase conducts through another SCR, through the armature and back into yet another phase. The cycle keeps repeating 60 times per second due to 60 Hz line input. Shunt field winding power is supplied by a DC field exciter, which supplies a constant voltage to the field winding, thereby creating a constant field flux. Many field exciters have the ability to reduce supply voltage, used in above base speed operation.

## 10.7.2 Control of Speed and Torque

A "speed reference" is given to the drive's input, which is then fed to the "speed controller" (see Figure 10-11).

The "speed controller" determines the output voltage for desired motor speed. The "current controller" signals the SCRs in the "Firing Unit" to "gate on." The SCRs in the converter section convert fixed 3-phase voltage to a DC voltage and current output in relation to the desired speed. The "current measuring/scaling" section monitors the output current and makes current reference corrections based on the torque requirements of the motor. If precise speed is not an issue, the DC drive and motor could

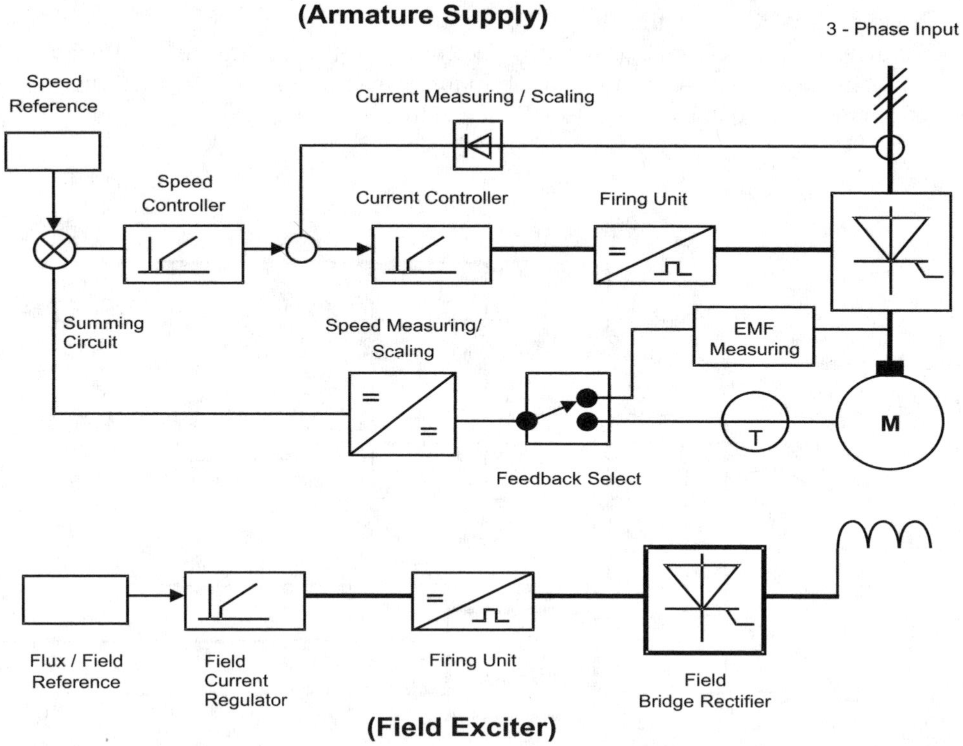


*Figure 10-11: Digital Speed & Current Controllers and Field Exciter*

operate "open loop." When more precise speed regulation is required, then the "speed measuring/scaling" circuit will be engaged by making the appropriate "feedback selection." If the "feedback" is using the "EMF measurement" circuit, then the "speed measuring/scaling" circuit will monitor the armature voltage output. The summing circuit will process the speed reference and feedback signal and create an "error" signal. This "error" signal is used by the "speed controller" as a new speed command—or a corrected speed command.

If tighter speed regulation is required (<1%), then "tachometer generator" feedback is required (e.g., tach feedback or tacho). A tachometer mounts on the end of the motor, opposite that of the shaft, and feeds back exact shaft speed to the speed controller. When operating in current regulation mode (controlling torque), the drive closely monitors values of the "current measuring/scaling circuit."

### 10.7.3 Single and 4-Quadrant Drives

There are (4) possible modes of motor operation, determined by the relationship of speed, torque and direction (Figure 10-12).

Operation in the first quadrant means the motor is actually driving the load, with torque and speed in the positive direction. This method of operation is usually accomplished with a single controller (one armature SCR bridge rectifier, and a field winding supply). The natural inertia of the system will bring the motor to a stop in acceptable time ("coast to stop").

If "coasting" to a stop is not acceptable, then a reverse polarity voltage is applied to the armature, reversing the magnetic field, and bringing the motor to a quick stop. This is called "plug" stopping. Using reverse torque to stop a motor is called "two quadrant operation."

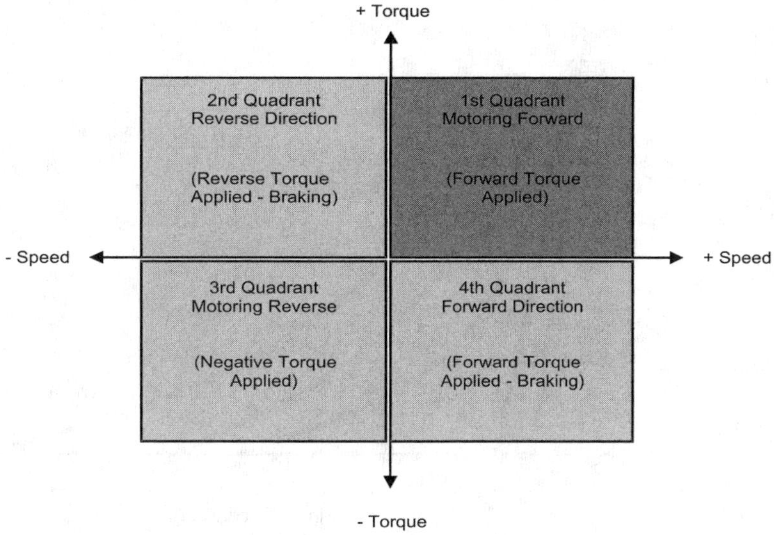


*Figure 10-12: Single, Two and Four-Quadrant Operation*

If true control throughout the positive and negative speed and torque range is desired, then a 12 SCR bridge controller is required (two armature SCR bridge rectifiers—one in the forward and one in the reverse direction), as is a field winding supply. The motor can be brought to a very fast stop by engaging the reverse armature supply bridge and "regenerating" the motor's energy back into the power line. This is called "regenerative braking." A 4-quadrant supply (4-quadrant operation) is used if "motoring" the load in the reverse speed and torque direction is required. The "reverse" SCR bridge is now used as the driving supply and the "forward" SCR bridge acts as the braking device to bring the motor quickly back to zero speed.

### 10.7.4 Braking Methods (Dynamic and Regenerative)

After "ramp to stop," the next fastest stopping time would be achieved by dynamic braking (see Figure 10-13).

This form of stopping uses a fixed, high wattage resistor (or bank of resistors) to transform the rotating energy into heat. This system uses a contactor to connect the resistor across the DC bus at the time needed to dissipate the voltage generated by the motor.

The fastest "electronic" stopping method is that of "regeneration." With "regenerative braking," all of the motor's energy is fed directly back into the AC power line. In order to accomplish this, a second set of "reverse connected" SCRs is required. This allows the drive to conduct current in the opposite direction (generating the motor's energy back to the line). A regenerative drive allows "motoring" and "regenerating" in both the forward and reverse directions.

## 10.8 Variable Speed Drives (Electronic AC)

### 10.8.1 Principles of Operation – Pulse Width Modulation (PWM)

There are several types of AC drives (VFDs—Variable Frequency Drives). All of them have one concept in common: they convert fixed voltage and frequency input into a variable voltage and frequency output to change the speed of a motor (see Figure 10-14).

3-phase power is applied to the input section of the drive, called the "converter." This section contains (6) diodes, arranged in an electrical "bridge." These diodes convert AC power to DC power. The "DC

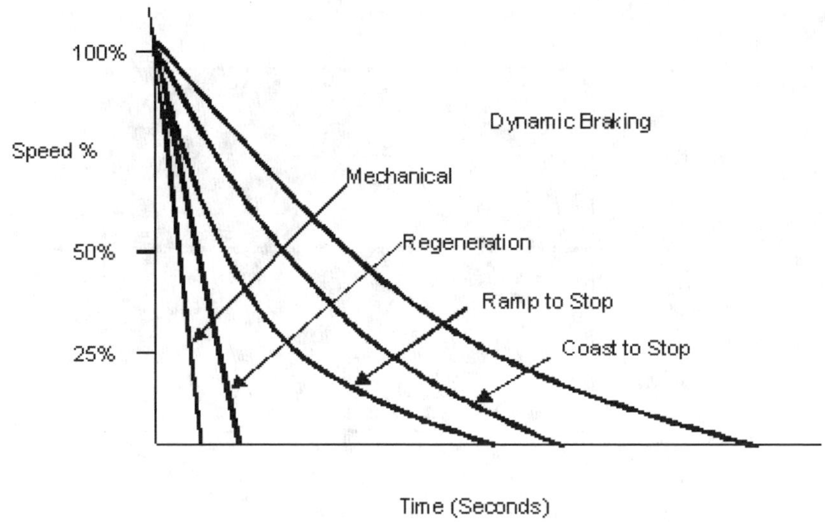


*Figure 10-13: Braking Methods for DC Drives*

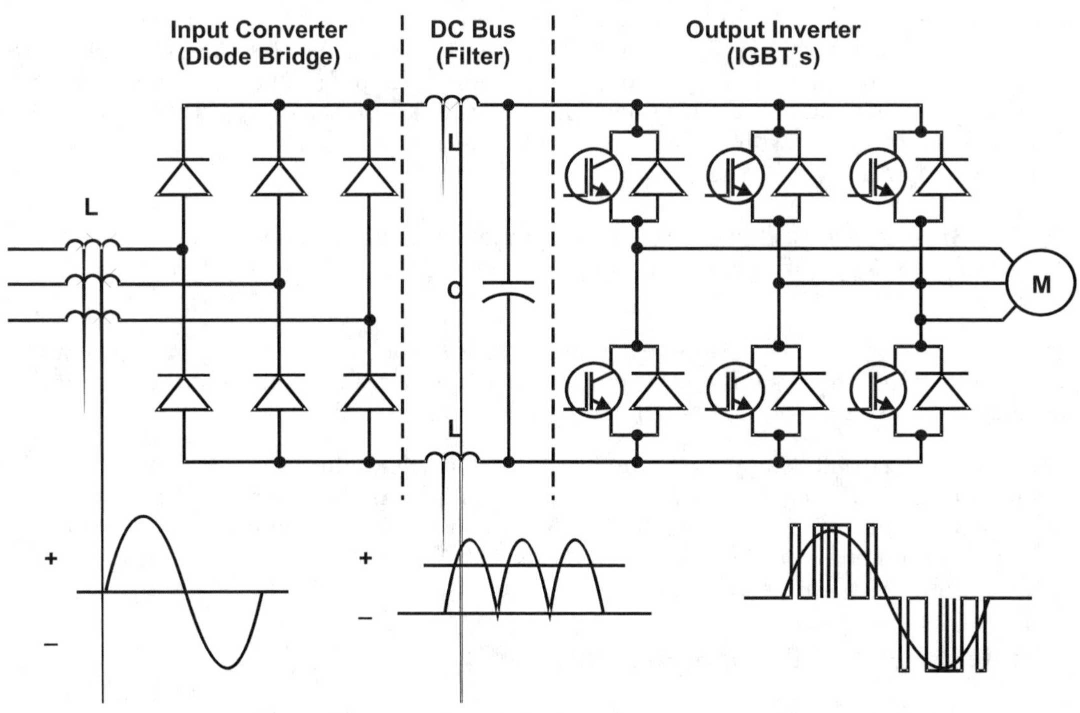


*Figure 10-14: PWM Drive (VFD) Block Diagram (Courtesy of ABB Inc.)*

bus" section accepts the now converted, AC to fixed voltage DC. The "DC bus" filters and smoothes the waveform, using "L" (inductors) and "C" (capacitors). The diodes reconstruct the negative halves of the waveform onto the positive half, with an average DC voltage of approximately 650-680 Volts (460VAC unit).

Once filtered, the DC bus voltage is delivered to the final section of the drive, called the "inverter" section. This section "inverts" DC voltage back to AC—but in a variable voltage and frequency output. Devices called insulated gate bipolar transistors (IGBTs) act as power switches that turn on and off the DC bus voltage, at specific intervals. Control circuits, called gate drivers, cause the control part of the IGBT (gate) to turn "on" and "off" as needed.

### 10.8.2 Control of Speed and Torque

The torque of a motor is determined by a basic characteristic—the volts per Hertz ratio (V/Hz). If an induction motor is connected to a 460 volt power source, at 60 Hz, the ratio is 7.67 V/Hz. As long as this ratio is kept in proportion, the motor will develop rated torque. The output of the drive doesn't actually provide an exact replica of the AC input sine waveform (see Figure 10-15).

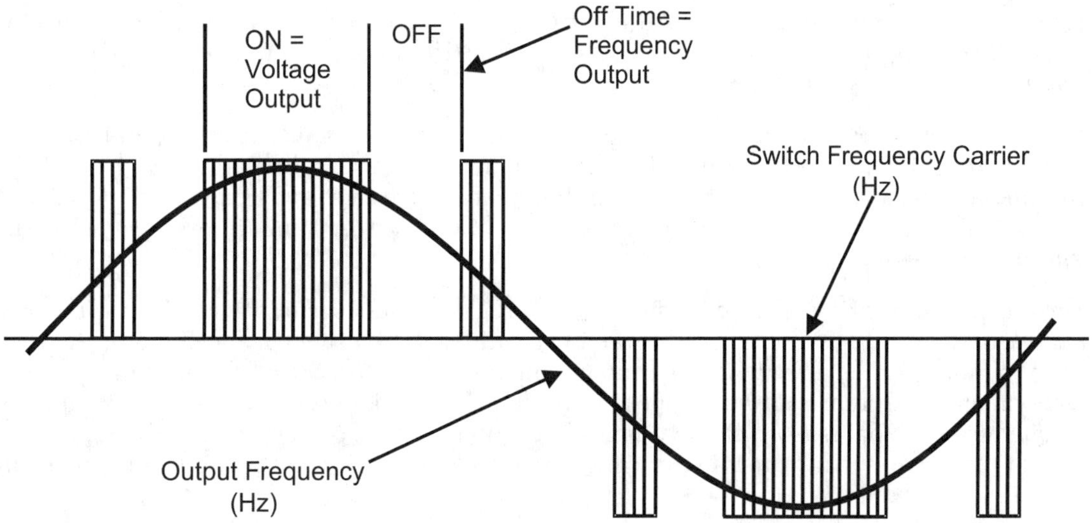


*Figure 10-15: Frequency and Voltage Creation from PWM*

It actually provides voltage pulses that are at a constant magnitude in height. The positive and negative switching of the IGBTs re-creates the 3-phase output. The speed at which IGBTs are switched is called the "carrier frequency" or "switch frequency." The higher the switch frequency, the more resolution each PWM pulse contains (typical carrier frequencies range from 3kHz to 16kHz).

### 10.8.3 Other VFD Types (Variable Voltage Inverter, Current Source Inverter, Flux Vector, Sensorless Vector, Torque Controlled)

A VVI design takes the supply voltage, rectifies it using controllable SCRs and sends the variable voltage to the DC bus and then to the inverter section. The inverter section then "inverts" the variable voltage DC to a variable voltage and frequency AC. The inverter section contains power semiconductors such as transistors or thyristors (SCRs).

A CSI drive has components similar to the VVI drive. The major difference is that it is more of a current-sensitive drive, as opposed to a VVI which is more voltage-sensitive. This design also takes the supply voltage, rectifies it, and sends the variable voltage to the DC bus and then to the inverter. The inverter section "inverts" variable voltage DC to a variable voltage and frequency AC. The inverter section is made up of power semiconductors such as transistors or thyristors (SCRs).

One of the basic principles of a flux vector drive is to simulate the torque produced by a DC motor (full torque at zero speed). Until the advent of flux vector drives, slip had to occur [30 to 50 revolutions per minute (RPM)], in order for motor torque to be developed (termed a V/Hz or scalar drive). With flux vector control, the drive forces the motor to generate torque, at zero speed.

A flux vector drive features field-oriented control similar to that of a DC drive, where the shunt field windings continuously have flux, even at zero speed. The motor's electrical characteristics are simulated in the drive controller circuitry called a "motor model." The motor model takes a mental impression of the motor's flux, voltage, and current requirements for every degree of shaft rotation. To emulate the magnetic operating conditions of a DC motor, the flux-vector drive needs to know the angular position of the rotor flux. The rotor status is fed back to the drive logic by a "pulse encoder" and a microprocessor is used to mathematically model and process the data. The advantages of this type of drive include: good torque response (<10 msec) and full torque at zero speed (at approximately 0.5 Hz output).

Sensorless flux vector control is similar to a DC drive's electromagnetic field (EMF) control. With sensorless flux vector control, a modulator is used to vary the strength of the field, which is in reality, the stator.

The role of sensorless flux vector fits generally in between the standard PWM open loop control method and a full flux vector, closed loop control method. This method provides higher starting and running torque, as well as smoother shaft rotation at low speed, compared with standard V/Hz, PWM drives. One of the main advantages of sensorless flux vector over standard PWM is higher starting torque on demand.

The direct torque control method is similar to an AC sensorless vector drive, which uses a *direct torque control* scheme. Field orientation is achieved without feedback using advanced motor theory to calculate the motor torque directly. There is no modulator used in direct torque control and no need for a tachometer or position encoder for speed or position feedback of the motor shaft. This drive has a torque response that is as much as 10 times faster than any AC or DC drive. The dynamic speed accuracy of this drive is many times better than any open loop AC drive. It is also comparable with a DC drive that uses feedback.

Direct torque control includes the basic building blocks upon which the drive does its calculations, based on a motor model (see Figure 10-16).

The two fundamental sections of direct torque control are the *torque control loop* and the *speed control loop*. During drive operation, two output-phase current values and the DC bus voltage value are monitored, along with the IGBT switch positions. This information is fed to the *adaptive motor model*. The motor model calculates the motor data on the basis of information it receives during a self-tuning process (motor identification). During this automatic tuning process, the drive's motor model gathers information, such as stator resistance, mutual inductance, and saturation coefficients, as well as the motor inertia.

The output of this motor model is the representation of actual motor torque and stator flux for every calculation of shaft speed. The values of actual torque and flux are fed to their respective comparators, where comparisons are performed every 25 μs. Every 25 μs, the inverter IGBTs are sent optimum pulse information for obtaining accurate motor torque. The correct IGBT switch combination is determined during every control cycle.

### 10.8.4 Braking Methods (Dynamic and Regenerative)

The DC bus of a typical AC drive will take on as much voltage as possible, without tripping. If an overvoltage trip occurs, the operator has three choices—increase the deceleration time, add "DC injection braking" (a parameter), or add an external dynamic braking package. If the deceleration time is extended, the DC bus has more time to dissipate the energy and stay below the trip point.

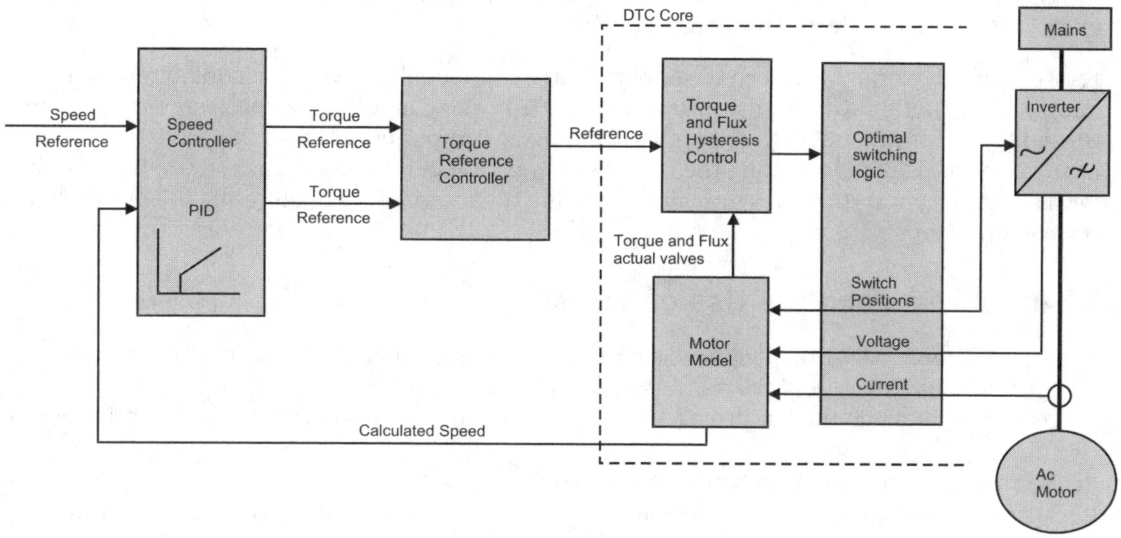


*Figure 10-16: The Direct Torque Control (DTC™) Method (Courtesy of ABB, Inc.)*

With DC injection braking, DC voltage is "injected" into the stator windings for a pre-set period of time. Braking torque (counter torque) brings the motor to a quicker stop, compared to "ramp." Dynamic braking (DB) uses an externally mounted fixed, high wattage resistor (or bank of resistors) to transform the rotating energy into heat (see Figure 10-17).

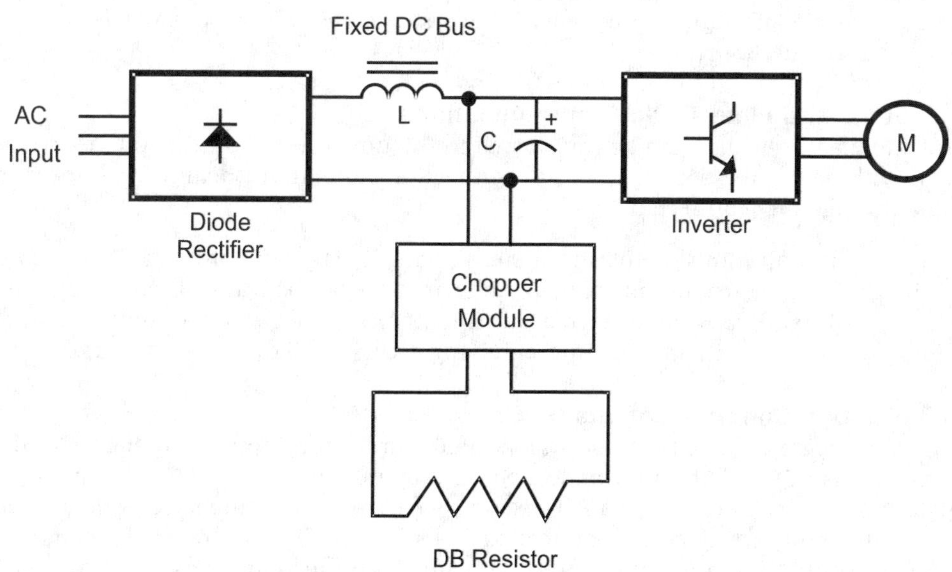


*Figure 10-17: AC Drive Dynamic Braking*

When the motor is going faster than commanded speed, the rotational energy is fed back to the DC bus. Once the bus level increases to a pre-determined point, the "chopper" module activates, and the excess voltage is transferred to the DB resistor.

For regenerative braking, a second set of "reverse connected" power semiconductors is required. The latest AC regenerative drives use two sets of IGBTs in the converter section (some manufacturers term this an "active front end"). The reverse set of power components allows the drive to conduct current in the opposite direction (taking the motor's energy and generating it back to the line). As expected with a 4-quadrant system, this unit allows driving the motor and regenerating in both the forward and reverse directions.

## 10.9. Automation and the Use of VFDs

The more complex AC drive applications are now accomplished with modifications in drive software, which some manufacturers call "firmware." IGBT technology and high-speed application chips and processors have made the AC drive a true competitor to that of the traditional DC drive system.

### 10.9.1 Intelligent and Compact Packaged Designs

Because of the use of microprocessors and IGBTs, a 1 HP drive of today is about one-third the size of a 1 HP drive 10 years ago. This size reduction is also attributed to "surface mount" technology used to assemble components to circuit boards. AC drive designs have fewer parts to replace, and include troubleshooting features available through "on-board" diagnostic or "maintenance assistant" software. In most cases, packaged AC drives of approximately 50 HP or less only use two circuit boards—control board and motor control board.

Programming is typically done with a removable touch keypad or remote operator panel. With the latest advancements in $E^2$PROMs and "flash PROMs," the programming panel can be removed from power after permanently storing parameter values. Drive panels guide the user through use of a "start-up assistant" and feature multi-language programming and "soft keys" similar to that of a cell phone. The functions of the keys change depending on the mode of the keypad. Keypads also feature pre-programmed default application values called "macros" such as PID (proportional-integral-derivative), as shown in Figure 10-18.

### 10.9.2 Serial and Fiber Optic Communications

Control and diagnostic data can be transferred to the upper level control system at a rate of 100 milliseconds. With only three wires for control connections, the drive "health" and operating statistics are available through any connected laptop.

Fiber optic communications use plastic or silica (glass fiber) and an intense light source to transmit data. With optical fiber, thousands of bits of information can be transmitted at a rate of 4 million bits per second (4M Baud). Several drive manufacturers offer serial and fiber optic software that installs directly onto a laptop or desktop computer, giving access to all drive parameters.

### 10.9.3 Fieldbus Communications (PLCs)

Data links to programmable logic controllers (PLCs) are common in many high-speed systems that process control and feedback information. Several manufacturers of PLCs offer a direct connection to many drive products. Because each PLC uses a specific programming language, drive manufacturers are required to build an "adapter" box (called a fieldbus module) to translate one language to another (called a protocol). Several manufacturers allow drive connections to existing internal network structures through the use of Ethernet modules. Modules that communicate through TCP and IP addresses allow high level controls through automated internal systems. Additional interfaces through wireless technologies and personal digital assistants (PDAs) are also making their way into programmable VFDs.

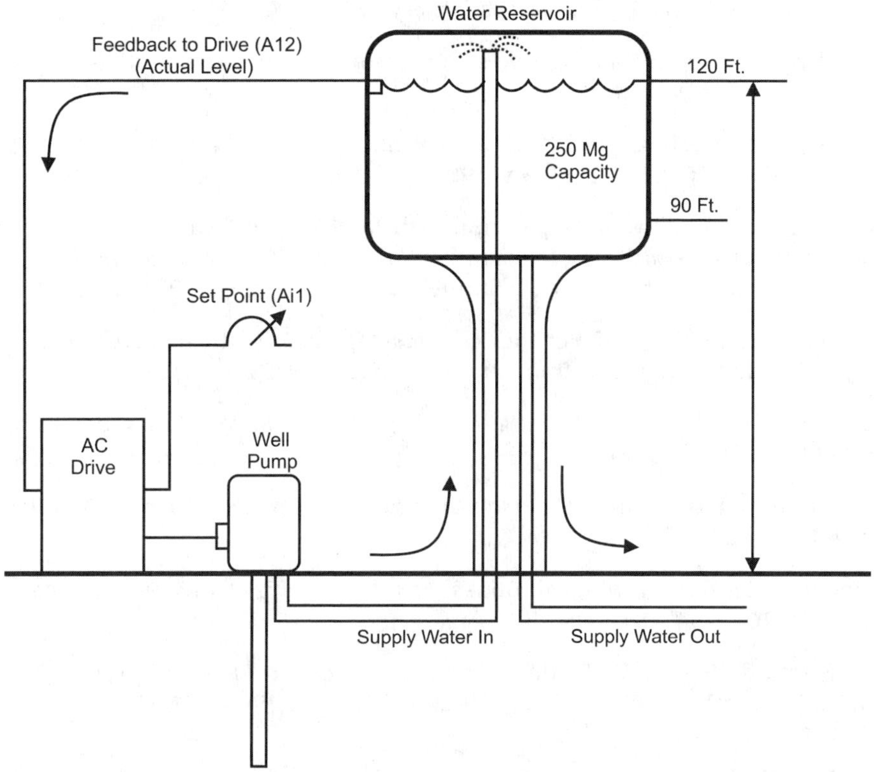


*Figure 10-18: PID Control in a Pumping Application*

### 10.9.4 Drive Configurations

Several manufacturers offer a variation of the standard 6-pulse drive. An AC drive that is termed "12-pulse ready" offers the optional feature of converting a standard 6-pulse drive to a 12-pulse drive (addition controls and a phase-shift transformer are required). The 12-pulse drive does an impressive job of reducing the highest contributors of harmonic distortion back to the power line.

One of the features of AC drive technology is the ability to "bypass" the drive if it stops operating for any reason. Known as "bypass," this configuration is used in many applications where a fan or pump must continue operating, even though it is at fixed speed. One manufacturer, ABB Inc., offers "E-bypass" circuitry. If required, a circuit board operates all the diagnostics and logic for an "automatic" transfer to bypass and feeds bypass information to the building automation system.

### 10.9.5 Chapter Summary

The use of electronic motor speed controls is expected to grow in non-traditional applications, such as automotive subsystems, household appliances, electric vehicles, people movers and marine propulsion units. As of this printing, VFDs comprise over 50% of the control methods used on standard induction AC motors. That trend will only increase in the years to come, due to increased focus on energy saving devices and micro-drive use (below 5 HP). Trends in AC drives appear to include: drives with increased intelligence; extensive use of communication options; and the use of external control devices for drive information, I/O and feedback status. Future technology will allow for an identification (ID) chip to be embedded into the motor, allowing for an almost "automated" start-up.

## 10.10 References

Polka, Dave. *Motors & Drives – A Practical Technology Guide*. ISA, 2003.

Polka, Dave. *What is a VFD?* Training Notes, P/N Training Notes 01-US-00. ABB Inc., June 2001. pp. 1-3.

"Power Transmission Design." *1993 Guide to PT Products*. Penton Publishing Inc., 1993. pp. A151-A155, A183, A235-A237, A270-A273, A337-A339.

Oliver, James A., prepared by, in cooperation with James N. Poole and Tejindar P. Singh, P.E., principal investigator. *Adjustable Speed Drives – Application Guide*. JARSCO Engineering Corp., December 1992. [Prepared for Electric Power Research Institute. Marek J. Samoty, EPRI Project Manager.] pp. 46-47.

Ebasco Services Inc. and EA-Mueller Inc. *Adjustable Speed Drive Applications Guidebook*, January 1990. [Prepared for Bonneville Power Administration] pp. 28–29, 32–33, 36-37.

ABB, Inc. Drive Operations, ST-223-1, *Basics of Polyphase AC Motors*, Reference Information, October 1998. pp. 6-29.

U.S. Electrical Motors, Division of Emerson Electric Co. *DC Motors Home Study Course*/No. HSC616-124. 1993. pp.12-18, 20-27.

Carrow, Robert S. *Electronic Drives*. Tab Books – a division of McGraw-Hill Companies Inc., 1996. pp. 96-100, 201-207, 254-255.

Patrick, Dale R. and Stephen W. Fardo. *Rotating Electrical Machines and Power Systems*. Second Edition. The Fairmont Press Inc., 1997. pp. 122, 249-250, 287-290, 296-297.

## About the Author

**Dave Polka** is Technical Instructor for ABB Inc., Automation Technologies, Low-Voltage Drives, New Berlin, Wis. He has been involved with AC and DC drive technology for more than 21 years, much of that time focusing on training and education efforts on AC drives. A technical writer, he has written user manuals and technical bulletins, along with several motor speed control articles published in major trade journals. He graduated from the University of Wisconsin – Stout, Menomonie, Wis., with a BS degree in Industrial Education, with an emphasis in Electronics and Controls.

# 11 Motion Control

*By Thomas B. Bullock and Lee A. Lane*

## Topic Highlights

*What is Motion Control?*
*Advantages of Motion Control*
*Feedback*
*Actuators*
*Electric Motors*
*Controllers*
*Servos*
*Feedback Placement*
*Multiple Axes*
*Leader/Follower*
*Interpolation*
*Performance*

## 11.1 What is Motion Control?

Motion control of machines and processes began with humans turning cranks or moving levers to actuate motion as they watched a measuring scale. Their brains were the control as they compared the desired position to the actual scale and took whatever corrective action was necessary to bring them into agreement. As automation was introduced, the scale was replaced with a feedback device; the human muscle was replaced with an actuator (motor); and the brain was replaced with electronics (controller). An operator (or input device) would enter the desired position; the controller would compare the feedback position to this desired position and decide the direction and speed needed to achieve the desired position. The controller would send these instructions to the motor on a continual basis until the desired position was achieved. Initially, machine tools were the major beneficiary of this automation. Today, packaging, material handling, food and beverage processing, and any industry that uses machines with movable members are enjoying the benefits of motion control.

## 11.2 Advantages of Motion Control

Saving time is a major benefit of motion control. It might take a person a minute or two to hand crank a machine a short distance and align it with the scale. A typical servo will do it in less than 0.5 seconds. Accuracy is another plus. In 0.5 seconds, the servo will get the machine to within the accuracy of the system. To achieve this accuracy manually may take verniers or other time-consuming means.

Coordinating two axes is impossible with manual hand cranks, but easy with servos. Servo clamping is another benefit. As will be seen shortly, a servo will return an axis to position when an outside force has moved the axis.

## 11.3 Feedback

The feedback device can be considered the eyes of the system. It determines the velocity and the position of the axis in a motion system. Motion control systems use many different types of feedback devices, which can be analog or digital and come in both incremental and absolute configurations. Both incremental and absolute types of devices can track position changes: the difference is in how they respond to a loss of power. Absolute devices can determine their position on power up, providing the axis was calibrated during start up. Incremental devices will lose their position and need to go through a homing sequence on power up. In this chapter we will briefly discuss several of the more popular types of feedback devices.

### 11.3.1 Resolvers

Resolvers are analog devices relying on magnetic coupling to determine position. They do this by looking at the magnetic coupling of a rotating winding, the rotor, compared to two stationary windings, stators (see Figure 11-1). This coupling varies with the angle of the shaft relative to the stators. As such, resolvers are rotary transformers and typically are interfaced with an A/D circuit to be used with a digital controller. These circuits typically provide 12 or 13 bits of resolution, although there are models with up to 16 bits of resolution. Resolvers are commonly used as velocity feedback devices in brushless AC servomotors. They are extremely rugged and provide absolute feedback for one revolution of their shafts. This makes them ideal for AC servomotors, as the absolute nature in one rev allows the drive to know where the motor shaft is. This allows it to commutate the motor. Although this works well for commutation, it is less than ideal for position absolute feedback. Typically an axis travels over more than 1 revolution of a motor shaft; therefore a single resolver loses its ability to act as an absolute device. If the application allows the axis to perform a homing routine on power up, this configuration offers the advantage of using a single feedback device. You can use the resolver for the drive for commutation and the controller for position and velocity. To achieve absolute positioning over multiple turns, use a dual resolver set or master vernier resolver set. Two resolvers are connected to the load, but each resolver is geared at a different ratio. By looking at the phase shift between the two resolvers it is possible to determine the absolute position of the axis over multiple turns.

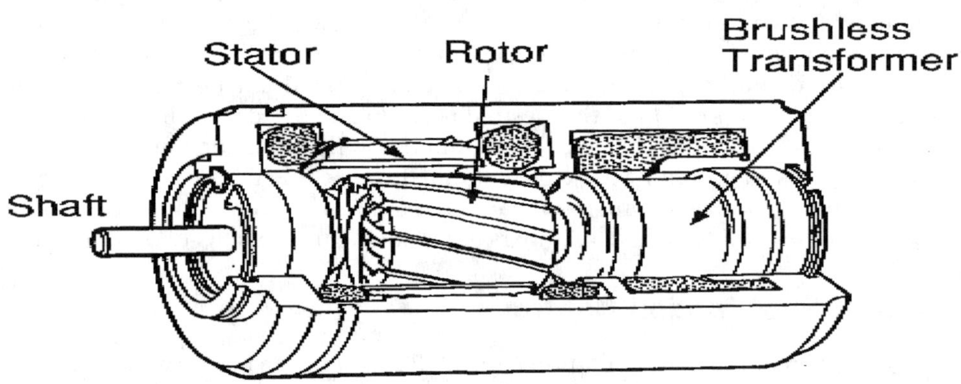


*Figure 11-1: Resolver*

### 11.3.2 Magnetorestrictive Transducers

Magnetorestrictive transducers are unique due to the noncontact nature of this type of feedback device, which makes them ideal for linear hydraulic applications. Magnetorestrictive transducers operate in a manner similar to sonar. A sensing magnet is placed on or, in the case of a hydraulic cylinder, inside the actual load. The magnetorestrictive transducer sends out a pulse to the moving magnet, which causes a mechanical strain that is conducted back to the unit. The time it takes for the strain to conduct back determines the position of the axis. These devices are absolute and must be used in linear applications. The downside to these devices is their limited resolution, typically several hundred counts per inch with moderate accuracy.

### 11.3.3 Encoders

Encoders are extremely popular in motion control applications. They are digital, relatively inexpensive, and have very high resolution and accuracy. Encoders come in both incremental and absolute configurations and are available from many vendors.

Incremental encoders are extremely popular and are used in many motion control applications. They are basically discs with slots cut in them and a through-beam photo sensor (see Figure 11-2). This configuration creates a pulse train as the encoder shaft is turned. The controller counts the pulse train and thereby determines how far the axis has traveled from a known position. This known position is determined at power up by going through a homing sequence. Typically there are two photo detectors, channel A and B, set 90 degrees apart. By looking at which channel rises first, the controller determines the direction of travel. The price of incremental encoders is primarily determined by their resolution. Incremental encoders of 1000 counts or less are very inexpensive. Incremental encoders with more than 10,000 counts are considerably more expensive. Specialty encoders with greater than 50,000 counts per rev are available but are very expensive.

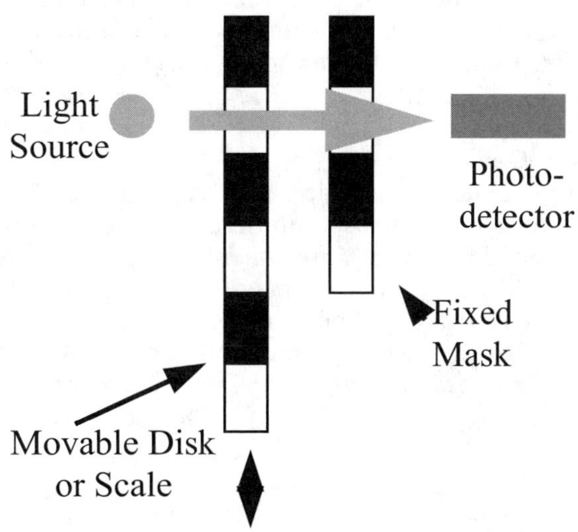


*Figure 11-2: Incremental Encoder*

Absolute encoders, as their name suggests, are absolute devices. Instead of creating a pulse train, an absolute encoder uses a disc that reveals a specific binary or gray code based on the position of the disc. Absolute encoders can be either single turn or multiturn. A single turn absolute encoder, like a resolver, gives an absolute position over one turn of its shaft. A multiturn absolute encoder incorporates an integrated gear that is encoded so that the number of turns can be recorded. Like incremental encoders, the higher the resolution, the more expensive the encoder is. Typical systems use absolute encoders with 12 bits (4096 counts) or 13 bits (8192 counts) per revolution. In addition, a multiturn absolute encoder will record 12 or 13 bits of revolutions. Typically a multiturn encoder will be described by the resolution bits plus the revolution bits; therefore a 13-bit (8192) resolution disc with the ability to track 12 bits (4092) revolutions will be referred to as a 25-bit multiturn absolute encoder. Many of today's absolute encoders are programmable and sit on a variety of industrial networks such as Asi, CAN, Devicenet, or Profinet.

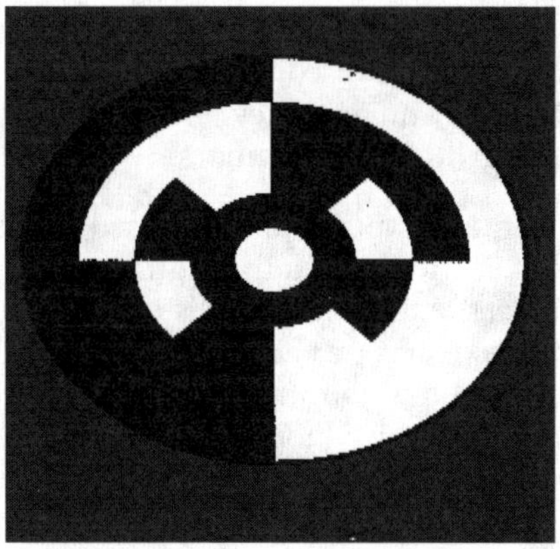

*Figure 11-3: A Three-bit Absolute Encoder Rotary Disc*

Linear encoders are incremental encoders that have been *rolled* out along the length of an axis. Also known as a glass scale, these encoders have been used for decades in machine tool applications. A reader is placed on the moving axis and picks up the pulse train generated by moving along the glass scale. Today, with the increasing use of linear motors, this form of encoder is seeing more and more use.

### 11.3.4 Other Feedback Devices

There are many other types of feedback devices that we have not discussed. These include linear variable displacement transducers (LVDTs), laser interferometers, synchros, and even Hall effect devices. These devices, while in use today, are not as prevalent as the previously described types of feedback in industrial systems. LVDTs still see a lot of use in aerospace applications, and laser interferometers are used in very high-precision applications. Today, using special interfaces and encoders with sine and cosine signals, it is possible to achieve 4 million counts per rev! Many servo drive and motor vendors offer this technology, which is patented by a company in Germany.

## 11.4 Actuators

The actuator takes the command from the controller and moves the axis. Based on the signal coming from the feedback device, the controller will command the actuator to move the axis at a particular velocity until it comes to the desired position. The actuator provides the means of accelerating and decelerating the axis and maintaining its velocity and position. The actuator can be considered the muscles of the motion control system and can be pneumatic, hydraulic, or electric.

### 11.4.1 Pneumatic

Pneumatic systems employ compressed gas under high pressure to move an axis. Compressed gas is held in a tank and released into an expandable chamber with a rod attached to it. The gas is released into the chamber through the use of an electrically operated valve. As the gas expands in the chamber it pushes the rod forward. The rod can be pulled back in either through a second chamber on the other side of the rod, or through a mechanical mechanism such as a spring. Due to the compressibility of a gas, pneumatic systems are usually not stiff enough for typical industrial motion control applications. They have limited use in specialty robotics or are used to position point-to-point systems, such as a flipper or diverger. Typically these systems are open loop, relying on a mechanical or prox switch to tell them that the axis is in position.

### 11.4.2 Hydraulic

Hydraulic systems are used when great force is required to move the axis and its load. Like pneumatic systems, hydraulic systems employ a pump and a valve, but in this case a liquid is used. The liquid is incompressible; thus the system is extremely stiff when tuned correctly. This liquid, hydraulic fluid, is made up of many different chemicals and is typically toxic to people and the environment. Proper equipment maintenance and proper storage and handling are therefore very important, which increases the ongoing costs of the system. However, there are applications that simply must use this actuator due to its capability to provide tremendous force on demand. Typical applications use a magnetorestrictive transducer for feedback and a hydraulic actuator to move large loads. An example is a transfer line in an automotive engine-machining center. The transfer bar, lifting several engine blocks from one station to the next, would be powered by a hydraulic actuator.

### 11.4.3 Electric

Electric systems consisting of a servomotor and drive are widely used in industry for motion control applications. This combination can consist of permanent magnet DC or brushless AC motors and drives. Today the performance of variable frequency drives and induction motors has increased tremendously. In some simple applications they have been used to control positioning of axis. The demanding high-performance applications still use servomotors over standard drives; however, the two types of drives are evolving together. Today many servo drives can use either synchronous (servo) or standard induction motors. There is also an increase in the use of linear motors. A linear motor is basically a synchronous motor rolled out on a flat plane. Linear motors are capable of incredible acceleration, high force, and very accurate positioning.

## 11.5 Electric Motors

The motion control industry uses several types of electrical motors. The stepper motor is used in simple applications, because it is less expensive than a servo. It is popular in its simplicity. You simply tell it how many steps you want it to move. Typically, it doesn't have feedback; you assume it moves the programmed amount.

Many motion control applications do not require positioning. A variable frequency drive runs a motor at a speed proportional to the input signal. For turning a drill or milling cutter, this is sufficient. When you need better speed regulation, use velocity feedback to close a velocity loop. The velocity regulation is determined by how good the velocity feedback device is.

More demanding applications use servo drives and motors, which are typically sold as a set from a particular vendor. This allows the motor to be precisely matched to the drive to extract the maximum performance from the system. Permanent magnet DC and brushless AC are the most common types of servomotors.

### 11.5.1 Permanent Magnet DC

Permanent magnet DC servos are DC motors that use a permanent magnet to create a field rather than a field winding with current as is used in most DC motors. This eliminates the need for a field winding, but increases cost because it requires high-quality rare-earth magnets. DC motors use a wound rotor. The rotor has windings in it to carry current; this current creates a second magnetic field that is acted upon by the field created by the permanent magnet. This produces torque on the motor shaft. To keep the rotor turning, the current must switch directions every half revolution. A modern DC motor is more complex than this, but this example demonstrates the theory. A DC motor uses brushes to switch current in the windings on the rotor. This is known as a brush commutator.

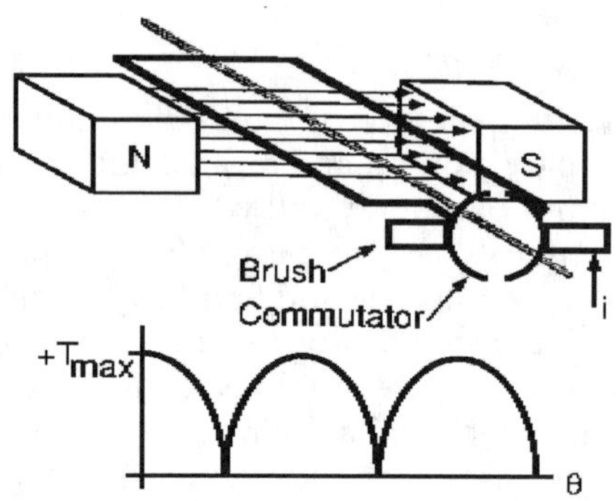


*Figure 11-4: Permanent Magnet DC Motor*

You must replace the brushes in DC servomotors after extended running time due to abrasion caused by the brushes contacting the moving rotor. The brush commutator also suffers from an effect called arc over. This occurs when the speed of the motor becomes great enough to induce a spark from one plate to another on the commutator. Due to the wound rotor, DC motors tend to have poor heat dissipation. Still, they are relatively inexpensive and require fewer electronics than AC servo motors do.

### 11.5.2 Brushless AC

Brushless AC motors are a permanent magnet DC motor turned inside out. The windings move to the stator, and the magnets move to the rotor. As the name suggests, the brushes are no longer used. Instead they employ electronic commutation and divide current into the appropriate windings to keep the motor turning. The windings are typically configured in a three-phase Y. Using a resolver on the rotor allows the drive to know where the rotor is, and therefore where the magnets are. Using this information allows the current in the windings to be adjusted to produce smooth torque. The division of the current in the windings produces sinusoidal commutation; thus these motors are referred to as synchronous AC motors. Heat dissipates better with this configuration, because the windings are in the stator and thus are able to use the body of the motor to dissipate heat. This allows for higher peak torques compared to the equivalent DC motor. With the elimination of the brushes there is no arc over; therefore AC motors are able to achieve higher speeds.

## 11.6 Controllers

The controller is the brain of the system. Its commands can be entered manually or downloaded. Manual entry occurs through the operator's station (keyboard, switches, etc.) The earliest controllers used punched paper tape to enter prepared commands, but almost any computer friendly media is possible now, as is downloading from a host computer and even a Web browser. The controller executes the program, produces the command signal, reads the feedback, compares it to the command, and decides the action that needs to be signaled to the amplifier/motor to bring the difference to zero.

The motion controller has evolved over time. It has gone from a dedicated unit that interfaced with other CPUs on the machine, typically a PLC, to becoming integrated with the PLC. Some vendors have added limited PLC functionality to their motion controllers, while others have integrated the motion controller into the PLC. This has led to many improvements to motion systems, such as eliminating the need to program two CPUs with different languages and to create handshaking routines to deal with two asynchronous processors. A single software package and single CPU make it easier to program and troubleshoot a modern system. The integrated controllers can be either resident on a PC or in a traditional dedicated hardware package. Either way, today's user will be able to use less space, have fewer points of failure, and use less wiring than the traditional systems of a PLC and dedicated motion controller.

## 11.7 Servos

A servo is the combination of the three components described above. Its basic block diagram has two elements as shown in Figure 11-5. The summing network that subtracts the feedback number from the digital command, thereby generating an error, is part of the controller.

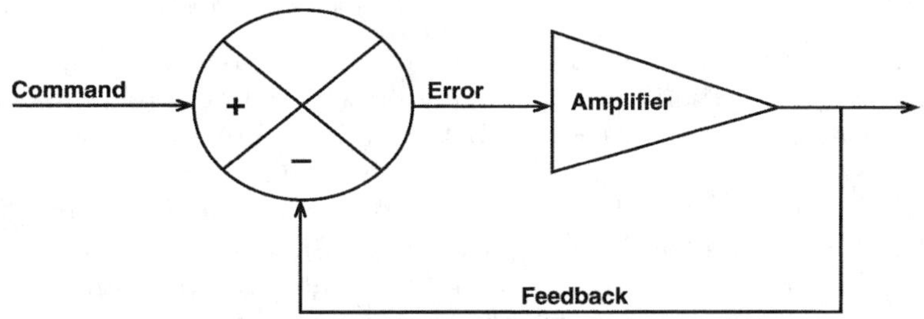


*Figure 11-5: Basic Servo Block Diagram*

The amplifier consists of the drive and motor. The feedback device that shows the movements is attached to the physical motor output or to the load. The error signal from the summing network to the drive can be either digital or analog, depending on the drive used. The drive is designed to run the motor at a velocity proportional to this error signal.

The beauty of a servo is the error must be zero for the motor to be at rest. This means the motor will continue to drive the load in the proper direction until the command and feedback are equal. Only then will it remain at rest. For positioning servos, this has the added benefit that if some external force disturbs the load, the resulting position loop error will force the motor back into position.

Servo motion control systems typically consist of three cascaded loops. The innermost loop is the current or torque loop. This loop is typically set by the vendor of the drive motor package and controls the amount of torque produced by the motor. Torque is produced by current: the more current in the motor, the more torque is produced. The amount of torque produced is based on the Kt constant of the motor. The Kt constant is a measure in units of torque per ampere. The amount of torque produced by the motor can be calculated by T=Kt*A, where *T* is torque and *A* is the amps going to the motor. This loop always resides in the drive.

The velocity loop is the second loop. This loop takes a velocity command and feeds it into the torque loop to control acceleration or deceleration. This loop can reside in either the drive or the motion controller depending on what mode you are operating the drive in. If the drive is set to torque mode, then the velocity loop resides in the controller. This means the controller will give the drive a torque command. If the drive is in velocity mode, then the drive handles the velocity loop. In this configuration the controller produces a velocity command for the drive. The velocity loop must be tuned by the user, and it determines the ability of the system to accurately follow the velocity command and to overcome disturbances to the velocity of the axis. The velocity loop must always be properly tuned before the user attempts to tune the position loop.

The position loop is the outermost loop, and it always resides in the motion controller. This loop is the final loop the application engineer must tune. The methods employed to tune a position loop are similar to the methods used to tune any other loop. In the end the user typically verifies performance by looking at following error. Following error is the difference between the command and actual position. Typically following error will jump when an axis accelerates or decelerates, and it tends to become smaller when the axis is at steady speed. By using velocity feedforward, you can minimize the jump in position error when changing speeds. The smaller the following error is at speed, the hotter the axis is tuned. Depending on the application this can be good or bad. In a point-to-point move an extremely hot system will tend to overshoot, which would not be desirable. For an application requiring two axes to be synchronized, an electronic gearing or camming situation, it is desirable to minimize following error. It is up to the user to determine the best method of tuning the axis and what level of performance is expected.

## 11.8 Feedback Placement

When the feedback is directly on the motor shaft (the usual case), the shaft is exactly where the servo is being commanded to go when the system comes to rest. Because there may be a coupling, gearbox, rotary-to-linear converter, or other element between the motor and the actual load, the load may not be perfectly in position. Backlash and wind-up are the two main culprits explaining the discrepancy, so the mechanical design has to keep these below the maximum error allowed. An alternative is to place the feedback on the load itself, but now the backlash and wind-up are within the servo loop and may present stability problems. There are notch filters and other compensation networks that can alleviate some of these stability problems, but they are not for the novice. A solid mechanical design is paramount.

## 11.9 Multiple Axes

Most applications require multiple axes, so the coordination of those axes becomes a major issue. Most "set-up" type applications allow the axes to run independently and reach their positions at different times. For instance, corrugated box making requires that dozens of axes be moved during setup to accommodate the different box sizes. These axes will determine where the cuts, creases, and printing occur. They can all be moved independently. However, once the machine starts running the boxes, any movements must be coordinated. Historically, gears and cams coordinated these motions mechanically, but electronics is taking over many functions.

## 11.10 Leader/Follower

Many machines (automotive transfer lines or cereal packaging) were originally designed with a line shaft that ran the length of the machine. Each station where particular operations occurred had one or more cams attached to the shaft to synchronize the operations within the station and with other stations. The shape of the cam dictated the movement that any particular axis had to make. The entire line was controlled by changing the speed of the line shaft. The electronic leader/follower is replacing these mechanisms. Each axis now has a servo, and its command is generated from a data table. The data table for each axis contains the position needed for that axis at each count position of the leader. The leader could be a counter that allows you to vary the count rate to simulate the line speed changing. This "electronic cam" arrangement allows many axes to be synchronized to a variable leader.

## 11.11 Interpolation

Linear/circular interpolation is a common way to coordinate two axes. It uses an algorithm that meters out movement increment commands to each axis servo on a fixed-time basis (1 millisecond is a typical time basis). Interpolation was incorporated in the earliest numerical controls for machine tools in the 1950s and is still in common use today. Parts that needed to be milled could get quite complex, but the theory was that any complex shape could be described with a series of straight lines and circles. The Electronic Industry Association developed a programming standard (RS-274) for instructing the machines in the early 1960s. The linear/circular algorithm executed those instructions.

For linear interpolation, it is only necessary to program the end point and velocity. The controller figures out how far each axis must move in every time increment to get it from its present location to the end point at the required speed. For circles, you must program the center of the circle from the present machine position, the end point on the circle, the velocity, and whether it should execute the circle in a clockwise or counterclockwise direction. The velocity can be changed by operator intervention without affecting the coordination, so the process can be slowed down or sped up as needed.

## 11.12 Performance

The performance of a motion control axis is typically specified in bandwidth or in response to a step input. Bandwidth refers to the frequency at which the servo loop output begins to fall off from its command. If you command a position loop servo with an AC function, the output will follow exactly at low frequencies. As you increase the frequency, a point is reached where the output begins to fall off, and it will do so rapidly with further frequency increases. The bandwidth is defined as the point where the output is 0.7071 of the input in amplitude. With typical industrial machinery, this is about 3 Hertz. Amplifier/drive builders also use bandwidth to rate the performance of their products. This is the bandwidth of the velocity loop that they provide, not the position loop. When tied to a real machine, you can expect this velocity loop bandwidth to be in the 30-Hertz ballpark. Many vendors might claim 100 Hertz or more, but that is not possible on industrial machinery.

Response to a step input is a measure of how fast the servo will get to its final state when a small command is initiated. You might recall that a spring/mass system gets to 63.6% of its final value in one time constant and follows the natural logarithmic curve. This time constant and the amount of overshoot (if any) are the values to consider for performance. A typical position loop servo on industrial machinery will have a time constant of 50 milliseconds. The amplifier/drive (velocity loop) will have about a 5-millisecond time constant.

## 11.13 Conclusion

Motion control has offered many benefits in automating factories. Its use continues to expand faster than the economy grows, as companies convert many applications to this better technology. The vendors are also making controllers easier to apply and tune, thereby taking some of the previous mystique away.

## 11.14 References

1. Younkin, G.W. *Industrial Servo Control Systems – Fundamentals and Applications.* Marcel Dekker, Inc., 1996.

2. Bullock, T. B. *Servo Basics for the Layman.* Bull's Eye Research, Inc., 1991.

## About the Authors

**Thomas B. Bullock** spent 31 years in industry before forming Bull's Eye Research, Inc., in 1990. He obtained a degree in engineering from the University of Wisconsin – Madison in 1959. He also has a Master's degree in engineering from the University of Pennsylvania and has 27 credits in business as a graduate student at the University of Wisconsin – Oshkosh. A registered professional engineer, he holds five patents. In addition, he has taught market research at Marian College in Fond du Lac, Wisconsin. He is a past Chairman of the NC Committee of the AMT (formerly the National Machine Tool Builders Association), and he has served as National Director, Area Vice President, and National Treasurer of NMA (National Management Association). He also served as the Steering Committee Chairman and Past President for the Industrial Computing Society (ICS). Bullock is also a fellow and honorary board member of ICS and a recipient of its distinguished service award. A free-lance writer, he has had dozens of articles published in the last five years.

**Lee Lane** has a BSEE from the University of Maine and has over 14 years of experience in Industrial Automation. He has moved into positions of increasing responsibility at Rockwell Automation where he started as an application engineer. Today Lee is a marketing manager in the Logix / Netlinx / Kinetix business of Rockwell Automation. Lee is an ISA member, a Certified Automation Professional (CAP), and is part of the CAP steering team. Lee has been very involved with the formation of the first CAP tests, the study guide, and in promoting CAP.

# III

# Advanced Control Topics

*Not all the topics in this category are really advanced control, but they are in some sense above and beyond basic plant floor controls. The first topic,* Process Modeling, *is not really control at all, but modeling is becoming increasingly important in automation, as systems become so complex that control design, checkout of configuration, and operator training all need to use process models.*

*The neural network type of modeling seems so academic to some people who meet it for the first time that it hardly seems worth a second glance. However, the successful application of this technology to virtual or soft sensors for things too complex to easily measure online, including stack gas composition, makes this technology something that every automation professional needs to understand.*

Advanced Process Control, *the second topic, is truly advanced control. Over time, what we used to call "advanced" becomes commonplace and some new things take on the advanced name. A few years ago, feedforward control was considered "advanced," but today we simply group that with the regulatory control in the* Continuous Control *topic. The model reference, model predictive control, and fuzzy logic are generally considered advanced today, but since these are built-in capabilities in current-generation DCS controllers, these will become so common that, in a few years, we will consider them part of basic control. Of this group, model predictive control is the most widely used and should be in the knowledge base of all automation professionals, whether or not they work in the process area.*

Batch *is one of the success stories of ISA standardization work, with the widespread acceptance and benefits of the ANSI/ISA88 Batch Control series of standards. In this sense, "batch" does not mean just a batch process where the steps are sequenced. Rather, batch in the ISA88 context means the use of recipes that are separated from the equipment capabilities. A general knowledge of the batch standard is fundamental to automation knowledge.*

Environmental *is not exactly an automation discipline, but it is so important in automation that it has been given its own topic. Monitoring of emissions is a very important task for automation professionals so that subject has been given strong emphasis.*

*Some years ago, heating, ventilating, and air-conditioning controls (HVAC) were considered too commercial and low cost to be a real part of true automation. However, today* Building Automation *is a rapidly growing technical field that we increasingly need to treat as a part of automation. The sophistication of controls and communications in this area now parallel the sophistication in any other area of automation.*

# 12 Process Modeling

*By Gregory K. McMillan*

## Topic Highlights

*Fundamentals*
*Linear Dynamic Estimators*
*Multivariate Statistical Process Control*
*Artificial Neural Networks*
*First Principal Models*
*Capabilities and Limitations*
*Costs and Benefits*
*References*

## 12.1 Fundamentals

Process models are used to provide important controlled variables ($CV1^*_n$), such as stream compositions, from process inputs ($X1_n$ and $X2_n$), such as flows, pressures, and temperatures. Process models help automation systems increase the efficiency and capacity of production systems despite changes in product grade, raw materials, utilities, environment, and equipment operating conditions by providing better controlled variables and finding more optimum setpoints. The change in a model output ($\Delta CV1^*_n$) with time, the dynamic response, for a change in process inputs ($\Delta X1_n$ and $\Delta X2_n$) as shown in Figure 12-1, is of greatest interest for determining how well a controlled variable can be predicted or moved to a better setpoint. The fundamental and practical ability of process models to provide the correct dynamic response in terms of three fundamental parameters called process gain ($K_{11}$ and $K_{12}$), total dead time ($\tau_{d1}$), and the open loop time constant ($\tau_{o1}$) is important, and the degree to which these parameters vary is a measure of the nonlinearity and difficulty of the system.

The conventional indication of model fidelity is based on process design where the emphasis is on the error ($E1_{ss}$) between the modeled controlled variable ($CV1_{ss}$) and the measured process variable ($PV1_{ss}$) for inputs ($X1_{ss}$ and $X2_{ss}$) all at steady state, as noted in Figure 12-1. Presently, "high fidelity" models for process design have "low fidelity" in terms of process dead time, cycling, and noise because transportation delays, sensors, valves, and transient behavior are not adequately modeled. The fidelity for control system, rather than process design, should be emphasized. For control system design, the dynamic delta error ($\Delta E1_n$) between the time response of a change in a modeled controlled variable ($\Delta CV1_n$) and a change in a measured process variable ($\Delta PV1_n$) is of greatest interest. Offsets or fixed errors between steady state values from the model and process can easily be corrected by a bias to the model output.

If one variable depends upon another variable, there is a cross correlation from cause and effect. If current values of an input or output variable depend upon past values of the same variable—as would be the case for a process variable with a time constant in its dynamic response, such as temperature—

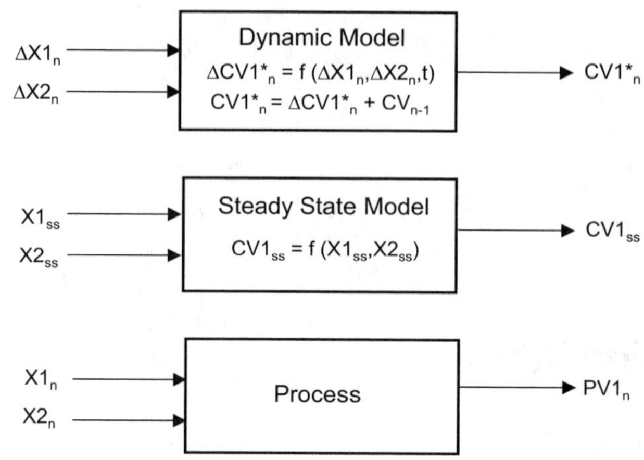


**Indication of Model Fidelity for Process Design is the Steady State Error $E1_{ss} = | PV1_{ss} - CV1_{ss} |$**

For a continuous process at a steady state, the values of successive measurement scans or model executions are equal

$$PV1_{ss} = PV1_n = PV1_{n-1} = PV1_{n-2} \ldots\ldots$$
$$CV1_{ss} = CV1_n = CV1_{n-1} = CV1_{n-2} \ldots\ldots$$
$$X1_{ss} = X1_n = X1_{n-1} = X1_{n-2} \ldots\ldots$$
$$X2_{ss} = X2_n = X2_{n-1} = X2_{n-2} \ldots\ldots$$

**Indication of Model Fidelity for Control System Design is a Dynamic Delta Error $\Delta E1_n = | \Delta PV1_n - \Delta CV1^*_n |$**

For a batch or disturbed process in a transient, the delta between successive measurement scans or model executions is important

$$\Delta PV1_n = PV1_n - PV1_{n-1}$$
$$\Delta CV1^*_n = CV1^*_n - CV1^*_{n-1}$$

*Figure 12-1: Dynamic and Steady State Model Fidelity*

there is an auto correlation. It is desirable that there be a significant cross correlation between the model output and each model input. Cross correlations between model inputs or between model outputs and auto correlations in model inputs are problematic.

Linear dynamic estimators, multivariate statistical process control, and artificial neural networks are generated from plant data, ideally plant tests, and are thus termed experimental models. First principal models are based on physical laws and require equations for physical properties and process outputs as functions of the process compositions and conditions of each stream.

Online, test skid, or laboratory measurements of the process variables must be used to verify and improve the accuracy of the model's prediction of the dynamic response of the controlled variable. As shown in Figure 12-1, the best method is by changing inputs to the process and model and computing the error between the controlled variable and the process variable. It is mistakenly thought that models can be generated from historical data that does not change, as might be the case with a process that is at steady state, or for process variables that are never measured.

## 12.2 Linear Dynamic Estimators

A Linear Dynamic Estimator (LDE), as shown in Figure 12-2, is an incremental model that sums the products of the changes in process inputs and their process gains ($K_{11}*\Delta X1_n$ and $K_{12}*\Delta X2_n$) and adds a feedback correction from a lab or online measurement of the process variable ($PV1_n$). Note that the subscripts n and n-1 denote the current and last measurement scan or model execution, respectively. In an LDE, the change in the controlled variable ($\Delta CV1_n$) is added to the last value of the controlled variable ($CV1_{n-1}$) to create an instantaneous value of the controlled variable ($CV1_n$) delayed by just the execution time.

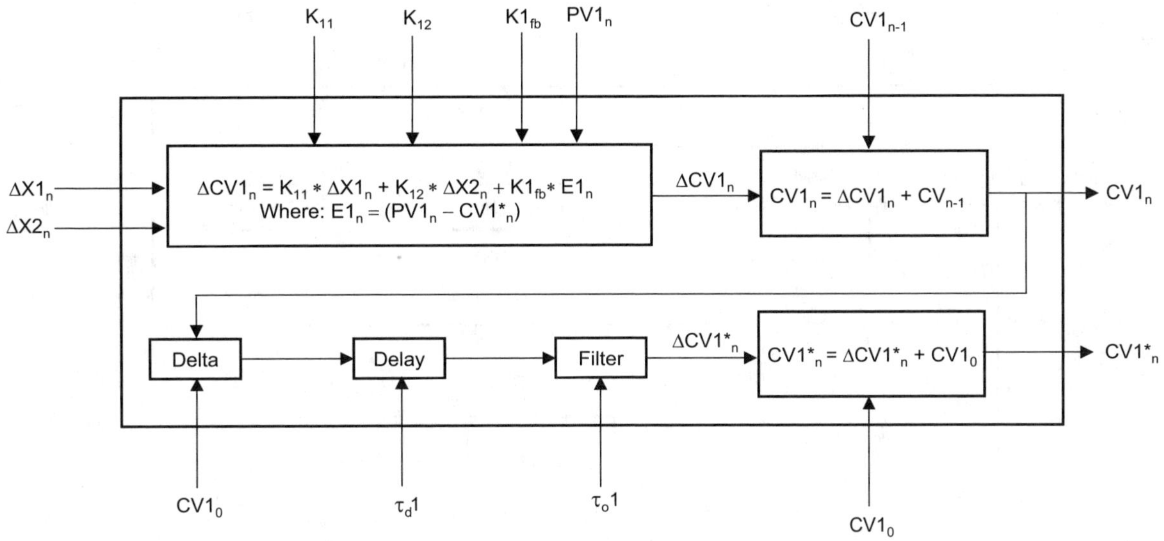


*Figure 12-2: Linear Dynamic Estimator*

The elimination of process dead time in the response provides considerable improvement in the performance of the control system and predicts where the controlled variable will be in the future. However, in order to get feedback correction, the controlled variable must be synchronized with the field or lab measurement. The delta between the instantaneous value ($CV1_n$) and an initial value ($CV1_0$) is delayed by a time equal to the process dead time, filtered with a filter time equal to the process time constant, and added back in the initial value ($CV1_0$). The resulting synchronized controlled variable ($CV1^*_n$) is subtracted from the measurement of the process variable ($PV1_n$) and multiplied by a fractional feedback gain ($K_{fb}$) and summed with the changes from the inputs to correct the LDE output.

The process gain, dead time, and time constant are obtained by manual or automatic tests and identification. The same software and techniques used for developing and deploying model predictive control (MPC) are used for linear dynamic estimators. The incremental nature of both the LDE and MPC is critical because the deltas capture the local slope and minimize the nonlinearities for small changes at an operating point to provide a better dynamic response.

In a tieback model, the percent manipulated variable ($\%MV1_n$), a PID controller output, is tied back to the percent controlled variable ($\%CV1_n$), a PID controller input, as shown in Figure 12-3. While these models may have provisions for a process gain, time constant, and dead time, these parameters are typically left at defaults or are at best manually set to provide a general type of response suggested by operations. Since the process gain normally changes significantly over the scale range, and a tieback model is not an incremental model, the model fidelity is usually poor even when the gain parameter is adjusted. The tieback model uses a single input and provides a single output and therefore does not show the effect of interactions and disturbances.

## 12.3 Multivariate Statistical Process Control

Multivariate statistical process control (MSPC) uses principal component analysis (PCA) to provide a small set of uncorrelated principal components, called latent variables, from a linear combination of a large set of possibly correlated process inputs. Consider a three-dimensional (3-D) plot of process output data versus three process inputs as shown in Figure 12-4. The first latent variable (PC1) is a line through the data in the direction of maximum variability. The second latent variable (PC2) is a line

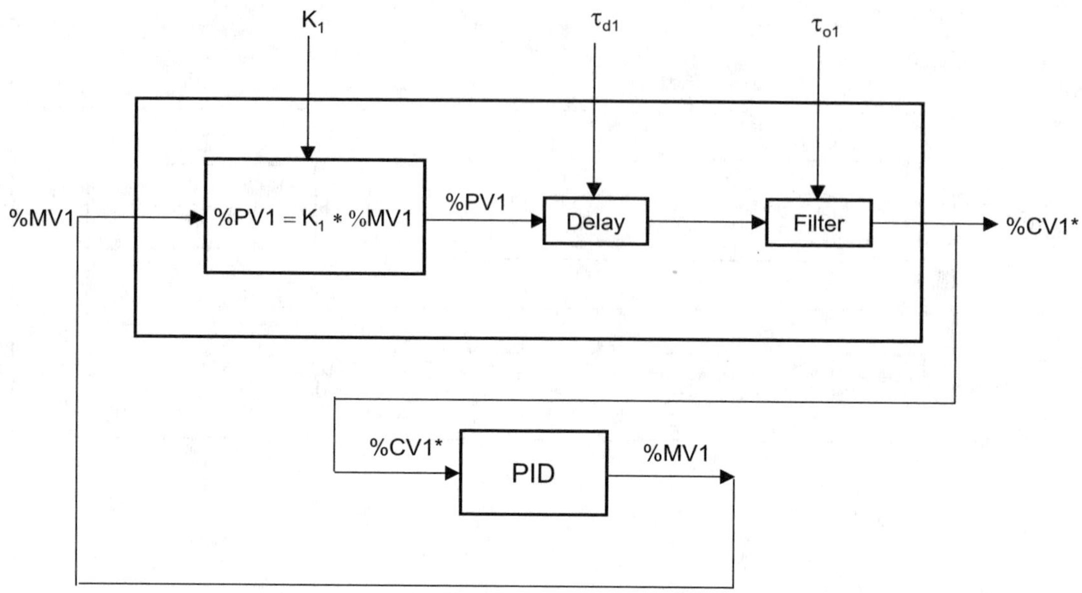


*Figure 12-3: Tieback Model*

perpendicular (orthogonal) to the first in the direction of second greatest variability. The data projected on this new plane is a "Scores" plot. While this example is for three process inputs reduced to two latent variables, MSPC can reduce hundreds of process inputs into a few principle components and still capture a significant amount of the variability in the dataset.

If the data points are connected in the proper time sequence, they form a "Worm" plot, as shown in Figure 12-5, where the head of the worm is the most recent data point ($PV1_n$). Outliers, such as data point at scan n-x ($PV1_{n-x}$), are screened out as extraneous values of a process variable, possibly because of a bad lab analysis or sample. When the data points are batch end points, the plot captures and predicts abnormal batch behavior. In Figure 12-5, the sequence of data points indicates that the process is headed out of the inner circle of good batches.

A Partial Least Squares (PLS) estimator predicts a controlled variable based on a linear combination of the latent variables that minimizes the sum of the squared errors in the model output. To synchronize the predicted with the observed process variable, each process input is delayed by a time approximately equal to the sum of the process dead time and time constant.

## 12.4 Artificial Neural Networks

An Artificial Neural Network (ANN) consists of a series of nodes in hidden layers where each node is a nonlinear sigmoidal function to mimic the behavior of the brain. The input to each node in the first hidden layer is the summation of the process inputs biased and multiplied by their respective weighting factors as shown in Figure 12-6. The outputs from nodes in a layer are the inputs to the nodes in a subsequent layer. The predicted controlled variable is the summation of the outputs of the nodes from the last layer. The weights of each node are automatically adjusted by software to minimize the error between the predicted and measured process variables in the training data set. The synchronization method for an ANN is similar to that for a PLS estimator, although an ANN may use multiple instances of the same input with accordingly set delays to simulate the exponential time response from a time constant.

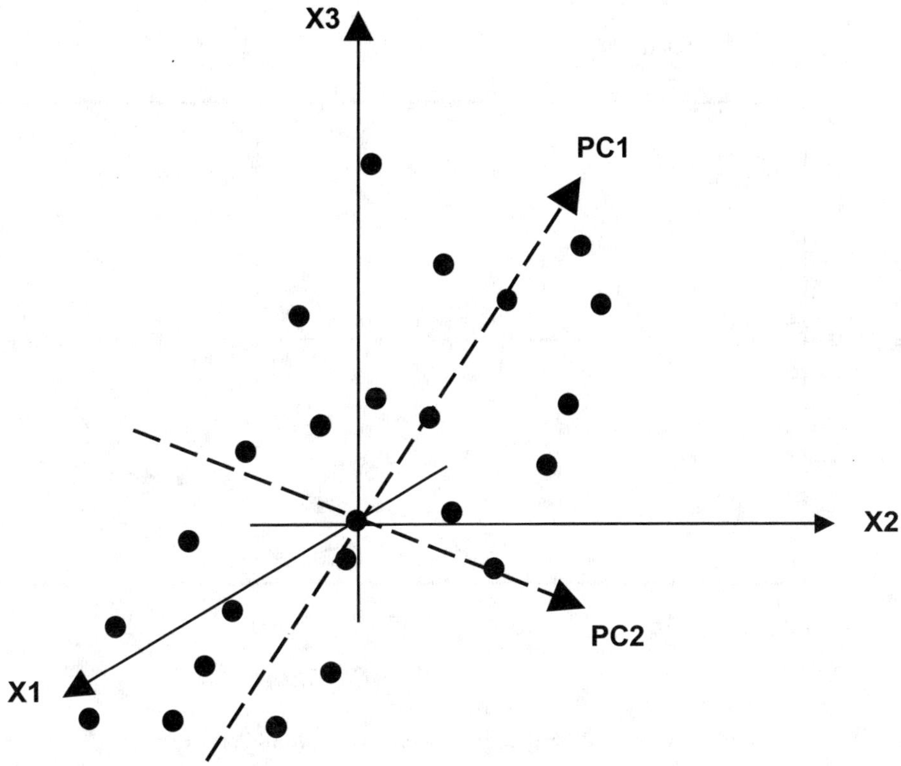


*Figure 12-4: Reduction from 3 Inputs to 2 Principal Components*

## 12.5 First Principal Models

First principal models use equations that obey the laws of physics, such as the conservation of quantities of mass, component, energy, charge, and momentum. The rate of accumulation of a given physical quantity is the rate of the quantity entering minus the rate of the quantity exiting the volume. The controlled variables are computed from the accumulation, equipment dimensions, physical properties, and equations of state. The mass balance for level and pressure is calculated by the integration of the mass flows entering and exiting the liquid and gas phase, respectively. Temperature is computed from an energy balance and pH from a charge balance.

Industry tends to use ordinary differential equations for dynamic first principal models where the physical quantity of interest is assumed to be evenly distributed throughout the volume. Profiles and transportation delays are modeled by the breaking of a process volume, such as the tube side of a heat exchanger, into several interconnected volumes. This lumped parameter method avoids the use of partial differential equations and the associated boundary value problems.

In steady state models, physical attributes, such as composition and temperature, of a stream are set by an iterative solution proceeding from input to output streams, or vice versa, to converge to a zero rate of accumulation of a quantity, such as the mass of a chemical component or energy, within the volume. In dynamic models, the rate of accumulation is nonzero and is integrated. However, most dynamic models have steady state models and thus iterative solutions for pressure-flow interrelationship of liquid streams because the integrations step sizes required for the millisecond momentum balance response time are too small. Special dedicated software is required to use momentum balances to

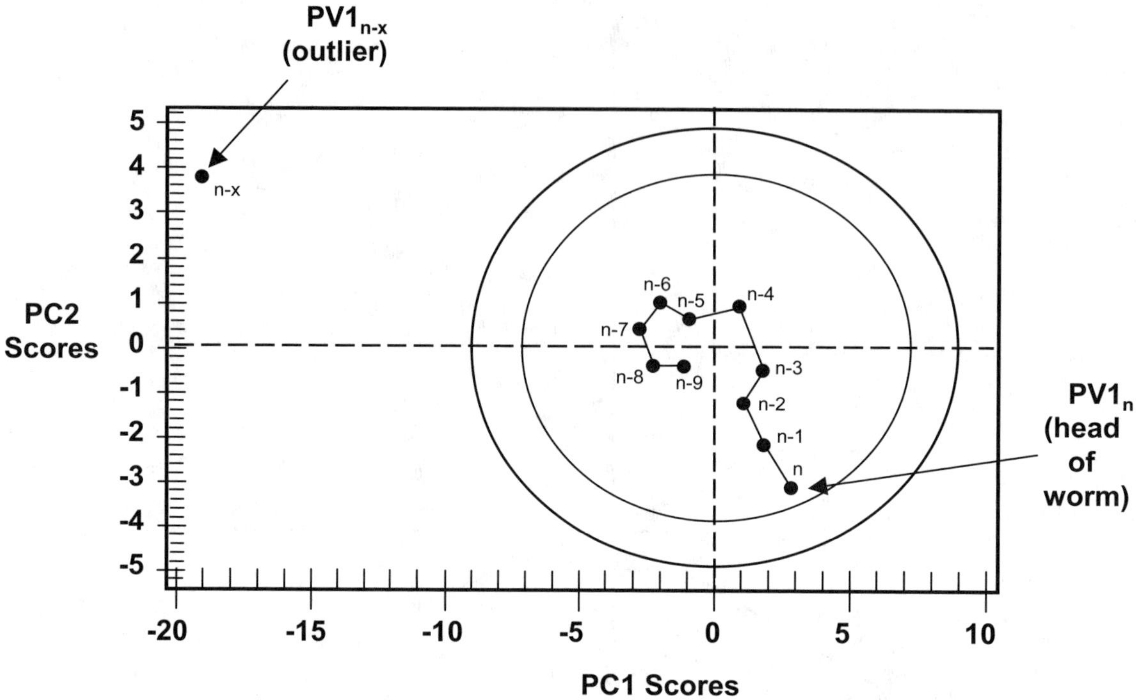


*Figure 12-5: Worm Plot for 2 Principal Components*

study hydraulics, hammer, and surge. Steady state models generally do not have a pressure-flow solver because it makes convergence of the overall model too difficult and lengthy. Consequently, a stream flow must be directly set in such models because a change in valve position or operating point on a pump curve does not change flow.

Figure 12-7 is an example of a dynamic and steady state model to compute level. In the dynamic model the rate of accumulation of liquid mass ($\Delta M_{Ln} / \Delta t$) is the net mass flows ($F1_n$ and $F2_n$) into and out of the volume ($F3_n$). The rate of accumulation multiplied by the integration step size ($\Delta t$) and added to the last accumulation ($\Delta M_{Ln-1}$) gives the new accumulation ($\Delta ML_n$). The liquid level ($L_{Ln}$) is the present accumulation of mass divided by the product of the liquid density ($\rho_L$) and cross sectional area ($A_v$) of the vessel. In the steady state model, the liquid mass ($M_{Ln}$) and level ($L_{Ln}$) is constant since the rate of accumulation is zero. Also, the mass flow out ($F3_n$) of the vessel is not a function of the pressure drop, maximum flow coefficient of the control valve, and the level controller output, but is set equal to sum of the mass flows into the vessel ($F1_n + F2_n$).

Dynamic models run real time if the time between successive executions of the model (execution time) is equal to the integration step size. A dynamic model will run faster than real time if the execution time is less than the integration step size. In order for the model to be synchronized with the control system, both must start at the same point in time and run at the same real time factor, which is difficult if the model and control system are in separate software packages. Frequently, complex dynamic models will slow down because of a loss of free time during an upset when the dynamic response of the model and control system is of greatest interest.

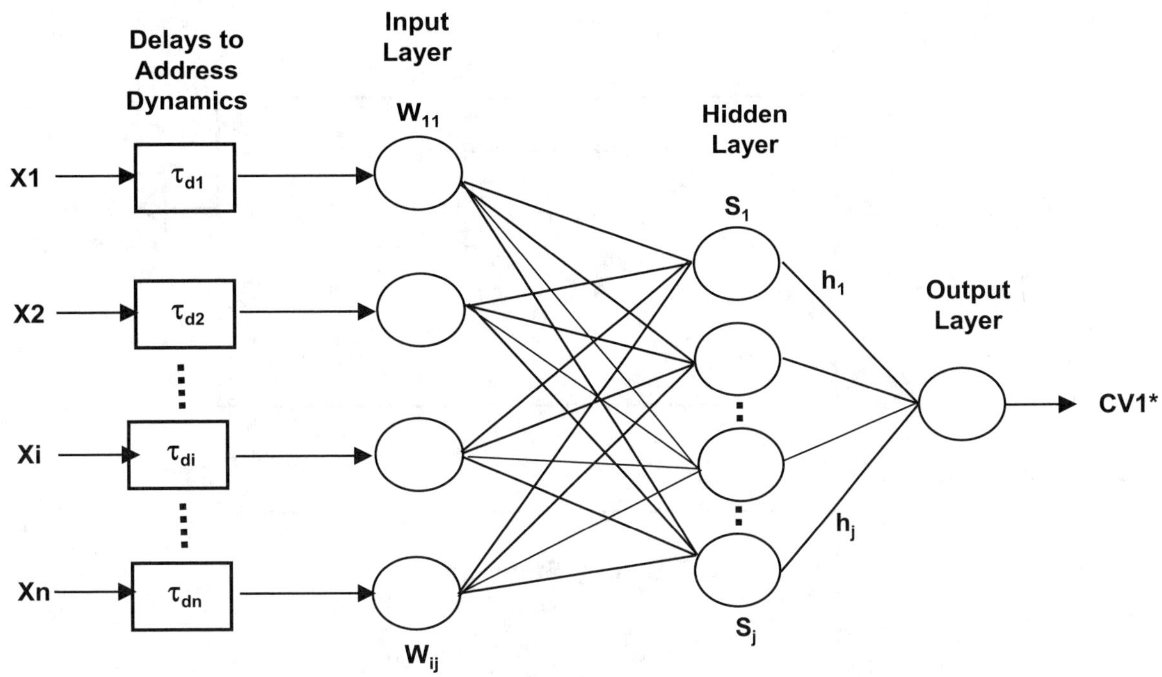


$$S_j = \sum_i w_{ij} x_i$$

$$h_j = \frac{1 - e^{-S_j}}{1 + e^{-S_j}}$$

*Figure 12-6: Neural Network with One Hidden Layer*

## 12.6 Capabilities and Limitations

A Linear Dynamic Estimator (LDE) and Multivariate Statistical Process Control (MSPC) use linear models. Both require that there be no correlations between process outputs. However, the MSPC is designed via PCA to handle correlated process inputs. An LDE by definition can accurately model process outputs with significant time constants and auto correlations. However, presently, LDE software tend to have practical limits of 50 process inputs, and most LDE have less than five process inputs, whereas MSPC and ANN software are designed to handle hundreds of inputs.

An ANN excels at interpolation of nonlinear relationships. However, extrapolation beyond the training set can lead to extremely erroneous process outputs because of the exponential nature of the response. Also, a large number of inputs and layers can lead to a bumpy response.

An LDE requires testing the process to accurately identify the dynamic response. Both MSPC and ANN advertise that you can develop models by just dumping historical data. Generally, the relationships between whole values rather than changes in the inputs and outputs are identified. Consequently, these models are at risk of not identifying the actual process gain at operating conditions. An MSPC and ANN also generally require the process to be self-regulating, meaning output variables reach steady states determined by measured process inputs. Non-stationary behavior (shifting process outputs) from an unmeasured disturbance or an integrating response is better handled by identification and feedback correction methods employed by an LDE.

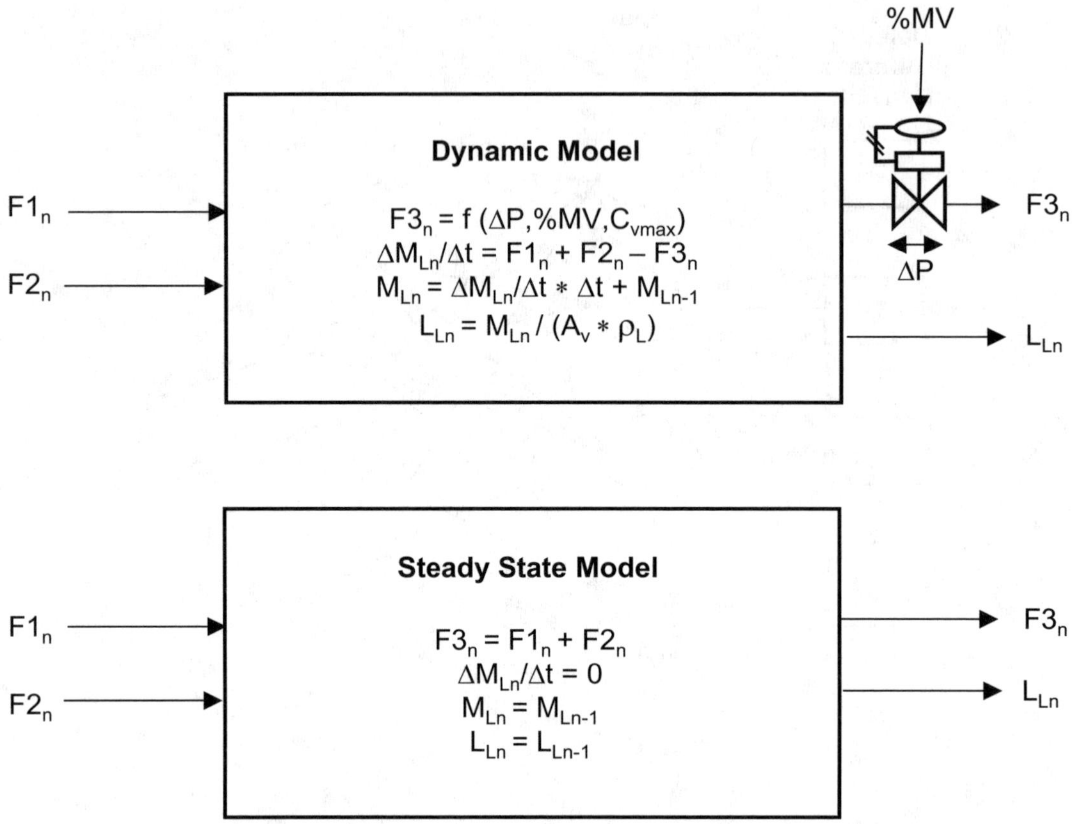


*Figure 12-7: First Principal Dynamic and Steady State Level Models*

First principal models can inherently handle all types of nonlinearities, correlations, and non-self-regulation, and show the compositions and conditions of streams. However, the physical property data is missing for many components, requiring the user to construct theoretical compounds. Also, these models tend to focus on process relationships and omit transportation and mixing delays, thermal lags, non-ideal control valve behavior, and sensor lags. Consequently, while first principal models potentially show nonlinear interrelationships better than experimental models, the errors in the process dead times and times constants are much larger than desired.

First principal models offer the significant opportunity to explore new operating regions, investigate abnormal situations, and provide online displays of the composition of all streams and every conceivable indicator of process performance.

An LDE requires step changes five times larger than the noise band for each process input with at least two steps held for the time to steady state, which is the dead time plus four time constants.

An MSPC and ANN should have at least five data points, preferably at different values significantly beyond the noise level, for each process input. A hundred process inputs would require at least 500 data points. A feedback correction from an online or laboratory measurement can be added to an MSPC and ANN similar to an LDE. If laboratory measurement is used, the synchronizing delay must be based on the time between when the sample was taken and the analysis entered. In most cases, the total number and frequency of lab samples is too low.

MSPC and ANN models that rely totally upon historical data, rather than a design of experiments, are particularly at risk at identifying a process input as a cause, when it is really an effect or a coincidence. For example, a rooster crowing at dawn is not the cause of the sunrise and the dark background of enemy tanks at night is not an indicator that the tanks are a legitimate target. Each relationship should be verified that it makes sense, based on process principals. A sure sign of a significantly weighted extraneous input is an MSPC or ANN model that initially looks good but is inaccurate for a slight change in process or equipment conditions.

The LDE, MSPC, and ANN all offer future values of controlled variables by the use of the model output without the process time delay and the time constant for synchronization with the process measurement needed for model development and feedback correction. The synchronization for an LDE is more accurate than for an MSPC or ANN because it includes the process time constant.

Frequently the process gain, dead time, and time constant of controlled variables, such as temperature and composition, are inversely related to flow rate. Consequently, an experimental model is only valid for limited changes in production rate. For large changes in throughput, different models should be developed and switched if the process gain, dead time, and time constant cannot be computed and updated in the model.

Steady state first principal models offer projected steady states. However, it may take 30 minutes or more for a large steady state model to converge. Dynamic first principal models can run faster than real time to rapidly show the dynamic response into the future, including response as it moves between steady states. It is not limited to self-regulating processes and can show integrating and runaway responses. The step size should be less than one fifth of the smallest time constant to avoid numerical stability, so faster than real time execution is preferably achieved by a reduction in the execution time rather than an increase in integration step size. The fastest real time multiple is the largest stable integration step size divided by the original step size or the original execution time divided by the calculation time (execution time minus the wait time).

Dynamic models can be run in a computer with a downloaded configuration and displays of the actual automation system to form a virtual plant. A virtual plant provides inherent synchronization and integration of the model and the automation system, eliminates the need for emulation of the control strategy and the operator interface, and enables migration of actual configurations.

Steady state first principal models are limited to continuous processes and steady states. Thus, the first principal model needs to be dynamic to predict the process outputs for chemical reaction or biochemical cell kinetics, behavior during product or grade transitions, and for batch and non-self-regulating processes. Parameters, such as heat exchanger coefficients, must be corrected within the models based on the mismatch between the process model and the actual plant. For steady state models, this is done by a reconciliation step where the process model is solved for the model parameter. For online dynamic models in a virtual plant synchronized with actual plant, a model predictive controller has been shown to be effective in adjusting model parameters.

Only dynamic first principal models are capable of simulating valve resolution (stick-slip) and deadband (backlash). However, most to date do not include this dynamic behavior of control valves and consequently will not show the associated limit cycle or dead time.

Often the compositions of raw materials are not measured comprehensively enough or frequently enough. Most of the variability of well automated systems is typically caused by raw materials.

It is important to distinguish fidelity based on the purpose. In general, process models for control system optimization require the most fidelity. Process models for configuration testing require the least fidelity. Operator training for familiarization with displays can be accomplished by low fidelity models. However, process training, control strategy prototyping, and abnormal situation management require at least medium fidelity models. It is difficult to achieve more than low fidelity in the modeling of start-ups and shutdowns and many first principal models will crash for zero volumes.

## 12.7 Costs and Benefits

Process models provide process knowledge and identify more optimum operating points whose benefits generally pay for the cost of the LDE, MSPC, and ANN software in less than a year. The cost of comprehensive first principal modeling software with physical property packages is generally several times greater, and consequently require much larger applications and longer payoff periods.

The process knowledge needed to implement first principal models is greater but correspondingly, the process knowledge gain is more extensive and deeper. Some LDE, MSPC, and ANN software require little, if any, outside support after the first application, but require some process understanding to verify cause and effects. The cost of inside application engineering to configure, identify, and run these experimental models is usually less than the software cost after the learning curve. However, the systems engineering cost to integrate, interface, and historize data between new external software and an existing automation system can be several times greater than application engineering cost.

First principal models presently require outside support or internal simulation experts and a total engineering cost that generally exceeds the software cost. All of the models require an on-going yearly maintenance cost that is about 10-to-20% of the initial installed cost, or else the benefits will steadily diminish and eventually disappear.

The total cost of an LDE, MSPC, and ANN process model is generally less than the installed cost of a field analyzer with a sample system. However, the cost of improving the accuracy of lab analysis and increasing the frequency of lab samples must be considered. Often overlooked are the benefits from reducing the effect of noise, dead time, and failures on existing analyzers, and taking advantage of the composition information in density measurements from Coriolis meters.

## 12.8 References

1. McMillan, Gregory K., Terrance L. Blevins, and Willy K. Wojsznis. "Advanced Control Smorgasbord." *Control*, Vol. 17 No. 5 (May 2004), pp. 41 – 48.

2. Blevins, Terrence, Gregory K. McMillan, Willy K. Wojsznis, and Michael Brown. *Advanced Control Unleashed – Plant Performance Management for Optimum Benefits*. ISA, 2003.

3. McMillan, Gregory K. and Robert A. Cameron. *Models Unleashed – Applications of the Virtual Plant and Model Predictive Control – A Pocket Guide*. ISA, 2003.

4. Mansy, Michael M., Gregory K. McMillan, and Mark S. Sowell. "Step into the Virtual Plant." *Chemical Engineering Progress*, Vol. 98, No. 2 (February 2002), p. 56+.

## About the Author

**Gregory K. McMillan** is a retired Senior Fellow from Solutia Inc. and an ISA Fellow. McMillan received the ISA "Kermit Fischer Environmental" Award for pH control in 1991 and *Control* magazine's "Engineer of the Year" award for the process industry in 1994. Greg was inducted into *Control* magazine's "Process Automation Hall of Fame" in 2001 and honored as one of *InTech* magazine's most influential innovators in 2003. He received a BS from Kansas University in 1969 in Engineering Physics and an MS from University of Missouri – Rolla in 1976 in Electrical Engineering (Control Theory).

# 13 Advanced Process Control

*By Gregory K. McMillan*

## Topic Highlights

*Fundamentals*
*Fuzzy Logic Control*
*Adaptive Control*
*Model Predictive Control*
*Real Time Optimization*
*Capabilities and Limitations*
*Costs and Benefits*
*References*

## 13.1 Fundamentals

In advanced process control, process knowledge by way of process models is used to make the control system more intelligent. The Process Modeling topic shows how quantitative process models can provide inferred controlled variables, such as stream compositions, that can be less expensive, faster, and more reliable than the measurements from field analyzers. In this section, the quantitative models from the previous section and qualitative models based on fuzzy logic are used to provide better controller tuning settings, setpoints, and algorithms for feedback and feedforward control.

## 13.2 Fuzzy Logic Control

Fuzzy logic uses qualitative measures in linguistic rules that mimic the rapid nonlinear and synergistic characteristics of human logic. The qualitative measures are signs, such as "negative" and "positive," and relative sizes, such as "small" and "moderate." The linguistic rules usually take the form of "If, Then" statements. The relative value scaled from -1 to +1 of each antecedent (condition) and each consequent (result) is determined by membership functions that are often geometrically represented by triangles and bars. The technique and terminology is not commonly understood and is best explained by looking at a simple but significant example in industry.

In the process industry, a fuzzy logic controller that only requires four rules has successfully replaced a proportional-integral (PI) controller. For a reverse acting controller, these rules are:

- *Rule 1:* If the error is negative and the change in error is negative, then the change in output is positive.
- *Rule 2:* If the error is negative and the change in error is positive, then the change in output is zero.

- *Rule 3:* If the error is positive and the change in error is negative, then the change in output is zero.
- *Rule 4:* If the error is positive and the change in error is positive, then the change in output is negative.

A PI controller works to keep an output from the process termed the controlled variable (CV) at a desired operating point called the setpoint (SP) by adjusting an input to the process known as the manipulated variable (MV). The control error (E) is the controlled variable minus the setpoint. The subscripts "n" and "n-1" denote the value of the variable at scans, n and n-1, respectively. The CV, SP, and E in a PI control algorithm are converted to percent of measurement scale, and the MV is the percent of the scale of whatever is manipulated, which could be a valve, speed, or setpoint. In a fuzzy logic algorithm, these variables are converted to fractional value from -1 to +1 based on scale factors that the user must enter for each variable. A PI controller is tuned by adjusting the gain or proportional band and integral time settings. A fuzzy logic controller is tuned by adjusting the scale factors.

The rules for the fuzzy logic replacement for a PI controller have two antecedents (conditions) and one consequent (result). The fuzzy inputs to the first and second antecedents are a scaled error ($E_s$), and a scaled change in error ($\Delta E_s$), respectively. The fuzzification of the PI controller inputs, which is the controlled variable (CV) and the setpoint (SP), to provide these fuzzy inputs with values that range from -1 to +1 is computed from the scale factors for error ($S_E$) and change in error ($S_{\Delta E}$).

$$E_s = (CV_n - SP_n) \ / \ S_E \qquad (13\text{-}1)$$

$$\Delta E_s = [(CV_E - SP_n) - (CV_{n-1} - SP_{n-1})] \ / \ SD_E \qquad (13\text{-}2)$$

To better understand how the algorithm works, consider an error that is -80% of the error scale ($E = -0.8*S_E$) and a change in error that is -20% of the change in error scale ($\Delta E = -0.2*DS_E$).

The value of the "positive" membership set for the error ($P_E$) and the "negative" membership set for the error ($N_E$) is the intersection of the scaled error input with the respective right triangle in Figure 13-1 and can be computed as follows for each antecedent:

$$P_E = 0.5 + 0.5*E_s \qquad (13\text{-}3)$$

$$P_E = 0.5 + 0.5*(-0.8) = 0.1$$

$$N_E = 0.5 - 0.5*E_s \qquad (13\text{-}4)$$

$$N_E = 0.5 - 0.5*(-0.8) = 0.9$$

The value of the "positive" membership set for the change in error ($P_{\Delta E}$) and the "negative" membership set for the change in error ($N_{\Delta E}$) is the intersection of the scaled change in error input with the respective right triangle in Figure 13-2 and can be computed as follows for antecedent:

$$P_{\Delta E} = 0.5 + 0.5*\Delta E_s \qquad (13\text{-}5)$$

$$P_{\Delta E} = 0.5 + 0.5*(-0.2) = 0.4$$

$$N_{\Delta E} = 0.5 - 0.5*\Delta E_s \qquad (13\text{-}6)$$

$$N_{\Delta E} = 0.5 - 0.5*(-0.2) = 0.6$$

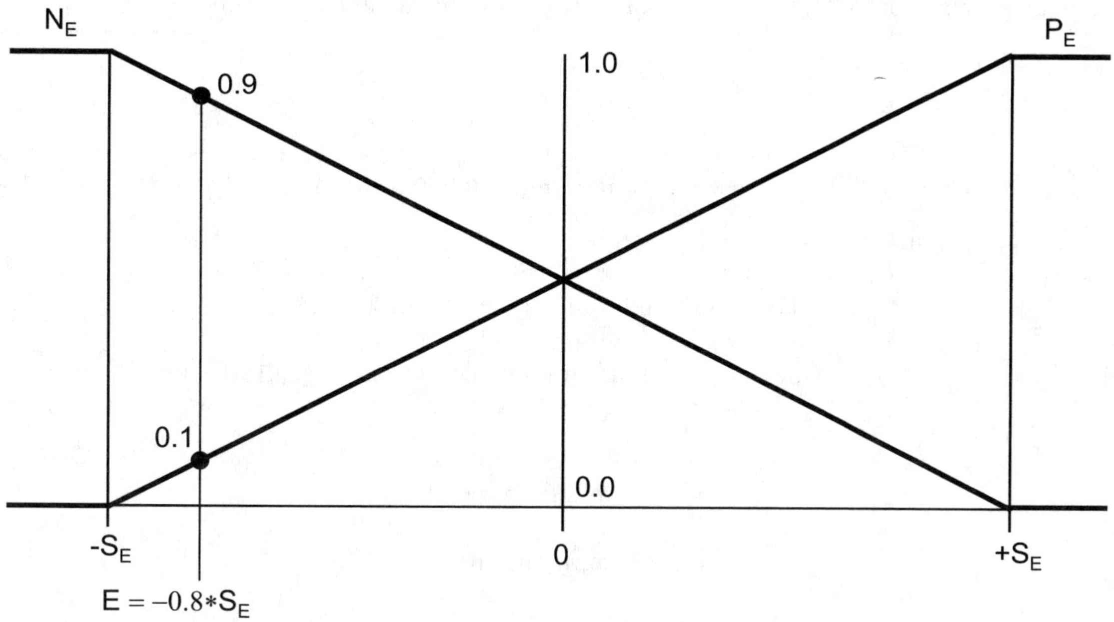


*Figure 13-1: FLC Antecedent Membership for Error*

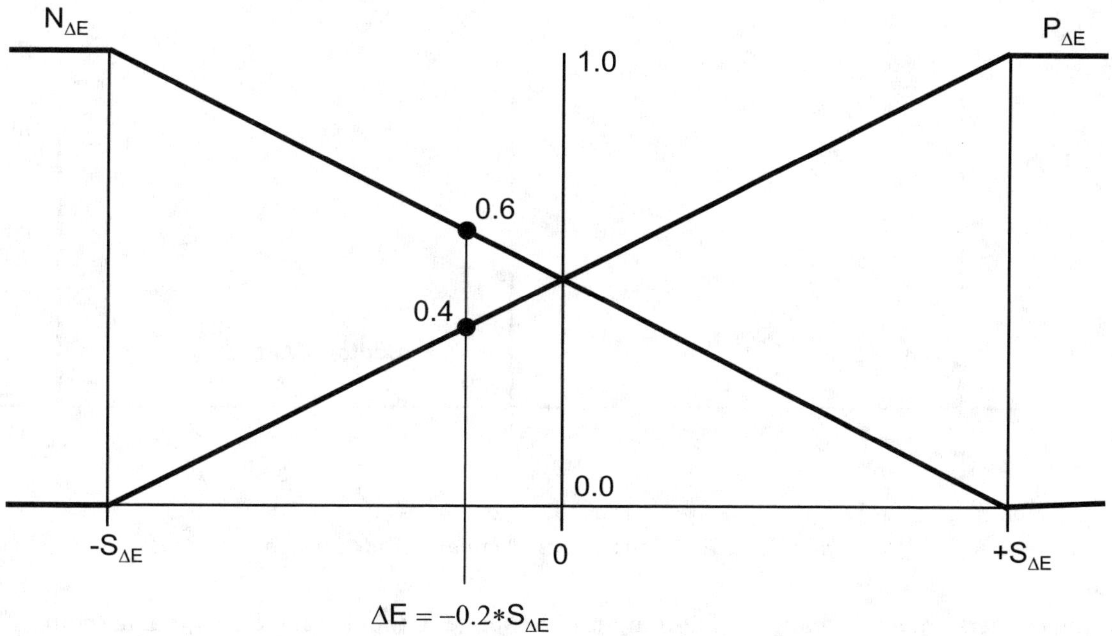


*Figure 13-2: FLC Antecedent Membership for Change*

The FLC computes a change in output that becomes the change in the manipulated variable ($\Delta MV$) for feedback control. The value of the "positive" membership set for the change in output ($P_{\Delta MV}$), the "negative" membership set for the change in output ($N_{\Delta MV}$), and the "zero" membership set for the change in output ($Z_{\Delta MV}$) are the height of the bars in Figure 13-3 and can be computed as follows for a reverse acting controller by simply taking the minimum value of each antecedent:

Rule 1:

$$P_{\Delta MV} = \text{minimum}\ (N_E, N_{\Delta E}) \tag{13-7}$$

$$P_{\Delta MV} = \text{minimum}\ (0.9, 0.6) = 0.6$$

Rules 2 and 3:

$$Z_{\Delta MV} = \text{maximum}\ (\text{minimum}\ (N_E, P_{\Delta E}), \text{minimum}\ (P_E, N_{\Delta E})) \tag{13-8}$$

$$Z_{\Delta MV} = \text{maximum}\ (\text{minimum}\ (0.9, 0.4), \text{minimum}\ (0.1, 0.6)) = 0.4$$

Rule 4:

$$N_{\Delta MV} = \text{minimum}\ (P_E, P_{\Delta E}) \tag{13-9}$$

$$N_{\Delta MV} = \text{minimum}\ (0.1, 0.4) = 0.1$$

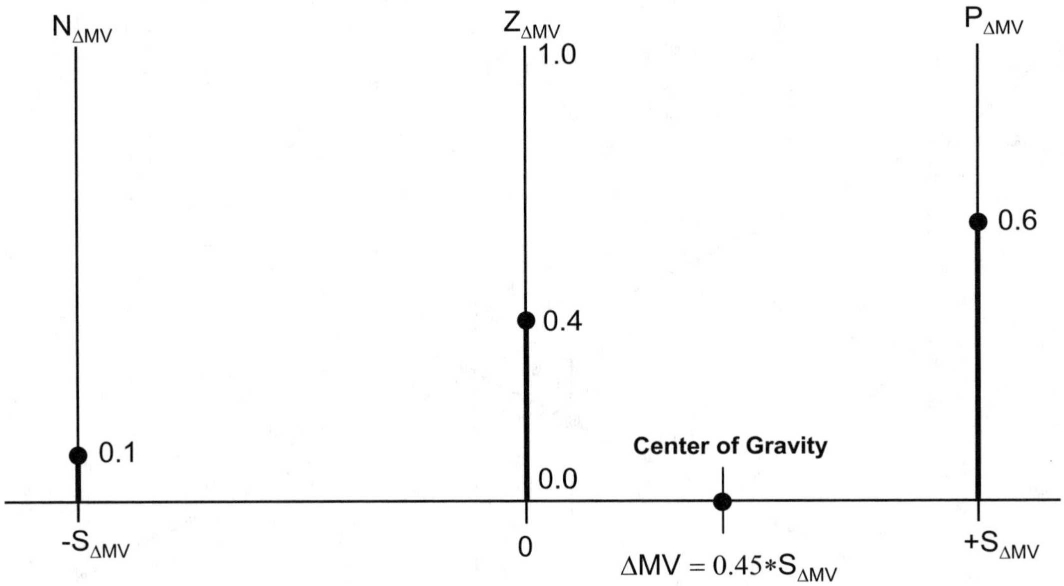


*Figure 13-3: FLC Consequent Membership for Change in Output*

The conversion of the change in FLC output to a change in the manipulated variable (defuzzification) is done by computing the center of gravity (centroid) from the location and height of the bar for each of the 3 membership sets for the output multiplied by the scale factor ($S_{\Delta MV}$).

$$\Delta MV = (P_{\Delta MV} - N_{\Delta MV})\ /\ (P_{\Delta MV} + Z_{\Delta MV} + N_{\Delta MV})\ *\ S_{\Delta MV} \tag{13-10}$$

$$\Delta MV = (0.6 - 0.1)\ /\ (0.6 + 0.4 + 0.1)\ *\ S_{\Delta MV} = 0.45\ *\ S_{\Delta MV}$$

The FLC scale factor for the error, change in error, and change in output can be computed from a process model, existing PI controller tuning settings, or the ultimate period and gain identified by an auto tuner for a particular range of dead time to time constant ratios. If the nominal size of setpoint changes is included in the calculations, the setpoint response can be improved. Derivative action on changes in the process variable as implemented in an industrial proportional-integral-derivative (PID) can be added to the FLC.

While this was an example of how fuzzy logic has been successfully deployed on a wide scale, application specific fuzzy logic algorithms have been developed to deal with variable and severe unknown nonlinearities, particularly as would be encountered in the pH control of plant waste.

## 13.3 Adaptive Control

The term adaptive control has been liberally used to describe any application where the controller tuning was changed. For example, error squared and notch gain controllers have been called adaptive controllers because the controller gain is changed as a function of error. Also, "on demand" auto tuners that calculate tuning settings from a manual or automated test sequence initiated by a person have been sometimes termed adaptive.

This section will focus on the adaptation of the tuning of the PID controller, where quantitative process models are continuously identified online and used to compute tuning settings. Figure 13-4 shows the general structure of these "self-tuning" controllers, where there are a process model, supervisor, and a set of tuning rules. Users may not be aware what is at work behind the scene because the model parameters and tuning rules may not be visible or accessible.

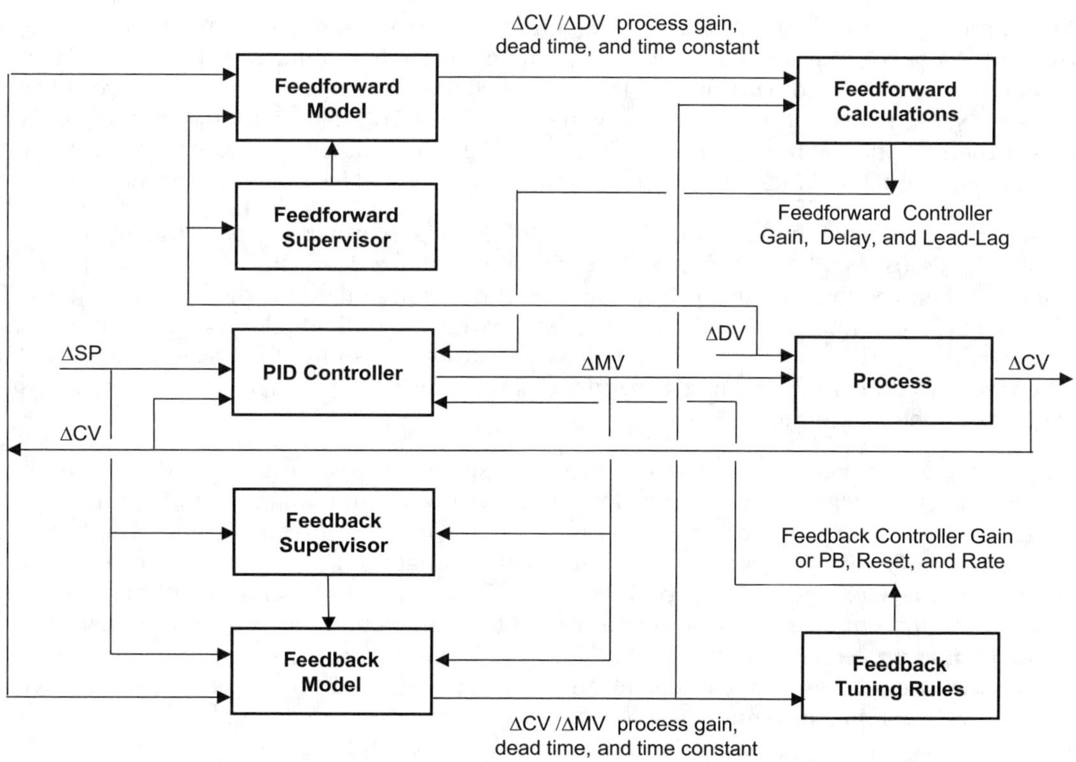


*Figure 13-4: Adaptive Feedback and Feedforward PID Controller*

The most widely applied self tuning controller in the process industry in the 1980s used a heuristic model based on pattern recognition. The controller gain was increased until the loop oscillates. The height and time interval of peaks were used to compute the overshoot, period, and damping (decay) of the oscillation. The Ziegler-Nichols type tuning rules were then applied but with the ratios of the reset and derivative time to the period adjusted by the algorithm.

A new generation of adaptive controllers has been developed that identifies parameters that minimize the squared error between the response of the model and the process to known changes in the set-point or output of the controller. The algorithm uses model switching (fastest known adaptation method) with parameter interpolation and re-centering (proven optimal approach) to provide the best value for the process gain, dead time, and the time constant. The model parameters can be historized and stored into user defined operating regions as a function of any process or manipulated variable. The stored model parameters can be used to provide a preemptive scheduling of the tuning settings when the controller enters a region so that controller does not have to wait for an excitation and adaptation. Tuning rules (Lambda, Internal Model Control, and modified Ziegler-Nichols), the degree of performance versus robustness, modes (monitoring, learning, scheduling, and full adaptive), and triggers can be set by the user.

The same identification procedure used to find the model for the process response to changes in the manipulated variable for feedback control can be used to find the model for the process response to measured disturbances. The feedback and feedforward process models are then used to compute the feedforward gain to give the correct magnitude and the feedforward delay and lead-lag time to provide the correct timing of the correction. These models can also be historized and stored for preemptive scheduling of feedforward settings.

## 13.4 Model Predictive Control

Model predictive control (MPC) uses incremental models of the process where the change in a controlled (CV) or constraint variable (AV) for a change in the manipulated (MV) or disturbance variable (DV) is predicted. The initial values of the controlled, constraint, manipulated, and disturbance variables are set to match those in the plant when the MPC is in the initialization mode. The MPC can be run in manual and the trajectories monitored. While in manual, a back calculated signal from the downstream function blocks is used so the manipulated variables track the appropriate setpoints.

A simplified review of the functionality of the MPC helps to provide a better understanding important for a later discussion of its capabilities and limitations. Figure 13-5 shows the response of a change in the controlled variable to a step change in each of two manipulated variables at time zero. If the step change in the manipulated variables was twice as large, the individual responses of the controlled variable would be predicted to be twice as large. The bottom plot shows the linear combination of the two responses. Nonlinearities and interdependencies can cause this principal of linear superposition of responses to be inaccurate.

Any errors in the modeled versus actual process response shows up as an error between the predicted value and actual valve of the controlled variable as shown in the upper plot of Figure 13-6. A portion of this error is then used to bias the process vector as shown in the middle plot. The MPC algorithm then calculates a series of moves in the manipulated variables that will provide a control vector that is the mirror image of the process vector about the setpoint as shown in the bottom plot of Figure 13-6. If there are no nonlinearities, load upsets, or model mismatch, the predicted response and its mirror image should cancel out with the controlled variable ending up at its setpoint. How quickly the controlled variable reaches setpoint depends upon the process dead time, time constant, move suppression, move size limit, and number of moves set for the manipulated variables.

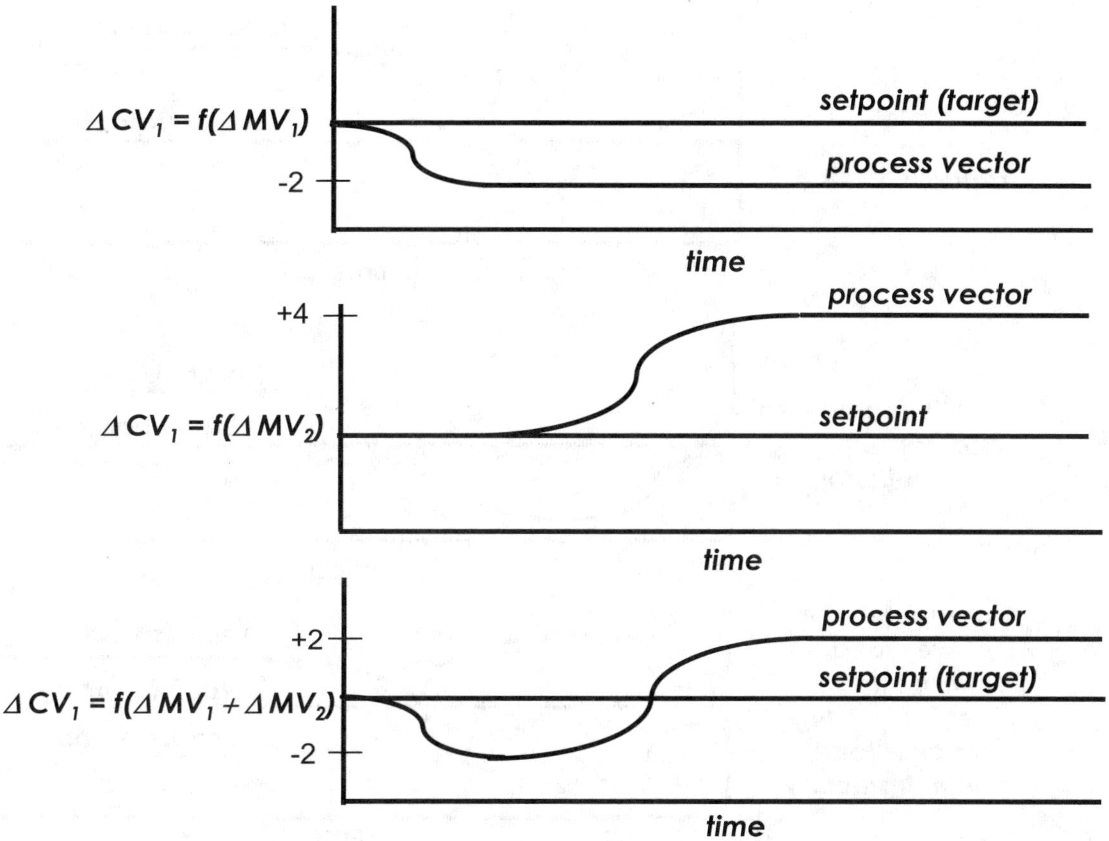


*Figure 13-5: Linear Superposition of MPC Response*

The first move in the manipulated variable is actually the linear summation of all the first moves based on controlled, disturbance, and constraint variables. Only the first move is executed because the whole algorithm is revaluated in the next controller execution.

As an example of an emerging MPC application, consider a feed-batch reactor. The process objectives in this case are to minimize batch cycle time and losses of reactant and product in the overhead system. Online first principal estimators are first developed and commissioned to provide online concentrations of the reactant in the overheads and product in the reactor that match periodic lab samples. The MPC setup uses these concentrations as controlled variables, the condenser temperature setpoint and reactant ratio as manipulated variables, the reactor jacket and condenser coolant valve positions as constraint variables, and the reactor temperature and feed flow as optimization variables. The reactor temperature and reactant feed rate is maximized to reduce the batch cycle time, if the projected coolant valve positions are below their high limit. The MPC setup is shown in Figure 13-7 with the relative trajectories for each pairing of a controlled and constraint variable with a manipulated variable.

MPC Models must be able to provide a reasonably accurate time response of the change in each process output (controlled or constraint variable) for a change in each process input (manipulated or disturbance variable). Any control loops that use these MPC variables must be in manual while the process is tested, otherwise it is difficult to impossible to separate the response of the controller algorithms and tuning from the process. The smallest size MPC should be sought that meets the process objectives to minimize the number of loops in manual and the test time.

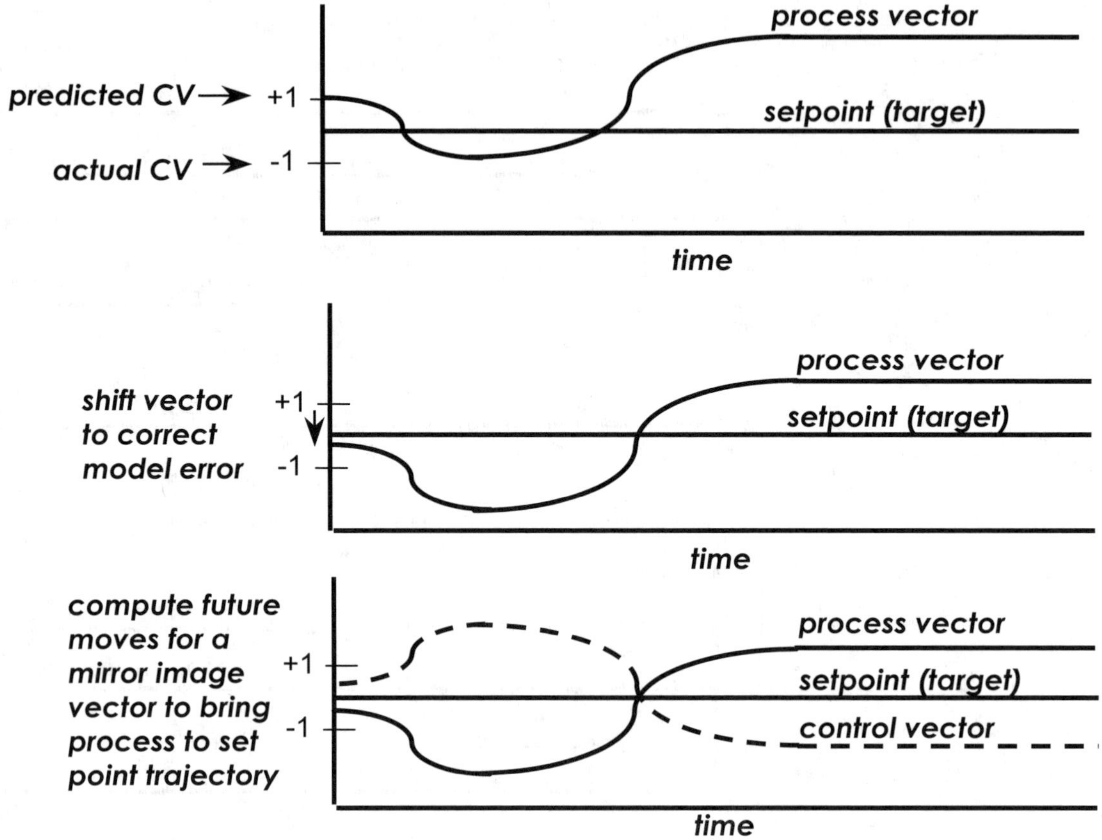


*Figure 13-6: Shift of Process Vector and Mirror Image Control*

The first process test makes a simple step change in each MV and DV and is commonly known as a bump test. This initial test provides an estimate of the model parameters as well as an evaluation of the step size and time to steady state. The parameters provide an estimate of the condition number of the matrix that is critical for evaluation of the variables selected.

The next test is either a more numerous series of steps each held for the longest time to steady state or a pseudo random binary sequence (PRBS). The PRBS test is favored because it excites more of the frequencies of the process and when combined with a noise model can eliminate the effects of noise and load upsets. In a PRBS test, each subsequent step is in opposite directions like the bump test, but the time between successive steps is random (a coin toss). However, one or more of the steps must be held to steady state and the minimum time between steps, called the flip time, must be larger than a fraction of the time lag. Theoretically the flip time could be as small as 1/8 of the time lag, but in practice it has been found that industrial processes generally require flip times larger than 1/2 of the time lag.

The number of flips and consequently the duration of the PRBS test are increased for processes with extensive noise and unmeasured upsets. A normal PRBS test time is about 10 times the longest time to steady state multiplied by the number of process inputs. For example, a typical PRBS test time would be four hours for a process with four manipulated variables and a maximum $T_{98}$ of six minutes. For distillation columns, the test can easily span several shifts and is susceptible to interruption by abnormal operation. While the PRBS test can theoretically make moves in all of the manipulated variables, PRBS tests are often broken up into individual tests for each manipulated variable to reduce the risk

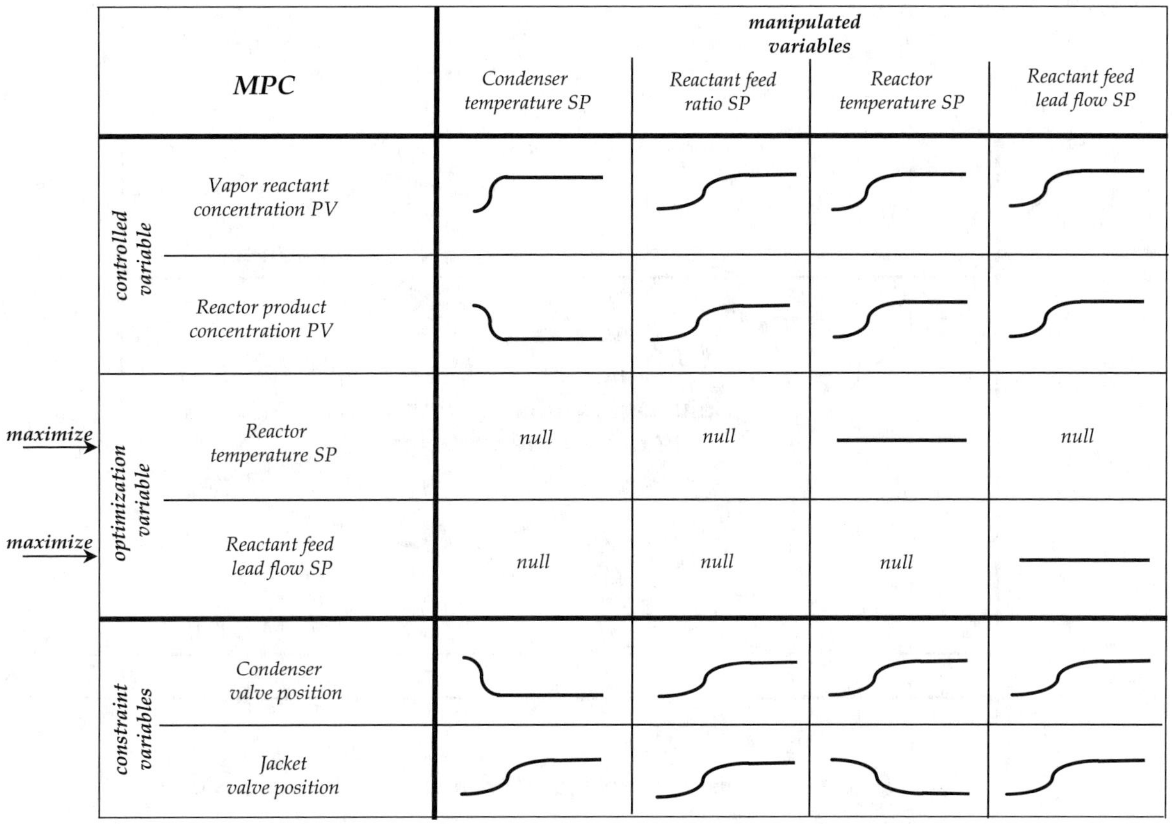


*Figure 13-7: MPC Setup for a Batch Reactor*

from an interruption and to make the identification process easier. If more than one manipulated variable is moved, it is important that the moves not be correlated. The PRBS sequence is designed to provide random step durations and uncorrelated moves to insure the software will identify the process rather than the operator.

Sometimes even the most sophisticated software gets confused and can cause gross errors in parameters and even a response in the wrong direction. The engineer should estimate the process gain, time delay, and time lag from the simple bump test and verify that the model direction and parameter estimates are consistent with these observations and process fundamentals. The manually estimated values can be used when the software has failed to find a model that fits the data. The rule is "if you can see the model, and it makes sense, use it."

## 13.5 Real Time Optimization

If a simple maximization of feed or minimization of utility or reagent flow is needed, a controlled variable that is the setpoint of the manipulated flow is added to the matrix. The setpoint of the flow loop is ramped towards its limit until there is a projected violation of a constraint or an excessive error of a controlled variable. While the primary implementation has been for flow, it can also used for the maximization of other manipulated variables.

It is important to realize that the optimum always lies at the intersection of constraints. This can be best visualized by looking at the plot of the lines for the minimum and maximum values of controlled, constraint, and manipulated variables plotted on a common axis of the manipulated variables as shown in Figure 13-8.

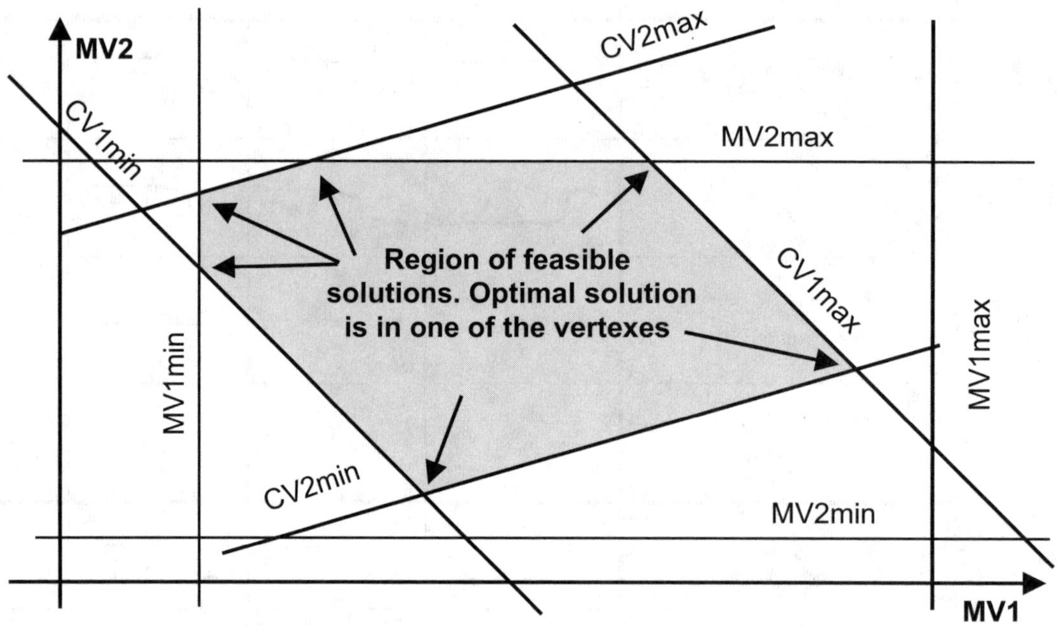


*Figure 13-8: Linear Program Optimization of a 2 Dimensional System*

If the best targets for controlled variables have a fixed value or if the same intersection is always the optimum one in Figure 13-8, the targets can be manually set based on the process knowledge gained from development and operation of the MPC and RTO. In this case the linear program (LP) or RTO can be run in the advisory mode. If the optimum targets of controlled variables move to different intersections based on costs, price, or product mix, then a LP can continuously find the new targets.

If the lines for the variables plotted in Figure 13-8 shift from changes in process or equipment conditions and cause the location of intersections to vary, then a high fidelity process model is needed to find the optimum. Steady state simulations are run for continuous processes. The model is first reconciled to better match the plant by solving for model parameters, such as heat transfer coefficients and column tray efficiencies. The model is then converged to find the optimums. This procedure is repeated several times and the results are averaged when the plant is not at a steady state. For batch operations and where kinetics and unsteady operation is important, dynamic models whose model parameters have been adapted by the use of an MPC and a virtual plant as shown in Figure 13-9 can be run faster than real time to find optimums.

Any controlled variable whose setpoint should be optimized is a prime candidate for an MPC, because the MPC excels at responding to setpoint changes and handling constraints and interactions. The best results of real time optimization are achieved in multivariable control when the setpoints are sent to an MPC rather than PID controllers.

## 13.6 Capabilities and Limitations

An FLC can provide a nonlinear feedback correction of disturbances and setpoint changes where the gain and reset action increases as the feedback control error get larger. If the scale factors have been

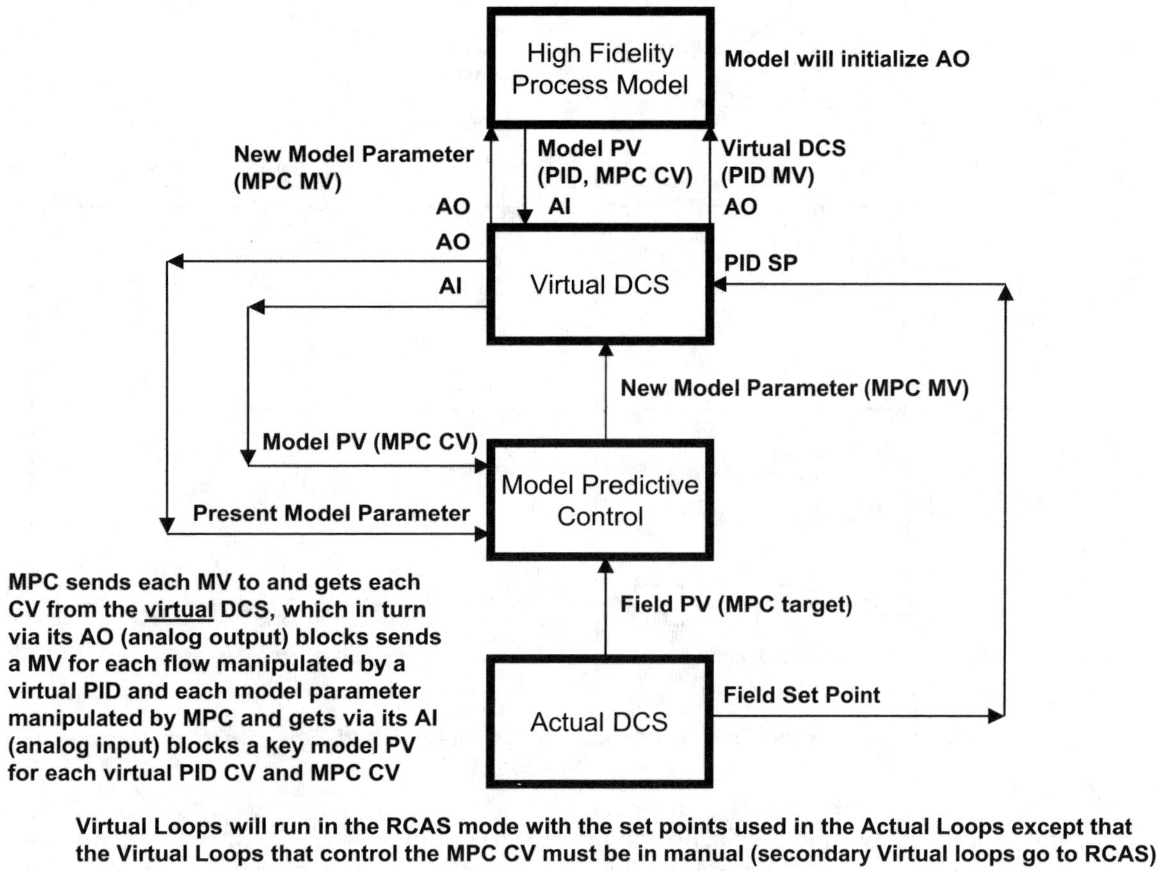


*Figure 13-9: Non-intrusive Adaptation of a High Fidelity Process Model by an MPC*

adjusted for a range of low dead time to time constant ratios and the size of a typical change in setpoint, an FLC has been shown to provide both a better load rejection and setpoint response than a PID controller for a process with a large time constant. The improvements have been most noteworthy in temperature loops, which have a large thermal lag. For this reason, manufacturers of extruder temperature control systems offer an FLC.

The order of execution of rules is not important for FLC and new rules can be readily added based on the decision logic from a proficient and knowledgeable process engineer. However, the scaling factors and the performance of a heuristic FLC algorithm are difficult to compute a priori, which makes the behavior of a special purpose FLC unpredictable. Thorough testing for abnormal situations and unmeasured upsets in an adapted virtual plant can alleviate these concerns.

All model based control is based on a simplification of the process dynamics. Figure 13-10 of the block diagram of a loop and a more realistic understanding of model parameters reveals the potential problems in the implementation of adaptive and model predictive control.

The process gain is really an open loop or static gain that is the product of the gains associated with the manipulated variable, the process variable, and controlled variable. For a manipulated variable that is a control valve, the gain is the slope of the installed characteristic at a given valve position for a perfect valve. Valve deadband from backlash and resolution from stiction reduces the gain to a degree that is dependent upon the direction and size of the change in controller output. A slip larger than the stick

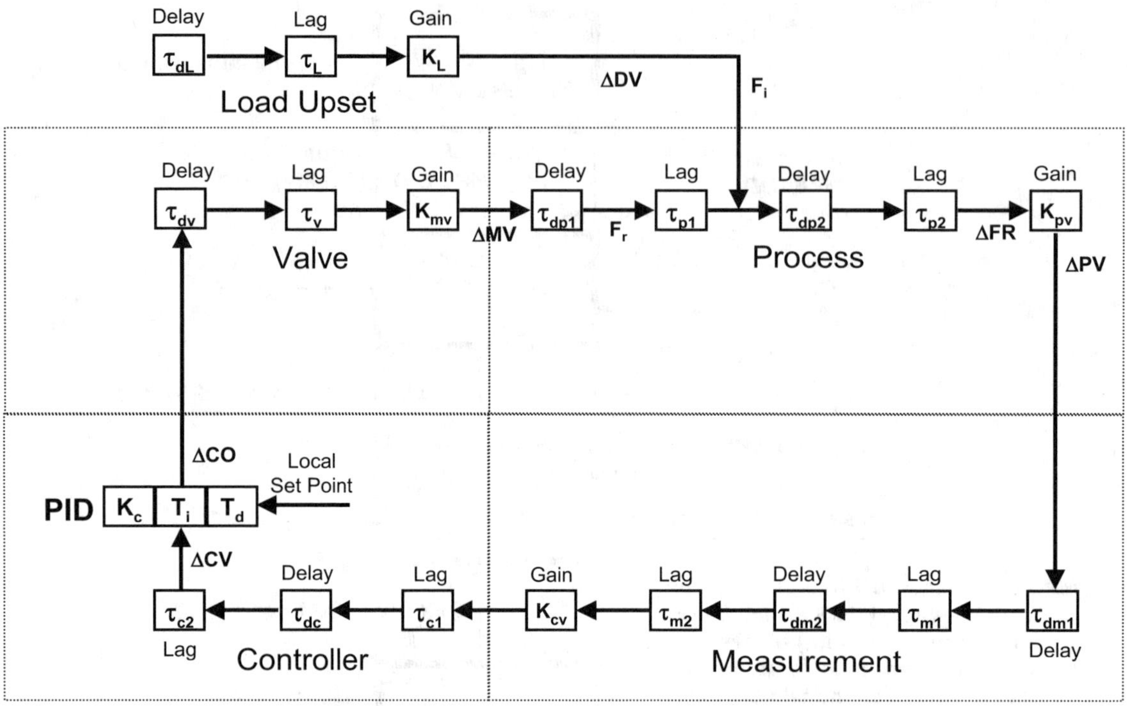


*Figure 13-10: Block Diagram of the Gains, Dead Times (Delays), and Time Constants (Lags) in a Control Loop*

will increase the valve gain. Deadband will result in a continuous cycle in integrating processes, such as level, and in cascade control systems where both controllers have integral action. Stick-slip will cause a limit cycle in all control systems. Excessive deadband and stick-slip has been the primary cause of the failure of adaptive controllers. For the more important process variables, such as temperature and composition, the process gain is a nonlinear function of the ratio of the manipulated flow to the feed flow and is thus inversely proportional to feed flow. For all controller algorithms, such as the PID with a percent input, the controlled variable gain is inversely proportional to measurement calibration span.

The process time constant is essentially the largest time constant in the loop; it does not necessarily have to be in the process but can be in the valve, measurement, and controller beside the process. Control valve time constants from large actuators are extremely difficult to predict and depend upon the size of the change in controller output. Process time constants may be interactive lags that depend upon differences in temperature and composition or are derived from residence times that are inversely proportional to throughput. Temperature sensor and electrode time constants are greatly affected by velocities, process conditions, sensor location, and sensor construction.

The process dead time is really a total loop dead time that is the sum of all the pure delays from the pre-stroke dead time of the actuator, the dead time from valve backlash and sticktion, process and sensor transportation delays that are inversely proportional to flow, unpredictable process delays from non-ideal mixing, and the execution time of digital devices and algorithms. All time constants smaller than the largest time constant add an equivalent dead time as a portion of the small time constant that gets larger as its size get smaller compared to the largest time constant.

Adaptive controllers require that the process be excited by a known change to identify the process gain, dead time, and time constant. When the controller is in manual, changes in valve position made by the operator trigger the identification of the process model. When the controller is in automatic, small pulses automatically injected into the controller output or changes in the setpoint initiate the identification. The models reportedly developed from closed loop control operation without any known excitation is really a combination of the process and controller. Studies have shown that it is not feasible to reliably extract the complete process model from the combined model of the controller and process for unknown disturbances and quiescent operation.

Unless the process exhibits or mimics an integrating or runaway response, an adaptive controller whose model is identified from excitations will wait for the time to steady state to find the process gain, dead time, and time constant. For a self-regulating process, which is a process that will go to steady state, the time to reach steady state is the total loop dead time plus four time constants. For processes with a large time constant, the time required for adaptation is slow. In the case where the time constant is much larger than the dead time, specifying the process as a pseudo integrator enables the identification to be completed in about four dead times, which could easily be an order of magnitude faster. This is particularly important for batch operations since there is often no steady state.

Pattern recognition controllers must wait for several damped oscillations, each of which is at least four or more times the total loop dead time. Thus, for processes where the dead time is very large, the identification is slow. Noise and limit cycles from non-ideal valves can lead to erroneous results. Integrators and runaway responses with windows of allowable gain, where too low besides too high of a controller gain causes oscillations, can cause a downward spiral in the gain.

Figure 13-11 shows how an MPC has a view of the trajectory of each controlled and constraint variable from changes in the manipulated and disturbance variables.

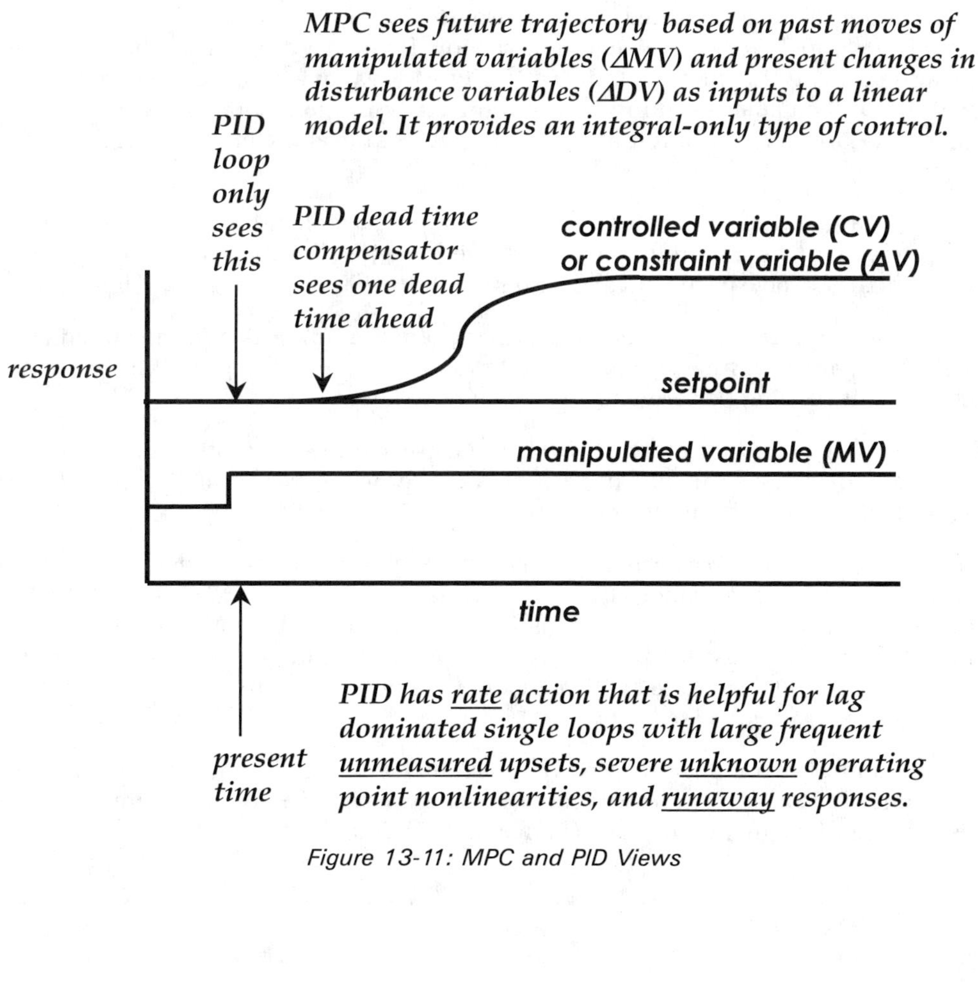


*Figure 13-11: MPC and PID Views*

In contrast, a PID controller only sees the current value and the rate of change of its controlled variable. The addition of a dead time compensator only extends its view of the future to the end of the dead time. However, besides the integral (reset) mode, it has a derivative (rate) mode that provides some anticipation based on the slope of response of the controlled variable and a proportional (gain) mode that result in immediate and abrupt action. Feedforward and decoupling can be added, but the addition of these signals to controller output is again based on current values, has no projected future effect, and is designed to achieve the goal of returning a single controlled variable back to its setpoint.

These fundamental differences between the MPC and the PID are a key to their relative advantages for applications. The MPC offers performance advantages to meet process objectives and deal with interactions. Since it also computes the trajectories of constrained variables and has built-in capabilities for move suppression and the maximization or minimization of a manipulated variable, it is well suited to multivariable control problems and optimization.

The PID algorithm assumes nothing about the future and is tuned to provide immediate action based on change and rate of change and a driving action via reset to eliminate offset. The PID offers performance advantages for runaway and nonlinear responses and unmeasured, and hence unknown, load disturbances where the degree and speed of the change in the process variable is the essential clue. The key to whether a loop should be left as a PID controller is the degree that the proportional and derivative mode is needed. For well tuned controllers with large gain settings (> 4) or rate settings (> 60 seconds), it may be inadvisable to move to MPC. Such settings are frequently seen in loops for tight column and reactor temperature, pressure, and level control. PID controllers thrive on the smooth and gradual response of a large time constant (low integrator gain) and can achieve unmeasured load disturbance rejection that is hard to duplicate.

The A/D chatter and resolution limit of large scale ranges of temperature inputs brought in through DCS cards rather than via dedicated smart transmitters severely reduces the amount of rate action that a PID can use without creating valve dither. The low frequency noise from the scatter of an analyzer reading also prohibits the full use of rate action. Process variable filters can help if judiciously set, based on the DCS module execution time and the analysis update time. An MPC is less sensitive to measurement noise and sensor resolution because it looking at the error over a time horizon and does not compute a derivative.

Valve deadband (backlash) and resolution (stick-slip) is a problem for both the PID and MPC. In the MPC, an increase in the minimum move size limit to be just less than the resolution will help reduce the dead time from valve deadband and resolution but will not eliminate the limit cycles.

In general, there is a tradeoff between performance (minimum peak and integrated error in the controlled variable) and robustness (maximum allowable unknown change in the process gain, dead time, or time constant). Higher performance corresponds to lower robustness.

An increase in the process dead time of 50% can cause damped oscillations in a typically tuned PID, but a decrease in process dead time of 50% can cause growing oscillations in an MPC with default tuning that initially has a better performance than the PID.

An MPC is more sensitive to a decrease than an increase in dead time. A decrease in process dead time rapidly leads to growing oscillations that are much faster than the ultimate period. An increase in dead times shows up as much slower oscillations with a superimposed high frequency limit cycle. A PID goes unstable for an increase in process dead time. A decrease in process dead time for a PID just translates to lost opportunity associated with a greater than optimal controller reset time and a smaller than optimal controller gain increased.

For a single controlled and manipulated variable, model predictive control shows the greatest improvement over PID for a process where the dead time is larger than the time constant. However, MPC is more sensitive to an unknown change in dead time.

For measured disturbances, the MPC generally has a better dynamic disturbance model than a PID controller with feedforward control, primarily because of the difficulty in properly identifying the feedforward lead-lag times. Often the feedforward dynamic compensation for PID controllers is omitted or tuned by trial and error.

For constraints, the MPC anticipates a future violation by looking at the final value of a trajectory versus the limit. MPC can simultaneously handle multiple constraints. PID override controllers, however, handle constraints one at a time through the low or high signal selection of PID controller outputs.

For interactions, the MPC is much better than PID controller. The addition of decoupling to a PID is generally just based on steady state gains. However, the benefits of the MPC over detuned or decoupled PID controllers deteriorate as the condition number of the matrix increases.

The steady state gains in the 2x2 matrix in Equation 13-11 show that each manipulated variable has the same effect on the controlled variables. The inputs to the process are linearly related. The determinant is nearly zero and provides a warning that MPC is not a viable solution.

$$\begin{bmatrix} \Delta CV_1 \\ \Delta CV_2 \end{bmatrix} = \begin{bmatrix} 4.1 & 6.0 \\ 4.4 & 6.2 \end{bmatrix} * \begin{bmatrix} \Delta MV_1 \\ \Delta MV_2 \end{bmatrix} \tag{13-11}$$

The steady state gains of a controlled variable for each manipulated variable in Equation 13-12 are not equal but exhibit a ratio. The outputs of the process are linearly related. Such systems are called *stiff* because the controlled variables move together. The system lacks the flexibility to move them independently to achieve their respective setpoints. Again, the determinant is nearly zero and provides a warning that MPC is not a viable solution.

$$\begin{bmatrix} \Delta CV_1 \\ \Delta CV_2 \end{bmatrix} = \begin{bmatrix} 4.1 & 6.0 \\ 2.2 & 3.1 \end{bmatrix} * \begin{bmatrix} \Delta MV_1 \\ \Delta MV_2 \end{bmatrix} \tag{13-12}$$

The steady state gains for the first manipulated variable ($MV_1$) are several orders of magnitude larger than for the second manipulated variable ($MV_2$) in Equation 13-13. Essentially there is just one manipulated variable $MV_1$ since the effect of $MV_2$ is negligible in comparison. Unfortunately, the determinant is 0.9, which is far enough above zero to provide a false sense of security. The condition number of the matrix provides a more universal indication of a potential problem than either the determinant or relative gain matrix. A higher condition number indicates a greater problem. For Equation 13-13, the condition number exceeds 10,000.

$$\begin{bmatrix} \Delta CV_1 \\ \Delta CV_2 \end{bmatrix} = \begin{bmatrix} 1 & 0.001 \\ 100 & 1 \end{bmatrix} * \begin{bmatrix} \Delta MV_1 \\ \Delta MV_2 \end{bmatrix} \tag{13-13}$$

The condition number should be calculated by the software and reviewed before an MPC is commissioned. The matrix can be visually inspected for indications of possible MPC performance problems by looking for gains in a column with the same sign and size, gains that differ by an order of magnitude or more, and gains in a row that are a ratio of gains in another row. Very high process gains may cause the change in the MV to be too close to the dead band and resolution limits of a control valve and very low process gains may cause an MV to hit its output limit.

## 13.7 Costs and Benefits

The cost of an industrially proven FLC replacement for the PID is about the same as a PID. The cost of an adaptive controller varies from $2K to $20K per loop. The cost of MPC software varies from $10K to $100K, depending upon the number of manipulated variables. The cost of high fidelity process modeling software for real time optimization varies from $20K to $200K. The installed cost of FLC and an adaptive controller is similar to a PID controller. The installed cost of MPC and RTO varies from about 2 to 20 times the cost of the software depending upon the condition of the plant and the knowledge and complexity of the process and its disturbances.

Process tests and model identification reveal measurements that are missing or non-repeatable and control valves that are sloppy or improperly sized. Simple preliminary bump tests should be conducted to provide project estimates of the cost of upgrades and testing time.

Often a plant is running beyond nameplate capacity or at conditions and products never intended. An MPC or RTO applied to a plant that is continually rocked by unmeasured disturbances or where abnormal situations are the norm, require a huge amount of time for testing and commissioning.

The proper use of advanced control can reduce the variability in a key concentration or quality measurement. A reduction in variability is essential to the minimization of product that is downgraded, recycled, returned, or scrapped. Less obvious is the product given away in terms of extra purity or quantity in anticipation of variability. Other benefits from a reduction in variability often manifest themselves as a minimization of fuel, reactant, reagent, reflux, steam, coolant, recycle, or purge flow and a more optimum choice of setpoints. Significant benefits are derived from the improvements made to the basic regulatory control system identified during testing. New benefits in the area of abnormal situation management are being explored from monitoring the adaptive control models as indicators of changes in the instrumentation, valve, and equipment.

The benefits for MPC generally range from 1 to 4% of the cost of goods for continuous processes with an average of around 2%. The benefits of MPC for fed-batch processes are potentially 10 times larger because the manipulated variables are constant or sequenced despite varying conditions as the batch progresses. Other advanced control technologies average significantly less benefits. RTO has had the most spectacular failures but also the greatest future potential.

## 13.8 References

1. McMillan, Gregory K., Terrance L. Blevins, and Willy K. Wojsznis. "Advanced Control Smorgasbord." *Control*, Vol. 17 No. 5 (May 2004), pp. 41 – 48.

2. Blevins, Terrence, Gregory K. McMillan, Willy K. Wojsznis and Michael Brown. *Advanced Control Unleashed – Plant Performance Management for Optimum Benefits*. ISA, 2003.

3. McMillan, Gregory K., Mark S. Sowell, and Peter W. Wojsznis. "The Next Generation Adaptive Control Takes a Leap Forward." *Chemical Processing*, Vol. 67. No. 9 (September 2004).

4. Kane, Les A. (editor). *Advanced Process Control and Information Systems for the Process Industries*. Gulf Publishing, 1999.

## About the Author

**Gregory K. McMillan** is a retired Senior Fellow from Solutia Inc. and an ISA Fellow. McMillan received the ISA "Kermit Fischer Environmental" Award for pH control in 1991 and *Control* magazine's "Engineer of the Year" award for the process industry in 1994. Greg was inducted into *Control* magazine's "Process Automation Hall of Fame" in 2001 and honored as one of *InTech* magazine's most influential innovators in 2003. He received a B.S. from Kansas University in 1969 in Engineering Physics and a M.S. from University of Missouri – Rolla in 1976 in Electrical Engineering (Control Theory).

# 14 Control of Batch Processes

By Lynn W. Craig

## Topic Highlights



## 14.1 What Is a Batch Process?

A batch process is generally regarded as one in which a *discrete* amount of material is operated on (or processed) in a stepwise fashion. While the equipment can vary greatly, a batch chemical processing facility is usually comprised of vessels and other processing equipment tied together with an appropriate piping network. The piping network is generally segmented with valves, hoses, or other diversion equipment. Such forces as pumps, process pressure, or gravity power material movement. Energy is introduced or removed through equipment like agitators, heat exchangers, vessel jackets, etc.

Some batch processes are dedicated to a single product or family of products. Others are purposely flexible enough to allow a variety of different products to be processed. Some have fixed sets of processing equipment, while others optionally include or exclude one or more pieces of processing equipment. Some are routinely reconfigured by changing the piping network or by resetting the valves in a fixed piping network. There is considerable variety in batch processing, which is at once its strength and the key to its complexity.

### 14.1.1 Why Have Batch Processes?

The general view of the process industry has been heavily influenced by the efficiency and economics of large continuous chemical processing plants. The ability to process large amounts of material, to control and optimize the process, to minimize energy and other operating costs, to maximize productivity, and to precisely predict material requirements all make continuous processing options attractive. However, the need for batch or stepwise processing remains—in spite of the manifest advantages of continuous processing technologies.

Batch can be the technology of choice for several reasons. The amount of material to be manufactured may be small, tilting economics toward multi-purpose equipment that can make many different products. In other situations it is just better to use batch technology. Wine is fermented as a lot (in some cases a very large lot). Beans are washed, cut, cooked, and canned in sequential steps, but as one or more lots. Special colors of paint are produced one lot at a time. Time to market is generally less if products can be introduced in existing flexible equipment. There are a multitude of other forcing functions that can make batch methods attractive. In spite of the diversity of reasons, however, most of the forces at work can be traced to business need as opposed to technology.

### 14.1.2 Typical Batch Processing Configurations

Batch plants are typically classified by plant topology and by number of products manufactured. It is convenient to divide those plants into three general categories, according to type of processing activity. The simplest is one which repetitively goes through the same sequence the same way every time it is called on to make another batch of the single product it is designed to produce.

The second type of plant is more common. In this case, the same general procedures are followed, but parameters are varied to produce a family of products that differ but have common characteristics. Typical of this kind of plant would be a paint manufacturing operation in which many different colors or grades of paint can be produced using essentially the same sequence of operations.

The third type of plant is flexible and often reconfigurable. It can make many differing products; it is the most common, as well as the most complex, of the three types. It tends to use both formula information and a specific procedure to describe the processing that is unique to a product. The product manufacturing rules are usually defined as a recipe that specifies both the sequence of operations to be followed and the parameters that define materials, quantities, rates, temperatures, etc. There is no reason the procedure for one product has to resemble the procedure for another. This type of process is an order of magnitude more difficult to control than either of the other two—and, for many reasons, is the most usual type.

### 14.1.3 Batch versus Continuous Process Control

The fundamental difference between batch control and more traditional control of continuous processes is the need to control procedure in a batch process. Of course there is a startup and a shutdown procedure inherent in any continuous process, but that is usually done manually and more or less disappears from a control agenda. There, the process may spend more than 95% of its time in a steady state, and the startup and shutdown procedures are viewed as a trivial part of the overall operation. In batch, the procedure is active essentially 100% of the time the process is operating. It is hard to ignore.

In the past, manual control of procedure was a standard part of most batch manufacturing. Simple batch systems were occasionally "automated" with fixed sequences, implemented with programmable logic controllers (PLCs), drum programmers, or other fixed sequence devices. Systems that could deal with variable sequences for flexible processes were not broadly available until 1985 or later. In that time frame, several proprietary batch control products became available, but terminology, flexibility, and features varied widely. In spite of the interest, no broadly accepted approach to batch control emerged.

The first part of the ISA-88 Batch Control standard, named ISA-S88.01, was published in 1995 and changed all that. Later, it was adopted by the American National Standards Institute and renamed ANSI/ISA-88.01-1995. However, it is still known broadly as S88. It provides standard terminology, an internally consistent set of principles and a set of hierarchical models that can be applied to virtually any batch process. Actually, it can be (and has been) applied to many other manufacturing processes that require procedural control.

## 14.2 What Is the ANSI/ISA-88 Standard?

Commonly called S88, it is an ANSI standard in five parts. The first part, Models and Terminology, defines most of the important principles and is the subject of this discussion. However, two other application-oriented parts have been completed. They are Part 2: Data Structures and Guidelines for Language, and Part 3: General and Site Recipe Models and Representation. As of this writing (mid-2005), Part 4: Batch Production Record has not yet been published. Neither has Part 5: Recipe to Equipment Interface, which will follow Part 4. In addition to the ANSI/ISA version, the International Electrotechnical Commission (IEC) has published an international equivalent—IEC 61512-1.

The standard provides a common batch control language via models and terminology and defines the framework for hierarchical recipe management and process segmentation. It does not supplant traditional control but defines two or more layers of control and functionality above the traditional technologies focused on control of individual equipment. Traditional control has an almost total focus on loops and devices. Batch control is focused on the procedural and coordinating functions that set the states or target conditions of those loops and devices.

Batch control has another function that differs. While collection and transmission of manufacturing data has been a common feature of traditional control for many years, batch control makes a significant step toward connecting the entire process to the business. There was a significant change in the role of control when it became obvious that it is necessary to control procedure. The addition of procedure as a major type of control changed everything. All of a sudden it was no longer possible to ignore how to make the product. That meant the control functionality also had to "know" when to make what product. The disturbing aspect of this realization was that control engineers could no longer stay in a comfortable cave, concerned only with the control of single pieces or small groups of equipment, or regulating a few process variables at a time. In addition to automation and control, S88 describes a business process because automation works best with good connection of the process to the business. Like it or not, with the advent of ANSI/ISA-95.00.01-2000 - *Enterprise-Control System Integration* (commonly known as ISA S95) to complete the connection, control (automation) is no longer an isolated technology, but is in the business mainstream. It is possible to do control well or poorly but, if business requirements are not recognized properly in a batch process, poorly is certain.

### 14.2.1 Principles

The S88 standard set out to define, in a consistent way, the control functionality and associated terminology necessary to control a batch process. In the process, several principles emerged. It is worthwhile to review some of those principles before digging into the structure of batch control.

#### Principle #1

The first and most important principle is that *control is a function*. Functions happen—regardless of the mechanism that makes it happen. To understand control, what happens is initially more important than how it is made to happen. Control is a function that makes equipment do things. Control happens—it doesn't matter how. It can be manual. It can be automatic. It can be a little of both. Regardless of how it is implemented, it must exist in some form to make a product. Given that understanding, it is obvious that batch control is not new. It has been around for many years. It just didn't necessarily look like control because most of it was done manually. It is helpful, then, to understand the control functions that have been implemented with manual control, if automatic control is to accomplish some of the things a well-trained person can do.

#### Principle #2

*Separate the recipe from equipment control.* All batch processes have some sort of recipe that defines how a product is to be made. Many early attempts at batch automation included some or all of the recipe information as part of the sequential control that manipulated equipment. ISA S88 recognizes that equipment control can be designed to carry out process-oriented tasks and that the recipe can then designate which tasks need to be done, their order, and the values or parameters required for the task.

By separating the recipe information from equipment control, it is possible to let the recipe drive the process. This also means that, by changing only the recipe, many different products or versions of a product can be produced with no change in the equipment control. This enables major benefits in flexibility, cost, and reliability.

**Principle #3**
*Both the recipe and the process are made up of smaller pieces*. The process is made up of units. The recipe is made up of Unit Recipes. A Unit Recipe directs initiation of process-oriented tasks in a unit. The result is a view of control in which a process is made up of modules with dedicated procedural control for each module such that each module can independently do a number of process tasks—tasks that are usually not product-specific. This allows a separate recipe that is product-specific to set the order modules should execute— and provide the values to use when they execute.

**Principle #4**
*Equipment Modularity.* A process cell contains all of the equipment in a process, but recipes direct the units. Units are a major module but are made up of other subordinate modules called equipment modules and control modules. The result is a hierarchy of modules as shown in Figure 14-1, each made up of subordinate modules. Each level in the equipment hierarchy has an associated hierarchy of control functionality, also shown in Figure 14-1. Each level in the associated hierarchies has a different functional purpose.

**Principle #5**
*Equipment Entities.* It has been traditional to look at control and equipment separately, much as it is depicted in Figure 14-1. When most control was associated with single pieces or small groupings of

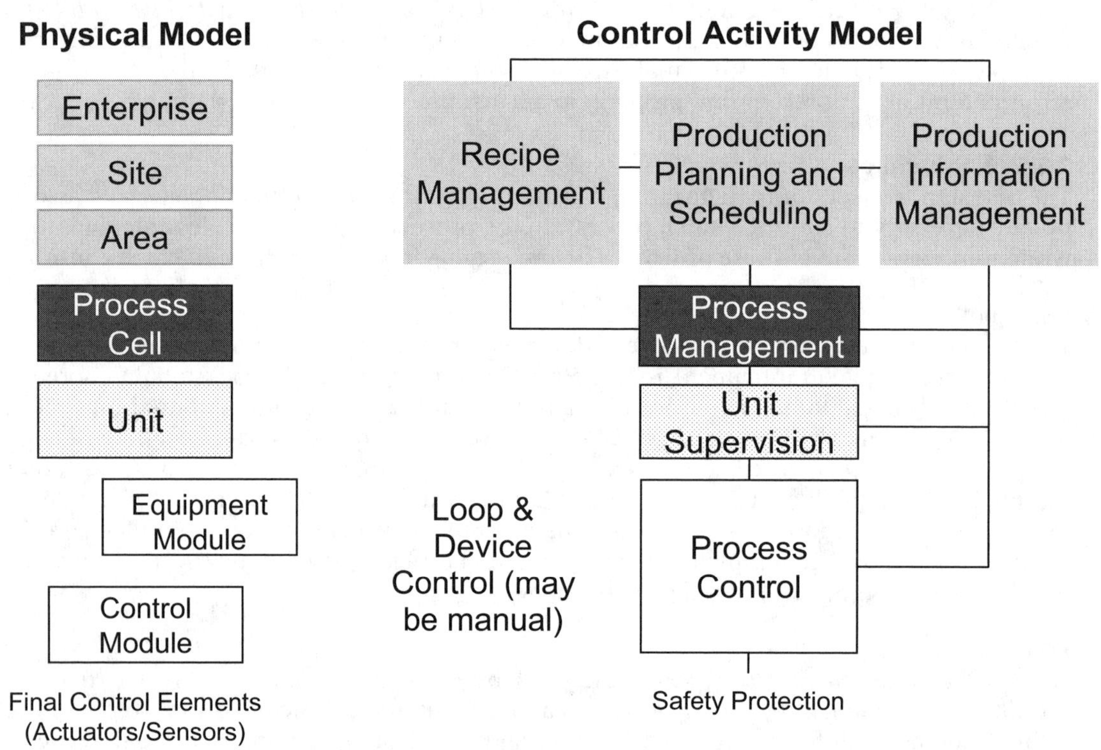


*Figure 14-1: Physical and Control Activity Model*

equipment, there was, in essence, a single level of automatic equipment control throughout a process. It was possible to think of a process as equipment that does the work and some separate control that sets or regulates final control elements, like valves and motors, to help get the work done and keep things stable. However, with hierarchies of equipment and control that require that each level in the hierarchy has unique function and procedural control that affects large groupings of equipment, it is no longer possible to look at control as a single kind of thing separate from the equipment. This led to the concept of the equipment entity. If one thinks of combining equipment and control into a single combined entity, then one can think of that combined entity as equipment that can be commanded, can carry out the commanded functionality, and can report on what it has done.

At each level in the equipment hierarchy, then, one can think of the defined equipment combined with all of the necessary control for that level as a single object that is capable of carrying out designated tasks or functions. We then have, as the standard defines, an equipment entity. This applies at each level in the hierarchy, so there is a hierarchy of equipment entities—a process cell entity, a unit entity, an equipment module entity, and a control module entity. (Please note this is a concept, not an implementation. The control may well be implemented manually or in a remote box, but the functionality that is destined to control a specific grouping of equipment is implemented in a way that allows the control to be mapped directly to the equipment grouping in question.)

An equipment entity at all levels has independent function and is stand-alone. It has equipment. It has function defined by that control regardless of how the control is implemented. It may be manual. It may be automatic. It may be some of both, but both control and equipment are needed to have an equipment entity.

### 14.2.2 The Modules

The entire ISA S88 standard is built on the presumption that any process can be broken down into the modules described. Actually defining the boundaries of modules in an optimally segmented process requires good understanding of both process and control issues, is most easily done with some experience, and is unfortunately beyond the scope of this overview. However, a good understanding of the modules in the hierarchy is the necessary first step. The hierarchy of equipment entities is shown in Figure 14-2.

#### The Process Cell

The *process cell* is the whole thing. It contains all equipment entities (modules) necessary to make one or more products. It is the interface to the business. ANSI/ISA-95.00.01-2000 - *Enterprise-Control System Integration* calls it a work center as a general term for any kind of manufacturing. When used specifically for batch, the term process cell is still the one to use. The process cell is made up mostly of units, along with their components parts. A process cell may have more than one batch active at a time and each of those batches may be following the same or different paths through the equipment. A batch does not need to use all the equipment in a process cell, and all batches of the same product do not need to follow the same path through the process cell. A process cell has many functions, but is mostly concerned with getting the proper unit recipe information to the proper unit at the proper time and keeping the batches and units sorted out into an orderly process.

#### The Subordinate Modules

The unit is the only grouping of equipment that can make something all by itself. It is the main S88 module for automatic control. Making a batch usually requires multiple units, although a batch can be completed in a single unit. It is made up of equipment modules and control modules. It is important to recognize that, while control modules are required, equipment modules are optional and, in many cases, may not be required at all. If used, equipment modules may be a permanent part of a unit, or a common resource that is temporarily attached to a unit and released for use by another unit when it has done its job. However, once a common resource is attached to a unit, it functions like any other permanent object in the unit.

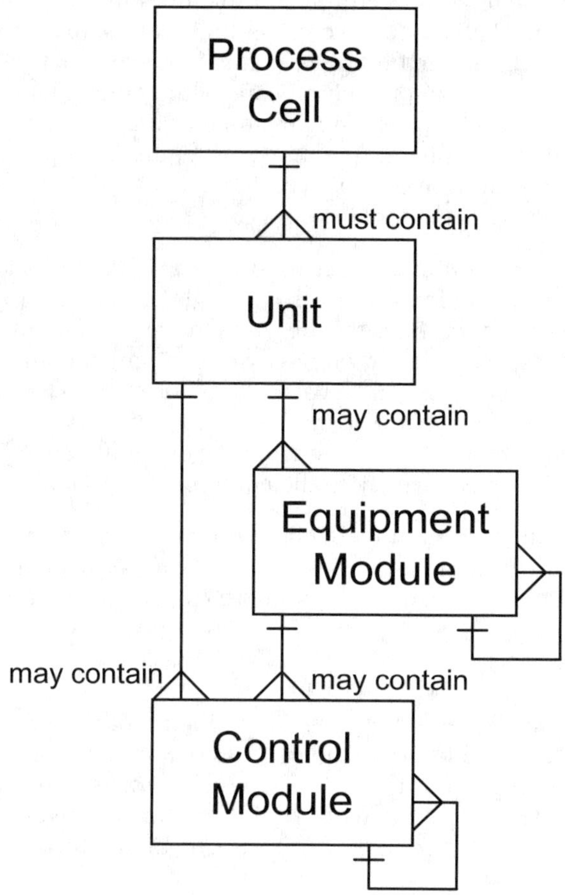


*Figure 14-2: Equipment Entities*

**The Unit**

The unit is the primary S88 module for automatic control. Nothing can be made without at least one unit. The primary function of a unit is to organize and execute procedural elements such as a unit procedure, operations, and phases. A unit procedure is made up of a hierarchy of operations and phases (see Figure 14-3). An operation is made up of an ordered set of phases. It executes or causes the execution of the phases to carry out process tasks. Only one batch at a time can be active in a unit. However, it can carry out several tasks at one time on that single batch because it can execute more than one phase at a time. The unit is the main module.

How can you tell if a grouping of equipment is a unit? If a grouping of equipment requires a variable sequence of procedures or tasks in order to operate, it is a unit. If not, most likely it is part of a unit. A unit is generally centered on a major piece of processing equipment, such as a mixing tank or reactor. It carries out product-related processing procedures such as combining material, initiating a chemical reaction, cooling a batch, etc. A unit can only operate on one batch of material, or portion of a single batch, at a time because it can execute only one unit procedure at a time. It gets a portion of a recipe (a unit recipe) that turns it into a mini process for making a single product.

**The Parts and Pieces of a Unit**

The unit is all the recipe cares about, but a unit is made up of smaller modules. Building a unit up from many smaller pieces makes all of the equipment modular so they can be combined like snap-together

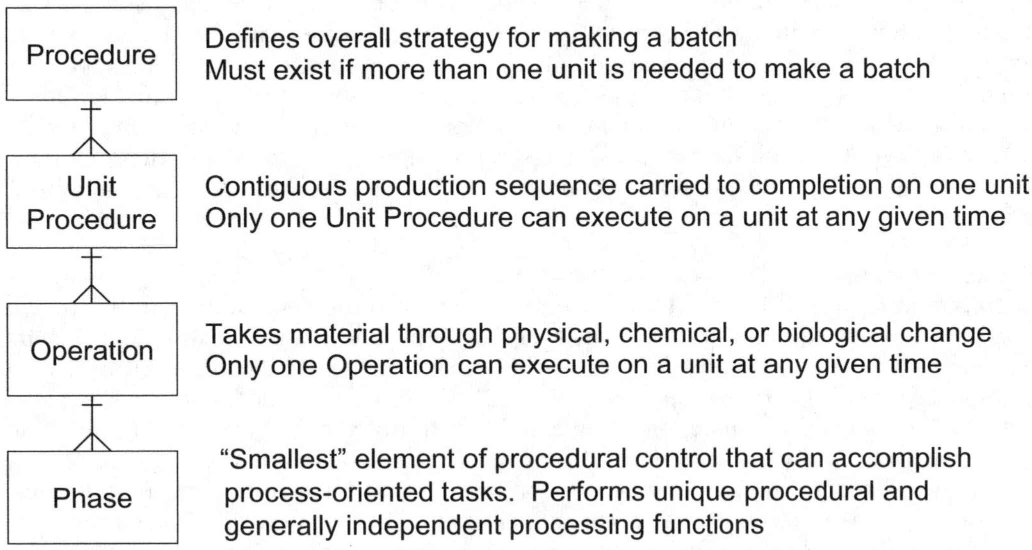


*Figure 14-3: Procedure Model*

blocks to "build" a unit. Having modular equipment entities not only makes it easier to define a unit, it also enables reusability. This adds enormous value and can result in lots of reusable work and much more uniformity of function. Subordinate parts of a unit are necessary because units are all about controlling procedure. Units tell their equipment modules, if they exist, when to execute their phase(s) and tell control modules what states or condition to set in what order. Control modules actually control equipment such as valves and control loops. The unit (or equipment module) commands a control module to set a condition. It is the job of the control module to set the condition and strive to maintain it. Control modules contain the physical actuators and sensors that manipulate equipment and measure process conditions.

### Control Modules

There is nothing new about control modules. The process control domain is where most traditional control has always been done. The only difference is that S88 treats the control module entity as a module so all of the basic control that goes on at that level can be encapsulated in reusable modules. Control modules directly connect to and contain process actuators and sensors as well as other simpler control modules. The simpler control modules are, in essence, the loops or devices that have long existed in traditional control. They are where essentially all basic control takes place—where the rubber meets the road. All elements encapsulated in a control module are treated as a single entity. The total purpose of a control module is to set a process state or condition and hold it, if at all possible, until commanded to change. In a loop configuration, the job is to set and maintain a process value. In a device configuration, the job is to set and maintain an equipment state. Control modules can contain other control modules. Whether they are simple or compound, their function is the same—they set a state and hold it. A typical example of a compound control module might be a valve header containing four valves that select a destination. Each of the valves could be a standard simple control module wrapped together in a control module that simply sets the proper valve to open and the others to close. The command to the overall control module might be the destination, and the conditions reported back could be "in transition" or "at state x" or "alarm."

Only a control module can directly manipulate a final control element. All other modules can affect equipment only by commanding one or more control modules. Every physical piece of equipment is controlled by one (and only one) control module. Sensors are treated differently. Regardless of which control module contains measurement instrumentation, all modules can share that information. Control modules do not execute phases, so they do not carry out process-oriented tasks. They are all about setting and holding a state or condition. However, a control module can take its equipment through a sequence of intermediate steps required in the transition from one commanded state to the next commanded state. Although it is a sequence, it is not a procedure. It is state transition logic necessary to the setting of a state or condition. In a typical situation, a control module is commanded to go to a state. It will take the equipment through steps required to get there. It will then maintain the commanded state, if at all possible, report its condition, and trigger an alarm if it cannot maintain the commanded state.

**Equipment Modules**

The equipment module is left until last because it is easier to understand once units and control modules are in mind. An equipment module is essentially a partition of a unit complete with its own phase(s) and its own control modules. This sort of module is required to allow a unit to acquire a common resource that can carry out a process-oriented task. Typical of this might be a raw material supply system that can deliver raw materials to several different units, but only one at a time. Since it can't belong exclusively to any one of the units, it must be the kind of animal that can be acquired, complete with its procedure (phase), and become part of the unit for a time. It can then be "told" to execute its phase and the amounts of a specific material to deliver to a specific destination; it can do that and then wait until it is told to deliver another charge, or is released. Once released, it belongs to no unit and can't contribute to the making of a batch until it is once again acquired. It is made up of control module entities and its own control, which is one or more phase(s). An equipment module must have at least one phase to function. If it has no phase, it is a control module. Equipment modules can also be configured as a permanent part of a unit.

Whether they are temporarily acquired, or are a permanent part of a unit, they exist to provide services to a unit. They do relatively simple, usually dedicated, procedural tasks using their own control modules. Since a unit can execute a phase, and contains control modules, the existence of a permanently configured equipment module is, in essence, optional. For that reason, many units may have no equipment modules at all and are still able to carry out the same process tasks. Their use as a permanently configured part of a unit is sometimes useful to provide partitions that prevent the phase that is not part of the equipment module from commanding control modules within the equipment module.

## 14.3 Recipe

Units carry out process-oriented tasks. The recipe turns the process into a machine to make a specific product by specifying which tasks are to be executed, the order in which they are to be executed, and the parameters or values to be used during that execution. The recipe is a little like the cartridge for a video game in that inserting a new cartridge totally changes the game even though no change at all was made to the control circuitry in the game controller. Recipes are easier to write than control code and are much more bulletproof in that safeguards can be built into the control code to keep really bad things from happening if a novice recipe author chooses to do something dumb. Usually, equipment control is the domain of control engineers, while process engineers and others familiar with the process, but not the control strategy, can write recipes.

ISA S88 defines a recipe with five parts:

*Header* — administrative information: recipe ID, description, author name, dates, source document references, revision history, etc.

*Procedure* — the sequence of operations and phases needed to make a product.

*Formula* — defines material inputs, expected outputs, and other parameters needed by the phases as they execute.

*Equipment Requirements* — defines equipment (or equipment class) needed to make the product.

*Other Information* — such product specific information as safety and regulatory information (does not contain global information like material safety data sheets, standard shipping requirements, etc.).

The standard also specifies four types of recipes:

*General Recipe* — the form of the processing plan that exists at the highest level of the organization or corporation. It is product and chemistry specific but defines the parameters and processing steps in a general and nonrestrictive way. Such items of information as batch size or specific selection of processing equipment are left for more focused versions of the recipe.

*Site Recipe* — a recipe version similar to the general recipe but changed to make it specific to an identified plant site. In some cases there will be no change in the recipe. In others, items of information will be added, such as local regulations, the language of choice for reports, and operator displays.

*Master Recipe* — a form of the recipe that has been made specific to either a process cell or a set of essentially identical process cells. Raw materials are designated by local code and quantities in engineering units. It is the template for a recipe that can be executed in the appropriate process cell.

*Control Recipe* — the executable recipe. It begins as a copy of the master recipe and is further modified to make it specific to a batch. Such information as batch ID and raw material lots may be identified at this level. If the control is manual, it might be called a "batch card" or something analogous. The difference between the control recipe and the master recipe is that options in the master recipe become absolute realities in the control recipe.

### 14.3.1 Schedule

In traditional control, schedules are something that happen somewhere in the operations department. In batch control, the schedule may come from the same place but is a vital part of control. Much as the recipe directs the tasks that must execute in order to make a product, the schedule directs the recipes that must execute in order to meet business requirements. It may also contain information that modifies the control recipe to make it specific to a single batch. The schedule is the primary business connection to the process and contains most of the information that ties the process to the business. It is very important. In a batch environment it has major impact on how well the process runs. A well-crafted schedule can directly and significantly improve quality, cost, throughput, inventory and other production metrics. A poorly crafted schedule can swing those metrics as much as 50% in the wrong direction.

### 14.3.2 Recipe Linkage

Recipes link to units through high-level control activities in the process cell. The procedural control model is split. Part of the procedure (the part for a product) is in the recipe. The rest is in the units (the product independent part). Although linkage can occur at any level in the procedure model (see Figure 14-3), the most usual level is at the phase level as illustrated in Figure 14-4. In that example, recipe phases link to equipment phases. The equipment phase is control that is embedded in equipment control and carries out a procedure. The control can be implemented as control code or can be implemented manually, but it is still an equipment phase carrying out a process-oriented task.

### 14.3.3 Tying It All Together

There is a logical sequence of events that takes place at the process cell level to start making a batch. First, the batch must be scheduled, and that information must be conveyed to the process management function that is part of the process cell. The process cell then requests a copy of the appropriate master recipe. It assigns a unique batch ID to the recipe copy, turning it into a control recipe. It then

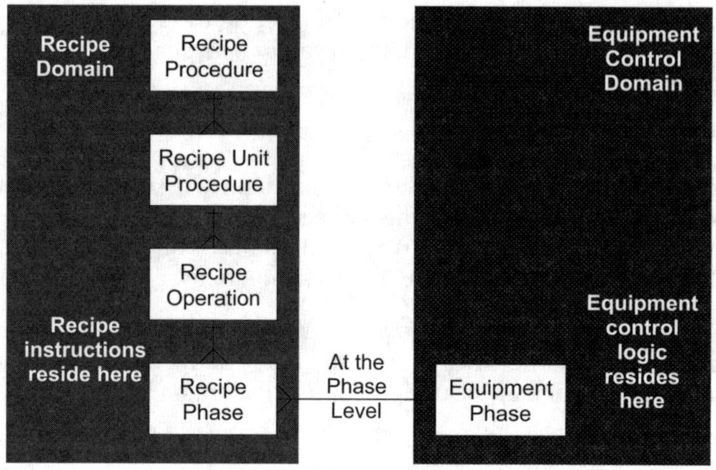


*Figure 14-4: Recipe Procedure to Equipment Procedure Linkage*

makes any further changes to the control recipe needed to make it specific for the batch to be made. It then separates the control recipe into unit recipes and, when the correct unit is available, hands the unit recipe off to that unit. At that point, the actual processing takes place in the unit. When the unit has totally finished the required processing, it becomes idle and available for the process cell to assign to another batch.

## 14.4 Summary

Automatic control of a batch process involves making the process operate the way it should—automatically. That requires knowledge of how the process should be operated. As a result, the first step in batch automation requires fundamental understanding of exactly how it should be run—in excruciating detail, recognizing that business requirements may change and the way the process is operated may also have to change. This goes beyond the typical engineering control problem. Focusing on control loops and valves, etc., can't solve it. The solution must focus on required functionality, recognizing that batch control is connected to the "business" and must be flexible enough to meet business requirements.

Regulatory control is a technology that has been around since the 1980s. It helps people who operate a plant do a better job. It is a necessary component of today's processing plants and has become so ubiquitous that it is more a safety and hygiene issue than an optional technology. Batch control is a technology that relies on, and directs, regulatory control, and it defines the way a plant will operate. Batch control is automation.

## 14.5 References

1. ANSI/ISA-88 series of batch control standards.

2. Parshall, J.H. and L.B. Lamb. *Applying S88: Batch Control from a User's Perspective.* ISA, 2000.

3. Fleming, D.W. and V. Pillia. *S88 Implementation Guide: Strategic Automation for the Process Industries.* McGraw-Hill, 1999.

4. Fisher, T.F. and W. Hawkins. *Batch Control Systems.* 2$^{nd}$ Edition. ISA, Forthcoming.

5. WBF.org - "The Forum for Manufacturing and Automation Professionals" (formally World Batch Forum). A wealth of papers and tutorials on batch related topics.

6. www.batchcontrol.com - Technical information with a sense of humor.

7. http://www.batchcentre.tudelft.nl/ - A batch knowledge center maintained by Delft University in the Netherlands.

## About the Author

**Lynn W. Craig** is one of the world's foremost authorities on batch processing technologies. He is president of Manufacturing Automation Associates, Inc., a consulting firm specializing in batch processing methods widely used by food, pharmaceutical, and specialty chemical companies. He is past chairman of the World Batch Forum, chairman of the ISA SP88 Batch Control Committee, and serves as convener of an international standards group addressing the same subject. His honors include induction to *Control* magazine's Hall of Fame, listing by ISA's *InTech* magazine as one of the 50 most influential people in control, and the WBF (World Batch Forum) Thomas G. Fisher award. His career spans more than 40 years, primarily with Rohm and Haas Co.

# 15 Environmental

*By Ian Verhappen*

## Topic Highlights

*Risk Reduction*
*Economic Incentives*
*Building Controls*
*Environmental Control Issues*

## 15.1 Introduction

Environmental controls are installed for a variety of reasons and incentives. They are like the traditional "carrot and stick" approach to modify behavior. There are significant rewards (carrots) possible as a result of environmental controls, but there are also significant penalties (sticks) that can result from poor implementation of environmental controls.

## 15.2 Risk Reduction

Environmental controls are one way of limiting liability and penalties such as fines and, potentially, incarceration. Environmental monitoring such as a Continuous Emission Monitoring System (CEMS) and other environmental systems for monitoring controlled and fugitive emissions from a facility are the most common way of ensuring a facility is in compliance with the associated operating licenses and regulations. In the United States, it is the responsibility of the Environmental Protection Agency (EPA) to mandate, through the appropriate Code of Federal Regulations (CFR), the rules and regulations regarding acceptable limits and measurement methods for a variety of substances considered harmful to the environment. In many cases, these operating limits form part of the license to operate a facility, meaning it is possible that lack of compliance with the regulations stipulated in the license can result in the temporary or permanent closure of a facility.

Leak Detection and Repair (LDAR) programs are another legislated tool used to minimize fugitive emissions from a facility. This program is not just a "stick" since it comes with the "carrot" that, if a facility consistently demonstrates low levels of fugitive emissions, the frequency at which the potential leak points need to be monitored can be reduced. A consistently "clean" environmental record not only reduces the potential of the "stick," but it is also the first of many "carrots." The broader community in which a company operates recognizes environmental efforts as those of a good corporate citizen adding value to a region through employment opportunities with negligible negative side effects.

## 15.3 Economic Incentives

Many companies have found that by tracking and minimizing their losses to the environment of a variety of substances, including their desired product, they are able to improve the operation of their facility. Obviously, every unit of product not lost to the environment results in increased revenue, because it is now being sold as product. In addition, facilities are finding that, by concentrating their effluent stream, it is possible to sell the resulting concentrate as a specialty product or, alternately, as a feed stock to a nearby facility to convert to a higher value product of their own.

The economics, or feasibility, of each recovery project can only be determined locally. However, what is "given" is that you cannot control what you cannot measure, so proper measurement and control of the process, and especially the stream constituents of interest, is critical. Fortunately, the analytical technology that is under development and becoming available continues to improve in reliability, repeatability, sensitivity, and accuracy.

What is commonly accepted and has been demonstrated is that converting a process loop from open to closed loop control results in a halving of the standard deviation of the system. This is demonstrated in Figure 15-1 that shows how a process will continue to have the same number of incidents when it will exceed the operating constraint but because the standard deviation is less, it is possible to move the mean, representing the process setpoint closer to that limit.

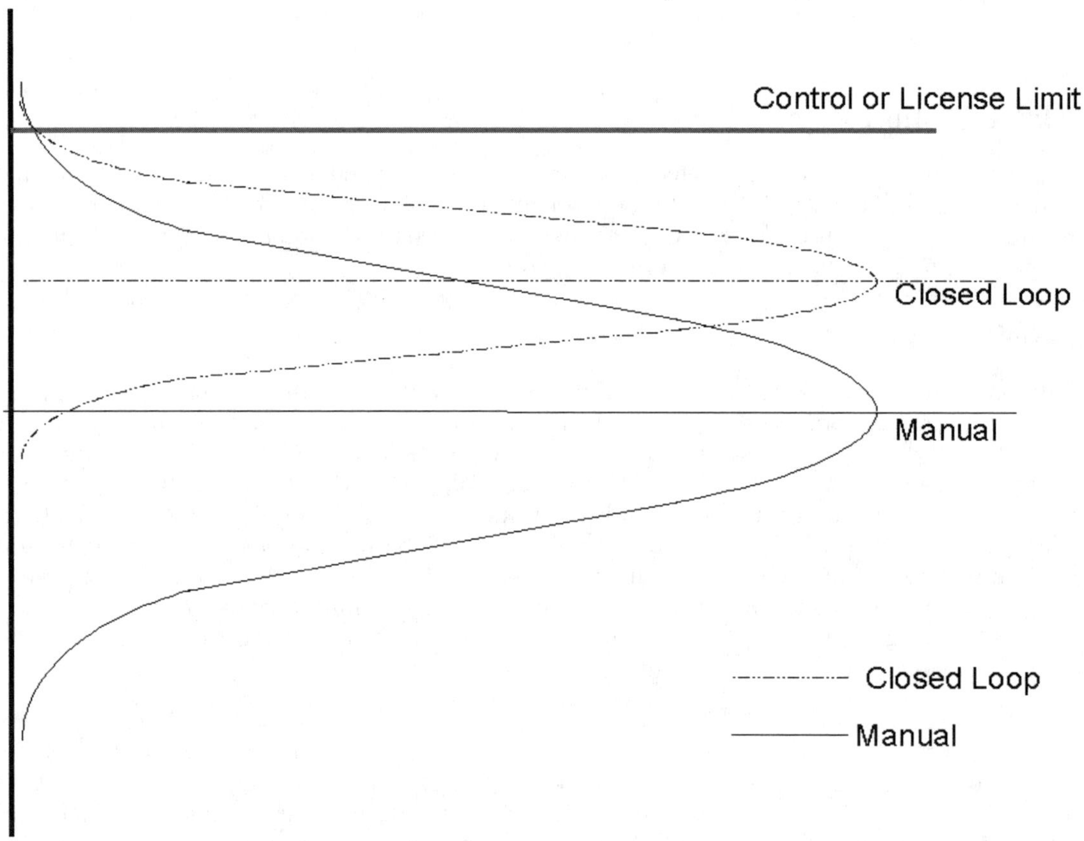


*Figure 15-1: Impact of Tighter Loop Control*

## 15.4 Building Controls

An often overlooked component of environment controls is the control of buildings in which workers and, in many cases, processes are housed.

In the case of workers, it is important to maintain a comfortable environment with a limited range of temperature and fresh air. "Sick Building Syndrome" was a periodic problem in the past because—in the quest for energy efficiency that included ensuring a building was "air tight"—a number of facilities were built that did not have enough fresh air being exchanged to prevent the concentration of gases, such as carbon dioxide and vapors from the building components themselves. This resulted in people having headaches and other symptoms that negatively impacted their ability to work.

Building controls are also taking on a more important role in industries as well, especially in the pharmaceutical and electronics industries. Both types of facilities have requirements for "clean environments" and, in the case of pharmaceutical facilities, the temperature must be maintained within a limited range. Electronics, of course, also require maintenance of building humidity, as well as a means to control static electricity.

For this reason, many facilities now have control systems dedicated to maintenance of the building environment/envelope that are comparable to the ones used for plant automation, often sharing the same components used to control the operating facility itself.

## 15.5 Environmental Control Issues

Environmental controls share many commonalities with other control systems and components of a facility. In addition, many analyzers used for detecting environmental constituents require regular maintenance to replenish consumables and maintain/verify the accuracy of measurements required of them by legislation.

### 15.5.1 Accuracy

In most control systems, repeatability is more important than accuracy. However, in the case of environmental monitoring, accuracy is also important. Accuracy of environmental systems, as verified by calibration and/or reference checking, must include the entire system from sample probe through to the signal, as received by the control system.

A Statistical Quality Control (SQC) process should be used to monitor any change, with the introduction of each reference gas sample and a change to the system's calibration factor made *only* when it is statistically significant *and* the reason for the change has been identified and resolved. Doing otherwise will result in the system continuing to "chase" the calibration gas and will not improve system accuracy at all.

### 15.5.2 Maintenance

All equipment and software applications require maintenance; environmental systems are no different. Without maintenance, the equipment will deteriorate and cease to function. Because of the specialized nature of the equipment used for environmental measurement, especially analyzers, the most effective way of ensuring continued reliable signals is to have a dedicated support team in place. This team should be made up of maintenance technicians/mechanics, engineers/technicians, and chemists working together to keep systems operating as they should. A poorly maintained system is a liability to a company, since it costs money to install and does not provide the benefit used to offset this cost.

## 15.6 References

1. Sherman, R.E., ed. *Analytical Instrumentation*. Practical Guides for Measurement and Control Series. ISA, 1996.

2. Pevoto, L.F. & Converse, J.G. "Decisions to Change Analyzer Calibration Based on Statistical Quality Control Charts." *ISA Transactions* Vol. 30, No. 1 (First Quarter, 1991).

3. ISA SP71 Series of Standards. *Environmental Conditions for Process Measurement and Control Systems.*

## About the Author

**Ian Verhappen** is an ISA Fellow and Director at ICE-Pros, Inc. an independent instrument and control engineering consulting firm specializing in process analyzer systems, fieldbus, and oil sands automation. He is a past ISA Analysis Division director and has been editor of the division newsletter since 1990. He has also contributed to the *Analytical Instrumentation* volume of ISA's Practical Guides for Measurement and Control Series, as well as the most recent editions of the *Instrument Engineer's Handbook,* jointly published by ISA and CRC Press.

# 16 Environmental Monitoring

*By Joe Bingham*

## Topic Highlights

*Introduction*
*Origins*
*Affected Companies*
*Extractive CEMS hardware*
  *Stack Probes*
  *Sample Lines*
  *Sample Conditioning Systems*
  *Analyzers*
*Calibrations*
*DAS/RTU Systems*
*Chart Recorders*
*System Design and Integration*
*Writing a Request for Proposal*
*Writing a Contract*
*Testing/Certification*
*Maintenance, QA/QC*

## 16.1 Introduction

Today's automation professional is not only involved with his company's plant monitoring and control systems, but also the air pollution control systems that the EPA (Environmental Protection Agency) has mandated. He will be responsible for many aspects of the environmental monitoring system design, installation, operations, maintenance, and upgrades. Currently within some companies, the air pollution monitoring systems have a huge impact on operations, maintenance, and profitability (due to possible fines).

## 16.2 Origins

In 1955, after many state and local governments had passed legislation dealing with air pollution, the federal government decided that this problem needed to be dealt with on a national level. Congress passed the Air Pollution Control Act, the nation's first piece of federal legislation. The bill identified air pollution as a national problem and called for research and additional steps to improve the situation. It was an act to make the nation more aware of this environmental hazard.

Although important legislative precedents had been set, the existing laws were deemed inadequate. By 1970, the issue of air pollution was addressed again with the introduction of the Clean Air Act of 1970. Though this act was technically an amendment, the Clean Air Act of 1970 was a major revision and set much more demanding standards.

In 1990, after a lengthy period of inactivity, the federal government believed that they should again revise the Clean Air Act due to growing environmental concerns. The Clean Air Act of 1990 emerged, addressing five main areas: air-quality standards, motor vehicle emissions and alternative fuels, toxic air pollutants, acid rain, and stratospheric ozone depletion. In many ways, this law set out to strengthen and improve existing regulations.

## 16.3 Affected Companies

Though an automation professional will face a great deal of environmental monitoring issues, stack gas monitoring and its affects, will be a primary and frequent concern.

The Clean Air Act mandates that the oil, gas, petrochemical, power generation, utilities, and industrial facilities monitor their stack emissions for certain non-attainment pollutants. In addition, these facilities are required to maintain their emission levels at the state, local, and/or federal limits, utilizing a Continuous Emission Monitoring System (CEMS). A CEMS refers to a packaged system of gas analyzers, gas sampling system, flow, and temperature monitors. Technical requirements for CEMS can be found in 40 CFR (Code of Federal Regulations) part 60 and CFR part 75. State implementation plans are based on the EPA's regulatory authority in 40 CFR, though some technical variations exist from each state and local Air Pollution Control Districts (APCDs). Here is a basic list of industries and the pollutant(s) that they are required to monitor:

- Refineries – $No_x/SO_2/CO/$ or $CO_2$ and flow;
- Chemical Plants – $NO/NO_x/$ $NH_3$ (as $CO_2$);
- Gasification & Gas Plants – $NO_x/O_2/CO/CO_2/$ and Flow;
- Gas Turbine Power Plants – $NO_x/CO/CO_2/O_2/$ and Flow;
- Cogeneration -- $NO_x/CO/CO_2/O_2/$ and Flow;
- Coal Fired Power Plants -- $NO_x/CO/CO_2/O_2/$Flow and Opacity;
- Steel and Cement -- $NO_x/CO/CO_2/O_2/$Flow and Opacity;
- Copper Smelting – $SO_2$; and
- Semiconductor – VOC (Volatile Organic Compounds);

These are but the most common industries and pollutants to be monitored. State and local APCD's may require additional monitoring requirements.

## 16.4 Extractive CEMS Hardware

All of the monitoring requirements for CEMS have similar configurations. (See Figure 16-1.) A typical extractive CEMS will be comprised of a stack probe, sample line, sample conditioning system, analyzers, PLC or other controller for calibrations, and Data Acquisition System (DAS).

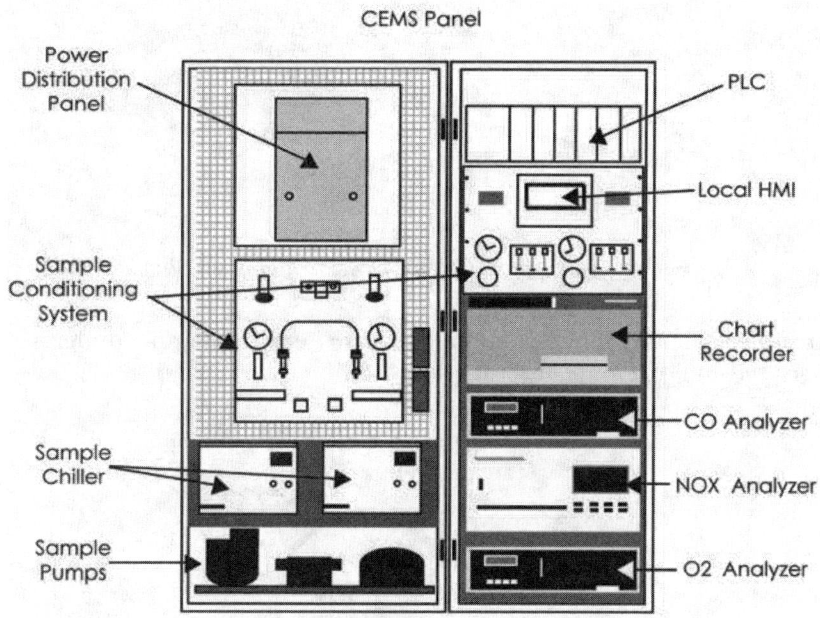


*Figure 16-1: Typical CEMS Enclosure*

### 16.4.1 Stack Probes

A stack probe is the primary connection to the source being monitored and is typically made out of 316 stainless steel. Some special materials include:

- Inconel,
- Hastalloy, and
- Monel.

An ANSI Class 150, 4-inch RF flange is used to attach the probe to the process or exhaust stream. Most stack probes are heated to keep the sample gas above dew point. A typical stack probe will have a primary filter around 10 microns and made up of 316 Stainless Steel to keep any large particulate from clogging the sample line, sample conditioning system and/or the analyzers. The basic probe will also have a one-way check valve connection to allow for calibration gas to flood the probe (also known as a dilution calibration) or a solenoid in order for the system to perform a calibration. (See Figure 16-2.)

Some processes are extremely dirty and require an additional connection to the probe in order to provide a regularly scheduled cleaning of the probe. Instrument air is normally used for this process, and controlled by a solenoid valve.

### 16.4.2 Sample Lines

The sample line transports the gas that is being sampled to the sample conditioning system and analyzers. (See Figure 16-3.) The line is heated to prevent the formation of moisture and acid gas condensation during sample extraction and transport, preserving the integrity of the gas sample for accurate measurement. A sample line is made up of many pieces and is typically three to four inches in diameter. It contains the sample line, calibration line, blow back line, possible electrical signal and a heating element running the length of the sample line. The sample line and the calibration line are usually made of PFA or FEP tubing, which has a high resistance to chemical absorption. If absorption occurs due to moisture or containments, it will cause the sample gas to read incorrectly. There are two meth-

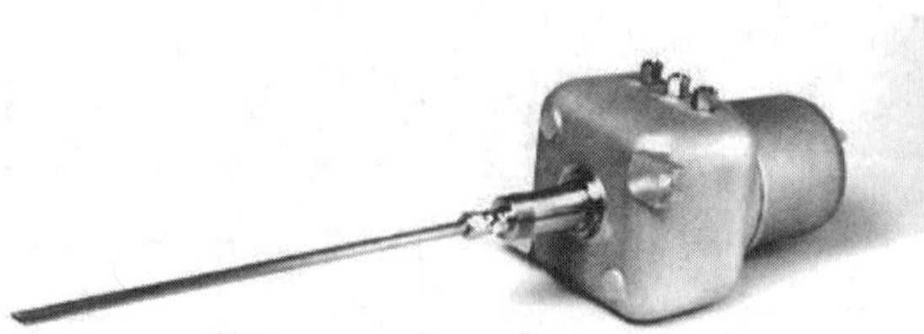

*Figure 16-2: Typical Class 1 Division 1 Heated Sample Probe*

ods of heat tracing available for heated sample lines to keep the sample under dew point prior to sample gas analysis: self-limiting and constant wattage.

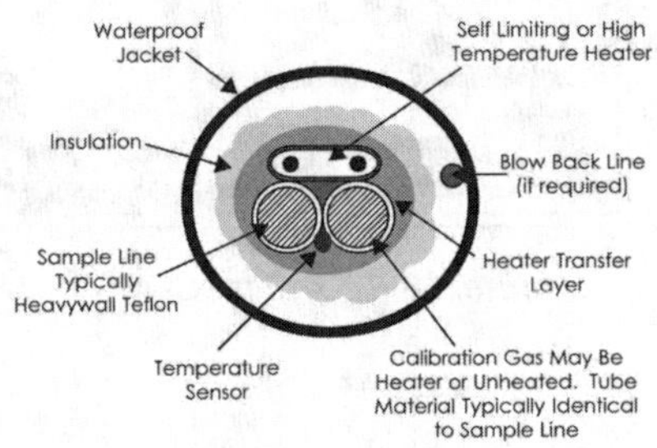


*Figure16-3: Schematic Diagram of a Sample Bundle*

Self-Limiting (has no temperature controller) heat tracing becomes more resistive as its temperature rises, limiting power delivery. It is only acceptable for temperatures up to 250 degrees Fahrenheit.

Constant wattage (has a temperature controller) heat tracing allows for higher operational temperatures. It is used for sample systems requiring temperatures above 250 degrees Fahrenheit, or in applications where temperature must be closely controlled.

### 16.4.3 Sample Conditioning System

The gas sample must be delivered to gas analyzers in a timely manner and must be clear of all contamination from condensed liquids, condensable water vapor, and particulate matter. The sample conditioning system is designed to remove moisture and particulate matter and to handle the calibration gases. A sample chiller is used to drop the sample gas below dew point and condense out the moisture. The moisture is then removed by a peristaltic pump. Inline filters are used to remove particulate matter, as well as any moisture the sample chiller was unable to remove.

The sample conditioning system also handles the calibration gases, using solenoids to select which calibration gas is sent to the probe through the calibration line. This provides a bias calibration. This type of calibration sends the calibration gas through the calibration line to the sample probe and floods the probe so that only the calibration gas is drawn through the sample line and the sample conditioning system. Bias calibrations indirectly check the entire sample gas transport system. If there are any leaks, the analyzers are unable to pass their individual calibrations. An internal calibration is also available to allow for multiple calibrations without sending the calibration gas all the way to the probe and back. This type of calibration is typically performed during maintenance or repair periods.

### 16.4.4 Analyzers

There are many brands of analyzers that employ multiple types of methods to detect a single, or several different pollutants (Figure 16-4). Analyzers are becoming more and more sophisticated. Today's modern analyzers are almost like your typical plant monitoring system. They have Human Machine Interfaces (HMI's), internal networks between each component, control valves, and flow meters.

*Figure 16-4: Example of Typical Multi-pollutant Analyzer*

Below is a list of the different types of pollutants and the best and/or most common detecting methods:

#### 16.4.4.1 $NO_X$

Chemiluminescence analyzers are more accurate and widely used over infrared analyzers. This method has been developed for measurement of NO and $O_3$, using a specific reagent which reacts with NO or $O_3$ to produce a light (chemiluminescence).

#### 16.4.4.2 CO

Infrared absorption is the most common method of detecting carbon monoxide. Gaseous materials normally contain absorption molecules peculiar to the infrared region. CO is one of the infrared active gases. A non-dispersive infrared gas analyzer (NDIR), with a reference cell sealed with the measuring component or other gases having absorption spectrum identical with the measuring component, and detects the change in the absorption of infrared rays at particular wavelength in a sample cell. To measure the concentration of gases in ambient air or in flue gas this non-dispersive infrared gas analyzer is normally used.

#### 16.4.4.3 $SO_2$

There are three categories of analyzers for measuring sulfur dioxide. They are solution conductivity, infrared absorption, and constant potential electrolysis. Currently, infrared absorption systems are the most commonly used.

#### 16.4.4.4 $O_2$

Oxygen behaves as a paramagnetic gas in the presence of an uneven field. When an uneven magnetic field is applied to a paramagnetic gas, the gas is attracted towards the strongest region of the field, raising the pressure in that region. A sample gas containing a large volume of oxygen will move vigorously in the presence of a strong oscillating magnetic field. A sample containing a relatively small volume of oxygen will move relatively less, and a sample containing no oxygen will not change pressure in a magnetic field. A paramagnetic analyzer excites an electromagnet to create pressure changes in the measurement call. A capacitor microphone detects the pressure changes and converts them to an electrical signal. The output of the detector is linear with the concentration of oxygen.

#### 16.4.4.6 Flow

A mass flow meter, or orifice meter, is typically used to measure fuel flow and is used for the calculation for stack flow. Optical flow meters are independent of the pressure, temperature, and density of the flue gas. They have an automatic calibration check that runs at a regular interval and require very little, or no calibration.

## 16.5 Calibrations

The EPA requires all CEM systems to pass a daily calibration. This guarantees that the facility CEMS will always be accurate and within a reasonable degree of error. The CEMS will be equipped with a Programmable Logic Controller (PLC) or the analyzer will have a built-in advanced calibration controller. Each pollutant will have to be calibrated for zero and span, and pass within five percent of each. If a calibration fails, the system is seen as being out of control, and the DAS system will substitute missing data until the pollutant finally passes its required calibration.

## 16.6 DAS/RTU Systems

Each CEM System is required to have a Data Acquisition System/Remote Terminal Unit (DAS/RTU) recording all of the CEMS monitored pollutants, the CEMS calibrations, and all other required monitoring (i.e. fuel flows, fuel temperature, fuel temperature, stack flow – calculated or derived, and equipment on/off status, etc.). The DAS/RTU's HMI (Human Machine Interface) will give the automation professional the ability to view all of the monitored data, CEMS calibrations, and required reports.

The RTU portion of the HMI is responsible for transmitting the required monitored data to the local APCD or the EPA. The HMI will also provide the operator the ability to view the reports, see when they were transmitted, and retransmit them as necessary.

## 16.7 Chart Recorders

A paper (also known as a strip chart recorder) or digital based chart recorder is required as a backup to the DAS system and is considered a critical part of the overall DAS system. Recorder failure is seen as a DAS failure, and missing data is required to be used during the time the recorder was down.

The recorder is required to monitor the pollutants, pressures, stack flows, fuel flows, temperatures, and status signals (for calibrations start/stop, alarms, equipment on/off, etc.) in parallel with the DAS/RTU.

## 16.8 System Design & Integration

Some CEMS/DAS/RTU systems are integrated into the facility's already existing plant monitoring/controls system and/or existing industrial data highway. Automation professionals use the existing systems to help reduce the installation costs of the system or systems, and to provide operations and engineering personnel additional data for plant operations and predictive maintenance. For example, monitoring a boilers Carbon Monoxide (CO) can help you determine the boiler's efficiency (the more efficient the boiler, the less fuel per BTU of heat out, providing a possible significant cost savings).

## 16.9 Writing a Request for Proposals

One of the keys to a successful installation is to have a strong Request For Proposal (RFP). The main purpose of an RFP is to fully document your company's needs. The RFP should provide a short history of your company's monitoring needs and detail the type of monitoring system that you currently employ. This will help the vendor determine whether it can use your existing hardware in order to reduce costs, or if it will be necessary to replace everything.

You should also include:

- A detailed listing of your equipment's I/O (analog and digital Inputs and outputs).
- The pertinent sections from your facilities air quality permit involving your monitoring and reporting requirements.

- A copy of all of the rules and regulations that your facility falls under (this way they cannot say that they did not know, or have the information).
- The date when the proposal is due, as well as what time frame in which the project needs to be started and completed.
- A list of specific personnel who may serve as a point of contact for technical or purchasing questions.

In writing an RFP, it is important that you involve certain key company personnel. The purchasing agent will send out the proposals and handle the responses. He should create two reports for review. One package will consist of the received reports based on their merits and offerings; the second report should be organized based on cost.

Accounting will need to create a capital work order and chart of accounts for the project. They will also be responsible for making sure that the vendors get paid.

Facility management and operations involvement will be crucial to the success of the project, as they will be responsible for maintaining and operating the system. They may also have special concerns or want custom reports. If you have an Environmental Department and/or Engineering Department, they will either be involved in the project management or deployment of the system. In rare cases the Information Technology (IT) Department will also be involved.

## 16.10 Writing Contracts

Most CEMS/DAS/RTU systems are considered a capital expenditure and require more than a basic Purchase Order. In writing a CEMS/DAS contract, it is important that you use basic, straight forward language and refrain from using technical jargon and uncommon acronyms. Most of the data required will come from you RFP and the vendors' proposals. Making sure that everyone - from the plant manger to the operator - is able to understand what the contractual obligations are for both parties. This information will be one of the keys to the CEMS installation success and differs from company to company. Some companies have project managers or project engineers who are more than capable of handling the technical requirements, while other companies may relegate the responsibility to the plant manager, operations supervisor or even the local operator (although this rarely happens).

## 16.11 Testing/Certifications

After your CEMS/DAS/RTU system has been installed, you will be required to perform a compliance verification test: a Relative Accuracy Test Audit (RATA).

A RATA (CFR 40, part 75, App. A) test requires a third party testing company to come out to your facility in a large box van (also known as the Reference Method) and monitor your exhaust stack side by side with your existing system. See Figure 16-5.

There are many qualification checks that a system must pass in order to be certified. Here are the basic requirements:

- 7-day calibration error check (CFR 40 part 75, App. A, Section 3.1 for $NO_x$ and $O_2$, and CFR 40 part 60 for CO)
- Linearity check (CFR 40, part 75 App. A)
- Cycle/Response time test (CFR 40, part 75 App. A, Section 6.4)
- Gaseous Stratification (CFR 40, part 75, App. A)

The source testing company will perform 12 runs monitoring the same pollutants that your CEMS is required to monitor. They will monitor your stack flow using an s-type probe, stack temperature, and moisture (see CFR 40 Part 60, Method 1 – 4). The information from each run will be compared to your DAS system and a final Bias Adjustment Factor (BAF) will be tabulated for your monitored pollutants. The bias test will be applied to the $NO_x$ concentration, $NO_x$ mass emission rate and volume flow rate relative accuracy test audit results. If the mean difference between the reference method (RM) average and the source CEM average is less than the confidence coefficient, then the parameter passes the bias test and is assigned a Bias Adjustment Factor (BAF) of 1.000. If the mean difference is greater than the confidence coefficient, than the parameter fails the bias test and is assigned a BAF according to the equation below:

$$BAF = 1 + |d| / |CEM|$$

In the event that the parameter fails the bias test, but the CEM average is greater than the RM value, no adjustment is necessary and a BAF of 1.000 is assigned.

*Figure 16-5: Source Test Van at a Power Plant*

## 16.12 Maintenance, Quality Assurance/Quality Control

To reduce operational costs and extend the life of your CEM/DAS/RTU system, a proper QA/QC (Quality Assurance/Quality Control) program should be put in place. The QA/QC program will involve the operations, maintenance, and automation professionals. The QA/QC program is also a requirement of the EPA. Not having and/or performing a QA/QC program could put your facility in violation. The facility may be fined per incident, per day from when the violation first occurred.

Operation personnel are usually tasked with the daily, weekly, and monthly inspections and replacement of basic consumables (i.e. filters, chemical type $NO_2$ converters, etc.).

Maintenance personnel will usually be tasked to change out pumps, chemically cleaning the sample lines, cleaning the probe, and fixing the air conditioning system.

Automation professionals will be tasked with making sure that the QA/QC program is being properly followed and performing high level maintenance on the systems. They will analyze problems with CEMS and DAS systems, work on the analyzers, and perform needed repair. They will also perform backups of the computer systems, conduct routine inspections, and recommend any needed system upgrades or replacements.

## 16.13 Chapter Summary

Congressional legislation regulatory air quality demands that companies modify their process to reduce air pollution, install monitoring systems, and report the mass emissions to the local APCD or the EPA on a daily basis. These requirements are costly and do not typically provide a return on investment. However, forward thinking companies have been able to take advantage of the regulatory monitoring requirements to aid in the maintenance of their equipment and to increase equipment efficiency.

## 16.14 References


1. www.EPA.gov (Environmental Protection Agency)
2. Code of Federal Regulations 40 Part 60 (appendices)
3. Code of Federal Regulations 40 Part 72 to 80
4. www.AMETSOC.org (American Meteorological Society)
5. www.bei-reno.com (Baldwin)
6. www.Horiba.com (Horiba Instruments)
7. www.apec-vc.or.jp (Environmental Technology Exchange)
8. www.SCEC.com (Southern California Environmental Company)


## About the Author

**Joe Bingham** has been responsible for the installation, management, and replacement of many CEMS and DAS/RTU systems for several Fortune 500 companies. The founder and president of AES Automation, Bingham holds a patent for DAS/RTU systems. He is a senior member of ISA, having served on the ISA Executive Board and as Vice President for District 11. He has also been instrumental in the development of the ISA Certified Automation Professional Program (CAP).

# 17 Building Automation

*By Ken Sinclair*

## Topic Highlights

*History of the Evolution to Direct Digital Control (DDC)*
*Example of the Application of DDC*
*Open Protocols Use in Building Automation*
*How to Specify Building Automation Systems*
*Future of Web Services in Building Automation*
*Web-Based Facilities Operations Guide*

## 17.1 Introduction & Overview

This chapter is designed to provide insight into the industry that automates large buildings. Each large building has a custom-designed heating, ventilating, and cooling (HVAC) air conditioning system to which automated controls are applied. Access control, security, fire, life safety, lighting control, and other building systems are also automated as part of the building automation system. These systems are viewed by building management on-site and off-site with standard computer technology. In the past, this was done by proprietary communication systems using telephone modems, but now the trend is to browser-based presentation using Internet access with open communication standards.

The names given to building automation systems are varied:

- Building Automation Systems or BAS
- Building Management Systems or BMS
- Facility Management Systems or FMS
- Energy Management Systems or EMS
- Energy Management Control Systems or EMCS
- Client Comfort System or CCS

plus others, based on building-based functions such as lighting control, fire/life safety, security, video, digital signage, etc.

Basic climate control systems (building HVAC) vary widely, depending on owner and building requirements and the mechanical engineer's design. To provide insight into the complete process, this chapter includes an example of how the instrument design for a typical air handing system is laid out. The building zoning is determined by the design engineer and owner of the building and is based on cost versus comfort. Thermal zones usually reflect the internal thermal barriers of the building.

Most controls are based on simple feedback of space temperature. However, the complexity comes from optimizing the interaction of the central equipment—such as the chillers, boilers, pumps, and air-handling equipment—with the great number of individual zones that make up the HVAC system.

Solar control is sometimes used to operate window shades, but most North American building designs depend on brute strength to overcome heating and cooling requirements. The green building movement is slowly changing this type of thinking, encouraging passive designs and interactive control of the building envelope. Outdoor air is used when possible for free cooling, but many buildings have only minimum outside air to provide for ventilation. Temperature is generally sensed with low-cost thermistors.

The energy management control for a building can be as simple as time clock control or as complex as Gridwise, which provides connections based on interactive Web information that allow the building to be a controllable load on the nation's electrical grid. In smaller buildings and homes, the controls are often packaged and provided with the heating or cooling equipment.

## 17.2 History of the Evolution to Direct Digital Control or DDC

A brief history of control evolution provides insight to the development of direct digital control, or DDC, an industry foundation. The automated buildings evolution followed the growth of the pneumatic control industry for a century, seeing the gradual introduction of pneumatic transmission, followed by electric and electronic control. In the mid-1960s, electronic control evolved to multiplex control systems that evolved into the first computerized systems using head-end computers. This quickly gave way to minicomputers, then ended with microcomputers and the DDC revolution. In the late 1970s and early 80s, the use of DDC exploded, greatly expanding the scope of the traditional building automation control market while displacing the traditional pneumatic control industry. The automated building control market growth curve went almost vertical. The decade between mid-1980 and mid-1990 saw the growth of many new control companies able to bring low cost, high functionality, proprietary protocol control systems rapidly to the marketplace. By the mid-1990s, the cost of DDC was much lower than the cost of pneumatics. This rapidly fueled the replacement of pneumatic controls and the expansion of traditional and non-traditional automation markets.

During the mid-1990s, the BACnet movement, "one of the first open communication protocols," began to gain momentum, and a few systems were built around the evolving standard. About the same time, the industry started to follow the revolution that was occurring in other industries with products based on the Echelon chip. Standard protocols became something the industry wanted to achieve. Slowly, but surely, the number of proprietary communication standards were reduced and replaced by open-standard communication protocols.

The next revolution to hit was the Internet/intranet and browser-based presentations of everything. This forced all companies to look hard at what the information technology (IT) industry was doing to standardize presentation models. Some companies had built their new systems around IT thinking and were able to rapidly follow, and even lead, this new trend. Once DDC concepts were embedded in browser-based presentations, their convergence fueled rapid evolution with wireless cell phones, personal digital assistant (PDA) interfaces, e-mailed alarms, etc. With this followed the rapid evolution that further explored Internet capabilities and increased the understanding of the complete power of the newly inherited tools from the Web environment; with this, building automation functionality exploded. Figure 17-1 provides a visual portrayal of all these events.

### 17.2.1 Example of the Application of DDC

An example of the points list for a basic DDC for an air handling system—shown in Table 17-1 on the following page, with accompanying Figure 17-2 following—is reprinted here with permission from its source, BC Buildings Corp. (BCBC), Victoria, British Columbia, Canada.

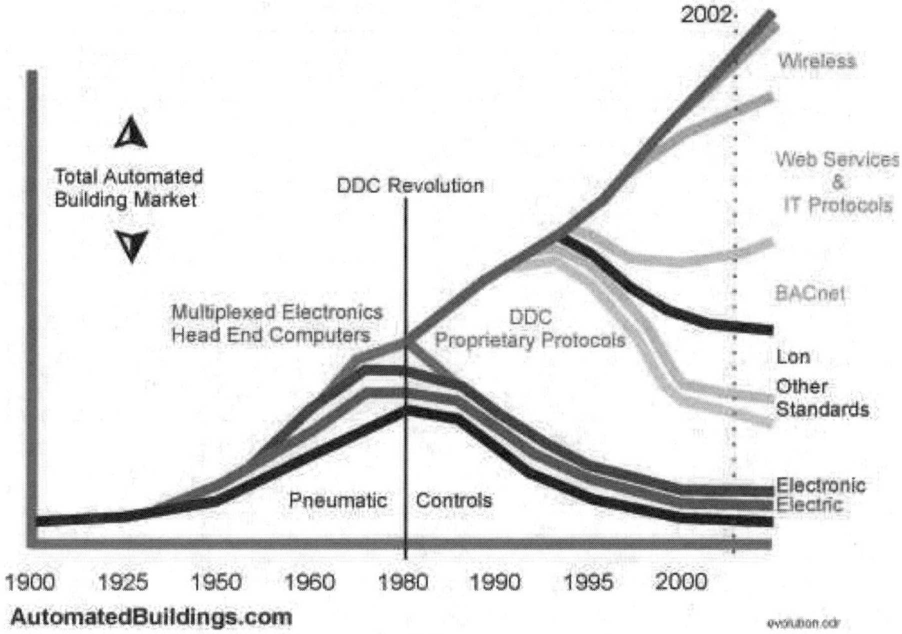


*Figure 17-1: Automated Buildings Evolution*

The *BCBC Client Comfort System Design Manual* is primarily intended to instruct and assist designers who are specifying computerized building controls for BC Buildings Corp. It starts with an introduction that describes how to use the manual for the procurement process.

Once you conceptually understand how this is done, you will see how automation is applied to other building components. Sensing and control points are provided for each air system, plus all other major pieces of equipment provided in the mechanical building design. Both hardware and software points are specified.

The points list provides each hardware and software point with a unique number, shown on Figure 17-2. The point type is indexed in a specification which provides a complete detailed specification for each end sensing and control device as well as all software points.

The graphic of how this system should be represented on the computer screen is part of the specification and shows the exact location of each of the required sensing points and the actuated devices. Software points, which are detailed in the specification, are also shown. The actual location on the graphic where the operator would click to get trend information, time schedules, access to control programs, etc., is also shown.

The graphic is followed by the control language, which shows the exact relationship between the real and virtual points in the automation system presented in a basic If, Then, Else, control language.

This is an example of one air system. In an extremely large building, there could be hundreds of air systems, typically one per floor in high-rises, plus thousands of individual terminal zones—each with their own DDC control. In addition, central equipment such as chillers and boilers to provide heating or cooling all have their own control systems. These DDC control points are all networked together to allow interaction of all points. This is done in the control language and the presentation to the operator is via a browser-based interface.

*Table 17-1: Example Points List for a Basic DDC Air Handling System (Source: BC Buildings Corp. Reprinted with permission.)*

| Project Number: | | Project Name: | | | | | | | | | |
|---|---|---|---|---|---|---|---|---|---|---|---|
| Graphic Logic Location | | Point Name Mnemonic | | Point | Hardware Point | | | | Virtual | Notes See Page # | Note |
| | Point Description | System | Point | Type | DO | DI | AO | AI | Point | Comments | # |
| 1.1 | Supply Fan Start/Stop | | SF | CR1 | 1 | | | | | | |
| 1.2 | Supply Fan Status | | SF_S | V | | | | | 1 | | |
| 1.3 | Supply Fan Program | | SF_PG | PG | | | | | 1 | | |
| 1.4 | Supply Fan Trendlog | | SF_TL | TL | | | | | 1 | SF,SF_ASD,SAT,SAP | |
| 1.5 | Supply Fan Run Hours | | SF_TZ | TZ | | | | | 1 | | |
| 1.6 | Weekly Schedule | | WS | WS | | | | | 1 | | |
| 1.7 | Supply Fan ASD | | SF_ASD | DDC | | | 1 | | | | |
| 2.1 | Return Fan Start/Stop | | RF | CR1 | 1 | | | | | | |
| 2.2 | Return Fan Program | | RF_PG | PG | | | | | 1 | | |
| 2.3 | Return Fan Trendlog | | RF_TL | TL | | | | | 1 | RF,RAT,BSP,ASD | |
| 2.5 | Return Fan ASD | | RF_ASD | DDC | | | 1 | | | | |
| 3.1 | Mixed Air Damper | | MAD | DA2 | | | 1 | | | | |
| | Mixed Air Damper Controller | | MAD_CO | CO | | | | | 1 | use SAT:SAT_SP | |
| | Mixed Air Damper Minimum | | MAD_MIN | V | | | | | 1 | | |
| | Mixed Air Damper Ramp | | RAMP | V | | | | | 1 | | |
| 3.2 | Mixed Air Program | | MAD_PG | PG | | | | | 1 | | |
| 3.3 | Mixed Air Trendlog | | MAD_TL | TL | | | | | 1 | SF_S,FRZ,MAD,MAT | |
| | Cooling Mode | | CLG_MODE | V | | | | | 1 | | |
| 4.1 | Cooling Coil Valve | | CCV | CV3 | | | 1 | | | | |
| | Cooling Coil Controller | | CCV_CO | CO | | | | | 1 | use SAT:SAT_SP | |
| 4.2 | Cooling Coil Trendlog | | CLG_TL | TL | | | | | 1 | CCV,SAT_SP,SAT,EC | |
| 4.3 | Cooling Coil Program | | CCV_PG | PG | | | | | 1 | | |
| | Heating Mode | | HTG_MODE | V | | | | | 1 | | |
| 5.1 | Heating Coil Valve | | HCV | CV3 | | | 1 | | | | |
| | Heating Coil Controller | | HCV_CO | CO | | | | | 1 | use SAT:SP = 12 | |
| 5.2 | Heating Program | | HTG_PG | PG | | | | | 1 | | |
| 5.3 | Heating Coil Valve Trendlog | | HCV_TL | TL | | | | | 1 | HCV,SAT_SP,SAT,HCP | |
| 5.4 | Heating Coil Pump | | HCP | CR1 | 1 | | | | | | |
| 5.5 | Heating Coil Pump Run Hours | | HCP_TZ | TZ | | | | | 1 | | |
| 6.1 | Evaporative Cooling | | EC | CR1 | 1 | | | | | | |
| 6.2 | Evaporative Cooling Program | | EC_PG | PG | | | | | 1 | | |
| 6.3 | Evaporative Cooling Run Hours | | EC_TZ | TZ | | | | | 1 | | |
| 7.1 | Return Air Temperature | | RAT | DTS2 | | | | 1 | | | |
| 8.1 | Mixed Air Temperature | | MAT | DTS1 | | | | 1 | | | |
| 9.1 | Freeze Control | | FRZ | FRZ | | 1 | | | | | |
| 10.1 | Room Temperature Minimum | | RT_MIN | V | | | | | 1 | | |
| 10.2 | Room Temperature Average | | RT_AVG | V | | | | | 1 | | |
| 10.3 | Room Temperature Maximum | | RT_MAX | V | | | | | 1 | | |
| 10.4 | Room Temperature Setpoint | | RT_SP | V | | | | | 1 | | |
| 10.5 | Room Temperature Trendlog | | RT_TL | TL | | | | | 1 | SAT,MIN,AVG,MAX | |
| 10.6 | Building Pressure | | BSP | DPS | | | | 1 | | | |
| 10.7 | Building Pressure Setpoint | | BSP_SP | V | | | | | 1 | | |
| 11.1 | Supply Air Temperature | | SAT | DTS2 | | | | 1 | | | |
| 11.2 | Supply Air Temp Setpoint | | SAT_SP | V | | | | | 1 | | |
| 11.3 | Supply Air Temp Program | | SAT_PG | PG | | | | | 1 | | |
| 12.1 | Supply Air Pressure | | SAP | DPS | | | | 1 | | | |
| 12.2 | Supply Air Pressure Setpoint | | SAP_SP | V | | | | | 1 | | |
| 12.3 | Supply Air Pressure Program | | SAP_PG | PG | | | | | 1 | | |
| | Supply Air Pressure Controller | | SAP_CO | CO | | | | | 1 | use SAP:SAP_SP | |
| | **Total** | | | | 4 | 1 | 5 | 5 | 34 | | |

## 17.3 Open Protocols used in Building Automation

In the past, proprietary protocols were used to network DDC points. However, today it is more likely open protocols such as BACnet, LonWorks, and/or TCP/IP, oBIX , and Niagara standards would be used.

**BACnet** is an acronym for a data communication protocol for Building Automation and Control Networks. The American Society of Heating, Refrigerating and Air-Conditioning Engineers (ASHRAE) maintains a BACnet website. See: http://www.bacnet.org/.

**BCBC VARIABLE AIR VOLUME TYPE L (VAV:L) - GRAPHIC**

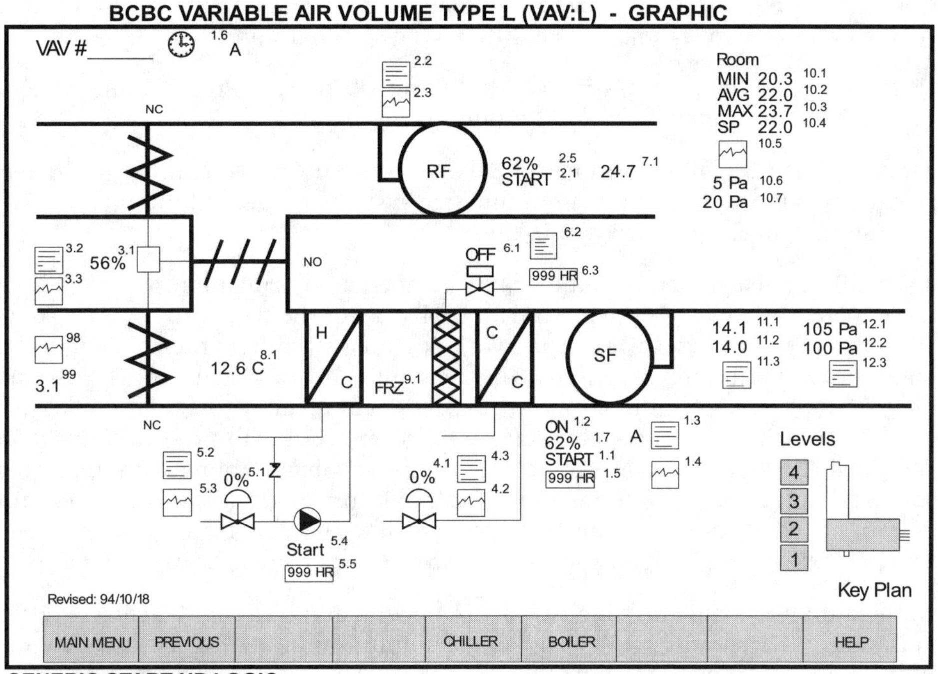


**GENERIC START UP LOGIC**

```
1.3 Supply Fan (SF_PG)
    IF RF ON-FOR 30 SEC THEN
        START SF
    ELSE STOP SF
    IF SAP > 30 THEN SF_S = ON
        ELSE SF_S = OFF
    [Supply Fan ASD]
    IF SF ON THEN
        SF_ASD = SAP_CO
    ELSE SF_ASD = 0
2.2 Return Fan (RF_PG)
    IF WS ON OR RT_AVG < 14 THEN
        START RF
    IF WS OFF AND RT_AVG > 16 THEN
        STOP RF
    [Return Fan ASD]
    BSP_SP = _____ DEFAULT = 20 Pa
    IF RF ON THEN
        RF_ASD = SF_ASD - (BSP_SP - BSP)
    ELSE RF_ASD = 0
3.2 Mixed Air Damper (MAD_PG)
    MAD_MIN = _____ DEFAULT = 30%
    DO EVERY 3 S RAMP = RAMP + 1%
    IF SF OFF THEN
        RAMP = 0%
    IF RAMP > 100% THEN
        RAMP = 100% END DO
    IF SF_S ON AND WS ON AND FRZ OFF THEN
        IF CLG_MODE ON OR HTG_MODE ON THEN
            IF RAT < OAT OR HTG_MODE ON THEN
                MAD = MAD_MIN
            ELSE MAD = 100%
        ELSE
            MAD = MAX(MAD_MIN, MAD_CO)
            MAD = MIN(MAD, RAMP)
    ELSE MAD = 0
```

```
4.3 Cooling (CLG_PG)
    IF SF_S ON THEN
        IF RT_AVG > RT_SP + 1.3 OR
        RT_MAX > RT_SP + 2 THEN
            CLG_MODE = ON, CCV = CCV_CO
        IF RT_AVG < RT_SP + 0.8 AND
        RT_MAX < RT_SP + 1.5 THEN
            CLG_MODE = OFF, CCV = 0
    ELSE
        CLG_MODE = OFF, CCV = 0
5.2 Heating (HTG_PG)
    IF SAT < 10 THEN
        HTG_MODE = ON, HCV = HCV_CO
    IF SAT > 14 THEN
        HTG_MODE = OFF, HCV = 0
    [Heating Coil Pump]
    IF HCV > 15 OR MAT < 3 THEN
        START HCP
    IF HCV < 2 AND MAT > 7 THEN
        STOP HCP
6.2 Evaporative Cooling (EC_PG)
    IF SF_S ON AND CLG_MODE OFF THEN
        IF RT_AVG > RT_SP + 0.5 OR
        RT_HIGH > RT_SP + 1 THEN
            START EC
        IF RT_AVG < RT_SP + 0.1 THEN
            STOP EC
    ELSE STOP EC
11.3 Supply Air Temp (SAT_PG)
    RT_SP = ____ DEFAULT = 22
    SAT_SP = 18 - ((RT_AVG - RT_SP) * 4)
12.3 Supply Air Pressure (SAP_PG)
    SAP_SP = ____ DEFAULT = 100 Pa
```

5_VL1.CDR    Updated: 94/10/31 by KWS    Page _____

*Figure 17-2: Unique Number Provided for Each Hardware and Software Point (Source: BC Buildings Corp. Reprinted with permission.)*

**LonMark International** is a global membership organization created to promote and advance the business of efficient and effective integration of open, multi-vendor control systems utilizing ANSI/EIA/CEA 709.1 and related standards. See: http://www.lonmark.org/.

**TCP/IP** is the acronym for Transmission Control Protocol/Internet Protocol, the suite of communications protocols used to connect hosts on the Internet.

**oBIX** stands for Open Building Information Xchange, an industry wide initiative to define XML- and Web Services-based mechanisms to present building systems-related information on TCP/IP networks such as the Internet. See: http://www.obix.org/.

**Niagara** is a different way to address the challenge of creating true interoperable and open systems. Niagara takes data from diverse systems and "morphs" all of the data into uniform software "components." It's a comprehensive software framework designed to address the challenges of integrating diverse smart devices into unified systems. These components can then easily be assembled into applications—for example, dynamic displays, control sequences (even across different systems and devices), alarms, schedules, reports, etc. Niagara makes it possible to expose all this data to the enterprise in a unified IT-friendly way. Niagara combines the capabilities of network management, protocol conversion, distributed control, and the user interface into a single software solution that can operate on a wide range of hardware platforms from very small to the very large. It can be used on PCs, on servers and in small, embedded controllers. See: http://www.tridium.com/.

The art of the automated buildings industry is connecting and creating the correct relationships of these thousands of DDC points, while providing communication to a Web browser with open protocols. Rapid evolution in the automated buildings industry has been fuelled with the standardization of communication protocols.

Figure 17-3 is from the Reliable Controls Web site at http://www.reliablecontrols.com/, and is reprinted here with permission. It provides a good example of how an open protocol operating system using BACnet protocol takes the DDC points and creates a relationship for each point in a control language (BASIC Code) and displays them on a standard internet browser.

## 17.4 How to Specify Building Automation Systems

### What is CtrlSpecBuilder?

CtrlSpecBuilder™ is a free online productivity tool for preparing bid specifications for HVAC control systems. CtrlSpecBuilder can prepare BACnet specifications or specs that allow other protocols. You can view the spec online and download it as a Microsoft Word file.

### CtrlSpecBuilder provides tool for preparing quicker, easier specifications

A new Website has been launched providing HVAC engineers with a new productivity tool. Previously, engineers wrote specifications from scratch, cut and pasted from previous projects, or relied on proprietary vendor "canned" specifications. CtrlSpecBuilder is a free, online tool that prepares non-proprietary, custom specifications. It follows ASHRAE Guideline 13-2000, specifying DDC systems, and CSI MasterFormat for section 15900. This tool also generates point lists and sequences of control for all HVAC equipment in the project, and can provide specifications in U.S. or metric units.

CtrlSpecBuilder can be accessed at http://www.ctrlspecbuilder.com/.

## 17.5 Future of Web Services in Building Automation

The following is an excerpt from an article published January 2002 on AutomatedBuildings.com's Web site. The article, titled *Information Model: The Key to Integration* at http://www.automatedbuildings.com/news/jan02/art/alc/alc.htm, has been widely read and is a cornerstone for Web-based evolu-

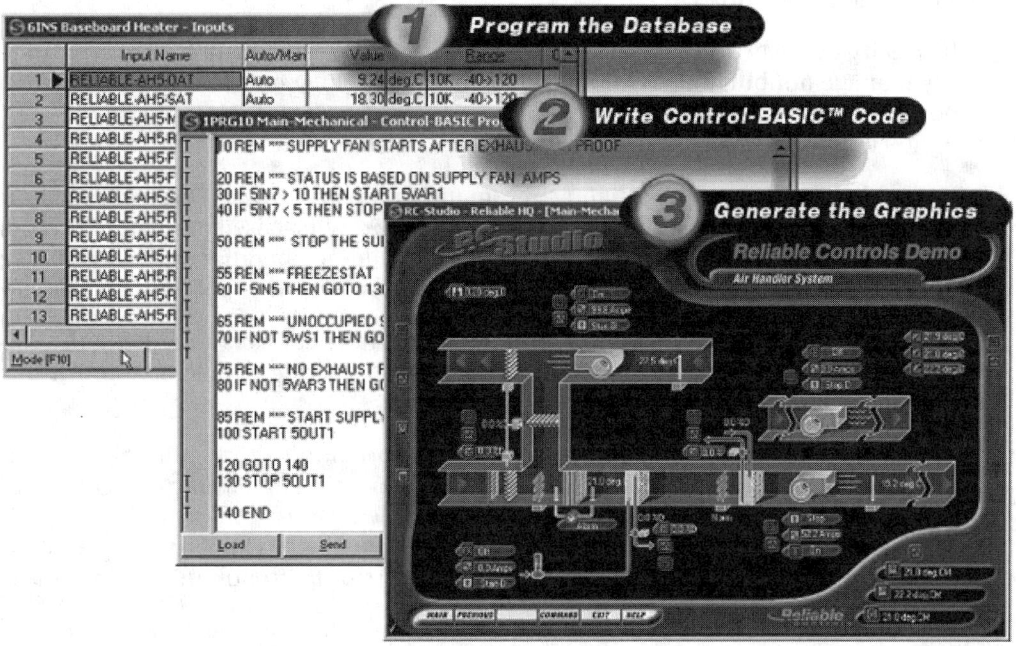


*Figure 17-3: Display Showing How BACnet Protocol Is Applied*
*(Source: Reliable Controls Corp., Victoria, British Columbia, Canada. Reprinted with permission.)*

tion. Its authors are Eric Craton, head of product development, and Dave Robin, head of software development, at Automated Logic Corp., Atlanta, Ga.

Let's look at the four trends in terms of Web services:

1. **Content is becoming dynamic** - A Web service has to be able to combine content from many different sources. That may include furniture inventories, maintenance schedules and work orders, energy consumption and forecasts, as well as traditional building automation information.

2. **Bandwidth is getting cheaper** - A Web service can now deliver types of content (streaming video or audio, for example) unthinkable a few years ago. As bandwidth continues to grow, Web services must adapt to new content types.

3. **Storage is getting cheaper** - A Web service must be able to deal with massive amounts of data intelligently. That means using technologies such as database replication, LDAP directories, caches, and load balancing software to make sure that scalability isn't an issue.

4. **Enterprise computing is becoming more important** - A Web service can't require that users run a traditional browser on some version of Windows. Web services have to serve all sorts of devices, platforms, and browser types, delivering content over a wide variety of connection types for a wide variety of purposes.

## 17.6 Web-Based Facilities Operations Guide

In August of 2002, I prepared a supplement for *Engineered Systems Magazine*, called *A Guide to Web-Based Facilities Operations*. See: http://www.automatedbuildings.com/news/aug02/articles/ksin/ksin.htm. Here's an excerpt:

"Doing more with less by using Web-based anywhere information to amplify your existing building operational resources.

"The reality of today and tomorrow's economy is that we will be doing more with less to effectively manage our buildings. Our saviour is that technology in the form of Web-based everything is now providing us a path to improved communications, while simplifying the movement of complex building information to the evolving Building Operations Specialist.

"The rapid movement of the building automation industry towards Web-based allows us to interweave critical dynamic building information into a browser-based anywhere presentation. This allows us to concentrate and amplify our existing building operation resources and operators into virtual operational centres, in which all critical information is exposed to all stakeholders.

"This visible-from-anywhere information allows authorized users to provide the correct input, management and dollar accountability skills that will provide excellent comfort/energy performance with total accountability.

"A Web-based presentation of dynamic building information allows not only operators to operate from anywhere with full functionality, it allows interaction of contractors, equipment suppliers, and consultants to provide valuable feedback and feed-forward information to the building operating equation. Upper management can also participate by having browser-based bottom line screens that provide the dynamic proof of the success or failure of building performance.

"This new concentration and amplification of existing building operational resources and personnel will provide a strong re-focus on the values of good operating principles.

"This focus will cause management and operations to re-establish communications on what is important and what is required to achieve cost effective excellent building operations."

## 17.7 Summary

We are witnessing a large building automation industry converging with corporate enterprises. As technologies converge, clients' expectations are fueled by the ease of access and freedom of information on the Internet. The scope has increased to encompass environmental control, energy metering/accounting systems, lighting control, life safety/fire, security, communications, high tech tools, Web resources, and interactive information systems. With the trend towards IT-type solutions, video, card readers, and enterprise-based solutions are all becoming an active part of what is now called building automation.

## 17.8 Resources to Learn More

Below are some of the resources available free with the online magazine, AutomatedBuildings.com (http://www.automatedbuildings.com), that will help you learn more about the industry. The magazine is updated monthly and provides connection to almost all the dot coms in the evolving automated building industry.

**Automated Building Systems Education**
http://www.automatedbuildings.com/frame_education.htm

The rapid changing Automated Building industry has always required that all players be constantly re-educating themselves to keep current, but never has there been a time when this is so important.

**Technology Roadmap for Intelligent Buildings**
http://www.automatedbuildings.com/news/jan03/review/roadmap.htm

This 60-page glossy roadmap also available in a (66 pages) Adobe Acrobat PDF, 2.3 MB file is an excellent starting point for our large building automation industry to begin its reinvention and repackaging and to get on with the task of helping create an even better roadmap.

**Intelligent Building Ranking System**
This Task Force is developing an online tool intended to assist building owners/managers, the commercial real estate industry, and other industry stakeholders to assess the level of integrated systems within a building (a Building Intelligence Quotient - BIQ). A comprehensive list of intelligent building criteria has been developed as well as a "ranking matrix." The next phase of this project is to develop content detail on each line item in the matrix to be available as part of the online tool for ease of use by all industry stakeholders.

**Middleware White Paper**
The paper defines middleware and describes a number of case studies where middleware has provided a solution to integrate new intelligent building technology implementations with legacy systems.

**Building Control Network Protocols White Paper**
Communications protocols are simply a means by which different systems may communicate. They are the message formats and procedures used to transfer information, in an understandable form, from one device, or array of devices, to another. They permit products from different vendors to communicate with each other and interact to produce intelligent integrated building systems and manage and interface with these products as if the same vendor supplied them all. This paper, prepared by the Continental Automated Building Association (CABA) Intelligent & Integrated Buildings Council Building Protocol Task Group explores four of the most common protocols used today and compares a number of the parameters that "assist the large building industry to understand the strengths and overall features of the building control communication protocols that are available for use in designing and implementing an "Intelligent Building." CABA Resources can be found at: http://www.caba.org/councils/council-pubs.html

**Best-Practices Guide for Evaluating IBT**
This guide, authored by Kenneth P. Wacks, Ph.D., builds upon the Technology Roadmap for Intelligent Building Technology. This paper consists of criteria by which intelligent building technologies can be evaluated. The topics in this guide are important for various audiences, such as building owners and managers, intelligent building designers, and installers.

**The Builconn Story**
http://www.automatedbuildings.com/news/mar03/articles/builconn/bcon.htm

BuilConn intends to help integrators deliver a pragmatic approach to integration, not product or technology based, but solutions oriented.

**Bring the Power of the Internet**
http://www.automatedbuildings.com/news/jul00/articles/tridium/trid.htm

Perhaps no other technology has had the profound impact on virtually everything we do than the Internet. It is changing the way we make purchases, the way transactions are conducted between businesses, the way in which we obtain information, and the way we communicate with one another. The above article discusses how the power of the Internet is being adapted to real-time control and automation systems.

## About the Author

**Ken Sinclair** is owner and editor of AutomatedBuildings.com, an online magazine and Web resource providing news and Web connection to the rapidly evolving industry that automates and implements intelligent, integrated buildings. He is a founding member and a past president of both the local chapter of Association of Energy Engineers (AEE) and the Vancouver Island Chapter of the American Society of Heating, Refrigerating and Air-Conditioning Engineers (ASHRAE). He has overseen 100+ computerized control conversions, working closely with building operators to insure acceptance of this new technology.

# IV

# Reliability, Safety and Electrical

Alarm Management *has become a very important topic in the safety area. The press continues to report plant incidents caused by poorly designed alarms, alarm flooding, and alarms being bypassed. Every automation professional should understand the basic concepts of this topic.*

*A basic knowledge of* Reliability *is fundamental to the concepts of safety and safety instrumented systems.* Process Safety and Safety Instrumented Systems *(SIS) is an increasingly important topic. Safety is important in all industries—but especially in large industrial processes such as petroleum refineries, chemicals and petrochemicals, pulp and paper mills, and food and pharmaceutical manufacturing, for example. Even in areas where the materials being handled are not inherently hazardous, personnel safety and property loss are important concerns. SIS is simple in concept, but requires a lot of engineering to apply well.*

*The topic on* Electrical Installations *underscores the reality that the correct installation of electrical equipment is required to design and implement almost any automation system. This is a very large topic with rigid codes; admittedly, we can only scratch the surface here.*

*While the safety of electrical installations is closely related to correct electrical installations in general, electrical safety is treated separately here to ensure that sufficient emphasis is placed on the safety aspects of these installations.*

# 18 Alarm Management

By Nick Sands

**Topic Highlights**

*Alarm System Practices*
- *Alarm Philosophy*
- *Rationalization*
- *Design*
- *Training*
- *Monitoring*
- *Management of Change*

*Alarm System Problems*
- *Nuisance Alarms*
- *Stale Alarms*
- *Alarm Floods*
- *Alarm Clarity*

*Alarms for Safety*

## 18.1 Introduction

The term "alarm management" refers to processes and practices for determining, documenting, designing, monitoring, and maintaining alarm messages from process automation and safety systems. Alarm system performance issues have contributed to many significant incidents in the process industries, with an estimated cost over $13B USD each year in the U.S. alone [Ref. 1].

The issues with alarm systems are well known, as are the practices to address those issues. Practices will be discussed first, followed by the main issues of alarm management and the application of the practices to those issues. The last section mentions the limitations of alarms for risk reduction.

## 18.2 Alarm System Practices

The following practices are often cited as essential steps to improve alarm system performance: alarm philosophy, rationalization, design, training, monitoring, and management of change.

### 18.2.1 Alarm Philosophy

The foundation of an alarm management system is the development of an alarm philosophy—a document that establishes the principles and procedures to consistently manage an alarm system over time. The philosophy does not specify the details of any one alarm, but defines each of the key processes used to manage alarm systems: rationalization, design, training, monitoring, and management of change. Alarm system improvement projects can be implemented without a philosophy, but the systems tend to drift back toward the previous performance. Maintaining an effective alarm system requires the discipline to follow these practices.

The philosophy begins with the basic definitions and extends those to operational definitions with the principles of the alarm system. The philosophy should define such things as the number of levels of alarm, the types of alarms allowed, and the assigned alarm priorities.

***Alarm:*** An audible or visible means of indicating to the plant operator an equipment or process malfunction of abnormal condition [Ref. 2].

The following are examples of principles:

- Each alarm must have a defined operator action.
- Each alarm must be rationalized prior to installation.
- Each alarm will be designed in accordance with site guidelines.
- Operator training is required for each alarm prior to installation.
- Each safety related alarm must be tested prior to start-up and, thereafter, at an explicitly documented frequency.
- Alarm system performance must be monitored on a daily basis and corrective action taken when performance limits are not met.
- All additions, modifications, and deletions of alarms must follow a "management of change" procedure.

Principles like these are critical to an alarm philosophy. They provide the standards against which all potential alarms are tested. A well-defined set of principles will yield a consistent and useful set of alarms.

### 18.2.2 Rationalization

Rationalization is the process of examining one alarm at a time against the principles and criteria defined in the alarm philosophy. The product of rationalization is a set of consistent, well-documented alarms. The documentation supports both the design process and operator training.

Rationalization begins with identifying the signal, the rationale for the alarm and the associated action. If the alarm is consistent with the philosophy, it is prioritized based on consequences and response time. Any further requirements for the alarm design are captured as well.

The alarm philosophy will capture information for each alarm, such as the basic control system information:

- Tag
- Alarm type
- Description
- Units/states
- Setting/alarm state

The tag is the tag number of the alarm in the database. The alarm type describes the alarm as high, low, or a discrete state. The description is for the tag, from the same tag database. The units are the engineering units for an analog type value, and the states are the discrete states of a digital value. The setting is the analog alarm limit or the discrete state that generates the alarm.

Some information is necessary to document the alarm for procedures and training:

- Consequence of deviation
- Corrective action
- Time for response
- Consequence category
- Basis

This information is required to train operators to respond to the alarm—specifically what action is necessary, and how fast must it be completed before the consequence results. Documenting the basis for the alarm allows re-evaluation of the consequences, especially with process changes.

Other information is required to complete the requirement specifications:

- Priority
- Retention period
- Report requirements
- Notification requirements

This information specifies properties of the alarm. The priority in the operator interface is a critical way to designate the importance of the alarm. The alarm record may need to be kept for a certain period of time, included in certain reports, or the alarm may be set up to trigger e-mail, pager, or voice mail messages. These functions are defined in the philosophy, and the rationalization identifies individual alarms that require these functions.

**Example**
A new tank containing flammable materials has the following alarms identified:

| | **Alarm 1** | **Alarm 2** | **Alarm 3** | **Alarm 4** |
|---|---|---|---|---|
| **Tag** | LIG502 | LIG502 | PIG502 | PIG502 |
| **Alarm type** | LL | HH | LL | HH |
| **Description** | T502 Level | T502 Level | T502 Pressure | T502 Pressure |
| **Units/states** | % | % | INWC | INWC |
| **Setting/alarm state** | 10 | 90 | 1 | 10 |
| **Consequence of deviation** | Cavitate pump | Overflow tank | Air intrusion | Excess venting |
| **Corrective action** | Stop pump | Close inlet valve | Stop pump | Close inlet valve |
| **Response time** | 2 min | 2 min | 10 min | 10 min |
| **Consequence category** | Equipment | Safety | Safety | Environmental |
| **Basis** | Pump cavitation at 2% | Tank overflow at 107% | Vacuum breaker setting | Conservation vent setting |
| **Priority** | Low | Emergency | High | High |
| **Retention period** | 1 year | 5 years | 5 years | 5 years |
| **Report requirements** | Pump report | Safety report | Safety report | Environmental report |
| **Notification requirements** | None | None | None | Environmental coordinator |

### 18.2.3 Design

The design phase utilizes the rationalized alarms and design guidance. Design practices are often documented in a separate guidance document specific to the type and generation of the control system. As systems change, the guidance should be updated to reflect features and limitations of the control system. Design practices fall into three areas: the basic configuration of alarms, the human-machine interface (HMI), and advanced techniques for managing alarms.

The guidance on basic configuration may include default settings for alarm deadbands, alarm practices for redundant transmitters, timing periods for discrete valves, alarm practices for motor control logic, and the methods for handling alarms on bad signal values. Many alarm system problems can be eliminated with good basic configuration practices.

***Deadband:*** the change in process value from the alarm point in the reverse direction of the alarm necessary to clear the alarm state.

The guidance on the HMI may include alarm priority definitions, alarm color codes, alarm tones, alarm groups, alarm summary configuration, and graphic symbols for alarm states. Alarm functions are only one part of the HMI, so it is important that these requirements fit into the overall HMI design philosophy. The consistent use of color for alarms is often listed as a principle.

A common component of the HMI design guide is a table of alarm priorities, alarm colors, and alarm tones. Some systems have the capability to show shapes or letters next to alarms. This is a useful technique for assisting color blind operators in recognizing alarm priorities.

Example of alarm priority features:

| Priority | Color | Tone | Shape |
|---|---|---|---|
| Emergency | Red | Tone 1 | Red triangle, point up |
| High | Yellow | Tone 2 | Yellow Diamond |
| Low | Orange | Tone 3 | Orange triangle, point right |

Beyond the basic configuration and HMI design, there are many techniques to reduce the alarm load on the operator and improve the clarity of the alarm messages. These techniques range from first-out alarming to state-based alarming to expert systems for fault diagnosis. The techniques allowed should be defined in the alarm philosophy, along with the implementation practices in the design guide.

***First-out (First-up):*** A sequence feature that indicates which of a group of alarm points operated first [Ref. 3].

***Alarm suppression***: Use of condition-based logic to determine that an alarm should not occur when the base alarm condition is present.

***State-based alarming:*** Use of measurements or models of the equipment or plant operating state to suppress alarms when they are not needed and activate alarms in the operating states to which they are relevant.

***Dynamic prioritization:*** Use of measurements or models of the equipment or plant operating state to change alarm priority based on the current operating state.

Testing is a common requirement when the design is implemented. Testing requirements vary with the type of alarms. Initial and periodic testing requirements should be documented in the rationalization so the accommodations for testing can be made in the design step.

### 18.2.4 Training

Training is an essential step in developing an alarm system. Since an alarm exists only to notify the operator to take an action, the operator must know the corresponding action for each alarm, as defined in the alarm rationalization. A program should be in place to train operators on these actions. Documentation on all alarms should be easily accessible to the operator. Beyond the alarm specific training, the operator should be trained on the alarm philosophy and the HMI design. A complete training program includes initial training and periodic refresher training.

### 18.2.5 Monitoring

Monitoring alarm systems is a critical step in alarm management. Since each alarm requires operator action for success, overloading the operator reduces the effectiveness of the alarm system. Instrument problems, controller performance issues, and changing operating conditions will cause the performance of the alarm system to degrade over time. Monitoring and taking action to address bad actors can maintain a system at the desired level of performance.

The alarm philosophy should define report frequencies, metrics, and thresholds for action. Common measurements include:

- Frequency of alarms, such as total number of alarms per day.
- Frequency of alarm by tag, such as the number of times a tag alarms per day.
- Time in alarm by tag, such as the number of minutes a tag is in alarm.
- Rate of alarms, such as alarms per ten-minute interval.
- The number of alarm floods (more than 10 alarms per 10 minutes) per day.

Measurement tools allow reporting of the metrics at different frequencies. Typically, there are daily reports to personnel responsible to take action, and weekly or monthly reports to management. The type of data reported varies, depending on the control system or safety system and the measurement tool.

Distinct limits to trigger action should be set on the measurements. These limits are dependent on the type of process and the resources to take corrective action. If the action limits are too relaxed, they will not be effective. If they are too aggressive, they will be ignored. The performance metrics are usually calculated per operator position or operator console.

**Example**

- Alarm measurement triggers points and actions:
- Frequency of alarm by tag greater than 10 alarms/day.
- Time in alarm by tag greater than 24 hours.
- Rate of alarms greater than 10/minute.
- Rate of alarms greater than 300/day.

The Engineering Equipment Materials and Users Association (EEMUA) Publication 191, *Alarm Systems: A Guide to Design, Management, and Procurement,* provides guidance on metrics for performance classification. As above, these metrics are calculated per operator since they are related to the operators' ability to process alarms.

*Table 18-1: Benchmark for Assessing Average Alarm Rates [Ref. 4]*

| Long term average alarm rate in steady operation | Acceptability |
|---|---|
| More than 1 per minute | Very likely to be unacceptable |
| One per 2 minutes | Likely to be over-demanding |
| One per 5 minutes | Manageable |
| Less than one per 10 minutes | Very likely to be acceptable |

*Table 18-2: Guidance on Alarm Rate Following an Upset [Ref. 5]*

| Number of alarms displayed in 10 minutes following a major plant upset | Acceptability |
|---|---|
| More than 100 | Definitely excessive and very likely to lead to the operator abandoning use of the system |
| 20-100 | Hard to cope with |
| Under 10 | Should be manageable – but may be difficult if several of the alarms require a complex operator response |

### 18.2.6 Management of Change

Another key procedure for maintaining an alarm system is management of change. Usually there are one or more management of change processes already established for Process Safety Management (PSM) or current Good Manufacturing Practices (cGMP) which would encompass changes for alarms. The alarm philosophy will define the change processes and the steps necessary to change alarms. These steps are usually the same steps, though the scope may be smaller, as a project.

## 18.3 Alarm System Problems

The main problems in alarm management are nuisance alarms, stale alarms, alarm floods, and clarity of the alarm to the operator. The processes defined in the alarm philosophy, implemented with operational discipline, can address these problems.

### 18.3.1 Nuisance Alarms

Nuisance alarms are alarms that indicate an abnormal condition when none exists, or when no change in process condition has occurred. Nuisance alarms desensitize the operator, reducing the response to all alarms. Instrument problems or alarms set within the normal operating range often cause nuisance alarms. Measurement of the alarm frequency by tag is used to detect nuisance alarms at a threshold defined in the alarm philosophy—for example, 10 alarms per day. Once detected, nuisance alarms should be investigated and corrected as soon as possible. Typical alarm reports show a very small percentage of tags are responsible for the majority of alarms. Without monitoring and prompt follow-up, nuisance alarms can quickly deteriorate the performance of an alarm system to the point where tens of thousands of alarms are recorded per day.

### 18.3.2 Stale Alarms

Stale alarms are alarms that remain in the alarm state when no abnormal condition exists or no operator action is required. Stale alarms form a baseline of alarms that require no action and train the operator to ignore certain alarms. These alarms are often caused by alarm configuration problems or alarms set within the normal operating range. Measurement of the time in alarm by tag is used to detect stale alarms at a threshold defined in the alarm philosophy—for example, 24 hours. Without monitoring and follow-up, the number of stale alarms slowly increases, decreasing the effectiveness of the alarm system.

### 18.3.3 Alarm Floods

Alarm floods are a temporary high rate of alarms, usually associated with an event like a process upset. Alarm floods overwhelm the operator, masking the important alarms and reducing the operator's ability to correctly respond to the upset. Alarm floods are often caused by configuring multiple alarms for a given event. Alarm floods are detected by measuring the rate of alarms in a given time interval with a threshold defined in the alarm philosophy—for example, 10 alarms per 10 minutes. Alarm floods are one of the more difficult problems to solve, but a problem closely linked with plant disasters. Monitoring can detect and report alarm floods, but reducing floods takes detailed process understanding and good alarm practices. Rationalization can help reduce duplicate alarms. Advanced alarming techniques can reduce the number of alarms during an upset.

### 18.3.4 Alarm Clarity

Clarity of alarms is an issue related both to configuring the alarms and to training the operator to respond to the alarm. Alarm documentation generated during rationalization provides the information for training. Alarm clarity problems are a difficult thing to measure. Operator training can provide the opportunity to identify clarity problems. Sometimes the problems can be resolved with changes to the basic alarm configuration. Advanced alarm techniques are also often employed to produce fewer alarms that have clear meaning.

## 18.4 Alarms for Safety

Alarms mark the boundary between normal and abnormal conditions in the process. They alert the operator to take action to return the process to normal conditions. Because alarms are linked to operator intervention, they are sometime used as a layer of protection for hazardous events. Care should be used when evaluating an alarm in the process automation system as a safety layer.

While some alarms provide safety warnings, there is a key difference between an alarm system and a safety system. The alarm function always requires an operator to take action. The safety function is almost always designed to function without the operator. One consequence of this difference is that the alarm system's effectiveness is limited by the operator's ability to respond correctly to each alarm. An operator can be overwhelmed as the rate of alarms or the complexity of the response increases. When the process control system is used for safety related alarms, monitoring can maintain the alarm system performance. Even with monitoring, the risk reduction factor for the basic process control system (BPCS), including the process alarms, is limited to 10 unless the system is treated as a safety instrumented system [Ref. 6].

## 18.5 References

1. Nimmo, Ian. *Abnormal Situation Management.*
2. ISA-RP77.60.02-2000, *Fossil Fuel Power Plant Human-Machine Interface: Alarms,* p. 9.
3. ANSI/ISA-18.01-1979 (R2004), *Annunciator Sequences and Specifications,* p. 9.
4. *Alarm Systems: A Guide to Design, Management and Procurement.* EEMUA, p. 105.
5. *Alarm Systems: A Guide to Design, Management and Procurement.* EEMUA, p. 107.
6. ANSI/ISA-84.00.01-2004-Part 2 (IEC 61511-2 Mod) - *Functional Safety: Safety Instrumented Systems for the Process Industry Sector - Part 2: Guidelines for the Application of ANSI/ISA-84.00.01-2004 Part 1 (IEC 61511-1 Mod) - Informative.* Sections 9.4.2 and 9.4.3.

## About the Author

**Nick Sands** has worked in various process control assignments at DuPont for the past 15 years, after graduating from Virginia Tech. He is a Chemical Solutions Process Technology Manager at DuPont. He is an active ISA member, serving as a section, division, and standards committee volunteer, and a contributor to the new ISA Certified Automation Professional program. Nick is a Certified Automation Professional.

# 19 Reliability

*By William Goble*

## Topic Highlights

*Measurements of Successful Operation—No Repair*
*Useful Approximations*
*Measurements of Successful Operation—Repairable*
*Average Unavailability with Periodic Inspection and Test*
*Periodic Restoration and Imperfect Testing*
*Equipment Failure Modes*
*SIF Modeling of Failure Modes*
*Redundancy*

## 19.1 Introduction

There are a number of common metrics used within the field of reliability engineering. Primary ones include reliability, unreliability, availability, unavailability and mean time to failure (MTTF). But, when different failure modes are considered—as they are when doing safety integrated functions (SIF) verification—then new metrics are needed. These include probability of failing safely (PFS), probability of failure on demand (PFD), probability of failure on demand average (PFDavg), mean-time-to-failure spurious ($MTTF_S$), and mean time to dangerous failure ($MTTF_D$).

## 19.2 Measurements of Successful Operation – No Repair

*Probability of Success*—This is often defined as the probability that a system will perform its intended function, when needed, and when operated within its specified limits. The phrase at the end of the last sentence tells the user of the equipment that the published failure rates only apply when the system is not abused or otherwise operated outside its specified limits.

Using the rules of reliability engineering, one can calculate probability of successful operation for a particular set of circumstances. Depending on the circumstances, that probability is called "reliability" or "availability" (or, on occasion, some other name).

*Reliability*—a measure of successful operation for a specified interval of time. Reliability, R(t), is defined as "the probability that a system will perform its intended function when required to do so if operated within its specified limits for a specified operating time interval." The definition includes five important aspects:

1. The system's "intended function" must be known.
2. "When the system is required to function" must be judged.

3. "Satisfactory performance" must be determined.

4. The "specified design limits" must be known.

5. An operating time interval is specified.

Consider a newly manufactured and successfully tested component. It operates properly when put into service (T = 0). As the operating time interval (T) increases, it becomes less likely that the component will remain successful. Since the component will eventually fail, the probability of success for an infinite time interval is zero. Thus, all reliability functions start at a probability of one and decrease to a probability of zero (Figure 19-1).

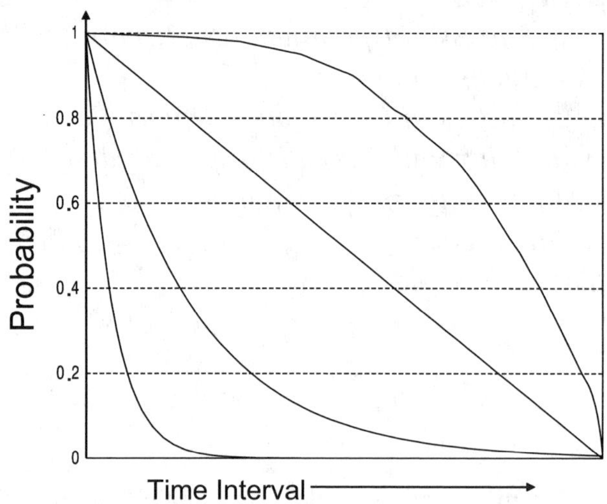


*Figure 19-1: Example Reliability Plots*

Reliability is a function of the operating time interval. A statement such as "system reliability is 0.95" is meaningless because the time interval is not known. The statement "The reliability equals 0.98 for a mission time of one hundred hours" makes perfect sense.

A reliability function can be derived directly from probability theory. Assume the probability of successful operation for a one-hour time interval is 0.999. What is the probability of successful operation for a two-hour time interval? The system will be successful only if it is successful for both the first hour and the second hour. Therefore the two hour probability of success equals:

$$0.999 \times 0.999 = 0.998 \tag{19-1}$$

The analysis can be continued for longer time intervals. For each time interval the probability can be calculated by the equation:

$$P(t) = 0.999^{t} \tag{19-2}$$

Figure 19-2 shows a plot of probability versus operating time using this equation. The plot is a reliability function.

Reliability is a metric originally developed to determine probability of successful operation for a specific "mission time." For example: if a flight time is 10 hours, a logical question is, "What is the probability of successful operation for the entire flight?" The answer would be the *Reliability* for the 10 hours

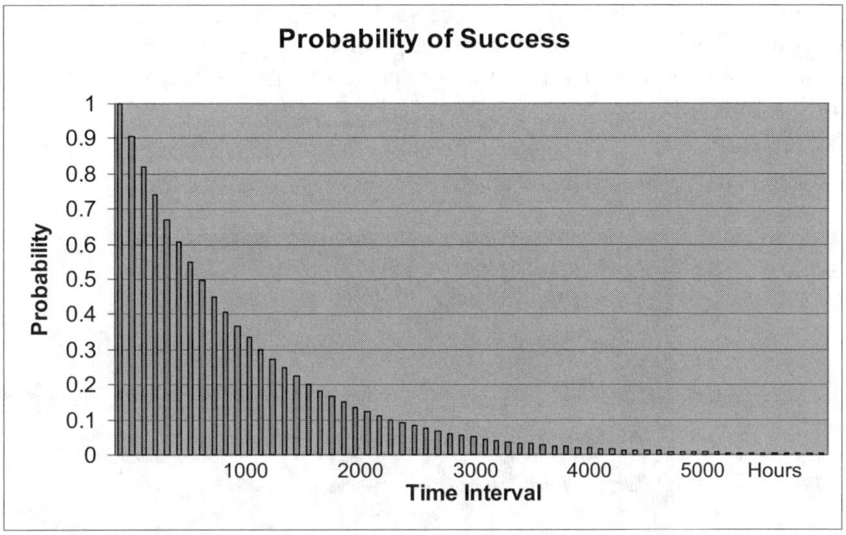


*Figure 19-2: Reliability Plot Using Constant Probability of Failure*

duration. It is generally a measurement applicable to situations where online repair is not possible, like an unmanned space flight or an airborne aircraft. Unreliability is the complement of reliability. It is defined as the probability of failure during a specific mission time.

*Mean Time To Failure (MTTF)*—One of the most widely used reliability parameters is the MTTF. It has been formally defined as the "expected value" of the random variable *Time To Fail, T.* Unfortunately, the metric has evolved into a confusing number. MTTF has been misused and misunderstood. It has been misinterpreted as "guaranteed minimum life."

Formulas for MTTF are derived and often used for products during the useful life period. This method excludes wearout failures. Ask an experienced plant engineer "What is the MTTF of a pressure transmitter?" This engineer would likely include wearout and might answer "35 years." Then the engineer would look at the specified MTTF of 300 years and think that the person who calculated that number should come out and stay with him for a few years and see the real world.

Generally, the term MTTF is being defined during the useful life of a device. "End of life" failures are generally not included in the number.

*Constant Failure Rate*—When a constant failure rate is assumed (which is valid during the useful life of a device) then the relationship between reliability, unreliability, and MTTF are straightforward. If the failure rate is constant then:

$$\lambda(t) = \lambda \quad (19\text{-}3)$$

For that assumption it can be shown that:

$$R(t) = e^{-\lambda t} \quad (19\text{-}4)$$

$$F(t) = 1 - e^{-\lambda t} \quad (19\text{-}5)$$

And

$$MTTF = 1/\lambda \quad (19\text{-}6)$$

Figure 19-3 shows the reliability and unreliability functions for a constant failure rate of 0.001 failures per hour. Note the plot for reliability looks the same as Figure 19-2, which shows the probability of successful operation given a probability of success for one hour of 0.999. It can be shown that a constant probability of success is equivalent to an exponential probability of success distribution as a function of operating time interval.

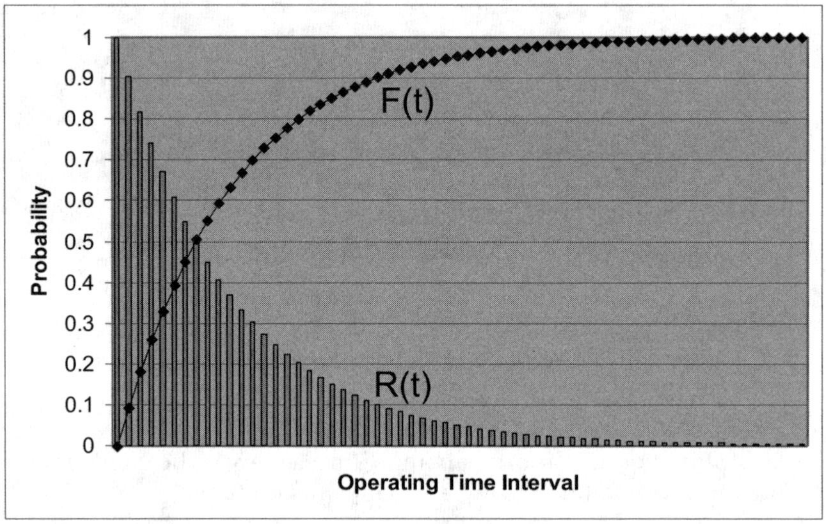


*Figure 19-3: Reliability/Unreliability Function for a Constant Failure Rate*

## 19.3 Useful Approximations

Mathematically, it can be shown certain functions can be approximated by a series of other functions. For all values of *x*, it can be shown that

$$e^x = 1 + x + x^2/2! + x^3/3! + x^4/4! + \ldots \qquad (19\text{-}7)$$

For a sufficiently small value of *x*, the exponential can be approximated with

$$e^x = 1 + x$$

Substituting $-\lambda t$ for x

$$e^{\lambda t} = 1 + \lambda t$$

Thus, there is an approximation for unreliability when $\lambda t$ is sufficiently small

$$F(t) = \lambda t \qquad (19\text{-}8)$$

Remember, this is only an approximation and not a fundamental equation. Often the notation for unreliability is PF (probability of failure) and the equation is shown as

$$PF(t) = \lambda t \qquad (19\text{-}9)$$

## 19.4 Measurements of Successful Operation—Repairable Systems

The measurement "reliability" requires that a system be successful for an interval of time. While this probability is a valuable estimate for situations where a system cannot be repaired during a mission, something different is needed for an industrial process control system where repairs can be made—often with the process operating.

*Mean Time To Restore (MTTR)*—MTTR is the "expected value" of the random variable "restore time" (or time to repair). The definition includes the time required to detect that a failure has occurred as well as the time required to make a repair once the failure has been detected and identified. Like MTTF, MTTR is an average value. MTTR is the average time required to move from unsuccessful operation to successful operation.

In the past, the acronym MTTR stood for "Mean Time To Repair." The term was changed in IEC 61508 because of confusion as to what was included. Some thought that "Mean Time To Repair" included only actual repair time. Others interpreted the term to include both time to detect a failure (diagnostic time) and actual repair time. The term "Mean Dead Time (MDT)" is commonly used in some parts of the world and means the same as Mean Time To Restore.

Mean Time To Restore (MTTR) is a term created to clearly include both diagnostic detection time and actual repair time. Of course, when actually estimating MTTR, one must include time to detect, recognize and identify the failure, time to obtain spare parts, time for repair team personnel to respond, actual time to do the repair, time to document all activities and time to get the equipment back in operation.

Reliability engineers often make the assumption that the probability of repair is an exponentially distributed function, in which case the "restore rate" is a constant. The lower case Greek letter mu is used to represent restore rate by convention. The equation for restore rate is:

$$\mu = 1/MTTR \tag{19-10}$$

Restore times can be difficult to estimate. This is especially true when periodic activities are involved. Imagine the situation where a failure in the safety instrumented system is not noticed until a periodic inspection and test is done. The failure may occur right before the inspection and test, in which case the detection time might be near zero. Or it may occur right after the inspection and test, in which case the detection time may get as large as the inspection period.

In such cases, it is probably best to model repair probability as a periodic function, not as a constant (see Ref. 1). This is discussed later in the section "Average Unavailability with Periodic Inspection and Test."

*Mean Time Between Failures (MTBF)*—MTBF is defined as the average time period of a failure/repair cycle. It includes time to failure, any time required to detect the failure and actual repair time. This implies that a component has failed and then has been successfully repaired. For a simple repairable component,

$$MTBF = MTTF + MTTR \tag{19-11}$$

The MTBF term can also be confusing. Since MTTR is usually much smaller than MTTF, MTBF is approximately equal to MTTF. The term MTBF is often substituted for MTTF and applies to both repairable systems and non-repairable systems.

*Availability*—The reliability measurement was not sufficiently useful for engineers who needed to know the average chance of success of a system when repairs are possible. Another measure of system success for repairable systems was needed. That metric is "availability." Availability is defined as "the probability that a device is successful at time t when needed and operated within specified limits." No operating time interval is directly involved. If a system is operating successfully, it is available. It does not matter whether it has failed in the past and has been repaired or has been operating continuously from startup without any failures. Availability is a measure of "uptime" in a system, unit, or module.

Availability and reliability are different metrics. Reliability is always a function of failure rates and operating time interval. Availability is a function of failure rates and repair rates. While instantaneous availability will vary during the operating time interval, this is due to changes in failure probabilities and repair situations. Often availability is calculated as an average over a long operating time interval. This is referred to as "steady state availability."

In some systems, especially safety instrumented systems (SIS), the repair situation is not constant. In safety instrumented systems the situation occurs when failures are discovered and repaired during a periodic inspection and test. For these systems, steady state availability is NOT a good measure of system success. Instead, average availability is calculated for the operating time interval between inspections. [Note: This is not the same measurement as steady state availability.]

*Unavailability*—A measure of failure that is used primarily for repairable systems. It is defined as "the probability that a device is not successful (is failed) at time t." Different metrics can be calculated including steady state unavailability and average unavailability over an operating time interval. Unavailability is the one's complement of availability; therefore,

$$U(t) = 1 - A(t) \tag{19-12}$$

*Steady State Availability*—Traditionally, reliability engineers have assumed a constant repair rate. When this is done, probability models can be solved for "steady state" or average probability of successful operation. The metric can be useful, but it has relevance only for long intervals of time.

Figure 19-4 shows a Markov probability model of a single component with a single failure mode. This model can be solved for steady state availability and steady state unavailability.

$$A = MTTF/(MTTF + MTTR) \tag{19-13}$$

$$U = MTTR/(MTTF + MTTR) \tag{19-14}$$

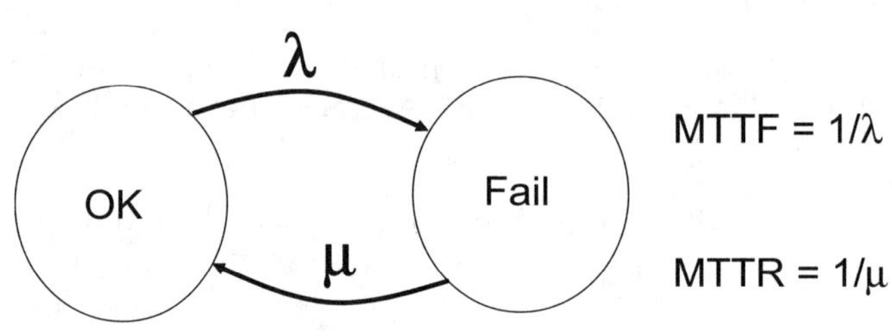


*Figure 19-4: Markov Model for Single Component and Single Failure Mode*

When the Markov model of Figure 19-4 is solved for availability as a function of operating time interval, the result is shown in Figure 19-5 labeled A(t). It can be seen that the availability reaches a "steady state" after some period of time.

Figure 19-6 shows a plot of Unavailability versus Unreliability. These plots are complementary to those shown in Figure 19-5.

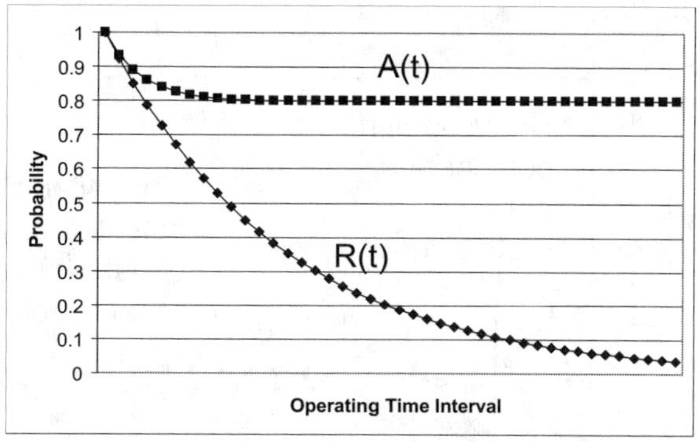


*Figure 19-5: Reliability Versus Availability*

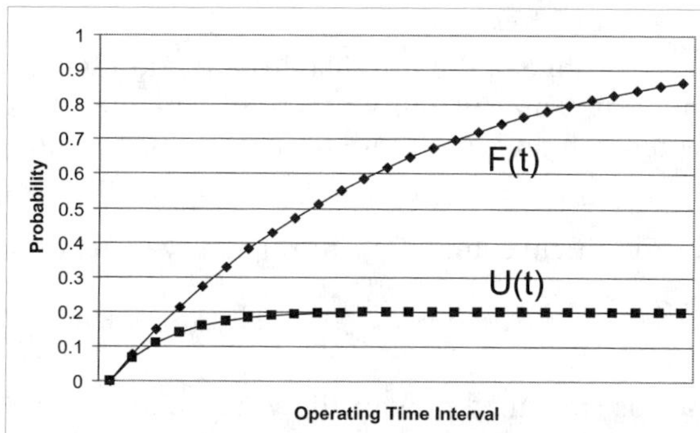


*Figure 19-6: Unreliability Versus Unavailability*

## 19.5 Average Unavailability with Periodic Inspection and Test

In low demand safety instrumented system applications, the restore rate is NOT constant. For failures not detected until a periodic inspection and test, the restore rate is zero until the time of the test. If it is discovered the system is operating successfully, then the probability of failure is set to zero. If it is discovered the system has a failure, it is repaired. In both cases the restore rate is high for a brief period of time. Dr. Julia V. Bukowski has described this situation and proposed modeling repair as a periodic impulse function (Ref. 1).

Figure 19-7 shows a plot of probability of failure in this situation. This can be compared with unavailability calculated with the constant restore rate model as a function of operating time. With the constant restore model, the unavailability reaches a steady state value. This value is clearly different than the result that would be obtained by averaging the unavailability calculated using a periodic restore period.

It is often assumed that periodic inspection and test will detect all failed components, and the system will be renewed to perfect condition. Therefore, the unreliability function is suitable for the problem. A mission time equal to the time between periodic inspection and test is used. In safety instrumented system applications, the objective is to find a model for the probability that a system will fail when a dangerous condition occurs. This dangerous condition is called a "demand."

Our objective, then, is to calculate the probability of failure on demand. If the system is operating in an environment where demands are infrequent (for example, once per 10 years) and independent from

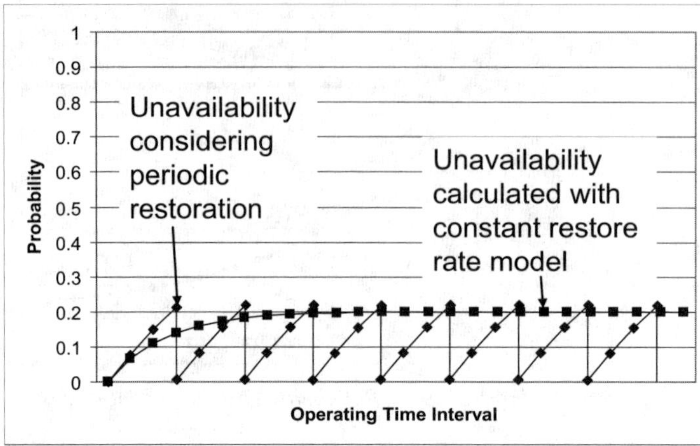


*Figure 19-7: Probability of Failure with Periodic Testing*

system proof tests; then an average of the unreliability function will provide the average probability of failure. This by definition is an "unavailability function" since repair is allowed. [Note: This averaging technique is not valid when demands are more frequent. Special modeling techniques are needed in that case.]

As an example, consider the single component unreliability function given in Equation 19-5.

$$F(t) = 1 - e^{-\lambda t}$$

This can be approximated as explained previously with Equation 19-8.

$$F(t) = \lambda t$$

The average can be obtained by using the expected value equation

$$PFavg = \frac{1}{T}\int_0^T PF(t)dt \tag{19-15}$$

with the result being an approximation equation

$$PFavg = \lambda t/2 \tag{19-16}$$

For a single component (non-redundant) or a single channel system; the approximation is shown in Figure 19-8. Note that the approximation is conservative and supplies a pessimistic value.

## 19.6 Periodic Restoration and Imperfect Testing

It is quite unrealistic to assume that inspection and testing processes will detect all failures. In the worst case, assume that testing is not done. In that situation what is the mission time? If the equipment is used for the life of an industrial facility then plant life is the mission time. Probability of failure would be modeled with the unreliability function using the plant life as the time interval.

If the equipment is required to operate only on demand, and the demand is independent of system failure, then the unreliability function can be averaged as explained in the preceding section.

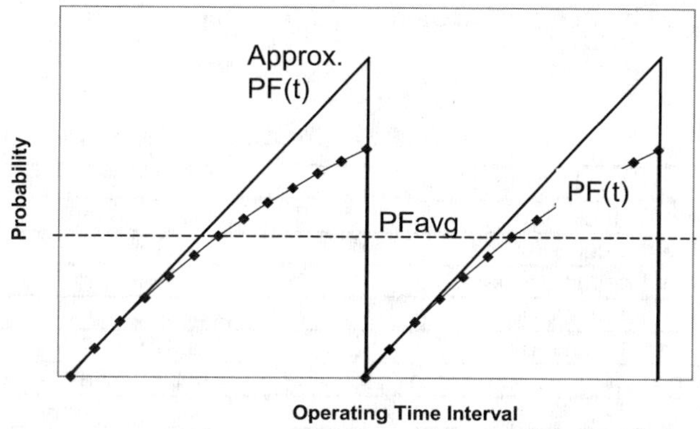


*Figure 19-8: Average Probability of Failure Approximation with Periodic Testing*

When only some failures are detected during the periodic inspection and test, then the average probability of failure can be calculated using an equation that combines the two types of failures—those detected by the test and those undetected by the test. One must estimate the percentage of failures detected by the test to make this split. The equation would be:

$$PFavg = C_{PT}\,\lambda\, TI/2 + (1 - C_{PT})\,\lambda\, LT/2 \qquad (19\text{-}17)$$

where

$\lambda$ = the failure rate
$C_{PT}$ = the percentage of failures detected by the proof test
TI = the periodic test interval
LT = lifetime of the process unit

## 19.7 Equipment Failure Modes

Instrumentation equipment can fail in different ways. We call these "failure modes." Consider a two wire pressure transmitter. This instrument is designed to provide a 4–20 milliamp electrical current signal in proportion to the pressure input. Detail failure modes, effects and diagnostic analysis of several of these devices reveals a number of failure modes; frozen output, current to upper limit, current to lower limit, diagnostic failure, communications failure and drifting/erratic output among perhaps others. These instrument failures can be classified into failure mode categories when the application is known.

If a single transmitter (no redundancy) were connected to a safety PLC programmed to trip when the current goes up (high trip), then the instrument failure modes could be classified as shown in Table 19-1.

*Table 19-1: Transmitter Failure Mode Categories*

| Instrument Failure Mode | SIF Failure Mode |
|---|---|
| Frozen output | Fail-Danger |
| Output to upper limit | Fail-Safe |
| Output to lower limit | Fail-Danger |
| Diagnostic failure | Annunciation |
| Communication failure | No Effect |
| Drifting / erratic output | Fail-Danger |

Consider possible failure modes of a PLC with a digital input and a digital output, both in a de-energize to trip (logic 0) design. The PLC failure modes can be categorized relative to the safety function as shown in Table 19-2.

*Table 19-2: PLC Failure Mode Categories*

| Instrument Failure Mode | SIF Failure mode |
|---|---|
| Input stuck High | Fail-Danger |
| Input stuck low | Fail-Safe |
| Input circuit oscillates | Fail-Danger* |
| Output stuck high | Fail-Danger |
| Output stuck low | Fail-Safe |
| Improper CPU execution | 50% Fail-Safe |
| | 50% Fail-Danger |
| Memory transient failure | 50% Fail-Safe |
| | 50% Fail-Danger |
| Memory permanent failure | 50% Fail-Safe |
| | 50% Fail-Danger |
| Power supply low (out of tolerance) | Fail-Danger* |
| Power supply high (out of tolerance) | Fail-Danger* |
| Power supply zero | Fail-Safe |
| Diagnostic timer failure | Annunciation |
| Loss of communication link | No Effect |
| Display panel failed | No Effect |
| * unpredictable - assume worst case | |

Final element components will fail also and, again, the specific failure modes of the components can be classified into relevant failure modes depending on the application. It is important to know if a valve will open or close on trip. Table 19-3 shows an example failure mode classification based on a close to trip configuration.

*Table 19-3: Final Element Failure Mode Categories*

| Instrument Failure Mode | SIF Failure mode |
|---|---|
| Solenoid plunger stuck | Fail-Danger |
| Solenoid coil burnout | Fail-Safe |
| Actuator shaft failure | Fail-Danger* |
| Actuator seal failure | Fail-Safe |
| Actuator spring failure | Fail-Danger |
| Actuator structure failure - air | Fail-Safe |
| Actuator structure failure - binding | Fail-Danger* |
| Valve shaft failure | Fail-Danger* |
| Valve external seal failure | No Effect |
| Valve internal seal damage | Fail-Danger |
| Valve ball stuck in position | Fail-Danger |
| * unpredictable - assume worst case | |

It should be noted the above failure mode categories apply to an individual instrument and may not apply to the set of equipment that performs a safety instrumented function, as the equipment set may contain redundancy. It should be also made clear that the above listings are not intended to be comprehensive or representative of all component types.

### Fail-Safe

Most practitioners use a definition of "Fail-Safe" for an instrument to define *a failure that causes a "false or spurious" trip of a safety instrumented function unless that trip is prevented by the architecture of the safety instrumented function.* Many formal definitions have been attempted that include "a failure which causes the system to go to a safe state or increases the probability of going to a safe state." This definition is useful at the system level and includes many cases where redundant architectures are used.

IEC 61508 uses the definition "failure which does not have the potential to put the safety-related system in a hazardous or fail-to-function state." This definition includes many failures that do not cause a false trip under any circumstances and is quite different from the definition practitioners need to calculate the false trip probability.

### Fail-Danger

Many practitioners define "Fail-Danger" as *a failure that prevents a safety instrumented function from performing its automatic protection function.* Variations of this definition exist in standards. IEC 61508 provides a definition similar to the one used in this book which reads "failure which has the potential to put the safety-related system in a hazardous or fail-to-function state." The definition from IEC 61508 goes on to add a note: "Whether or not the potential is realized may depend on the channel architecture of the system; in systems with multiple channels to improve safety, a dangerous hardware failure is less likely to lead to the overall dangerous or fail-to-function state." The note from IEC 61508 recognizes that a definition for a piece of equipment may not have the same meaning at the safety instrumented function level or the system level.

### Annunciation

Some practitioners recognize that certain failures within equipment used in a safety instrumented function prevent the automatic diagnostics from correct operation. When reliability models are built, many account for the automatic diagnostics ability to reduce the probability of failure. When these diagnostics stop working, the probability of dangerous failure or false trip is increased. While these effects may not be significant, unless they are modeled the effect is not known.

An annunciation failure is therefore defined as *a failure that prevents automatic diagnostics from detecting or annunciating that a failure has occurred inside the equipment.* Note the failure may be within the equipment that fails or inside an external piece of equipment designed for the purpose of automatic diagnostics. These failures would be classified as "Fail-Safe" in the definition provided in IEC 61508.

### No Effect

Some failures within a piece of equipment have no effect on the safety instrumented function nor cause a false trip nor prevent automatic diagnostics from working. Some functionality performed by the equipment is impaired but that functionality is not needed. These may simply be called "No Effect" failures. They are typically not used in any reliability model intended to obtain probability of a false trip or probability of a fail-danger. Per IEC61508, these would be classified as "Fail-Safe" or may be excluded completely from any analysis depending on interpretation of the analyst.

### Detected/Undetected

Failure modes can be further be classified as "detected" or "undetected" by automatic diagnostics performed somewhere in the safety instrumented system.

## 19.8 Safety Integrated Function (SIF) Modeling of Failure Modes

When evaluating safety instrumented function safety integrity, an engineer must examine more than the probability of successful operation. The failure modes of the system must be individually calculated. The normal metrics of reliability, availability and MTTF only suggest a measure of success. Additional metrics to measure safety integrity include probability of failure on demand (PFD), average probability of failure on demand (PFDavg), risk reduction factor (RRF) and mean time to fail dangerously ($MTTF_D$). Other related terms are probability of failing safely (PFS) and mean time to fail spurious ($MTTF_S$).

### PFS/PFD

There is a probability that a safety instrumented function will fail and cause a spurious/false trip of the process. This is called probability of failure safely (PFS). There is also a probability that a safety instrumented function will fail such that it cannot respond to potentially dangerous condition. This is called probability of failure on demand (PFD).

### PFDavg

PFD average (PFDavg) is a term used to describe the average probability of failure on demand. PFD will vary as a function of the operating time interval of the equipment. It will not reach a steady state value if any periodic inspection, test and repair is done. Therefore, the average value of PFD over a period of time can be a useful metric if it assumed that the potentially dangerous condition (also called hazard) is independent from equipment failures in the safety instrumented function.

The assumption of independence between hazards and safety instrumented function failures seems very realistic. (NOTE: If control functions and safety functions are performed by the same equipment, the assumption may not be valid! Detailed analysis must be done to ensure safety in such situations, and it is best to avoid such designs completely.) When hazards and equipments are independent, it is realized a hazard may come at any time. Therefore, international standards have specified that PFDavg is an appropriate metric for measuring the effectiveness of a safety instrumented function.

PFDavg is defined as the arithmetic mean over a defined time interval. For situations where a safety instrumented function is periodically inspected and tested, the test interval is correct time period. Therefore:

$$PFDavg(TI) = \frac{1}{TI}\int_0^{TI}(PFD)dt$$

This definition is used to obtain numerical results in several of the system modeling techniques. In a discrete time Markov model using numerical solution techniques, a direct average of the time dependent numerical values will provide the most accurate answer. When analytical equations for PFD are obtained using a fault tree, the above equation can be used to obtain equations for PFDavg.

## 19.9 Redundancy

There are applications where the reliability or safety integrity of a single instrument is not sufficient. In these cases more than one instrument is used in a design. Some arrangements of the instruments are designed to provide higher reliability (typically to protect against a single "safe" failure). Other arrangements of instruments are designed to provide higher safety integrity (typically to protect against a single "dangerous" failure). And some arrangements are designed to provide both high reliability and high safety integrity. When multiple instruments are wired (or configured) to provide redundancy to protect against one or more failure modes, these arrangements are known as "architectures." A listing of some common architectures is shown in Table 19-4. These architectures are described in detail in Chapter 14 of Reference 2.

*Table 19-4: Common Redundant Architectures*

| Architecture | Number of units | Output Switches | Objective |
|---|---|---|---|
| 1oo1 | 1 | 1 | Base unit |
| 1oo2 | 2 | 2 | High Safety |
| 2oo2 | 2 | 2 | Maintain output |
| 1oo1D | 1 | 2 | High Safety |
| 2oo3 | 3 | 6 | Safety and Availability |
| 2oo2D | 2 | 4 | Safety and Availability |
| 1oo2D | 2 | 4 | Safety and Availability |

The naming convention stands for X out of Y where Y is the number of equipment sets in the design and X is the number of equipment sets needed to perform the function. In some advanced architecture names, the term D is added to designate a switch that is controlled by diagnostics to reconfigure the equipment if a failure is detected in one equipment set.

## 19.10 References

1. Bukowski, J. V. "Modeling and Analyzing the Effects of Periodic Inspection on the Performance of Safety-Critical Systems." *IEEE Transactions of Reliability.* Vol. 50, No. 3 (September 2001).

2. Goble, W. M. *Control System Safety Evaluation and Reliability.* Second Edition. ISA, 1998.

3. Goble W. M. and H. Cheddie. *Safety Instrumented Systems Verification: Practical Probabilistic Calculations.* ISA, 2005.

4. Billinton, R. and R.N. Allan. *Reliability Evaluation of Engineering Systems: Concepts and Techniques.* Plenum Press, 1983.

## About the Author

**William M. Goble** is currently Principal Partner, exida.com, a company that does consulting, training and support for safety critical and high availability process automation. He has over 25 years of experience in control systems doing product development, engineering management, marketing, training and consulting. Dr. Goble has a BSEE from the Pennsylvania State University, an MSEE from Villanova and a PhD from Eindhoven University of Technology in Reliability Engineering. He is a registered professional engineer in the State of Pennsylvania and a Certified Functional Safety Expert (CFSE). He is a fellow member of ISA and author of several ISA books.

# 20 Process Safety and Safety Instrumented Systems

*By Paul Gruhn*

## Topic Highlights

*Safety Instrumented System Design Life Cycle*
*System Technologies*
*System Analysis*
*Abnormal Situation Management*

## 20.1 Introduction

Process plants produce products needed in today's society. The drawback of such plants is their operation involves some risk. While there is no such thing as zero risk, one goal is to design the process to keep the risk to as low a level as practical. This means we need to be able to evaluate and rank risk in order to make decisions on various design and cost options to control it.

Safety instrumented systems (SIS) are one means of maintaining the safety of process plants. These systems monitor a plant for potentially unsafe conditions and bring the equipment or the process to a safe state if certain conditions are violated. Today's safety instrumented system standards are performance-based, not prescriptive. In other words, they do not mandate technologies, levels of redundancy, test intervals, or system logic. Essentially they state "the greater the level of risk, the better the safety systems needed to control it."

There are a variety of methods of evaluating risk. There are also a variety of methods of equating risk to the performance required of a safety system. The overall design of safety instrumented systems is not a simple, straightforward matter. The total engineering knowledge and skills required are often beyond that of any single person. An understanding is required of the process, operations, instrumentation, control systems, and hazard analysis. This typically calls for the interaction of a multi-disciplined team.

Experience has shown a detailed, systematic, methodical, well documented design *process* is necessary in the design of safety instrumented systems. This starts with a safety review of the process, implementation of other safety layers, and systematic analysis, as well as detailed documentation and procedures. These steps are described in various regulations, standards, guidelines and recommended practices. The steps are referred to as the safety design life cycle. The intent is to leave a documented, auditable trail and make sure that nothing is neglected or falls between the inevitable cracks within every organization.

Hindsight is easy. Everyone always has 20/20 hindsight. *Fore*sight, however, is a bit more difficult. Foresight is required with today's large, high risk systems. We simply can't afford to design large petrochemical plants by trial and error. The risks are too great to learn that way. We have to try and prevent certain accidents, no matter how remote the possibility, even if they have never yet happened. This is the subject of *system safety.*

## 20.2 Safety Instrumented System Design Life Cycle

Safety instrumented systems require a methodical design process to prevent important items from falling through cracks. Figure 20-1 shows the life cycle steps as described in the ANSI/ISA-84.00.01-2004 Parts 1-3 (IEC 61511-1 through 3 Mod) - *Functional Safety: Safety Instrumented Systems for the Process Industry Sector* standards. This should be considered one example only. There are variations of the life cycle presented in other industry documents. A company may wish to develop its own variation of the life cycle based on its unique requirements.

Some will complain that performing all of the steps in the life cycle, like all other tasks designed to lower risk, will increase overall costs and result in lower productivity. One in-depth study conducted by a group including major engineering societies, 20 industries, and 60 product groups with a combined exposure of over 50 billion hours, concluded that *production increased as safety increased.* In the U.S., OSHA (Occupational Safety and Health Administration) documented that, since the adoption of their process safety management regulation (29 CFR 1910.119), the number of accidents has decreased over 20%, and companies are reporting their productivity is *higher.*

### 20.2.1 Hazard & Risk Analysis

One of the goals of process plant design is to have a facility that is inherently safe. Trevor Kletz, one of the pillars of the process safety community, has said many times, "What you don't have, can't leak." Hopefully the design of the process can eliminate many of the hazards, such as unnecessary storage of intermediate products, use of safer catalysts, etc.

One of the first steps in designing a safety system is developing an understanding of the hazards and risks associated with the process.

*Hazard analysis* consists of *identifying* the hazards and hazardous events. There are numerous techniques that can be used (e.g., HAZOP – HAZard and OPerability study, what-if, fault tree, checklist, etc.). Techniques such as checklists are useful for well-known processes where there is a large amount of accumulated knowledge. The accumulated knowledge can be condensed in a checklist of items that needs to be considered during the design phase. Other techniques, such as HAZOP or what-if, are more useful for processes that have less accumulated knowledge. These techniques are more systematic in their approach and typically require a multi-disciplined team. They typically require detailed review of design drawings and asking a series of questions intended to stimulate the team into thinking about potential problems, and what might cause them. For example, what if the flow is too high, too low, reverse, etc.? What might cause such a condition?

*Risk assessment* consists of *ranking* the risk of the hazardous events that have been identified in the hazard analysis. Risk is a function of the frequency, or probability, of an event, and the severity or consequences of the event. Risks may impact personnel, production, capital equipment, the environment, company image, etc. Risk assessment can either be qualitative or quantitative. Qualitative assessments subjectively rank the risks from low to high. Quantitative assessments attempt to assign numerical factors to the risk, such as death or accident rates, actual size of a release, etc. These studies are *not* the sole responsibility of the instrument or control system engineer. There are obviously a number of other disciplines required to perform these assessments, such as safety, operations, maintenance, process, mechanical design, electrical, etc.

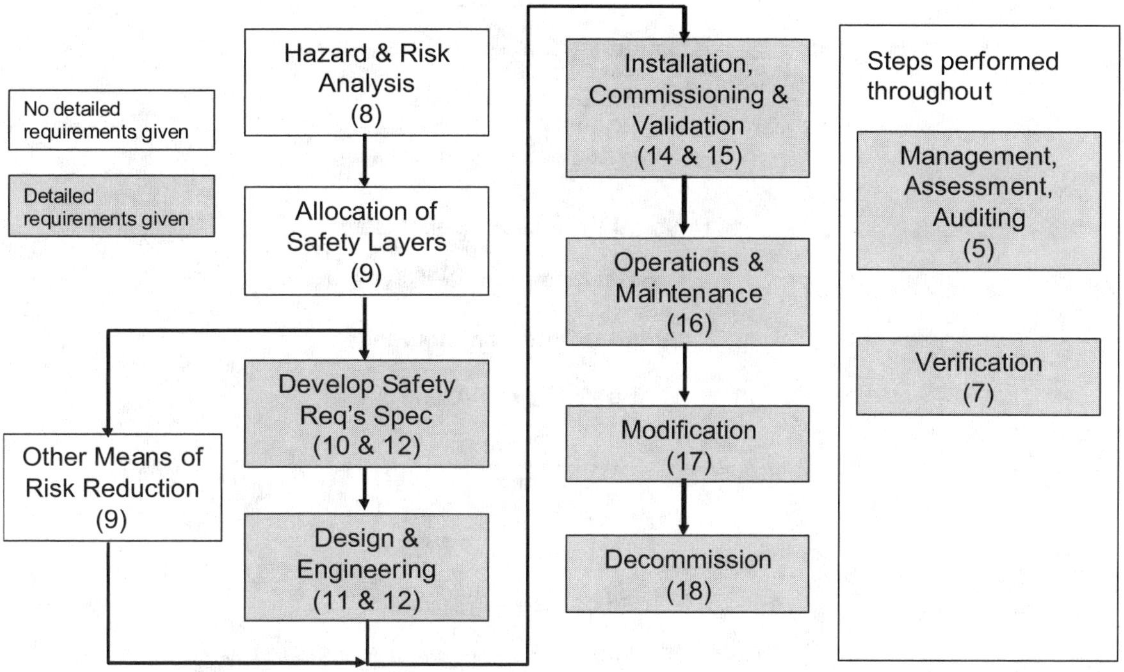


*Figure 20-1: SIS Design Life Cycle [with Clause Numbers from ANSI/ISA-84.00.01-2004 Parts 1-3 (IEC 61511 Parts 1 through 3 Mod) - Functional Safety: Safety Instrumented Systems for the Process Industry Sector]*

### 20.2.2 Allocation of Safety Functions to Protective Layers

Figure 20-2 shows an example of multiple independent protection layers that may be used in a plant. Various industry standards either mandate or strongly suggest that safety systems be completely separate and independent from control systems. Each layer helps reduce the overall level of risk. The inner layers help *prevent* a hazardous event (e.g., an explosion due to an over pressure condition) from ever occurring and are referred to as *protection layers*. The outer layers are used to *lessen the consequences* of a hazardous event once it has already occurred, and are referred to as *mitigation layers*.

Figure 20-3 is a graphical way of representing the risk reduction that each layer provides. The vertical line on the right side of the figure represents the level of risk inherent in the process. This would be determined from the safety review. Let's consider an example of the risk of a serious explosion to be once per year, assuming no safety layers were in place. Let's also assume our corporate safety target (i.e., the tolerable level of risk, shown as the vertical line on the left side of the figure) for such an event, is 1/10,000 per year. (Determining such targets is a significant subject all unto itself and is beyond the scope of this chapter.) The basic process control system (BPCS) maintains process variables within safe boundaries and therefore provides a level of protection. Standards state that one should not claim more than a risk reduction factor of 10 for the BPCS. If there are alarms separate from the control system, and assuming the operators have enough time to respond and have procedures to follow, one might assume a risk reduction factor of 10 for the operators. If relief valves could also prevent the overpressure condition, failure rates and test intervals could be used to calculate their risk reduction factor (also a significant subject all unto itself and beyond the scope of this chapter). Let's assume

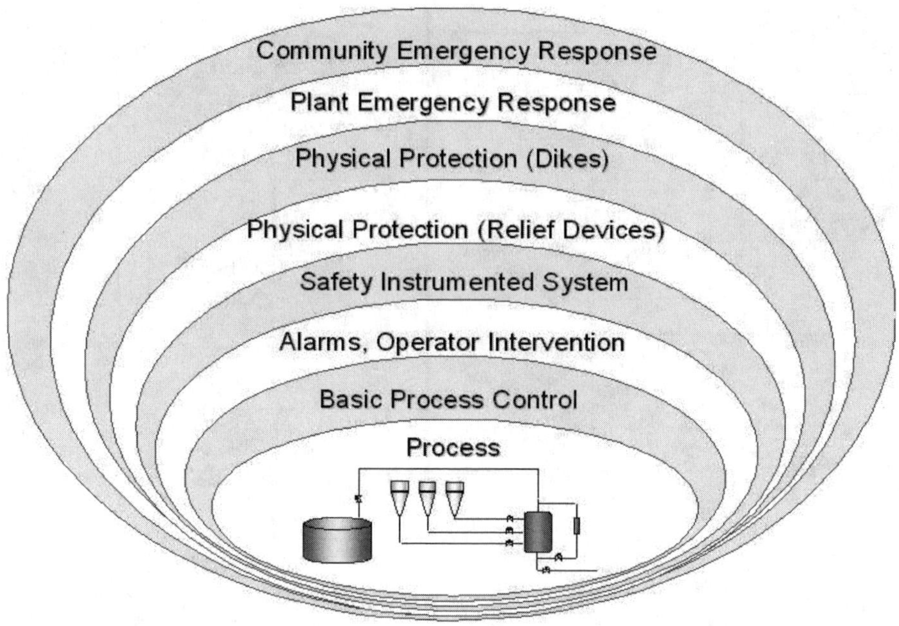


*Figure 20-2: Multiple Independent Protection Layers*

a risk reduction factor of 100 for the relief valves. Without a safety instrumented system, the level of overall risk is:

| 1 / year | * 1 / 10 | * 1 / 10 | * 1 / 100 | = 1 / 10,000 |
|---|---|---|---|---|
| Initiating event frequency | BPCS risk reduction | Operator risk reduction | Relief valve risk reduction | Overall risk reduction |

Without a safety system, the example above does *not* meet the corporate risk target of 1/10,000. However, adding a safety system that provides a level of risk reduction of at least 10 *will* result in meeting the corporate risk target. As shown in Table 20-1 below, this falls into the SIL (Safety Integrity Level) 1 range. This is an example of LOPA (Layer Of Protection Analysis), which is one of several techniques for determining the performance required of a safety system.

If the risks associated with a hazardous event can be prevented or mitigated with something other than instrumentation—which is complex, expensive, requires maintenance, and is prone to failure—so much the better. For example, a dike is a simple and reliable device that can easily contain a liquid spill. KISS (Keep It Simple, Stupid) should be an overriding theme.

For all safety functions assigned to instrumentation (i.e., safety instrumented functions), the level of performance required needs to be determined. The standards refer to this as Safety Integrity Level (SIL). This continues to be a difficult step for many organizations. Note that SIL is not directly a measure of process risk, but rather a measure of the safety system performance required in order to control the risks identified earlier to an acceptable level. The standards describe a variety of techniques on how safety integrity levels can be determined. This text will not attempt to summarize that material beyond the brief LOPA example given above.

Tables in the standards then show the performance requirements for each integrity level. Table 20-1 lists the performance requirements for "low demand" mode systems, which are most common in the

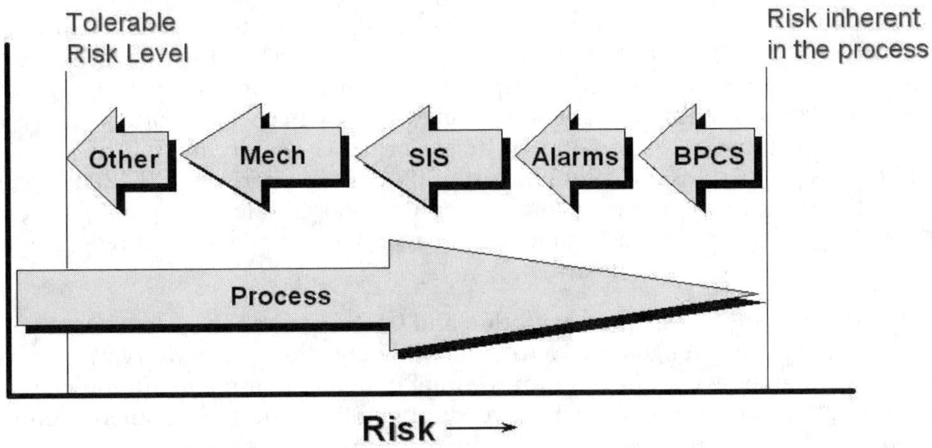


*Figure 20-3: Risk Reduction Provided by Each Protection Layer*

process industries. This shows how the standards are performance-oriented and not prescriptive (i.e., they do not mandate technologies, levels of redundancy, or test intervals).

*Table 20-1: Safety Integrity Levels and Required Safety System Performance for Low Demand Mode Systems*

| Safety Integrity Level (SIL) | Probability of Failure on Demand (PFD) | Safety Availability (1-PFD) | Risk Reduction Factor (1/PFD) |
|---|---|---|---|
| 4 | .0001 - .00001 | 99.99 – 99.999% | 10,000 – 100,000 |
| 3 | .001 - .0001 | 99.9 - 99.99% | 1,000 - 10,000 |
| 2 | .01 - .001 | 99 - 99.9% | 100 - 1,000 |
| 1 | .1 - .01 | 90 - 99% | 10 - 100 |

### 20.2.3 Develop Safety Requirements Specification

The next step consists of developing the safety requirements specification. This consists of documenting the I/O (Input & Output) requirements, functional logic, and the SIL of each safety function. This will naturally vary for each system. There is no general, across-the-board recommendation that can be made. One simple example might be: "If temperature sensor TT2301 exceeds 410 degrees, then close valves XV5301 and XV5302. This function must respond within three seconds and needs to meet SIL 2." It may also be beneficial to list reliability requirements if nuisance trips are a concern. For example, many different systems may be designed to meet SIL 2 requirements, but each will have different nuisance trip performance. Considering the costs associated with lost production downtime, as well as safety concerns, this may be an important issue. Also, one should include *all* operating conditions of the process, from start-up through shutdown, as well as maintenance. One may find that certain logic conditions conflict during different operating modes of the process.

The system will be programmed and tested according to the logic determined during this step. If an error is made here, it will carry through for the rest of the design. It won't matter how redundant or how often the system is manually tested. It simply won't work properly when required. These are referred to as systematic or functional failures.

### 20.2.4 SIS Design & Engineering

Any proposed *conceptual design* (i.e., a proposed implementation) needs to be analyzed to see if it meets the functional and performance requirements. One needs to initially select a technology, configuration, test interval, etc. This pertains to the field devices as well as the logic box. Factors to consider are overall size, budget, complexity, speed of response, communication requirements, interface requirements, method of implementing bypasses, testing, etc. One can then perform a simple quantitative analysis to see if the proposed system meets the performance requirements. The intent is to evaluate the system *before* one specifies the solution. Just as it's better to perform a HAZOP *before* you build the plant rather than afterwards, it's better to analyze the proposed safety system *before* you specify, build, and install it. The reason for both is simple. It's cheaper, faster, and easier to redesign on paper. This topic is discussed in greater detail below.

Detail design involves the actual documentation and fabrication of the system. Once a design has been chosen, the system must be engineered and built following strict and conservative procedures. This is the only realistic method we know of for preventing design and implementation errors. The process requires thorough documentation which serves as an auditable trail that someone else may follow for independent verification purposes. It's difficult to catch one's own mistakes.

After the system is constructed, the hardware and software should be fully tested at the integrator's facility. Any changes that may be required will be easier to implement at the factory rather than the installation site.

### 20.2.5 Installation, Commissioning and Validation

It is important to ensure the system is installed and started up according to the design requirements, and it performs per the safety requirements specification. The entire system must be checked, this time including the field devices. There should be detailed installation, commissioning, and testing documents outlining each procedure to be carried out. Completed checks should be signed off in writing documenting that each and every function has been checked and has satisfactorily passed all tests.

### 20.2.6 Operations and Maintenance

Not all faults are self-revealing. Therefore, *every* safety instrumented system *must* be periodically tested and maintained. This is necessary to make certain the system will respond properly to an actual demand. The frequency of inspection and testing will have been determined earlier in the life cycle. All testing must be documented. This will enable an audit to determine if the initial assumptions made during the design (e.g., failure rates, failure modes, test intervals, etc.) are valid based on actual experience.

### 20.2.7 Modifications

As process conditions change, it may be necessary to make modifications to the safety system. All proposed changes require returning to the appropriate phase of the life cycle in order to review the impact of the change. A change that may be considered minor by one individual may actually have a major impact to the overall process. This can only be determined if the change is documented and thoroughly reviewed by a qualified team. Hindsight has shown that many accidents have been caused by this lack of review. Changes that are made must be thoroughly tested.

### 20.2.8 Decommissioning

Decommissioning a system entails a review to make sure removing the system from service will not impact the process or any other surrounding units. Means must be available during the decommissioning process to protect the personnel, equipment, and the environment.

## 20.3 System Technologies

### 20.3.1 Logic Systems

There are a number of technologies available for use in safety systems—pneumatic, electromechanical relays, solid state, and PLCs (programmable logic controllers). There is no one overall "best" system, each has advantages and disadvantages. The decision over which system may be best suited for an application will depend on many factors, such as budget, size, level of risk, flexibility, maintenance, interface and communication requirements, security, etc.

Pneumatic systems are most suitable for small applications where there are concerns over simplicity, intrinsic safety, and lack of available electrical power.

Relay systems are fairly simple, relatively inexpensive to purchase, immune to most forms of EMI/RFI interference, and can be built for many different voltage ranges. They generally do not incorporate any form of interface or communications. Changes to logic require manually changing documentation. In general, relay systems are usually only used for relatively small applications.

Solid state systems (hardwired systems that do not incorporate software) are also available. Several of these systems were built specifically for safety applications and include features for testing, bypasses and communications. Logic changes still require manually changing documentation. These systems have fallen out of favor with many due to their high cost, along with the acceptance of software-based systems.

Software-based systems, generally industrial PLCs, offer software flexibility, self-documentation, communications, and higher level interfaces. Unfortunately, many general purpose systems were not designed specifically for safety and do not offer features required for more critical applications (such as effective self-diagnostics). However, certain specialized single, dual and triplicated systems were developed for more critical applications and have become firmly established in the process industries. These systems offer extensive diagnostics and better redundancy schemes and are often referred to as "Safety PLCs."

### 20.3.2 Field Devices

In the process industries, more hardware faults occur in the peripheral equipment—that is, the measuring instruments/transmitters and the control valves—than in the logic system itself. The overall reliability of a computerized control system may therefore not be significantly different than a conventional hard-wired relay or solid-state system.

#### Sensors

Sensors are used to measure process variables, such as temperature, pressure, flow, level, etc. They may consist of simple pneumatic or electric switches which change state when a setpoint is reached, or they may contain pneumatic or electric analog transmitters which give a variable output in relation to the strength or level of the process variable.

Sensors, like any other device, may fail in a number of different ways. They may cause nuisance trips (i.e., respond without any change of input signal). They may also fail to respond to an actual change of input condition. While these are the two failure modes of most concern for safety systems, there are additional failure modes as well, such as leaking, erratic output, responding at an incorrect level, etc.

Most safety systems are designed to be fail-safe. This usually means that, when power is lost, the safety system makes the process revert to a safe state, which usually means stopping production. (Nuisance trips should be avoided for safety reasons as well, since startup and shutdown operations are usually associated with the highest levels of risk.) Thought must be given to how the sensors should respond in order to be fail-safe.

**Final Elements**

Final elements generally have the highest failure rates of any component in the system. They are mechanical devices and subject to harsh process conditions. Safety shutoff valves also suffer from the fact that they are usually open and not activated for long periods of time, except for testing. One of the most common failure modes is a valve that is stuck, or frozen in place. Valves should be fail-safe upon loss of power, which usually entails the use of a spring loaded actuator.

Solenoids are one of the most critical components of final elements. It is important to use a good industrial grade solenoid valve. The valve must be able to withstand high temperatures, including the heat generated by the coil itself. In general, the reliability of solenoids is very low. One of the most common failures is a coil burning out, which causes a false trip.

## 20.4 System Analysis

What is suitable for SIL 1, for SIL 2, and SIL 3? (SIL 4 is now defined in the 2004 version of ANSI/ISA-84.00.01, but users are referred to IEC 61508, as such systems should be extremely rare in the process industry.) Which technology, what level of redundancy, what manual test interval, and what about the field devices are all questions that need to be answered. Things are not as intuitively obvious as they may seem. Dual is not always better than simplex, and triple is not always better than dual.

We do not design nuclear power plants or aircraft by gut feel or intuition. As engineers, we must rely on quantitative evaluations as the basis for our judgments. Quantitative analyses may be imprecise and imperfect, but it nevertheless is a valuable exercise for the following reasons:

- It provides an early indication of a system's potential to meet the design requirements.
- It enables one to determine the weak link in the system (and fix it, if necessary).

In order to predict the performance of a system, one needs performance data of all the components. Information is available from user records, vendor records, military style predictions, and commercially available data bases in different industries.

When modeling the performance of a safety system, one needs to consider two failure modes:

- ***Safe failures*** result in nuisance trips and lost production. Common terms used to describe this mode of performance are $MTBF_{sp}$ (Mean Time Between Failure, spurious) and nuisance trip rate.
- ***Dangerous failures*** result in hidden failures where the system will not respond when required. Common terms used to quantify performance in this mode are PFD (Probability of Failure on Demand), RRF (Risk Reduction Factor, which is 1/PFD), and SA (Safety Availability, which is 1-PFD).

Note that safety integrity levels *only* refer to dangerous system performance. There is *no* relationship between safe and dangerous system performance. A SIL 4 system may produce a nuisance trip every month, just as a SIL 1 system may produce a nuisance trip once in 20 years. Knowing the performance in one mode tells you *nothing* about the performance in the other.

There are a number of modeling techniques used to analyze and predict safety system performance. The ISA technical report ISA-TR84.00.02-2002 - Parts 1-5, *Safety Instrumented Functions (SIF) Safety Integrity Level (SIL) Evaluation Techniques Package* provides an overview of reliability block diagrams, fault trees, and Markov models. Each method has its pros and cons. No method is more "right" or "wrong" than any other. They are all simplifications and can account for different factors. Using such techniques, one can model different technologies, levels of redundancy, test intervals, and field device configurations. One can model systems using a hand calculator, or develop spreadsheets or stand-alone programs to automate and simplify the task. Table 20-2 is an example of a "cookbook" that one could develop using any of the modeling techniques.

*Table 20-2: General System Recommendations (see accompanying notes)*

| | Subsystem | | |
|---|---|---|---|
| **SIL** | **Sensors** | **Logic** | **Final Elements** |
| 1 | Simplex switches or transmitters | Relays<br>Solid state systems<br>General purpose PLCs | Simplex "dumb" |
| 2 | Redundant switches/ transmitters<br>Transmitters with comparison<br>Simplex safety transmitters | Relays<br>Fail-safe or fully tested solid state systems<br>Certified safety PLCs | Redundant "dumb"<br>Simplex "smart" (e.g., partial stroking valves) |
| 3 | Redundant transmitters | Relays<br>Fail-safe or fully tested solid state systems<br>Certified redundant safety PLCs | Redundant "smart" (e.g., partial stroking valves) |

**Table Notes:**
Such tables are, by their very nature, oversimplifications. It is not possible to show the impact of *all* design features (failure rates, failure mode splits, diagnostic levels, quantities, manual test intervals, common cause factors, etc.) in a single table. Users are urged to perform their own analysis in order to justify their design decisions. The above table should be considered an example only, based on the following assumptions:

1. Separate logic systems are assumed for safety applications. Safety functions should not be performed solely within the BPCS (Basic Process Control System).
2. One sensor and two final elements are assumed. Field devices are assumed to have an MTBF (Mean Time Between Failure) in both failure modes (safe and dangerous) of 50 years.
3. Simplex (non-redundant) transmitters are assumed to have 30% diagnostics, redundant transmitters >95%.
4. "Transmitters with comparison" means comparing the control transmitter with the safety transmitter and assuming 90% diagnostics.
5. "Dumb" valves offer no self-diagnostics, "smart" valves (e.g., automated partial stroking valves) are assumed to offer 80% diagnostics.
6. When considering solid state logic systems, only solid state systems specifically built for safety applications should be considered. These systems are either inherently fail-safe (like relays) or offer extensive self-diagnostics.
7. General purpose PLCs are not appropriate beyond SIL 1 applications. They do not offer effective enough diagnostic levels to meet the higher performance requirements. Check with your vendors for further details.
8. One-year manual testing is assumed for all devices. (More frequent testing would offer higher levels of safety performance.)
9. Redundant configurations are assumed to be either 1oo2 or 2oo3. [Editor's note: "1oo2" is ISA Standards speak for "one out of two"; "1oo2" means there are two devices making the decision and they both must be in a "go" mode before an output can be achieved. The elec-

trical equivalent of 1oo2 is two switches wired in series and connected to a load.] 1oo2 configurations are safe, at the expense of more nuisance trips. 2oo2 configurations are less safe than simplex and should only be used if it can be documented that they meet the overall safety requirements.

10. The above table does not categorize the nuisance trip performance of any of the systems.

## 20.5 Abnormal Situation Management

One subtopic of process plant safety is the handling of abnormal situations. The Abnormal Situation Management® (ASM®) Consortium (the phrases are U.S. registered trademarks of Honeywell, Inc.) is a research and development consortium of 11 companies and universities concerned about the negative effects of industrial plant incidents. The group was established in the early 1990s as an outgrowth of an effort to define improvements to DCS (distributed control system) alarm system technologies. The aim of the group is to identify problems facing industrial plant operations during abnormal conditions, and to develop solutions. The deliverables of the group consist of products and services, guidelines and other documents, and information-sharing workshops.

Abnormal situations are managed by prevention, early detection, and mitigation. The intent is to reduce unplanned outages and process variability that may reduce profits and place plant employees and local residents at risk.

The vision of the ASM consortium is to empower and enable organizations to proactively manage their plants to maximize safety and minimize environmental impact, while allowing the processes to be pushed to their optimal limits. The consortium conducts research that develops and advances the collective knowledge of its members. It also directs the development of tools, best practices, and services that facilitate the conversion of ASM knowledge into practice.

The consortium achieves its mission with three programs: research, development, and communications. The research program conducts investigations and shares experiences on factors contributing to the successful reduction of abnormal situations. The development program captures the knowledge represented in, and developed by, the consortium and returns it to customers in the form of products and services. The communication program disseminates the information within the consortium membership to enhance the understanding and use of effective ASM practices.

## 20.6 Key Points

- Follow the steps defined in the safety design life cycle.
- If you can't define it, you can't control it.
- Justify and *document* all of your decisions (i.e., leave an auditable trail).
- The goal is to have an inherently safe process (i.e., one where you don't even need an SIS).
- Don't put all of your eggs in one basket (i.e., have multiple, independent safety layers).
- The SIS should be fail-safe and/or fault-tolerant.
- Analyze the problem, *before* you specify the solution.
- *All* systems *must* be periodically tested.
- *Never* leave points in bypass during normal operation!

## 20.7 Rules of Thumb

- Maximize diagnostics. (This is the most critical factor in safety performance.)
- Any indication is better than none (e.g., transmitters have advantages over switches, systems should provide indications even when signals are in bypass, etc.).
- Minimize potential common cause problems.
- General purpose PLCs are not suitable for use beyond SIL 1.
- When possible, use independently approved and/or certified components/systems (e.g., FM, TÜV, etc.).

## 20.8 References

### 20.8.1 Standards, Guidelines, Recommended Practices, Technical Reports:

ANSI/ISA-84.00.01-2004 Parts 1-3 (IEC 61511-1 through 3 Mod) - *Functional Safety: Safety Instrumented Systems for the Process Industry Sector.*

IEC 61508-SER Ed. 1.0 b:2005. *Functional Safety of Electrical/Electronic/Programmable Electronic Safety-Related Systems.*

RP 554 - *Process Instrumentation and Control.* American Petroleum Institute, 1995.

*Guidelines for Safe Automation of Chemical Processes,* American Institute of Chemical Engineers, Center for Chemical Process Safety, 1993.

ISA-TR84.00.02-2002 - Parts 1-5, *Safety Instrumented Functions (SIF) Safety Integrity Level (SIL) Evaluation Techniques Package.*

### 20.8.2 Books:

Gruhn, Paul and Harry Cheddie. *Safety Shutdown Systems: Design, Analysis and Justification.* ISA, 1998.

American Institute of Chemical Engineers, Center for Chemical Process Safety. *Layer of Protection Analysis: Simplified Process Risk Assessment.* Wiley Publishing, 2001.

McMillan, Gregory K. and Douglas M. Considine. *Process/Industrial Instruments and Controls Handbook.* Fifth edition. McGraw-Hill Professional, 1999.

Leveson, Nancy G. *Safeware - System Safety and Computers.* Addison-Wesley, 1995.

American Institute of Chemical Engineers, Center for Chemical Process Safety. *Guidelines for Hazard Evaluation Procedures.* Second edition. Wiley-Interscience, 1992.

American Institute of Chemical Engineers, Center for Chemical Process Safety. *Guidelines for Chemical Process Quantitative Risk Analysis.* Wiley-Interscience, 1989.

Kletz, Trevor A. *What Went Wrong? Case Studies of Process Plant Disasters.* Gulf Publishing Co., 1994.

Perrow, Charles. *Normal Accidents.* Princeton University Press, 1999.

Chiles, James R. *Inviting Disaster: Lessons from the Edge of Technology.* HarperBusiness, 2001.

### 20.8.3 Papers:

Gruhn, Paul. "Safety Instrumented System Design: Valuable Lessons Learned." *Hydrocarbon Processing*, Aug. 2000.

Gruhn, Paul. "Separate Safety and Process Controls - Or Else." *InTech*, Aug. 1999, pp. 60-61.

Gruhn, Paul. "Accidents lead to modern safety instrumented systems." *InTech*, Jan. 1999, pp. 48-51.

### 20.8.4 Other:

ASM website: http://www.asmconsortium.com

## About the Author

**Paul Gruhn** is a safety product specialist with ICS Triplex in Houston, Texas. An ISA Fellow, he is a member of the ISA SP84 committee, which wrote the 1996 Application of Safety Instrumented Systems for the Process Industries and 2004 Functional Safety: Safety Instrumented Systems for the Process Industry Sector versions of the ISA 84 series standards. Paul is the developer and instructor for ISA's three-day course EC50, "Safety Instrumented Systems," along with the matching one-day course and three-part web seminar series. He is a member of the System Safety Society and the National Society of Professional Engineers. He has a BS degree in mechanical engineering from Illinois Institute of Technology in Chicago. He is a licensed professional engineer in Texas and a certified functional safety expert (a TÜV certification).

# 21 Electrical Installations

*By Victor Maggioli Sr., Graham Elvis, & Lawrence (Larry) M. Thompson*

## Topic Highlights

*Introduction*
*Scope*
*Grounding and Bonding*
*Ground Systems*
*Grounding Loops*
*Noise Reduction*
*Electrostatic Noise*
*Surge Suppressors*
*Power*
*Uninterruptible Power Systems*
*Electrical Installation Details*

Motto for Electrical Installation: *"It depends upon the application."*

## 21.1 Introduction

This chapter is about electrical installations (EI) in industrial facilities. It is designed to provide the necessary information to ensure automation processes do not fail due to faulty electrical installation practices.

Covered in this chapter are the basics of grounding, both for safety and signal noise reduction, including salient aspects such as power quality and uninterrupted power supply (UPS) systems, along with a cursory look at electrical circuit protection and enclosures.

## 21.2 Scope

This section addresses key electrical installation criteria necessary to achieve a reliable electrical installation in an industrial manufacturing facility utilizing automated processes. These EI criteria include power quality, plant-level electrical distribution, grounding, electromagnetic coupling (EMC), power line conditioning, and noise characteristics.

Please note that this section does not consider EI requirements for office buildings or large computer facilities, such as those found in system information departments.

## 21.3 Grounding and Bonding

Grounding plays an important role in the ability of EI to function safely and properly.

Properly grounded electrical installations allow protective devices—for example, fuses, circuit breakers, metering, ground fault interrupters (GFIs), lightning arrestors, and surge protectors to function properly. A properly grounded and designed EI will place the circuit in a safe, open (no power) condition during:

- overloads,
- short circuits (before equipment explodes),
- ground faults, and
- surges (caused by, for example, a lightning strike).

Personnel safety is outlined in the National Fire Protection Association (NFPA) National Electrical Code (NEC) and the Canadian Electrical Code (CEC) and should be completely understood and integrated into the EI design.

### 21.3.1 Grounding

Electrical systems that are grounded shall be connected to the earth in a manner that will limit the voltage imposed by lightning, line surges, or unintentional contact with higher voltage lines and that will stabilize the voltage to earth during normal operation. (NEC Article 250-4(A)(1))

### 21.3.2 Grounded

*Grounded* means connected to the earth or to some conducting body that serves in place of the earth. (NEC Article 100)

### 21.3.3 Grounding Conductor

The conductor used to connect non-current-carrying metal parts of equipment, raceways, and other enclosures to the system grounded conductor, the grounding electrode conductor, or both, at the service equipment or at the source of a separately derived system. (NEC Article 100)

### 21.3.4 Bonding

*Bonding* is the permanent joining of metallic parts to form an electrically conductive path which will assure electrical continuity and the capacity to conduct safely any current likely to be imposed. (NEC Article 100)

## 21.4 Grounding Systems

NEC requires the system ground be established (preferably at the service entrance), and is concerned primarily with personnel safety and fire protection. Establishing a signal zero volt reference must work in conjunction with the NEC. The zero volt reference is necessary to reduce noise and ground loops, which are not the goals NEC was designed to specify. Although the two purposes are generally in harmony, at times there are issues in satisfying both objectives. While we will cover some grounding considerations and practices in this chapter (this is not the definitive grounding reference) it should be understood that the chapter is primarily concerned with how the grounding system augments the performance of automation systems.

The main thrust of all signal grounding systems is to develop a workable ground plane. Due to the high frequency signals that abound in modern control systems, the aim is to establish an equipotential point—effectively a large area of the same impedance—with multiple grounds including the service entrance point. There are many ways to do this, which will be addressed at the appropriate point in the chapter.

### 21.4.1 Grounding Electrode

The primary grounding element for a small facilities ground is a single electrode consisting of a rod, pipe, or plate (made electrodes). Note that if a single ground electrode (at the service entrance or entry point of a separately derived ground) has a resistance greater than 25 ohms it should be augmented by an additional electrode. However, the NEC does not say what to do if the resistance is still above 25 ohms. In fact, most EI should have a ground resistance of 10 ohms (MIL HDBK 419A) and in practice, 5 ohms or less.

### 21.4.2 Grounding Triads

A single ground rod is scarcely able to establish an equipotential plane or even an effective ground. In fact, in most industrial systems a triad is used. A triad consists of three ground rods (a minimum of 3 meters [10 ft] long—longer in high resistive soils) driven in the ground in a triangular configuration a minimum of 3 meters (10 ft) apart. They are then bonded together and used as a single rod.

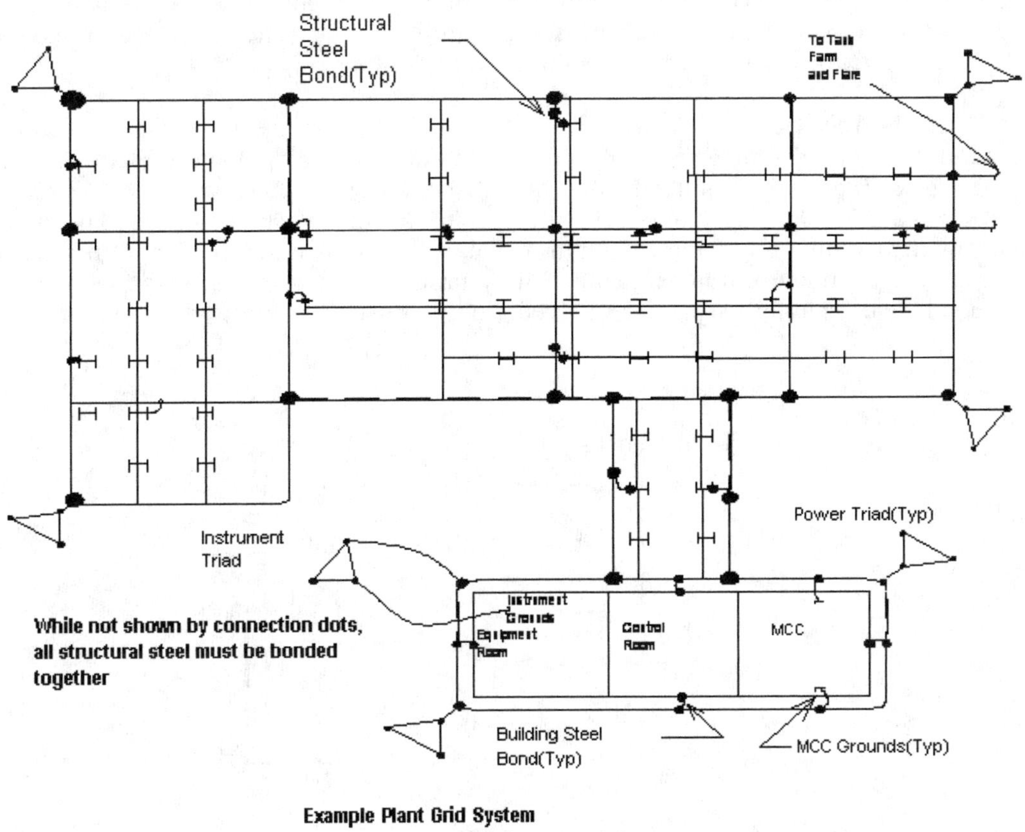


*Figure 21-1: Triads and Building Grounds*

### 21.4.3 The Equipotential Plane

The equipotential plane is a surface underneath (typically) the operating area consisting of conducting thin sheets or copper screen bonded together at multiple points typically by triads. It includes the service ground. This area has very low impedance and is effectively one large ground area.

The conducting media used for an equipotential plane can be (a) a copper grid embedded in the concrete floor, (b) a raised metal floor such as computer floor, (c) a subfloor of aluminum, copper, phosphor bronze screen, or sheet metal laid underneath the floor tile or carpet, or (d) a ceiling grid above tile equipment. The grid openings should not be larger than 1/20 wavelength at the highest frequency of concern, up to four inches. As a design objective the grid openings should not be larger than four inches." (Mil HDBK 419A)

At higher frequencies, this large conducting surface under the equipment to be grounded will present much lower characteristic impedance than a single wire. This is due to characteristic impedance (Z0) being a function of L/C. As the amount of capacitance to earth increases, Z0 will decrease.

Typically, the capacitance to ground of the conducting plane to earth is higher than that of a single conductor. If the size of the conducting plane is increased and allowed to encompass more area, the capacitance will increase. Also, the inductance generated by length of the conducting sheet will decrease with width further decreasing Z0.

Significantly increasing the dimensions of a conducting sheet (as it will be with the conducting plane) causes the characteristic impedance to approach a very low value. It will remain quite low throughout a large portion of the spectrum. This will establish an equipotential reference plane for all equipment bonded to it. Any "noisy" conductor through or along the equipotential ground plane will have its fields constrained between the conductor and the ground plane. (The material on the equipotential plane has been extracted and paraphrased from Mil HDBK 419A.)

### 21.4.4 Separately Derived Grounds

In a standard service entrance point solid ground, the neutral and safety ground are tied directly to the grounding electrode at this point, which is the only point they connect within the system. This is done so if there is a ground fault in the system, the circuit interrupter will open (circuit breaker or fuse), protecting the circuit. In some processes this could pose a problem, particularly if the ground fault causes an interruption to equipment controlling a process. Halting a process immediately is not generally a good idea. Typically a separately derived ground is used.

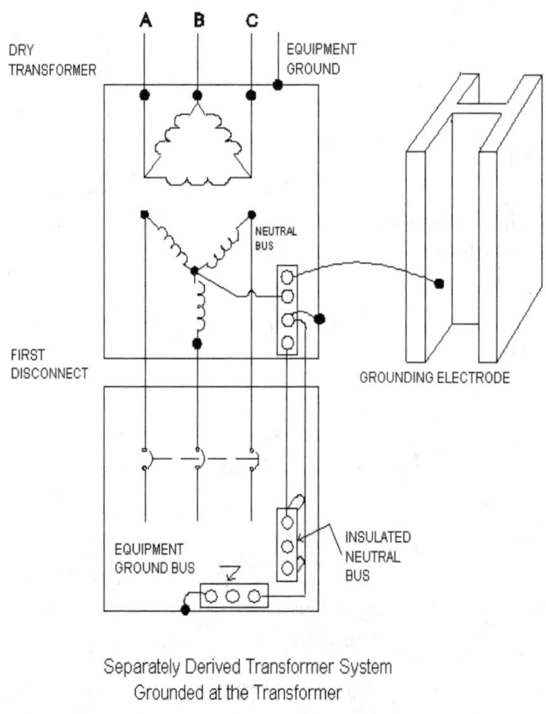


*Figure 21-2: Separately Derived System*

Note that the input from the delta transformer has no metallic connection to the grounded system (a generator could be used in place of a transformer), so the ground is separately derived. In place of the solid wire to the grounding conductor, an impedance or resistance could be used that would then allow a ground fault to exist without interrupting the system. The only requirement is a ground fault detection system (and a competent workforce) must be in place.

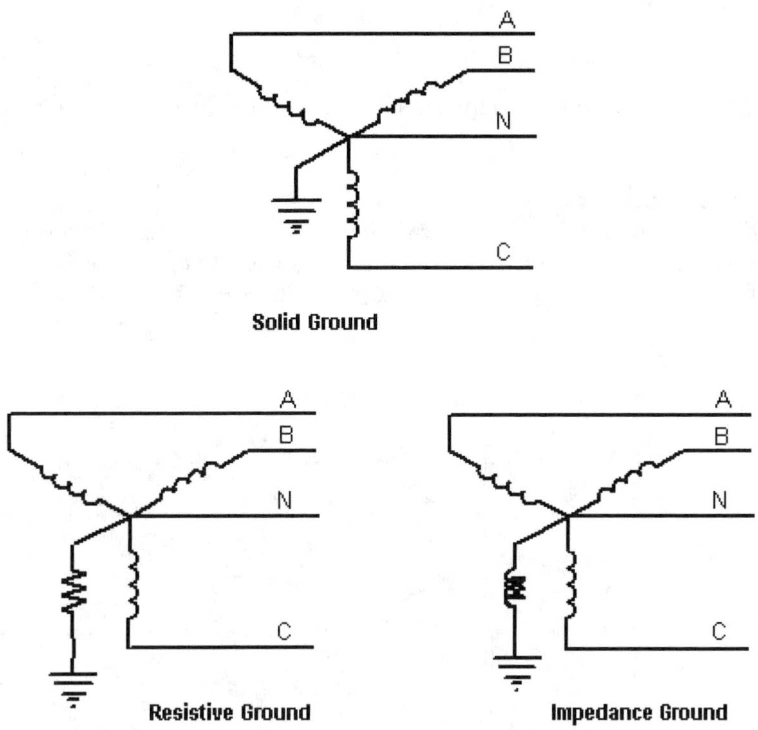


*Figure 21-3: Ground Types*

### 21.4.5 Other Recommendations

Steel framed buildings shall be designed so the steel is electrically continuous.

Single point grounds are recommended for instrument power panel(s) distribution transformer(s) servicing programmable electronic (PE) devices.

All communication wiring between ground planes, such as buildings and systems with different grounding approaches, shall be nonmetallic—for example, fiber optic with nonmetallic strength members.

All PE devices and related distribution equipment shall have surge protection sized to:

- protect the device or equipment immediately on the load side of the surge protector
- coordinate with the upstream device, such as substation lightning arrestors

### 21.4.6 Equipment

The EI shall use equipment suitable for the application (e.g., NE Code, Article 500). However, PE equipment requires additional considerations to achieve the desired safety and functionality. These considerations include:

- mounting to eliminate vibration and ground currents—e.g., PE device mounting on an enclosure backplane via nonconductive vibration isolation mounts
- assembly exactly per manufacturer's instructions to eliminate electromagnetic compatibility (EMC) problems
- wiring routing to allow proper heat dissipation in the cabinet housing PE devices
- single point grounds where possible
- isolation from EMC sources
- voltage level separation to eliminate EMC problems (see manufacturers installation instructions)

### 21.4.7 Enclosure Grounding

Cabinet grounding is illustrated in the diagram below. There is no need for an insulated (from the cabinet) grounding strip in the bottom of the cabinet, because the cabinet is bonded to the equipotential plane and is part of that plane. Signal returns and shields are connected as shown in the diagram.

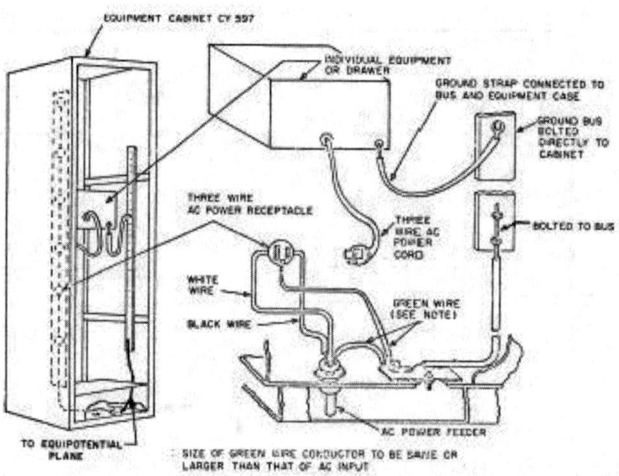


*Figure 21-4: Typical Cabinet Grounding (Graphic extracted from MIL HDBK 419A)*

## 21.5 Ground Loops

Ground loops are caused by a difference in potential between ground points in a circuit that causes undesired current to flow between the ground points, hence the term "ground loop." A ground loop does not have to involve current flowing in earth ground.

This is one reason why the shield should only be grounded at one end. There are other problems if the signal return itself is also grounded at both ends. The transmitter could very well be a thermocouple transmitter with the couple mechanically (and hence electrically) bottomed in the thermal well to make good thermal conductivity. There are many solutions, although the simplest is just to ensure the shield is only grounded in this case at one end. By grounding the shield at one end only we are using the "single point ground," which is quite effective for circuits with noise below 1 MHz.

In most cases grounding the shield at one end only effectively reduces the noise coupling acting as a Faraday shield between the signal line and ground. However, what about high frequency (HF)—above

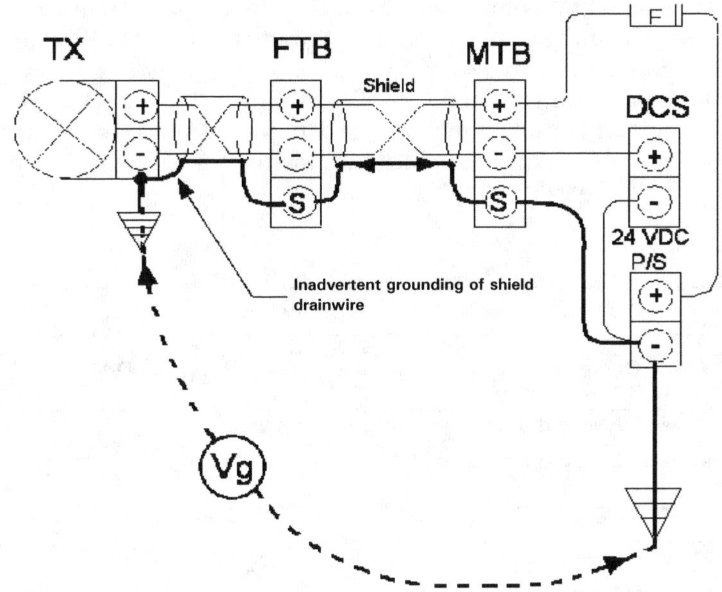


*Figure 21-5: Typical Ground Loop*

1 MHz—noise? The purpose of the equipotential plane is to have a 0 V reference over a large area, and in the case of HF a multiple-point ground is desired.

The use of shielded cable is a well-known technique for combating noise; however, when HF noise is a problem, it may be better to ground the shield at both ends.

A shielded cable, with both ends of the shield bonded to the relevant equipment chassis, acts as a 1:1 transformer at HF, the shield being the "primary" and the conductors being the "secondary." Any voltage applied between the two ends of the shield (the difference between the two pieces of equipment) will generate an identical voltage in the conductors. The polarity is such that the induced voltage exactly cancels the applied or common mode noise voltage. This effect is so strong that a 9-m (29-ft) length of feedback cable will typically reduce the common mode noise by a factor of 60dB (1000:1), and a 90-m (290-ft) length by 80dB (10000:1). When only one end of the shield is grounded this attenuation effect will be completely lost.

## 21.6 Noise Reduction

Although an entire treatise on shielding and other noise reduction techniques is beyond the scope of this chapter, there are certain installation details that if followed will go a long way toward reducing electrical noise.

### 21.6.1 The Noise Model

Noise is a communication, similar in characteristics to our desired communications. Noise has a source, a path, and a receptor. We have discussed grounding at some length, and having a good ground (0 volt reference) is essential to noise reduction.

Basically, if you can stop the noise at the source you are way ahead. Most of our efforts are placed on interrupting the path or protecting (shielding) the receptor (because it is very difficult to determine the source, when the source is offensive, and if we have the jurisdictional authority to move or diminish the source—lightning comes to mind).

### 21.6.2 Interrupting the Path

Physical separation is (usually) one of the least expensive methods for reducing noise. Radiated noise is a square law adherent, meaning the field strength of radiated noise falls off as the square of the distance from the source. If you are twice as far from the source, the radiated power reaching you is one-quarter of what it would be if you were only half that distance. Table 21-1 is a list of noise generators, their intensity of noise and what cabling to use. Please note that simply running the different level signals in different conduits will break up the paths by allowing less coupling of noise to your signal leads.

*Table 21-1: Noise Sources*

| | Zone Category | | | |
|---|---|---|---|---|
| **Cable** | **Very Noisy** | **Noisy** | **Quiet** | **Cable Shielded** |
| Three phase between drive and filter (shielded) | | X | | X |
| Three phase between drive and filter (unshielded) | X | | | |
| PWM drive/inverter – motor power (shielded) | | X | | X |
| PWM drive/inverter – motor power (unshielded) | X | | | |
| External dump resistor (unshielded) | X | | | |
| External dump resistor module (metal-clad) – bus connections | | X | | |
| External dump resistor module (metal-clad) – fan connections | | X | | |
| Motor feedback | | | X | |
| Three-phase supply power | | X | | |
| Single-phase supply power | | X | | |
| Hydraulic/pneumatic solenoids (suppressed) | | X | | |
| AC motor contactors | X | | | |
| Solid-state relay output (zero crossing) | | X | | |
| PLC digital I/O | | X | | |
| Dedicated drive inputs (except registration) | | X | | |
| Limit switches | | X | | |
| Pushbuttons | | X | | X |
| Proximity switches (except registration) | | X | | X |
| Photo electric cell | | X | | |
| 24 VDC relay contact (suppressed) | | X | | |
| Transformer indicator lamp | | X | | |
| Encoder buffer board | | | X | X |
| Data/communications (See note 2) | | | X | X |
| Encoder | | | X | X |
| High-speed registration | | | X | X |
| PLC analog I/O | | | X | X |
| PLC high-speed counter input | | | X | X |

The system builder's objective must be to minimize the coupling mechanism between noise sources and their receptors. There are five techniques commonly employed to deal with electrical noise in systems:

- High-frequency bonding
- Shielding
- Segregation
- Filtering
- Suppression (of mechanical contacts)

#### 21.6.2.1 High-frequency Bonding

Lengths of wire, however heavy, will exhibit significant impedance at HF, making them unsuitable for HF bonding. For example, a 500-mm (19.7-in) length of wire behaves as an efficient antenna at the Amateur Radio frequency of 144 MHz. It forms a quarter-wave antenna (a quarter of the 2-m wavelength) that matches the 50-ohm output impedance of the transmitter to that of free space, 377 ohms. Thus, a simple length of wire can appear as anything between 50 and 377 ohms, depending only on length and frequency.

The purpose of HF bonding is to ensure all the metalwork of a system is at the same voltage at HF; then, if two pieces of equipment are linked by a cable, the cable will not be carrying any "common mode" voltage (the voltage between the two ends of the cable).

The solution is to use the "Equipotential Ground Plane" principle: every point on a large, flat, conductive surface will be at the same potential at all frequencies, that is, at almost zero impedance between all points. This is a well-known practice in printed circuit board design. One layer is left largely intact and used as a reference for all the devices on the board.

We can easily achieve this within a control cabinet by using the panel itself, which is much easier if the panel is zinc-plated steel. If paint is used it must be scraped off under each mounting point and is then liable to corrode, thus negating the bonding effect.

Multiple panels, if adjacent, may be bonded top, middle, and bottom with wide, short straps to extend the ground plane over the whole suite.

PWM drive power cables should *always* be shielded if possible. At high powers this may become impractical, so conduit may be used to create a shield. However, it must be continuous, ferrous, and terminated at both the panel and motor frame.

#### 21.6.2.2 Separation

Clearly if the source and receptor are widely separated (the path is lengthened) they will not interact. A 200-mm (7.9-in) distance is enough to drastically reduce interaction, but even that amount is impractical in a typical, crowded panel. The solution is to define areas of the control panel as "quiet" and "noisy" zones. Each is reached via "quiet" and "noisy" wire ways that can be most easily identified by the use of different colors, say, black for noisy and gray for quiet. (See Table 21-1.)

### 21.6.3 Protecting the Receptor

If all else fails, we will fortify the receptor itself. We can actually shield the target (receptor). There are two forms of shielding: ferrous and nonferrous. *Ferrous* means the material is magnetic. Iron and steel are ferrous; aluminum is not. Surrounding equipment in nonferrous materials (such as with a cabinet made of aluminum) protects it from electromagnetic radiation—radiated signals. Reducing the electrostatic field of an electromagnetic radiation reduces the signal itself. However, in industry many fields are created by large currents, and these tend to be magnetic fields (H). Nonferrous shielding has very little effect on magnet fields. Ferrous materials, on the other hand, provide a lower reluctance path for magnetic fields and hence protect from them. Because most ferrous materials are also conductors, they protect against the electrostatic (E) field as well.

When protecting a piece of equipment or even a facility, all conductors in and out of the facility must have filters and other techniques applied to prevent noise from being conducted into (or in some cases, out of) the protected area.

## 21.7 Electrostatic Noise

Electrostatic noise is generated when the built up charges on insulating surfaces discharge. Lightning is a dramatic example. Wind currents in the updraft of a thundercloud charge the bottom of the cloud negatively (and hence the top of the cloud, where the ionized atoms end up, positively). The earth provides an oppositely charged surface. When the potential difference is high enough a leader stroke ionizes the path, and the main stroke then attempts to equalize the charges in the familiar lightning stroke, followed by a thunderclap due to the heated air around the stroke or strokes.

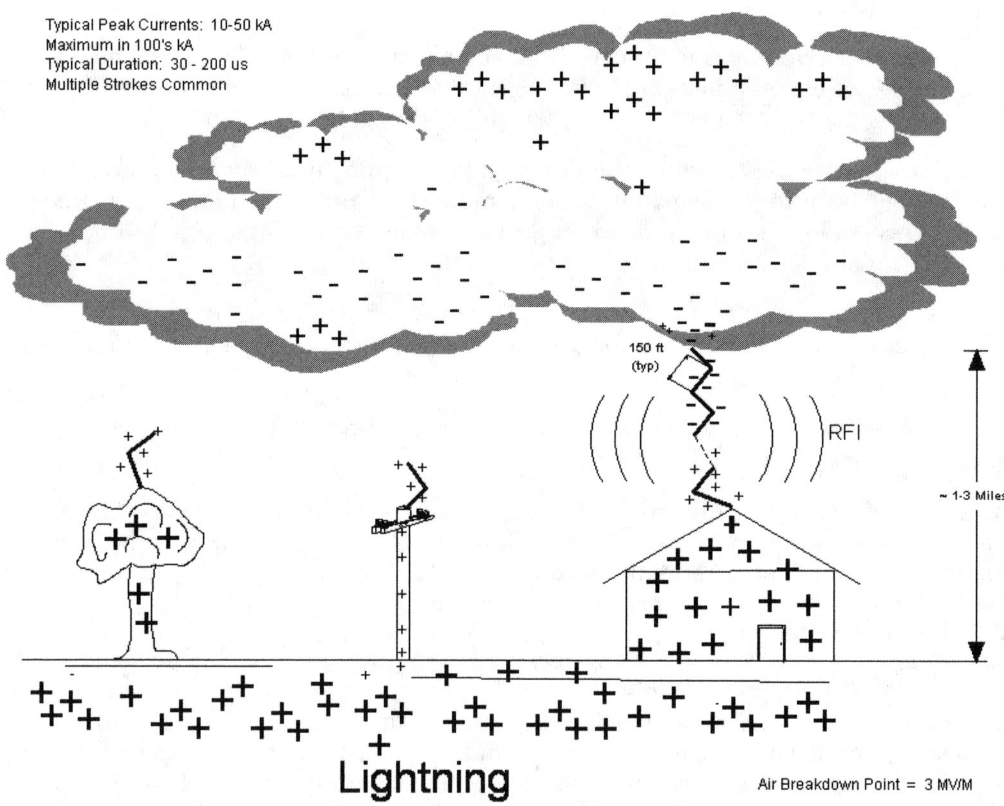


*Figure 21-6: Lightning as an ESD*

Similarly, on a much smaller scale, whenever there is a charge difference and it becomes high enough for the distance involved, an electrostatic discharge (ESD) will occur, destroying unprotected electronic circuits. The main concern in this chapter, though, is with lightning and its radiated and conducted effects.

### 21.7.1 Radiation from an ESD

When an ESD occurs, it generates extremely wideband electromagnetic energy with quite high (momentary) energy, resulting in a wideband pulse. This noise can enter control systems as a pulse, or spike, causing unanticipated results. Noise shielding and good grounding will reduce the susceptibility of circuitry to this noise.

### 21.7.2 Conducted Noise from an ESD

The major concern with lightning (other than a direct hit) is the noise spike introduced and conducted through the system. Long runs of conductors parallel to the earth's surface (unless encased in ferrous conduit) are subject to having ESD noise coupled due to induction. A cloud-to-cloud strike will produce a mirror earth current paralleling the cloud-to-cloud strike. If this is parallel to the conductor's, a current will be induced in the conductors and will have then entered the system. Other impulse noise is disruptive, but seldom destroys equipment.

### 21.7.3 Protection from ESD

There are a number of methods to protect systems from the effects of lightning: proper grounding, shielding, and active protection in the form of surge suppressors. Other concerns, such as noise spikes and random pulses produced by equipment are reduced by noise suppression techniques such as shielding and active noise reduction circuitry. Power quality concerns (of which ESD protection is one such) include a number of methods to ensure that the power quality will enable correct equipment operation and longevity. Surge suppressors are used for just that reason and are one of the primary protections from lightning as an ESD.

## 21.8 Surge Suppressors

Surge suppressors go by a number of names, dependent upon their construction and materials. Two of the best known names are:

- Transient Voltage Surge Suppressor (TVSS)
- Surge Protection Device (SPD)

### 21.8.1 TVSS

The major active components of transient voltage surge suppressors are:

- Gas tube, which is also called a gas discharge or gas surge arrestor
- Solid state
    - SAS are silicon avalanche diodes and go by commercial names such as Transorbs.
    - Silicon controlled rectifiers (SCRs) are used as crowbars in some surge suppressors to handle higher currents than SAS can.
- Metallic oxide semiconductors (MOVs) are also known as voltage-dependent resistors or varistors. MOVs are subject to deterioration from repeated surges and should have a fuse in series with the MOV and preferably some form of indication or alarm upon MOV failure.

*Table 21-2: General Capacities of TVSS*

| Type | Capacity | Response Time |
|---|---|---|
| Gas Tube | High | >5nS |
| Solid State | Low - Medium | <5nS |
| MOV | Medium | <5nS |

A commercial device called a RAV manufactured by Comm-Omni International is a combination of a gas tube and a varistor.

### 21.8.2 TVSS Circuitry

The active components alone do not make a complete TVSS but need additional components to work correctly. Most TVSS are hybrid devices (a combination of active components), and some are made up of only passive components. The following is an example of TVSS circuitry.

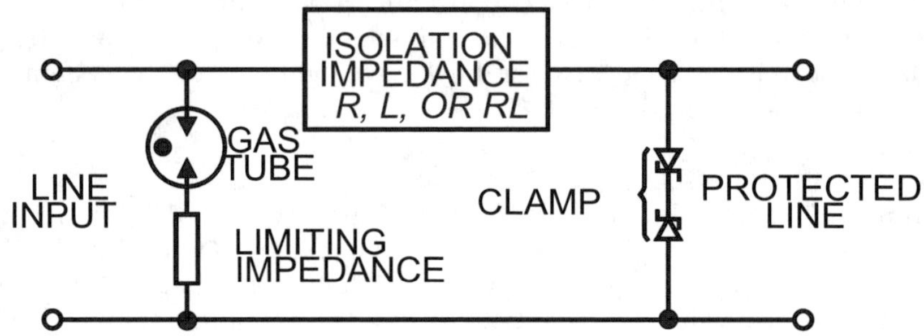


*Figure 21-7: A Hybrid TVSS*

Hybrid suppression circuits as illustrated in Figure 21-7 will respond to several types of line disturbances. A gas tube is combined with a silicon clamping device to provide two-stage suppression.

When a transient hits, the gas tube fires and crowbars the bulk of the energy. The clamping device catches the leading edge of the transient that the gas tube may have missed.

The series impedance may be a resistor or an inductor sized to provide sufficient impedance to ensure the pulse will produce a high enough voltage to fire the gas tube.

### 21.8.3 Installation

The location of the surge suppressors has a large impact upon their efficiency. It is best if they are distributed, generally along the power distribution network. Figure 21-8 is an example of this staged protection.

## 21.9 Power

### 21.9.1 Source

Site planning should include an analysis of power source options such as utility purchased, third-party purchased, self-generated, and hybrid (a combination of sources). You should analyze each option in a way that considers cost, availability, capacity, and operating history.

You should also analyze the site to define its exposure to environmental disturbances (e.g., thunderstorms, tornadoes, earthquakes, hurricanes, and humidity). The EI should include the necessary design features to address these conditions, as needed, to achieve the appropriate availability. Options, for example, include steel versus wood poles, a counterpoise system for all overhead lines, underground lines instead of overhead lines, and parallel lines using a different physical routing.

The electrical distribution configuration (e.g., ring, dual parallel, or star) shall be suitable for the desired availability and be compatible with the design of the substations that service the process. Distribution substation configuration—their number, location, and capacity—should be compatible with the process distribution to minimize common-cause process failures, such as one substation loss

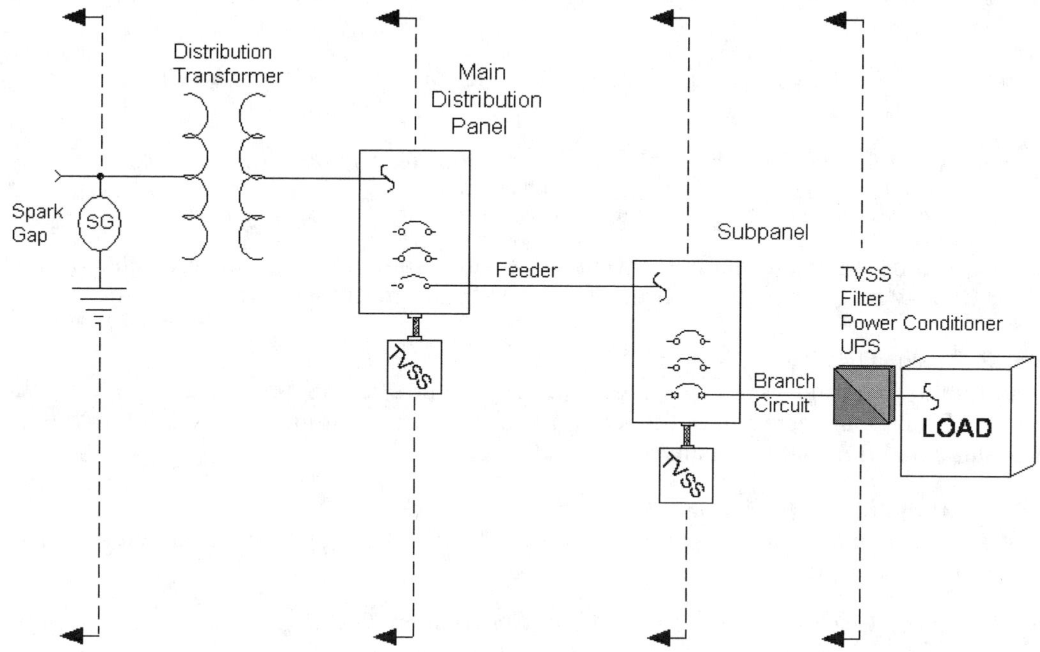


*Figure 21-8: Staged TVSS*

shutting down multiple process facilities. Electrical power voltage levels shall be compatible with the voltage required for the process electrical loads (e.g., three-phase motors/480 volts AC, single-phase motors/120 volt AC, PE device(s) power supply(s) input voltage/120 volt AC).

Electrical distribution metering is a key characteristic of any purchased power. For example, eight-cycle response time to transfer from one power source to another. What is the metering technology (e.g., carrier frequency on power lines, fiber optics, wireless)? What interface is required to the plant metering?

Is the metering provided compatible with the desired automation characteristics of the process? For example, will the process be able to withstand brownouts and voltage dips? Should process loads, such as pump motors, be provided with time-delay under-voltage protection to allow automatic start up after a voltage dip? What are the quantitative values used by the purchased power supplier for such things as voltage dips, voltage spikes, and brownouts? What impact will this have on the substation capacity? What impact will it have on the process?

A process may place heavy peak loads on the purchased power system. What effect does this have? For instance, the process may have a large motor (e.g., greater than 0.1500 HP) that starts up weekly. Energy needed to start the motor will lower the utility voltage. Will this level interfere with other customers? This analysis is required during preplanning to select the proper electrical power utility source.

Note that the contract agreed upon with the power supplier may stipulate maximum power demands the owner or user can place on the utility's power system. In any event, this value should be defined before EI design.

Electrical power systems for most PE automated processes require the ability to maintain power on portions of the control system for a fixed time during loss of utility power. Typical loads requiring this feature include:

- *PE logic solvers* such as programmable logic controllers (PLCs) and distributed control system controllers
- *sensor inputs* such as flow, pressure, temperature, and position switches
- *diagnostic outputs* such as indicating lights, alarms, and operator displays/keyboards

This power requires an alternative power source—for example, a fuel-powered electrical power generator or UPS—with the necessary speed of response and capacity to perform its function.

### 21.9.2 Power Quality

Power quality is a concept of powering equipment and devices in a manner suitable for the safe, reliable, and functional operation of that equipment. It requires monitoring and, if purchased, specifications that detail the amount of abnormality allowed.

#### 21.9.2.1 Definitions of Abnormal Power

**Sag** – Also known as a dip, a sag is a reduction in RMS voltage at power frequency for a duration of a half-cycle to a few seconds.

**Swell** – An increase in RMS voltage at the power frequency from a half-cycle to 1 minute

**Outage –** A complete loss of power for a period of time, also known as a blackout

**Impulse, Transient, or Surge** – A subcycle disturbance; a sharp, brief discontinuity of the waveform of either polarity that may be additive or subtractive to the nominal waveform

**Oscillatory Transient Or Ring Wave** – A subcycle disturbance that rings or has the appearance of a decaying sine wave

**Notch** – A switching or other disturbance of the normal power voltage lasting less than a half-cycle. Initially opposite in polarity to the normal waveform creating a notch.

**Noise –** Unwanted electrical signal component either superimposed upon or modulated onto the desired signal

**Harmonics –** Frequencies that may exist on a power line that are multiples (e.g., 120, 180, 240, 300) of the fundamental frequency (60 Hz in the U.S.)

#### 21.9.2.2 Harmonic characteristics

Nonlinear power devices cause harmonics. The measurement is total harmonic distortion (THD), and there are both voltage and current THDs. The ratio is the percent of the fundamental (line input) RMS value to the RMS value of the harmonic current or voltage.

Triplen harmonics are odd multiples of the third harmonic. These harmonics add when combined.

#### 21.9.2.3 Problems Caused by Harmonics

- overheating of neutral
- overheating of transformers
- overheating of panel boards
- premature fuse or circuit breaker blowing
- poor power quality leading to data error and equipment malfunction
- resonant conditions in power system leading to large voltage transients

## 21.10 Uninterruptible Power Systems (UPS)

UPS come in many shapes forms and sizes depending upon:

- power required
- duration of power required
- switchover time
- power quality
- costs

UPS can be offline, online, or in another arrangement.

### 21.10.1 Offline UPS

Figure 21-9 is a block diagram of an offline UPS.

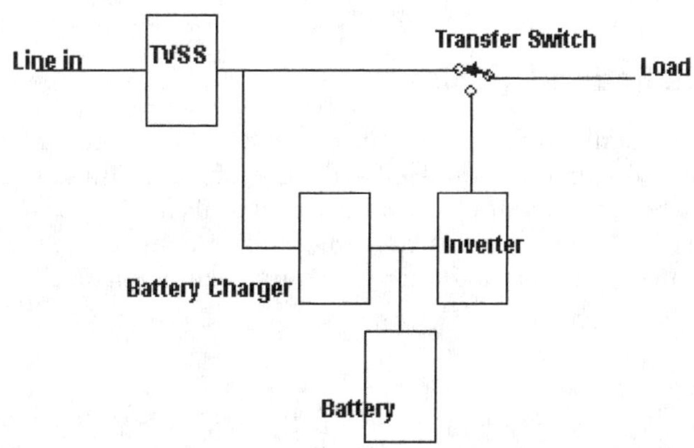


*Figure 21-9: Offline UPS*

The offline UPS is also called a standby or forward transferring UPS. It has a switching time of about 4 ms. This is generally not a problem with switch mode power supplies, because they have sufficient capacity to ride through the switching time. However, you will have to investigate the ability of linear power supplies to maintain voltage during the switching time; most industrial linear supplies should have little problem unless heavily loaded. A filter/TVSS must be present in the conduction path, because there is a direct path from line to load under normal operating conditions.

One consideration is the output wave shape. Many less expensive offline or standby UPSs output a square wave at about 90V (whose RMS value is 120V); it generates considerable noise, and you must consider the power quality factor. Generally, it is best to avoid a square wave output. On the way to a full sine wave, output is the stepped output allowing the load to integrate (smooth) into a quasi sine wave. More expensive units will produce a true sine wave output.

### 21.10.2 Online UPS

An online UPS is generally the supply for the process all the time. There are rotary models (using fly-wheels and motor/generators), and there are solid-state varieties. A block diagram of the solid-state type is illustrated in Figure 21-10. Note that this is an expensive solution, but one that ensures power quality, subcycle switchover, and (if solid state) no moving parts.

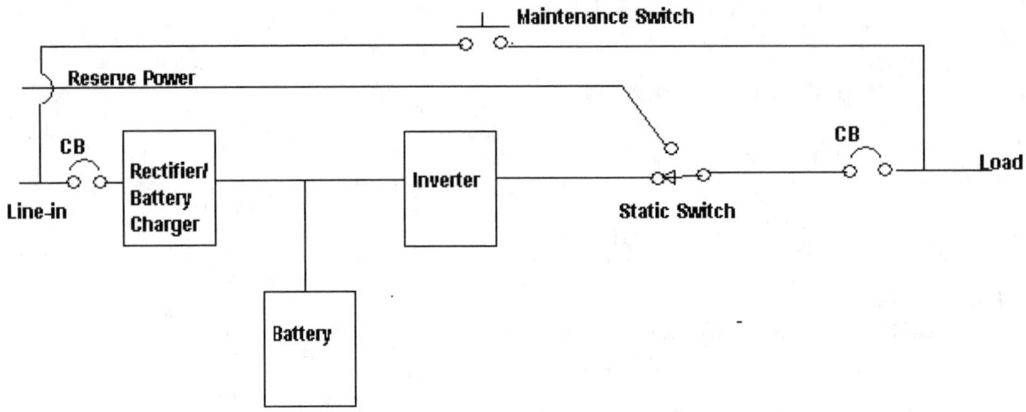


*Figure 21-10: Online UPS*

## 21.11 Electrical Installation Details

There are a myriad of installation details required for any EI. Pre-installation planning, process flow and control points, and documentation itself are absolute necessities. There are many differing requirements, and only some major electrical requirements are covered in the remainder of this chapter. The NEC and CEC dictate many of the wiring requirements that must be met for specific applications, including conductor capacity and sizing, conduit sizing, enclosure classifications, and protection characteristics and sizing.

### 21.11.1 Ampacity

NEC (2005) Section 310 *Conductors for General Wiring* defines and specifies the current carrying ability of conductors, their identification, and the derating necessary for specific conditions, while Annex C specifies how many conductors of which size can fit into conduits or raceways of a specified size.

### 21.11.2 Enclosures

Enclosures are required for various reasons, including protecting a person in the vicinity of electrical equipment that might prove hazardous.

Enclosures are required in hazardous areas to protect the area from the equipment. In hazardous areas equipment enclosures may be explosion proof. This means that even if there is an explosion in the cabinet, the cabinet is of sufficient strength and the portals to the areas outside the cabinet are of sufficient cooling length that ignition in the cabinet will not ignite the atmosphere outside the cabinet. Another cabinet use is that of a purge cabinet in which the cabinet is pressurized (usually with a non-incendive gas such as nitrogen), preventing ignition from taking place in the first place.

Enclosures protect equipment and personnel. See the National Electrical Manufacturers Association's (NEMA's) Web site (http://www.nema.com) for a list of standard enclosure specifications.

Enclosures are rated by their ability to protect. All provide protection against accidental exposure to the equipment inside (up to 1000V). However, they are then rated to withstand environmental conditions, such as indoors, outdoors, corrosion, or explosion proof. The IEC 60529 provides a set of specifications used in Europe and other countries, while in North America, NEMA has a set of standards. Please note that there is no direct correlation between the two specifications. However, in general, the more protections provided, the higher the cost of the enclosure.

One of the most popular types in the automation area is the NEMA Type 4. It is rated for indoors and outdoors; provides protection against dust, dirt, and water; and can even withstand external ice. Type 4X will provide all the above protections, as well as protect against corrosive agents. Type 6P provides all the protections of 4X and can also be submerged.

Types 7, 8, and 9 are for use in indoor hazardous locations, and type identifiers specify the class of materials. Type 10 is for methane gas (mines).

### 21.11.3 Protection

In this case, the term *protective devices* refers to devices such as circuit breakers, fuses, and ground fault current interrupters (GFCIs).

#### 21.11.3.1 Fuses

Fuses are one-time overcurrent protection devices. The active element melts when excessive current flows through the element. There are single element types (quick blow) and dual element types (slo-blo).

Fuses come in many forms and have four ratings: (1) voltage, (2) amperage, (3) current interrupting—AC Short Circuit [AIR]—rating, and (4) instantaneous rating. Please note that cartridge fuses above a certain amperage can no longer have the round end but must use blades for contacts.

#### 21.11.3.2 Fuse Rejection Feature

*Table 21-3: Rejection Lengths of Fuses*

| Amperage | 250V Length | 600V length |
|---|---|---|
| 0.1 – 30 | 2 in | 5 in |
| 35 – 60 | 3 in | 5.5 |
| 70 – 100 | 5 7/8 in | 7 7/8 |
| 110 – 200 | 7 1/8 in | 9 5/8 |
| 225 – 400 | 8 5/8 in | 11 5/8 |
| 450 – 600 | 10 3/8 in | 13 3/8 |

The fuse rejection feature is used on fuses so lower AIR rated fuses cannot be physically substituted for higher rated ones.

#### 2.11.3.3 Circuit Breakers

Circuit breakers are typically one of two types: bimetallic or magnetic.

Bimetallic types use a bimetallic strip (two metals of different temperature coefficients of expansion bonded together) to release or otherwise cause the breaker to trip. The bimetallic types have fair repeatability (relative to the other types), but ambient temperature must be considered because it causes lower trip points at higher temperatures.

Magnetic types are found in industry and use the line current to pull in and unseal a line experiencing overcurrent. They have trips that (in some cases) may be adjusted.

#### 2.11.3.4 GFCI

GFCI devices compare the current in the phase line and the neutral. Any imbalance means the current has found a different path to return, and at the trip rating of the GFCI will open the circuit, preventing possible electric shock.

These should be located in an industrial control system anywhere water is collocated with the power. Typically used where detection of minute (compared to a fault) current paths exist to ground to prevent electrical shock.

The typical residence type triggers on an imbalance of 4 mA, the industrial types at about 10 mA or, if adjustable, within their range.

## 21.12 References


### ISA

ISA-5.1-1984 - (R1992) – *Instrumentation Symbols and Identification.*

ISA-12 Series of Standards, Recommended Practices (RP), and Technical Reports (TR) related to Electrical Equipment for Hazardous Locations.

ISA-50 Series of Standards and Technical Reports related to Signal Compatibility of Electrical Instruments.

ISA-60 Series of Recommended Practices (RP) for Control Centers.

ANSI/ISA-84.00.01-2004, Parts 1-3 (IEC 61511-1-3 Mod) – *Functional Safety: Safety Instrumented Systems for the Process Industry Sector.*

### IEEE

BSR/IEEE 62 Series of Guides and Standards for Surge Protection.

IEEE 142-1991– *Recommended Practice for Grounding of Industrial and Commercial Power Systems.* (IEEE Green Book.)

IEEE 519-1992 – *Recommended Practices and Requirements for Harmonic Control in Electrical Power Systems.*

IEEE 1100-1999 – *Powering and Grounding Sensitive Electronic Equipment.*

IEEE 1159-1995 – *Recommended Practice for Monitoring Electric Power Quality.*

*IEEE Standards Reference* – An enhanced InfoBase version of the *IEEE Standard Dictionary of Electrical and Electronics Terms* on CD-ROM (formerly IEEE 100).

### National Fire Protection Association

NFPA 70 – *National Electric Code® - 2005 edition.*

NFPA 79 – *Electrical Standard for Industrial Machinery - 2002 edition.*

### Other

MIL-HDBK 419A – *Grounding, Bonding, and Shielding for Electronic Equipments and Facilities.*


## About the Authors

**Victor J. Maggioli Sr.**, a 38-year DuPont veteran and currently President, Feltronics Corp., Newark, Del., has made significant contributions to the process industry sector by applying new techniques in the areas of grounding, power line conditioning, programmable electronics, fiber optics, fail-to-safe configured control systems, and limited variability application software. A member of the European Workshop for Industrial Computers (EWICS), he co-authored a guideline on *Use of PLCs in Process Safety Applications,* which was issued by EWICS. Maggioli chaired ISA standards panel 84 (SP84) and led a committee of more than 100 engineers in the development of ANSI/ISA-84.00.01-1996, and ISA-TR84.00.02. He also participated in the International Electrotechnical Commission (IEC) development of the generic functional safety standard IEC 61508 and was selected to lead the process sector standard development (i.e., IEC 61511) for functional safety using PE technology, which resulted in approval of IEC 61511. He has been a member of ISA, IEEE, AIChE, and NFPA. Maggioli is a senior member in ISA and a life senior member in IEEE. Maggioli received the 1996 DuPont Engineering Excellence Award for outstanding contribution in the area of process control functional safety and *CONTROL* magazine's 1987 engineer of the year award for his contribution in process safety standards development. He serves as an ANSI-appointed technical representative on industrial computer safety to the IEC.

**Graham Elvis** earned a post-graduate diploma in Machine Tool Design and Technology from the University of Manchester (UK) Institute of Science and Technology. He has worked for Rockwell Automation for 27 years in various posts involving Drive System Technology and Industrial Automation. Currently, he is a senior Customer Success Team member.

**Lawrence (Larry) M. Thompson** has an extensive background as a technician, technical trainer, and course developer in electronics, measurement and control, and computer networking. A 20-year veteran of the U.S. Air Force, his military specialty included instruction in electronic encryption equipment and TEMPEST. He has maintained a General (formerly First Class) FCC Radiotelephone license since 1966. His industrial experience includes positions as technician, test engineer, and test engineering supervisor for numerous companies. He recently retired as Department Chair for Software Engineering Technology at Texas State Technical College, and now runs his own consulting business. He has served as an adjunct instructor for ISA for more than 20 years. He has a B.A.A.S. from Tarleton State University, Texas, and is the author of several books including ISA's *Industrial Data Communications* and *Basic Electricity and Electronics for Control: Fundamentals and Applications.*

# 22 Safe Use and Application of Electrical Apparatus

*By Ernie Magison*

## Topic Highlights

*Philosophy of General Purpose Requirements*
*Equipment for Use Where Explosive Concentrations of Gas, Vapor, or Dust Might be Present*
*Equipment for Use Where Combustible Dust May Be Present*

## 22.1 Introduction

This article discusses ways to ensure electrical equipment does not endanger personnel or plant. See another chapter for a discussion of Process Safety and Safety Instrumented Systems—protecting the plant against the risk of equipment failing to perform its function in a control system or in a safety instrumented loop (SIL).

Developments during the past half century have greatly eased the selection of safe equipment. Standardization of general purpose safety requirements has made it possible to design a single product that is acceptable with very minor modifications, if any, in all nations. Worldwide adoption of common standards progresses for constructing and selecting equipment for use in hazardous locations, but transition from long-accepted national practices to adopt a somewhat different international or harmonized practice is, of necessity, slowed by historical differences in philosophy. Emphasis in this article is on the common aspects of design and use. Present day differences in national standards, codes and practices will be reduced in coming years. To ensure safety, a user anywhere must select, install, and use equipment in accordance with local standards and codes.

General purpose safety standards address construction requirements that ensure personnel will not be injured by electrical shock, hot surfaces or moving parts—and that the equipment will not become a fire hazard. Requirements for constructing apparatus to ensure it does not become a source of ignition of flammable gases, vapors, or dusts are superimposed on the general purpose requirements in equipment to be used where potentially explosive atmospheres may be present. The practice in all industrialized countries, and in many developing countries, is that all electrical equipment must be certified as meeting these safety standards. An independent laboratory is mandated for explosion-protected apparatus but, in some cases, adherence to general purpose requirements may be claimed by the manufacturer, subject to strict oversight by a third party. Thus, once a user has selected the equipment he or she wants to use, the user can be assured it meets the applicable construction standards if it is certified or approved. It is the user's duty to install and use the equipment in a manner that ensures that safety designed into the equipment is not compromised in use.

## 22.2 Philosophy of General Purpose Requirements

Protection against electrical shock is provided by construction rules that recognize voltages below about 30V alternating current (AC), 42.4V peak, or 60 volts direct current (DC) don't pose danger of electrocution in normal industrial or domestic use, whereas contact with higher voltages may be life threatening. Design rules, therefore, specify insulation, minimum spacings, or partitions between low voltage circuits and higher voltage ones to prevent them from contacting each other so that an accessible extra low voltage circuit becomes hazardous. Construction must ensure higher voltage circuits cannot be touched in normal operation or by accidental exposure of live parts. Any exposed metallic parts must be grounded or otherwise protected from being energized from hazardous voltages.

Protection against contact with hot parts or moving parts is provided by the enclosure, or by guards and interlocks.

To prevent the apparatus from being the initiator of a fire, construction standards specify careful selection of materials with respect to temperature rises of parts, minimum clearances between conductive parts to prevent short circuiting, and enclosure to prevent any arcs or sparks from leaving the equipment.

Instructions, warnings, and installation diagrams are assessed by third party approval authorities, as part of the approval process, in addition to evaluating conformity to construction rules. The user must install and use apparatus in accordance with these specifications and documents to ensure safety.

## 22.3 Equipment for Use Where Explosive Concentrations of Gas, Vapor, or Dust Might be Present

Equipment intended for use in hazardous locations is always marked for the hazardous locations in which use is permitted and the kind of protection incorporated, and it is almost always certified by an independent approval authority. The user may depend on this marking when selecting equipment for use.

### 22.3.1 Area Classification

Any hazardous area classification system defines the *kind* of flammable material that could be present, and the *probability* that it will be present. In North America and some other locations, two nomenclatures are in use to denote type and probability: Class, Group and Division, and the more recent international usage of Material Class and Zone. A third system, introduced in the European Community (EC), also addresses the nature and degree of hazard, and is applied also to locations where hazards other than an explosion exist. Class and Group or Material Group defines the nature of the hazardous material that may be present. Division or Zone indicates the probability of the location having a flammable concentration of the material.

*Table 22-1: Material Classes and Groups and Divisions – National Electrical Code (NEC) Article 500*

| **Class I**<br>**Gases and Vapors** | **Class II**<br>**Combustible Dusts** | **Class III**<br>**Ignitable Flyings** |
|---|---|---|
| **Group A** –Acetylene<br>**Group B** – Hydrogen, butadiene, ethylene oxide, etc.<br>**Group C** – Ethyl ether, cyclopropane, ethylene, etc.<br>**Group D** – Gasoline, common solvents, natural gas, vinyl chloride. Propylene, styrene, etc. | **Group E** – Metallic dusts<br>**Group F** – Coal dust, carbon black and similar carbonaceous dusts<br>**Group G** – Grain dusts, plastics, sugar and other organic dusts | No Groups assigned. Typical materials are kapok, cotton linters, nylon, flax, wood chips, which are not normally in suspension |
| **Criteria for grouping** – Ease of ignition by electric arc and/or passage of explosion through narrow gap | | |
| **Division 1** | | |
| Continuous or intermittent hazard, as in below grade pits in Division 2 locations, near sources of release, such as packing glands, open kettles or vents, or where release of material and damage to electrical equipment may occur simultaneously. | Dust cloud of flammable concentration exists periodically, frequently, or intermittently, as near processing equipment.<br>Where conducting dust may accumulate. | Where cotton, Spanish moss, etc. are manufactured or processed. |
| **Division 2** | | |
| Areas below grade in unclassified location, or adjacent to Division 1 locations, or where flammable liquids are stored or piped in closed systems, or Div. 1 locations made nonhazardous by forced ventilation | Where failure of equipment may release a cloud or where dust layers deposited on floor or equipment may be resuspended as a cloud. | Areas where materials are stored or handled. |

In international practice, mining hazards are denoted as Group I, due to potential presence of both methane and dust. Equipment for use in mines is constructed to protect against both hazards in a physically and chemically arduous environment. Industrial facilities are denoted as Group II and the gases and vapors are classified as follows.

| | |
|---|---|
| Material group IIA - equivalent to Group D<br>Material group IIB - equivalent to Group C<br>Material group IIC - equivalent to Groups A and B | Alternatively:<br>Group A = IIC<br>Group B = IIB + Hydrogen<br>Group C = IIB<br>Group D = IIA |

The degree of hazard is given by the Zone designation:

- Zone 0, Zone 20 – hazardous atmosphere or dust cloud may be present continuously or a large percentage of the time
- Zone 1, Zone 21 – hazardous atmosphere or dust cloud may be present intermittently
- Zone 2, Zone 22 – hazardous atmosphere or dust cloud present only abnormally, as after equipment or containment failure.

Thus, Class I, Division 1 includes Zone 0 and Zone 1, and Class I, Div. 2 = Zone 2.

And, Class II, Division 1 includes Zone 20 and Zone 21, and Class II, Div. 2 = Zone 22.

### 22.3.2 Temperature Class

Electrical apparatus often develops areas of sufficiently high temperature to ignite a material it may contact. To allow the user to ensure equipment will not become a thermal source of ignition, the manufacturer must indicate the value of the highest temperature reached by a part that is accessible to the explosive or combustible mixture.

At present this is most often done by specifying a Temperature Class. These are defined in the table below. Classes without a suffix are internationally recognized. Those with suffixes are defined in North America, because many intermediate temperature limits were defined in electrical codes or standards prior to 1971. As a practical matter, a T4 class is safe for all but carbon disulphide and perhaps one or two other vapors. The maximum surface temperature for some equipment for use in dusty locations may be controlled by the testing standard and may not be marked on the equipment.

| Maximum Temperature | | Temperature Class |
|---|---|---|
| °C | °F | |
| 450 | 842 | T1 |
| 300 | 572 | T2 |
| 280 | 536 | T2A |
| 260 | 500 | T2B |
| 230 | 446 | T2C |
| 215 | 419 | T2D |
| 200 | 392 | T3 |
| 180 | 356 | T3A |
| 165 | 329 | T3B |
| 160 | 320 | T3C |
| 135 | 275 | T4 |
| 120 | 248 | T4A |
| 100 | 212 | T5 |
| 85 | 185 | T6 |

### 22.3.3 Selection of Apparatus

Once the classification of a location has been established, usually during plant design, one can select electrical apparatus which is safe to use in that location.

Figure 22-1 shows the types of protection against ignition of gases and vapors in common use, and their applicability. Ex d, Ex p, etc. are the International Electrotechnical Commission (IEC) designations for that type of protection, which are also recognized in the NEC (denoted AEx) and Canadian Electrical Code (denoted Ex or EEx) for marking apparatus for use in the several zones.

| TYPE OF PROTECTION (See Note 1) | | | | | | | | | | | |
|---|---|---|---|---|---|---|---|---|---|---|---|
| **Area Classification** | | **Intrinsic safety** | | **Explosionproof/ Flameproof** | **Pressurization (Note 2)** | | **Type N Protection** | **Encapsulation** | | **Increased Safety** | **Unprotected apparatus** |
| | | | | | Isolate | Alarm | | | | | |
| | | Ex ia | Ex ib | Ex d | Ex p | Ex p | Ex n | Ex ma | Ex mb | Ex e | |
| Zone 0 | Division 1 | Yes | | | | | | Yes | | | |
| Zone 1 | | Yes | Yes | Yes | Yes (Note 3) | Ex n apparatus | | Yes | Yes | Yes | |
| Zone 2/Division 2 | | Yes | Yes | Yes | Yes | Yes | Yes | Yes | Yes | Yes | If normally no ignition source |
| Unclassified | | Yes | Yes | Yes | Yes | Yes | yes | Yes | Yes | Yes | Yes |

**Note 1.** This figure illustrates the relationship among the types of protection and the Zone/Division notation. Only explosionproof, pressurized or intrinsically safe constructions are recognized for Class I, Division 1 use in North America. Few, if any, explosionproof and pressurized apparatus are installed in that small percentage of Division 1 locations that could be classified Zone 0. Types of Protection o (oil immersion) and q (powder filling) are Zone 1 techniques, but are not listed above because they are seldom used.

**Note 2.** "Pressurization" replaced the older term "purging" after it was recognized that flow more than that needed to assure a standardized minimum pressure inside an enclosure is unnecessary. "Purging" is now an initial deliberate flow through an enclosure to sweep out accumulated gases or vapors.

**Note 3.** In North American standards, a pressurization system which disconnects the power to ignition-capable apparatus in Division 1 when pressurization fails is Type X pressurization. In a Type Y system, loss of pressurization protecting apparatus suitable for Division 2 requires only an alarm. A Type Z system protecting ignition-capable equipment in Division 2 requires only an alarm. At present international standards recognize two constructions, with interlock and with alarm. Application is left to local codes of practice.

*Figure 22-1: Applicability of Common Types of Protection*

Table 22-2 describes the protection concept, salient construction features of each Type of Protection, and aspects of safe use specific to that Type of Protection. To ensure safe use of any Type of Protection, the user must install per instructions that accompany the apparatus—especially with regard to protection from the environment, shock, vibration, and, where specified, grounding and bonding. The user must inspect the installation frequently enough to ensure the integrity of the enclosure, and continuity of grounding and bonding, has not become impaired by accident or environmental attack.

*Table 22-2. Protection Concept, Salient Construction Features, and User Duties Specific to Each Type of Protection*

| Type of Protection | Protection Concept | Salient Construction Features | User Duties to Maintain Safe Installation |
|---|---|---|---|
| d – flameproof/explosionproof | An explosion inside the enclosure is not transmitted to the outside flammable atmosphere through gaps, shaft clearances, of by damage to the enclosure. | Heavy walls, usually cast metal, but may be plastic in small sizes.<br>Joints are tight fitting bolted flanges with narrow gaps or close fitting multiple threads. Shaft clearances very small. Small thick windows in reinforced mounting. Seals in wiring system complete the flameproof enclosure. | Ensure all bolts and threaded joints are tight and remain so. Ensure damage by corrosion or accident detected and corrected. Remove covers only if area non-hazardous, and ensure joints are clean, and undamaged before covers are replaced.<br>Ensure all seals are installed per codes. |
| p – pressurized | Positive pressure of air, or sometimes inert gas, inside the enclosure prevents entry of flammable gases, vapors and dusts. | Electrical equipment may be of standard design, though cost of supplying leaking pressurization medium may require sealing of openings. User may supply controls, interlocks, alarms, and warning labels required to ensure safe operation, but certified assemblies of apparatus and controls are available. | Periodic check to ensure that pressure or flow-actuated alarm or interlock works as designed. Enforce rules against energizing without ensuring enclosure is free of explosive materials if purging is not automatic. |
| i – intrinsic safety | Extremely conservative construction rules and limits on current and voltage available, even after failures, reduce probability to nearly zero that any failure in component or wiring has enough energy to ignite a flammable material. The only standard protection concept recognized safe in Zone 0.<br>Only protection concept generally suited for some servicing while energized without 'hot permit.'<br>ia – safe in Zone 0 or 1.<br>ib – safe in Zone 1. | Useful primarily for low power measurement and control applications (Note 1). Control of power ratings on components and stringent rules for spacings between conductors demands design for intrinsic safety from inception for apparatus located in Div. 1 or Zone 0/1 location. Isolating assemblies (barriers) can permit connection of general purpose equipment in a safe or Div. 2/Zone 2 location to intrinsically safe apparatus in the field. Stringent limitations on capacitance and inductance of field wiring and apparatus.<br>. | Essential that installation instructions, especially limits of inductance and capacitance of field apparatus and wiring, and grounding, be observed. Administrative controls must ensure that changes to the installation or addition of a new device do not violate installation rules and restrictions, and that i(b) apparatus is used only in Zone 1, not Zone 0. |

| | | | |
|---|---|---|---|
| e – increased safety | Originated as conservative design to reduce probability of ignition-capable arc developing in non-sparking rotating machinery, but same principles applied to terminals, connection facilities, resistance heaters, transformers, etc..<br>Type e construction often used for connection facilities of flameproof enclosures, and in conjunction with flameproof enclosures in switchgear. | In rotating machinery stator-rotor gaps are wider than normal to avoid frictional sparking. Coils are conservatively designed to reduce temperature rise and must be impregnated. Conservative rules for stating Temperature Class and selecting overcurrent protection devices to limiting temperature rise during under short circuit and stalled rotor conditions.<br>Internal connections are robust, soldered, welded, etc., and external connection terminals must not loosen under vibration, shock or temperature cycling and must maintain a low resistance connection. | Careful attention to specified ratings of protective devices, installation of external wiring. Safety is especially dependent on equipment being operated within all specified ratings. |
| m - encapsulation | Ignition capable elements sealed in encapsulant or potted to prevent exposure to flammable mixture.<br>ma – safe after two faults of encapsulated components, no switching contacts permitted, more robust construction than mb.<br>mb – safe after one fault. Less restrictive construction requirements. | Devices up to volume of 100cc encapsulated with specified minimum wall thickness except for volumes less than 1cc. Protection shall be maintained after fault(s). Creepage and clearances are defined which are presumed not to fault, as are certain components used at less than 2/3 rating.<br>Encapsulant is tested for water absorption and dielectric strength. Device tested over rated temperature range to ensure temperature rating of encapsulant and surface temperature class are not exceeded. | Primary duty of user, in addition to ensuring a device is used only in zone permitted, is to ensure that encapsulated device is housed in a protective enclosure if it is not rated for unprotected mounting, and that ambient limits and power ratings are not exceeded in use. |

| | | | |
|---|---|---|---|
| n – Apparatus which is nonsparking in normal operation,<br>-Normally sparking apparatus prevented from being an ignition source by one of several methods.<br>- Circuits in which opens, shorts, or arcing contacts are energy limited in normal operation ao ignition does not occur. | nA - nonsparking apparatus.<br>nC - sparking apparatus, isolated from flammable atmosphere by enclosed break device, hermetic sealing, sealing, non-incendive component, or encapsulation.<br>nR - restricted breathing enclosure.<br>nL - energy limited apparatus. | nA – rotating machines, low power measuring and control apparatus.<br>nC –enclosed break device is low volume enclosure strong and tight so explosion doesn't propagate outside if it should occur.<br>- hermetic seal is made by soldering, welding, etc.<br>- sealed device, a small enclosure made tight by means other than fusing metal or glass.<br>- nonincendive component, construction such that arcing is incapable of causing ignition.<br>- restricted breathing enclosure may be large and house ignition-capable contacts, but is so tight that outside flammable atmosphere in Zone 2 will not result in inside atmosphere becoming flammable. Internal temperature rise must be low.<br>- energy limited apparatus is assessed, in normal operation only, using ignition criteria for intrinsic safety. | User must be especially aware of the conditions and limitations of use required by the certification documents or instructions, as ambient conditions are very important in assessing safety of each design.<br>Techniques depending on maintaining nonflammable atmosphere inside enclosure demand care in use, and inspection often enough to reveal any degradation of the enclosure integrity. |
| **Note 1.** Single limiting values for voltage, power or energy cannot be stated, but most systems and devices operate at less than 30V, and a few watts, depending on the explosive material classification. However, electrostatic paint spray systems operating at kilovolts have been approved as intrinsically safe. Typical ignition energies in the standard test apparatus are 40 microjoules or higher. | | | |

All the protection techniques referred to in Figure 22-1 and Table 22-2, except intrinsic safety and energy limited Type n or Division 2 constructions, are device oriented. In principle, a user purchases the equipment, opens the box, and installs in accordance with the installation instructions and the applicable installation code—the NEC or CEC in North America. Intrinsic safety and energy limited design for Div. 2/Zone 2 applications are *system* oriented. Figure 22-2 illustrates an intrinsically safe system. A discussion of an energy limited system for Division 2/Zone 2 would be based on a similar diagram.

Intrinsically safe apparatus is composed entirely of intrinsically safe circuits. It can be described in terms of the maximum voltage, maximum current, and maximum power which may be impressed on its terminals; and, by the effective capacitance, inductance, or optionally, inductance to resistance (L/R) ratio, which can be seen looking into the terminals. Figure 22-2 shows only a single pair of terminals. There may be multiple pairs of terminals, each of which must be characterized by the characteristic values noted. The first symbol when two are shown is common in international standards. The second is common in North America, though both are permitted. Associated apparatus contains circuits that are intrinsically safe and circuits that are not intrinsically safe. If the nonintrinsically safe circuits are protected by some other technique, the apparatus may be located in a hazardous location. Otherwise, it must be located in an unclassified location. The intrinsically safe terminals are characterized by the maximum voltage, maximum current, and maximum power which can be delivered at the terminals, and the maximum values of capacitance, inductance, and optionally, L/R ratio, which may be connected safely to the terminals. Nonintrinsically safe terminals are characterized by the maxi-

**Intrinsically Safe Apparatus** | **Associated Apparatus**

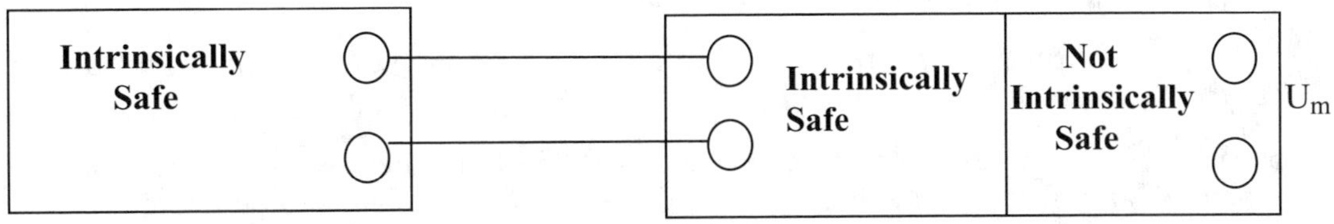


**Intrinsically Safe Apparatus**

$U_i$ or $V_{max}$ – max. safe applied voltage
$I_i$ or $I_{max}$ – max. safe input current
$P_i$ or $P_{max}$ – max. power input

$C_i$ – total equivalent capacitance seen at input terminals
$L_i$ – total equivalent inductance seen at input terminals
$L_i/R_i$ – effective ratio of inductance to resistance at input terminals

**Associated Apparatus**

$U_o$ or $V_{oc}$ – max. open circuit voltage
$I_o$ or $I_{sc}$ – max. output current
$P_o$ – max. deliverable power

$C_a$ or $C_o$ – maximum permitted connected capacitance
$L_a$ or $L_o$ – maximum permitted connected inductance
$L_a/R_a$ or $L_o/R_o$ – maximum permitted ratio of inductance to resistance in connected circuit

*Figure 22-2: Elements and Parameters of an Intrinsically Safe System*

mum rated voltage that may be applied to them. This voltage, $U_m$, in Figure 22-2, is usually 250V dc or rms for equipment for connection to the power line, but it could be 24V or other low voltage for other equipment.

In the former case, the design will be based on some minimum prospective current available from the power line. In the latter case, the assessment is carried out assuming presence of the low voltage, but the certificate or control drawing may demand that the 24V be supplied from a protective transformer, constructed in accordance with the provisions of the standard. The intent of this mandate is to ensure that the transformer, which is not a part of the certified apparatus, is of a quality of construction that will reduce to an acceptably low value the probability that its failure allows a voltage higher than 24V to appear at the terminals. In countries following the IEC pattern, apparatus is usually certified as an entity, without regard to the specific design of the other apparatus to which it is connected in a system. Intrinsically safe apparatus is defined by the parameters indicated in Figure 22-2. In principle, the use of these parameters to select devices to be used in a system is straightforward if the intrinsically safe device is a two-terminal device. It is only necessary to ensure that $U_o$ and $I_o$ are equal to or less than $U_i$ and $I_i$; and, that $L_i$, and $C_i$, are equal to or smaller than $L_o$ and $C_o$. If the intrinsically safe apparatus is a two-wire device, and both wires are isolated from ground, an associated apparatus (barrier) must be installed in each wire. In North America it has now become a requirement on the manufacturer of the intrinsically safe or associated apparatus to provide a "control drawing" which provides the details of the permissible interconnections and any special installation requirements peculiar to that apparatus. This control drawing is assessed and verified by the certifying agency as part of its examination of the product.

Standards following the IEC pattern define two levels of protection of intrinsic safety: level of protection ia apparatus and level of protection ib apparatus.

***Level of protection ia*** apparatus, suitable for Zone 0, will not cause ignition when the maximum permitted values are applied to its terminals:

- in normal operation with application of those non-countable faults which give the most onerous condition;
- in normal operation with application of one countable fault and those non-countable faults which give the most onerous condition; and
- in normal operation with application of two countable faults and those non-countable faults which give the most onerous condition.

Normal operation means that the apparatus conforms to the design specification supplied by the manufacturer, and is used within electrical, mechanical, and environmental limits specified by the manufacturer. Normal operation also includes open circuiting, shorting, and grounding of external wiring at connection facilities.

When assessing or testing for spark ignition, the safety factors to be applied to voltage or current are 1.5 in conditions a and b, and 1.0 in condition c. (These factors should properly be called "test factors." The real safety of intrinsic safety is inherent in the use of the sensitive IEC apparatus to attempt to ignite the most easily ignitable mixture of the test gas with hundreds of sparks. This combination of conditions is many times more onerous than any likely to occur in practice.)

North American Intrinsic Safety design standards are equivalent to ia intrinsic safety.

***Level of protection ib*** apparatus, suitable for Zone 1, is assessed or tested under the conditions of a and b above, with a safety factor on voltage or current of 1.5 in the condition of a and b.

It is likely that Level of protection ic apparatus, equivalent to and replacing Type of Protection nL, suitable for use in Zone 2, will be standardized.

Figure 22-3 shows typical grounded and ungrounded two wire intrinsically safe circuits.

Figure 22-3 illustrates the principle that every ungrounded conductor entering the Division 1/Zone 0, or 1 location in this case where the transmitter or transducer is located, must be protected against unsafe voltage and current by appropriate associated apparatus. The boxes with three terminals represent barriers, independently certified protective assemblies, certified and rated according to the national standard. Nonintrinsically safe devices connected to the barrier need only be suitable for their location, and must not contain voltages higher than the Um rating of the barrier.

Many barriers are passive, consisting of current limiting resistors and voltage limiting diodes in appropriate configuration and redundancy to meet the requirements of the standard. Others have active current or voltage limiting. Both types may be combined with other circuitry for regulating voltages, processing signals, etc. The user should follow the recommendation of the intrinsically safe apparatus manufacturer in the control drawing, or discuss the selection of appropriate barriers with the barrier vendor, all of whom have proven configurations for many field mounted devices.

## 22.4 Equipment for Use in Locations Where Combustible Dust May be Present

Table 22-3 illustrates the methods approved in 2005 for use in Class II locations in the NEC.

Definition of Zones 20, 21, and 22 in the Canadian Electrical Code is pending.

Dust ignitionproof construction is designed and tested to ensure no entry of dust under swirling test conditions while cycling equipment to create a pressure drop to draw dust into the enclosure. Temperature rise of the enclosure surface is determined after a thick layer of dust has accumulated on the enclosure. Dust tight construction may or may not be tested by a third party. If so, the test is essentially the same, but the equipment is cycled only once during the dust test.

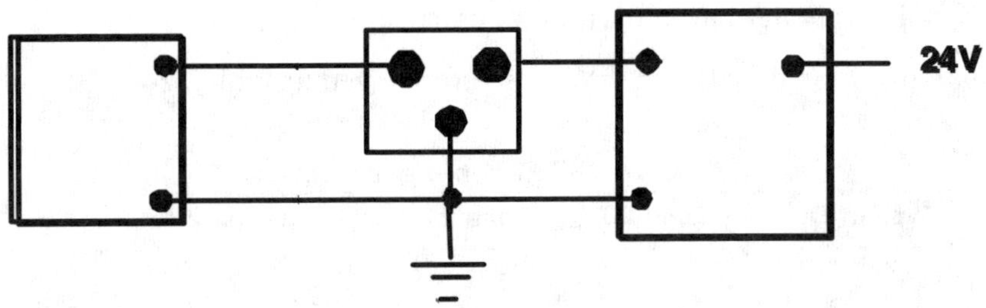


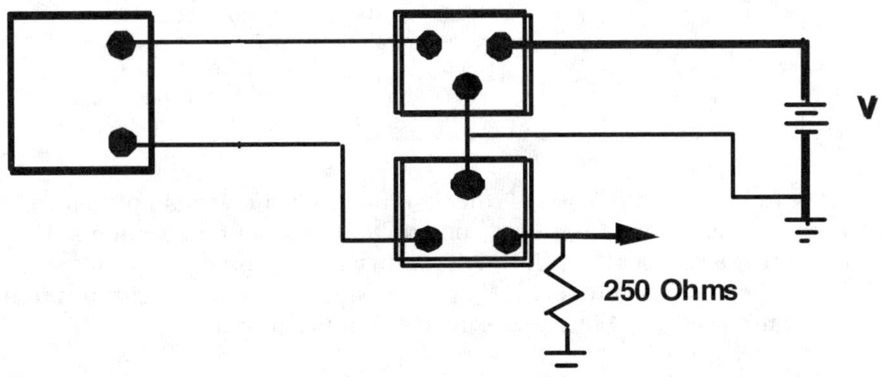


*Figure 22-3: Transmitter Input and Controller Output*

*Table 22-3: Methods of Protection for Equipment in NEC Class II Locations*

| Type of Protection | Division 1 | | Division 2 |
|---|---|---|---|
| | Zone 20 | Zone 21 | Zone 22 |
| Dust ignitionproof | No | Yes | Yes |
| Intrinsic Safety | Yes | Yes | Yes |
| Pressurized | No | Yes | Yes |
| Dusttight | No | No | Yes |
| Nonincendive circuit | No | No | Yes |
| Nonincendive equipment | No | No | Yes |

In European practice the user has responsibility for determining that equipment surface temperature under the expected accumulation of dust is safely below the ignition temperature of the dust involved. Equipment enclosure standards reflect this difference. The IEC standard for enclosures originally standardized constructions of both practices. Practice A is similar to American dust ignitionproof and dust tight enclosure. Practice B is European, in which enclosures for Zone 21 are tested under vacuum to be dust tight, i.e., degree of protection IP6X. Enclosures for Zone 22 are permitted to have entry of some dust, but not enough to impair function or decrease safety. However, it is expected that, in the near future, there will be greater harmonization of practice.

The recommendations adopted by IEC in 2004 are summarized in Table 22-4.

*Table 22-4: IEC Recommendations*

| Type of Dust | Zone 20 | Zone 21 | Zone 22 |
|---|---|---|---|
| Nonconductive | tDA20<br>tDB20<br>iaD<br>maD | tDA20 or<br>tDA21<br>tDB20 or tDB21<br>iaD or ibD<br>maD or mbD<br>pD | tDA20 or tDA21or<br>tDA22<br>tDB20 or tDB21<br>ortDB22<br>iaD or ibD<br>maD or mbD<br>pD |
| Conductive | tDA20<br>tDB20<br>iaD<br>maD | tDA20 or<br>tDA21<br>tDB20 or tDB21<br>iaD or ibD<br>maD or mbD<br>pD | tDA20 or tDA21or<br>tDA22<br>IP6X<br>tDB20 or tDB21<br>ortDB22<br>iaD or ibD<br>maD or mbD<br>pD |

The suffix D after the symbol for Type of Protection indicates the version of that technique intended for use with dusts, sometimes with reduced construction and test requirements. The symbol "t" refers to Protection by Enclosure, essentially dust tight or dust ignitionproof construction, and suffixes A and B are as discussed above. It is likely that IEC and U.S. standards for area classification and equipment selection in dusty areas will grow in closer agreement in the coming years.

### 22.4.1 The Label Tells about the Device

In North America, apparatus for use in Division 1 or 2 is marked with the Class and Group for which it is approved. It may be marked Division 1, but must be marked Division 2 if suitable for Div. 2 only. The temperature code, discussed below, is also shown.

Examples: Class I, Groups A-C, Division 1 T4

Class I, Groups C, D, Division 2 T6

In addition equipment approved for Class I, Division 1 or Division 2 may also be marked with Class, Zone, Gas Group, and temperature code. For the above examples the additional marking would be:

Class I, Zone 1, IIC, T4

Class I. Zone 2, IIA, T6

In the United States, apparatus approved for use specifically for use in Zone 0, Zone 1, or Zone 2 shall be marked with Class, Zone, AEx, the symbol for the method of protection, the applicable gas group, and temperature code.

Examples: Class I, Zone 0 AEx ia IIC, T4

Class I, Zone 1 AEx m IIC, T6

In Canada, the Class and Zone markings are optional and AEx is replaced by Ex or EEx, which are the symbols for explosion-protected apparatus conforming to IEC and CENELEC standards, respectively. In countries using only the zone classification system only marking such as those below are required.

Ex ia IIC, T4

EEx ia IIC, T4

If more than one Type of Protection is used, all symbols are shown: i.e., Ex d e mb IIC, T4.

In addition to the markings specifically relevant to explosion protection, a label will usually also provide additional information such as:

- Name and address of manufacturer
- Degree of protection afforded by the enclosure, NEMA or CSA enclosure rating in North America, or the IP code in countries following IEC patterns
- Symbol of certifying authority and an approval document number
- Voltage, current and power ratings
- Pressure ratings, if applicable.

If equipment is certified by a recognized approval agency, the user may be assured it is safe if it is installed and used according to the instructions supplied by the manufacturer. Conditions of use and installation instructions may be incorporated by reference in the certification documents or in drawings provided by the manufacturer. In all cases, it is essential the user install equipment in accordance with local codes, and operate it within its electrical supply and load specifications and rated ambient conditions.

## 22.5 For More Information

Few users need more detail about the fire and shock hazard protection that's built into the equipment they buy, but those who wish to dig deeper may consult the documents produced by the standards committees ISA-SP82, Electrical and Electronic Instrumentation, and IEC Technical Committee 66. ISA-SP82 developed ANSI/ISA-82.02.01-2004 (IEC 61010-1 Mod) - *Safety Requirements for Electrical Equipment for Measurement, Control, and Laboratory Use*, and the IEC group produced IEC 61010 - *Safety Requirements for Electrical Equipment for Measurement, Control, and Laboratory Use*. A broader understanding of area classification and the types of protection does, however, aid safe application of equipment in classified locations.

Publications by ISA-SP12, Electrical Equipment for Hazardous Locations, include ANSI/ISA-60079-0 (12.00.01)-2005 - *Electrical Apparatus for Use in Class I, Zones 0, 1 & 2 Hazardous (Classified) Locations*. That series includes information about design and use of electrical apparatus for classified locations, as do the U.S. versions of the IEC standards for the Types of Protection discussed above.

IEC publications form the basis of an increasing number of national standards of IEC member countries, the local versions of which can be purchased from the national member body of IEC.

The following is a list of standards addressing use and application of electrical apparatus rather than design:

### ISA

ANSI/ISA-12.01.01-1999 - *Definitions and Information Pertaining to Electrical Instruments in Hazardous (Classified) Locations*

ISA-TR12.2-1995 - *Intrinsically Safe System Assessment Using the Entity Concept*

ISA-RP12.2.02-1996 - *Recommendations for the Preparation, Content, and Organization of Intrinsic Safety Control Drawings*

ANSI/ISA-TR12.24.01-1998 (IEC 60079-10 Mod) - *Recommended Practice for Classification of Locations for Electrical Installations Classified as Class I, Zone 0, Zone 1, or Zone 2*

**IEC**

IEC 60079-10 *Electrical apparatus for explosive gas atmospheres - Part 10: Classification of hazardous areas*

IEC 60079-17 *Electrical apparatus for explosive gas atmospheres - Part 17: Inspection and maintenance of electrical installations in hazardous areas (other than mines)*

IEC 61285 *Industrial-process control - Safety of analyser houses*

IEC 61779-6 *Electrical apparatus for the detection and measurement of flammable gases - Part 6: Guide for the selection, installation, use and maintenance of apparatus for the detection and measurement of flammable gases*

**NFPA**

NFPA 70 *National Electrical Code®*

NFPA 70B *Recommended Practice for Electrical Equipment Maintenance*

NFPA 70E *Standard for Electrical Safety in the Workplace*

NFPA 496 *Standard for Purged and Pressurized Enclosures for Electrical Equipment*

NFPA 497 *Recommended Practice for the Classification of Flammable Liquids, Gases, or Vapors and of Hazardous (Classified) Locations for Electrical Installations in Chemical Process Areas*

NFPA 499 *Recommended Practice for the Classification of Combustible Dusts and of Hazardous (Classified) Locations for Electrical Installations in Chemical Process Areas*

The websites of the above-mentioned organizations are found at:

www.isa.org
www.iec.ch
www.nfpa.org

Many manufacturers provide free literature, often viewable on their Websites, which discuss the subjects of this article.

For an in-depth treatment of the science behind the types of protection and the history of their development, as well as design, installation, and inspection of installations one can consult the book *Electrical Instruments in Hazardous Locations*, published by ISA.

For discussion of installation of apparatus see:

Schram, P.J. and M. W. Earley. *Electrical Installations in Hazardous Locations*. NFPA, 1998.

McMillan, Alan. *Electrical Installations in Hazardous Areas.* Butterworth-Heinemann, 1998.

***Acknowledgment:*** The author is indebted to William G. Lawrence, P. E., Senior Engineering Specialist, Hazardous Locations, FM Approvals, for many valuable contributions to improve the accuracy of this article.

## About the Author

**Ernie Magison** was active in standards development in ISA, IEC, and NFPA for four decades. He has authored 40 articles, as well as papers and several books, most focusing on application of electrical apparatus in potentially explosive atmospheres. He has taught many short courses and consulted in the field.

# V

# Integration and Software

*Integrating systems and communications is now fundamental to automation. While some who work in a specific area of automation may have been able to avoid a good understanding of these topics, that isolation is rapidly coming to an end.*

*In fact, integration and communications are such an important part of automation that, in some companies, automation responsibility has been turned over to the information technology (IT) departments. While that may solve the integration issue, it usually does not deal with the unique real time and security issues in automation, and it totally ignores the plant floor issues.*

*What is really needed is a good coupling of IT know-how with a broad knowledge of plant floor automation—either by having IT systems specialists learn plant floor controls or by having automation professionals learn more about integration, or both.*

*Functionality and integration of the shorter time frame operating systems with both plant floor controls and with company business systems is called by several names. However,* manufacturing execution systems *(MES) is the most common.*

*The concepts of what functions to do where and when data flows occur has been wrestled with for a couple of decades since the computer integrated manufacturing (CIM) work of the 1980s. It has continued with the tireless work led by Theodore "Ted" J. Williams, the now retired Purdue University professor and 1968 ISA President. Now, with the ISA-95 - Manufacturing Enterprise Systems Standards series, real standardization has arrived in this area. The ISA-95 standards have been adopted by some of the biggest names in manufacturing and business systems. While a large percentage of automation professionals do not know really what MES is, this topic, like integration in general, cannot be ignored any longer.*

*No topic is hotter today than* Network Security*—including the Internet. Any automation professional who is working in any type of integration must pay attention to the security of the systems.*

Operator Interface, Data Management, *and other types of* Software *are also now basic topics for automation professionals, and they fit in this category better than anywhere else. Packaged automation software that is open with respect to OPC covers a significant portion of the needs of automation professionals; custom software is still needed in some cases. That custom software must be carefully designed and programmed to perform well and be easily maintained.*

# 23 Digital Communications

*By Dick Caro and Tom Phinney*

## Topic Highlights

*Protocol Concepts*
*Network Protocols*
*Network Topology*
*Wireless Networks*
*Bibliography*

## 23.1 Introduction

The terms "data communications" and "networking" are often used interchangeably, but both refer to the transmission of data in digital form from one place to other places. The most important factors are that the source of the data and the receivers of that data use the same electrical technology to encode and decode the digital data, and that they share the same scheme for formatting the data. These are universal truths for all networks used for information technology (IT) and for industrial automation.

The earliest automation networks were not considered networks at all but, instead, serial "buses." The term *fieldbus* stems from this concept. Naturally, each network was designed to solve one problem, then extended to solve other, perhaps related, problems. Because each supplier's business model was directed toward a slightly different business niche, the resulting *bus* turned out to be different from any other.

As long as industrial automation networks were slow and uncomplicated, they required no special components. For example, EIA[1]-232, EIA-422/423, and EIA-485 were often used for the physical layers, supported by commodity semiconductors. Early protocols, data formatting, and error handling procedures were simple enough to execute on eight-bit microprocessors such as the 8051, Z80, or 6809. When speeds became higher and protocols richer in functionality, custom silicon became necessary to implement these networks. Custom silicon is expensive to design due to the nonrecurring engineering or NRE costs, and because the volumes are usually small compared with volumes in the IT market, expensive to manufacture.

The first attempt to reduce these high costs was to standardize industrial automation networks through standards committees or establish de facto standards by opening the specifications to multivendor consortiums. The theory was that, if many system suppliers used the same chip set, there would be an economy of scale and lower manufacturing costs. It didn't work—there were *too many standards!* Due to the many niches of the industrial automation market, it was not possible to specify one chip set to meet the needs of all niches.

1. Originally noted as RS-232, RS-422, and RS-485, but the standards organization Electronics Industry Association prefers to designate it as EIA.

The industry is starting to shelve the idea that industrial automation networks are somehow different from other networks. The clear trend is to use commercial off-the-shelf (COTS) components and adapt them through software to industrial automation applications. Because Ethernet has been the clear winner in the IT market, it is no surprise that Ethernet is the basis for the newest evolution of industrial automation networks—at least at the high-performance end—which leaves the lowest-level networks used for connecting sensors and actuators with a different, and often more expensive, solution. These networks are also migrating and converging in both directions with commodity silicon as the basis. Some low-level networks will most likely use scaled-down versions of the higher-performance industrial automation networks; whereas others will use low-cost silicon developed for other markets. In some of the lowest-performance applications, sufficient volume may exist to produce custom silicon at low cost.

## 23.2 Protocol Concepts

Exchanging information through any communications method requires the transmitters and the receivers to have a common agreement about the type of electrical signals used, the organization of the data, and the processes used to ensure successful and error-free transmission of the information. Formalization of that agreement is called the communications *protocol.* Even simple point-to-point serial line transmission conforms to a protocol, although such a specification may be very brief and totally contained in the specifications of the Universal Synchronous/Asynchronous Receiver/Transmitter (USART) chip used to implement the communications.

### 23.2.1 Encapsulation and Layering

Network functionality is usually organized by a hierarchical set of *layers* called the communications *stack.* At the lowest layer of the stack is the electrical mechanism for sending/receiving data on the wire, optical, or radio transport media. At the highest layer of the stack is the interface to the user. In the middle are the layers necessary to perform the routing, segmentation (of long messages), and error detection and recovery. Each layer's protocol describes an action that results in a data field *appended to the message* as it is formed during transmission. The protocol's data field may be placed before the message as it was received from the previous layer, or after the message, or both. The message as it appears on the communications media is, therefore, the encapsulation of the original data field within the protocol data fields of all active protocol layers.

Upon reception, the receiver's corresponding layer removes its protocol appendage and processes the message accordingly. By the time the message arrives at the top of the communication stack, the original data stream is presented to the application with all communications protocols removed.

### 23.2.2 Addressing

Messages sent across a network begin at a *source address,* and are sent to some *destination address.* At the lowest layers of the network, all addresses are numbers, so the hardware can easily make sure the correct network node receives the message. At higher layers, the network address may be expressed as a human-readable name, commonly called a *tag.* Translation between the numerical address and the alphabetic expression of that address may be assigned to a network address translation device located somewhere in the network, or it may even be assigned to an end device itself as a higher layer function. However, in all data transmissions across the network, it is very common to have both the source address and the destination address included in the data transmission message.

There is a special type of network address called a broadcast or group address. Messages sent to a broadcast or group address are received by all network nodes or the network nodes in the numbered group.

### 23.2.3 Bit/Byte Ordering

Digital data is processed by computers in *words* that vary in length from eight binary bits called a *byte,* to longer concatenations of bytes that may be a total of 32, 64, or even 128 bits in length. Conventionally, the *most significant bit* is used to represent the sign, when the word represents a signed integer value.

There are two possible ways to send the data on a serial communications channel:

- big-endian – sign bit or most significant bit first
- little-endian – least significant bit first

### 23.2.4 Error Detection and Correction

Sending binary data via a communications network is never free of error. The most important factor is detecting errors, so an error recovery mechanism can restore the original data or the receiver can ignore the data. Therefore, we speak of reliability of a data communication network in terms of the probability of NOT delivering *undetected errors.* An acceptable error rate for delivery of voice might be expressed as one undetected error in $10^8$ data bits, or expressed as an error rate of $10^{-8}$. An industrial control system might require an error rate of $10^{-20}$.

Error detection for all communications protocols uses a mechanism called *parity,* which appends one or more extra bits to the communication frame. The extra bits are determined by the extent of the error detection *algorithm.* The simplest error detection algorithm inserts a parity bit with a value of "1" if the number of data bits with a value of 1 is odd; thus making the total number of 1 bits even. This is called *even parity.* Insertion of a data bit with the value of "0" in the same situation is called *odd parity.* Parity error detection is most often used for asynchronous communications and applies to each byte.

A slightly more efficient error detection method is called a *checksum.* Rather than adding together all of the data bytes of a message, the checksum is usually computed by exclusive or-ing (XOR) all of the bytes in the message that will yield the longitudinal redundancy check (LRC) byte. The receiver performs the same XOR operation and should get the same LRC byte. If not, then there was an error in the transmitted message.

A more complex form of error detection is called cyclic redundancy code (CRC). It is typically used on Ethernet and most local area networks. CRC with a length of 32 bits is commonly called CRC-32, and is used for Ethernet. CRC-16 is used for Foundation Fieldbus H1. The CRC algorithm is designed to detect all single-bit errors and usually some significant portion of multiple-bit errors. The CRC calculation uses a complex polynomial that results in a frame check sequence (FCS) that is appended to the message. The receiver makes the same computation and should get the same FCS value unless an error occurred during transmission.

The usual form of correcting a transmission error is to reject the data frame and request retransmission of the data. Sometimes, retransmission is not acceptable, because the time window would have expired, such as for control data in Foundation Fieldbus. In these cases, the data error is noted and the data is ignored as if it was not received, hoping that good data will be received the next time. In other cases, such as data integration, it may be necessary to retransmit.

There is a special case where retransmission is impractical, such as radio telemetry of data from deep space. For these applications, the most complex error recovery is available: error correction. It is possible to generate parity information with enough data redundancy to actually correct all single-bit errors and many multiple-bit errors. The transmitter complex computes the error correcting code (ECC) and appends it to the message. The receiver makes the same calculation, and if the answer is the same, then no errors in transmission occurred. If there is a difference, the receiver can correct the data errors. Error correction is highly wasteful of bandwidth, however. Simple error correction can use more than 25% of the available bandwidth just by inserting a large ECC.

### 23.2.5 Mastering

Before a network node sends data or a request for data, it must know that it alone controls the network and that its message will not be corrupted during transmission. This is done by temporarily transferring network mastership to that node. One of the most important parts of a network protocol is to determine which of the available network mastering schemes to use. Table 23-1 lists all four network mastering methods in actual use and an example of the networks using each method.

*Table 23-1: Network Mastering Methods*

| Mastering Method | Strategy | Example |
|---|---|---|
| Master/Slave | Network Master station polls all nodes and requests them to send data; node responds immediately | PROFIBUS, Modbus |
| Arbitration | Link scheduler assigns mastership on a regular schedule | FOUNDATION Fieldbus, WorldFIP |
| Contention | Node listens before transmit, on silence sends the data. If collisions occur, use a setback before retrying | 5BaseT Ethernet, DeviceNet |
| Token Passing | Peer nodes transmit then send token to next node. Lost tokens must be recovered. | IEEE 802.4 Token Bus, IEEE 802.5 Token Ring |

## 23.3 Network Protocols

All network architectures are described by the International Standards Organization (ISO) standard Open Systems Interconnection (OSI) basic reference model: standard ISO/IEC 7498-1:1994. This model, illustrated in Figure 23-1, is divided into seven parts that are independent of each other. Note that *network protocol* means the microprocessor firmware embedded in these layers. The end user only cares about the connection to the physical wires coming out the bottom and the features and functions made visible by the software for the layers above the ISO/OSI stack.

Notice there are two layers above the ISO/OSI seven layers shown in Figure 23-1. The object linking and embedding for process control (OPC) layer has the benefit of adapting the network layers to the host system. Thus the client-user layer can be created knowing only that it will be used with a server running compatible OPC server software. With OPC, the details of the network layers are effectively hidden from view. Also note that you can isolate the network application layer from the user layer software by using other network technologies incorporated in the user layer that do not use OPC. The figure illustrates this by the direct coupling of the user layer to the top of the communications protocol stack. Usually this is done to take advantage of the efficiency of the user layer connections and to make data transfers more efficient or deterministic than allowed by OPC.

Finally, in Figure 23-1, the network wiring is drawn with a little box between client and server. This box presents the physical network as more than just wire and cable. Very often, there are active switches and converters in these networks for a variety of reasons that we will discuss when each network is described in subsequent chapters. The word *cable* is often used, because it includes metallic wiring and fiber optics. The term also should include wireless connections, but that would require the oxymoron wireless cable, an expression actually used in the telecommunications market but not yet in industrial automation.

### 23.3.1 Physical Layer (PHY)

Layer 1 of the OSI seven-layer stack is the physical layer (PHY). Note the PHY does not include the communications media, only the electrical and logical interface to the media. Here, the type of media supported is presented with suitable information about electrical characteristics such as voltage levels, symbol representation (either a 1 or a 0) and encoding, signaling method, frequency bands, type of

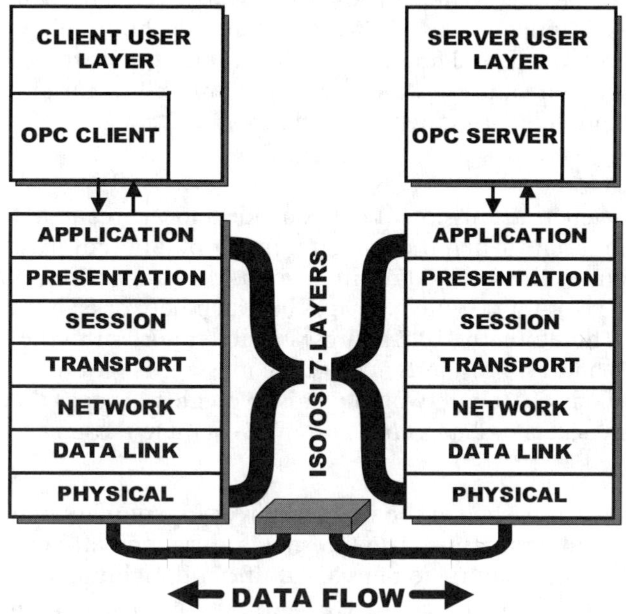


*Figure 23-1: Network Layers*

connectors supported, and the pin assignments for each connector type. For very simple networks, where contention for the media mastership can be determined electrically, that specification is included in the PHY including collision detection. Finally, this layer usually has some flow control capabilities to temporarily stop transmission if the receiver is too busy to accept data.

A very important concept is the choice of encoding method, which actually affects the capability of the receiver to discriminate between noise and a true signal. Once the signal is established as a voltage or current level, a frequency shift, or a phase shift, then it must be decided that each level or shift transition represents a data symbol (bit) or a more complex series of transitions will be required to represent each symbol. The PHY generally specifies the methods used to determine each bit of the transmission.

### 23.3.2 Data Link Layer (DLL)

Layer 2 of the OSI model is the data link layer (DLL). This layer specifies the basic message format including addressing and error detection and correction methods. The DLL is usually subdivided into the logical link control (LLC) sublayer and the Media Access Control (MAC) sublayer. The LLC provides services to higher-numbered layers while the MAC concentrates on the lower-level functions particular to the chosen network media and its characteristics. The MAC specifies what we usually think of when we say protocol, such as token passing, collision-based, peer-to-peer, or master/slave protocols. The address of the device follows the rules of MAC addressing, and is usually a number between four and 64 bits long.

Data framing is specified in the DLL. The frame is the "envelope" used to send data that includes the *from* and *to* addresses and the error checking code needed to detect and correct transmission errors. Exactly how the data is organized into the message frame is vital, because the receiving device must be able to interpret exactly how the data was organized in that frame. Irrespective of how the data in the host computers is organized, all data communications is organized in big endian order.

The most important function of the MAC is to determine which network device is allowed to send messages when the communications network is shared among many devices. When all network sharing is under the control of a single host device, then the MAC protocol is called master/slave or client/

server. When the network mastership passes from one device to another in a determined order, the protocol is called token passing. When any device may access the network if it is not busy, the protocol is called Carrier Sense Multiple Access (CSMA). The protocol must allow for new devices to be added to the network, devices to fail, and for errors to occur during transmission. Token-passing networks must provide for lost token failures. CSMA networks must allow for message collisions. Even master/slave networks must provide for masters that fail.

### 23.3.3 Network Layer

The network layer covers transmission of data outside a local area to a device on another network. This is particularly important when there are a number of different paths messages can take from the source to the destination. This is called routing, and is the primary responsibility of the network layer. Because the network layer must route messages between networks, network addressing is the primary feature of this layer. Therefore, the DLL address is only important to the local area network; while the Internet Protocol (IP) address is important once the message is sent to a device on another network on the Internet. The final task of the network layer is to divide messages that are longer than the limits of the DLL into shorter packets or datagrams, and to reassemble those packets in the correct order when the message is received.

Routing messages is accomplished at the network layer for complex networks with more than one path between source and destination. The Internet is a vast network of networks with very complex and dynamic routing rules adapting to network traffic and malfunctions. Simple local area networks are usually routed using the layer 2 address, because there is usually only a single route possible.

The most common network using a network layer is the Internet. For the Internet, this level of protocol is called IP, which is specified as an Internet standard by the Internet Engineering Task Force (IETF), the standards body of the Internet. Here is an excellent Web site describing all of the fields of the IP header: http://www.networksorcery.com/enp/protocol/ip.htm

Because of its use in all industries, the IP address is very important. Like MAC addresses, the IP address is a number. Using IP version 4, most common in 2005, the IP address is 32 bits long. In IP version 6, which perhaps will be popular in the future, the IP address is 128 bits long. Several IP addresses are reserved by Internet standards for local addressing purposes, and cannot be assigned to devices addressable on the Internet. These long numbers are difficult for humans to remember, so there is a naming system to express IP addresses in text strings that have more personal meaning. The Internet organization responsible for assigning text strings to Internet addresses is the Internet Corporation for Assigned Names and Numbers (ICANN). The full scope of IP addressing is beyond the scope of this book. See this excellent tutorial on the complexities of IP addressing: http://compnetworking.about.com/od/workingwithipaddresses/l/aa042400a.htm

Industrial automation networks are often quite simple local area networks and do not require the use of the network layer or IP addresses. Any routing is done within the DLL using MAC addresses, and segmentation, if needed, is done within the application layer or the application instead.

When there is no difference between the maximum message length of the source and destination networks (e.g., both are Ethernet), the network layer does no segmentation. However, where there is a difference, the network layer produces segments of the correct length for the destination. Segmentation at the network layer is a service for the underlying DLL of the receiving node.

### 23.3.4 Transport Layer

Although there are several standards for transport layers, many industrial networks do not specify a transport layer, and the corresponding services are assigned to other layers. When a transport layer is used, one of the two Internet standard transport layer protocols is usually specified:

- Transmission Control Protocol (TCP)

  TCP is a connected service guaranteeing end-to-end delivery across an IP network. Failed or unacknowledged segments are retransmitted to correct transmission errors. Another task of TCP is flow control. Because long messages can be sent along a route that is initially undefined, the first message frames are small to "probe" the network route. Once the route has been defined (connected), subsequent messages are usually longer with the expectation that they will be delivered without error or delay. Sending longer messages is usually more efficient than short messages.

- Universal Datagram Protocol (UDP)

  UDP is a simple service designed to deliver IP packets between two IP address using "best effort" methods. The application layer or the application itself supplies any error detection, correction, segmentation, or flow control.

The transport layer defines types of applications by the network "port" they use. The Internet Assigned Number Authority (IANA) has assigned many of the TCP/UDP ports, which are listed on this Web site: http://www.iana.org/assignments/port-numbers. This mechanism allows many applications to share a single communications line. For example, Modbus/TCP, ProfiNET, EtherNet/IP, and Foundation Fieldbus HSE all use TCP/IP and/or UDP/IP and have ports assigned in this group. Table 23-2 shows some of the ports assigned to common industrial automation networks:

*Table 23-2: IANA Port Assignments for Industrial Networks*

| Network Protocol | Port Number |
|---|---|
| Foundation Fieldbus HSE | 1089-1091 and 3622 |
| Modbus/TCP | 502 |
| EtherNet/IP | 2221-2223 and 44418 |
| ProfiNET | 34962-34964 |
| EtherCAT | 34980 |
| LonWorks | 2540-2541 |
| Worldwide Web http | 80 |

One of the duties of TCP is to ensure successful end-to-end delivery of every message; UDP has no such duty. At this point in the protocol, it is assumed that the DLL detects and corrects errors in the message contents. TCP only ensures the message is safely delivered to the end destination. Each message segment is acknowledged as it is received. The outgoing message segment header contains an acknowledgment number. When the acknowledgment frame is returned to the sender, it must contain the same acknowledgment number. If acknowledgment is not received before a time limit, it is automatically retransmitted. Because all segments are numbered, duplicate segments are discarded when the message is reassembled.

### 23.3.5 Session Layer

The session layer is not often used for industrial automation communications protocols. Unlike the lower four layers, the session layer provides sustaining connections between applications on different computers across the network. These connected services in industrial automation networks are usually implemented at the application layer or in the actual application itself. The layer originated with the use of sockets in UNIX software for connection between concurrent processes. The session layer extends the UNIX socket across the network.

It is also useful to think about the use of large all-digital telephone networks. Here, the session layer contains the protocol for establishing and holding open a virtual end-to-end connection for the voice grade telephone call.

### 23.3.6 Presentation Layer

Industrial automation networks also do not often use a specific presentation layer; the corresponding functionality is embedded into the application layer when needed. The three traditional functions of the presentation layer are:

- translation
- compression
- encryption

Because the computers at each end of the communications connection may be quite different, any differences in data formats and character encoding are resolved in the presentation layer. The classic function was the translation between ASCII and EBCDIC character encoding as defined by the American National Standards Institute (ANSI) and that defined by IBM for all of its mainframe computers.

Data transmissions are often compressed during the communications process to conserve bandwidth. The application can also compress files, but that does not define the dynamic compression used by modem communications standards. Compression was frequently used in the days of telephone-line modem transmission, but is used less frequently true today with broadband transmission.

Data encryption is widely used to establish virtual private networks (VPNs), because safeguarding data against interception is necessary when use of a very public carrier such as the Internet. Wireless data transmission also exposes the data to eavesdropping and is corrected with the robust Advanced Encryption Standard (AES) that has been adopted as part of the privacy standard for wireless data transmission, IEEE 802.11i (also known as Wireless Protected Access).

### 23.3.7 Application Layer

The application layer is the top layer of the formal communications stack, but it is not the software that we think of as an "application." Many times applications use features of the operating system that in turn use some of the services provided by the application layer. The following are typical services provided by the application layer:

- **HTTP –** Hypertext Transfer Protocol (Worldwide Web)
- **FTP** – File Transfer Protocol
- **SMTP** – Simple Mail Transfer Protocol (e-Mail)
- **DHCP** – Dynamic Host Communications Protocol
- **Telnet** – A text-based command and control language for remote computers
- **SNMP** – Simple Network Management Protocol
- **RMON** – Remote network monitoring protocol

All of these services are part of the TCP/IP "suite" of protocols. Layers 1-4 and 7 are defined by the Internet standards, a set of documents called Request for Comments (RFC) administered by the IETF (http://www.ietf.org/iesg/1rfc_index.txt.) Standards documents are too difficult to read unless you need to implement software to support that standard. A much better learning tool for discovering the details of the TCP/IP protocol suite is the TCP/IP Guide: http://www.tcpipguide.com/free/index.htm

## 23.4 Network Topology

All data communications networks connect devices positioned for their application, not for the convenience of the network installation. The network architecture, or wiring plan, is called the "topology" designed to interconnect network devices. The network protocol must be aware of, and compatible with, the topology selected—they are not independent.

### 23.4.1 Star Topology

The most popular of all network topologies is the star, illustrated in Figure 23-2. They are used in practically all office networks, especially where Ethernet and Token Ring are used. Star networks usually have an active server, switch, or hub at the central location. The spokes of the star are directly connected to the remote device with no branching. Centralization of the server, switch, or hub allows easy reconfiguration of the network as the devices move from one location to another. It also places most of the active network equipment at one central location for easier maintenance.

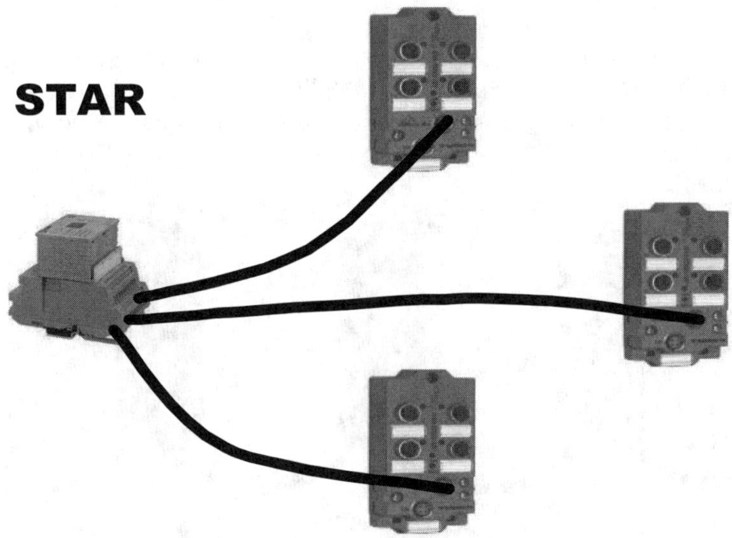


*Figure 23-2: Star Network Topology*

### 23.4.2 Multidrop Topology

Some manufacturing equipment is built as a long sequence of machines organized in a straight line. Accordingly, the sensors and actuators are positioned along the manufacturing line. The most natural method to wire these devices is with a "trunk cable" with "drops" to each device. Because many drops are required, this topology is called *multidrop*. The network protocol must support this topology, because, unlike a star with individual connections to a central location, all connected devices will receive any message placed on a multidropped network. The trunk wire, or *bus* as it is commonly called, must drive the entire line and all of the drops. A multidrop topology is sometimes used with a peer-to-peer protocol that allows direct communications between devices, but is often used with a master/slave protocol that only allows communications between the master and one slave device. A multidropped network is illustrated in Figure 23-3.

### 23.4.3 Daisy-chain Topology

Very similar to the multidropped network topology, a daisy-chain network is designed for devices laid out in a linearly distributed pattern. However, each device must drive the signal on the wire only as far as the next device, often requiring less power than a multidropped topology. Data not intended for

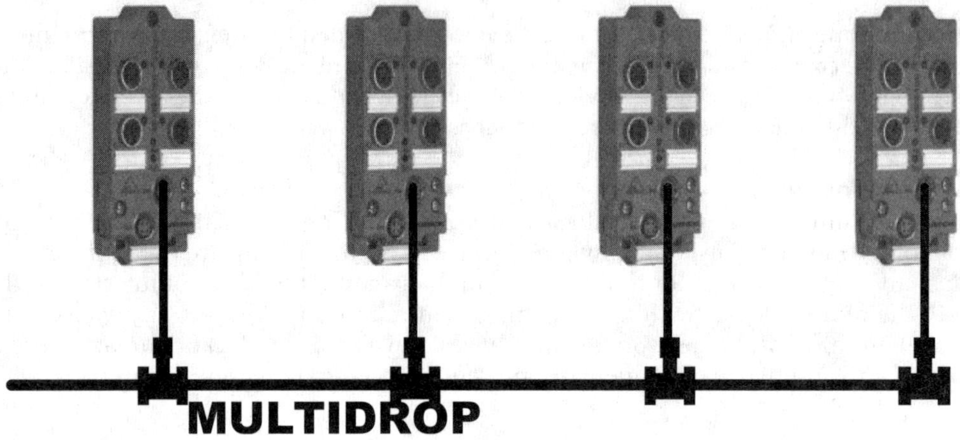


*Figure 23-3: Multidrop Network Topology*

a device is always forwarded to the next device. To use a daisy-chain topology for peer-to-peer devices, the data path must move data in both directions; therefore daisy chain is not often used for peer-to-peer protocol networks. Figure 23-4 illustrates a daisy-chain topology.

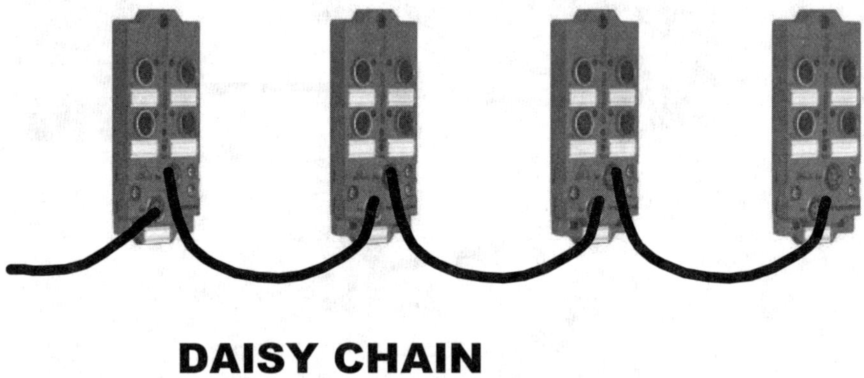


*Figure 23-4: Daisy-chain Topology*

### 23.4.4 Ring Topology

Ring networks are usually used when organizations desire high-reliability networking. The ring network is very similar to the daisy-chain topology except the last device is always wired back to the first device or network master device. Figure 23-5 illustrates a simple ring topology network. Most ring networks used in automation for high reliability actually use a dual ring with data moving in opposite directions: counter-rotating rings. Dual redundant counter-rotating rings allow all communications to continue uninterrupted when the network cable is cut at any one location. Figure 23-6 illustrates a dual redundant counter-rotating ring topology.

### 23.4.5 Mesh Topology

Mesh networks have more than one data path between network nodes to achieve path redundancy for reliability. However, mesh networks also usually have an increased protocol burden called "routing." In most other networks, there are no alternative paths—the one path is fully determined by the wiring. Mesh networks, because they have alternative paths, need to have the route defined by the

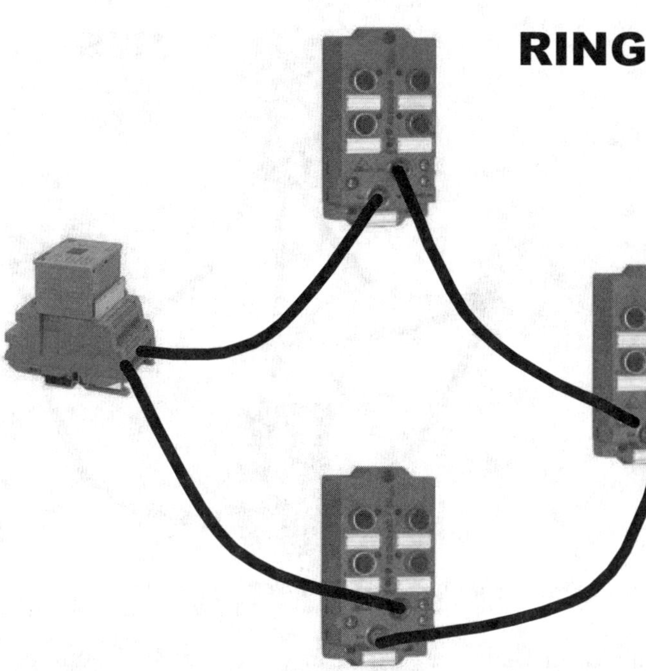


*Figure 23-5: Ring Network Topology*

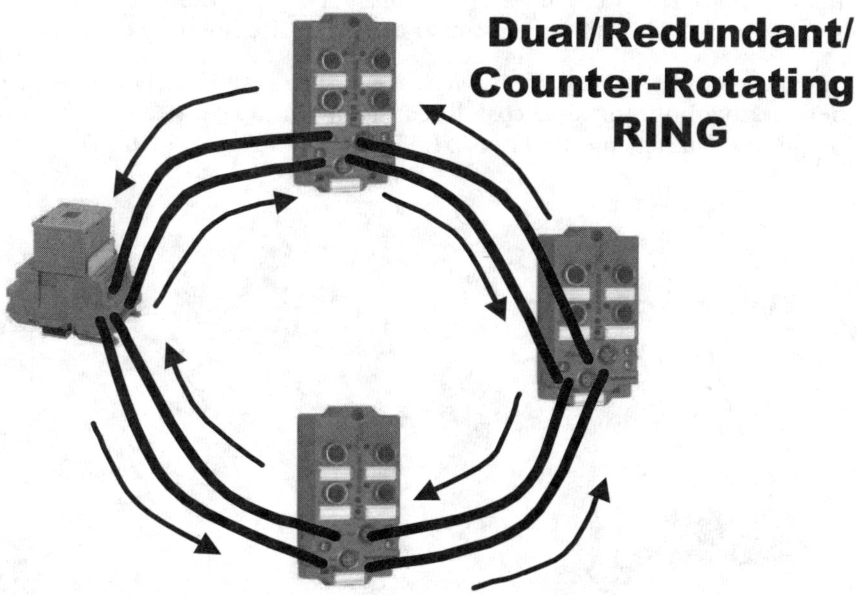


*Figure 23-6: Dual Redundant Counter-rotating Ring Topology*

destination address when a message enters the network. The Internet itself is a very large wired mesh network that was designed to survive loss of a major part of the network during a nuclear disaster. Figure 23-7 illustrates a wired mesh network topology in which each device is wired to every other device.

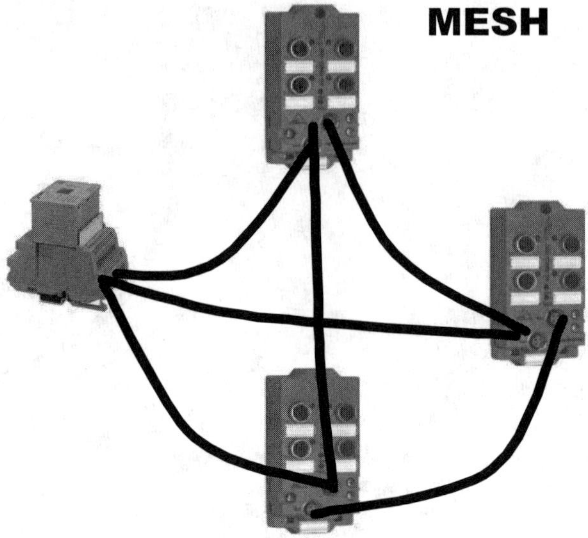


*Figure 23-7: Mesh Network Topology*

Mesh networks do not require that every device be directly connected to every other device. To reach devices not directly connected, routing tables must be constructed to supply sufficient information to pass messages to a directly connected device in the "direction" of the intended device. Each time a message is passed in such a direction is called a *hop*. The Internet generally reaches the final destination in less than 15 hops, for example. Routing tables for the Internet are exceedingly complex, but industrial mesh networks with a limited number of devices are not very complex.

One of the newest trends in networking is wireless data links. Mesh networking provides redundancy in wireless networks without the large cost of redundant wiring connections, although the cost of routing still applies. A wireless mesh network is illustrated in Figure 23-8.

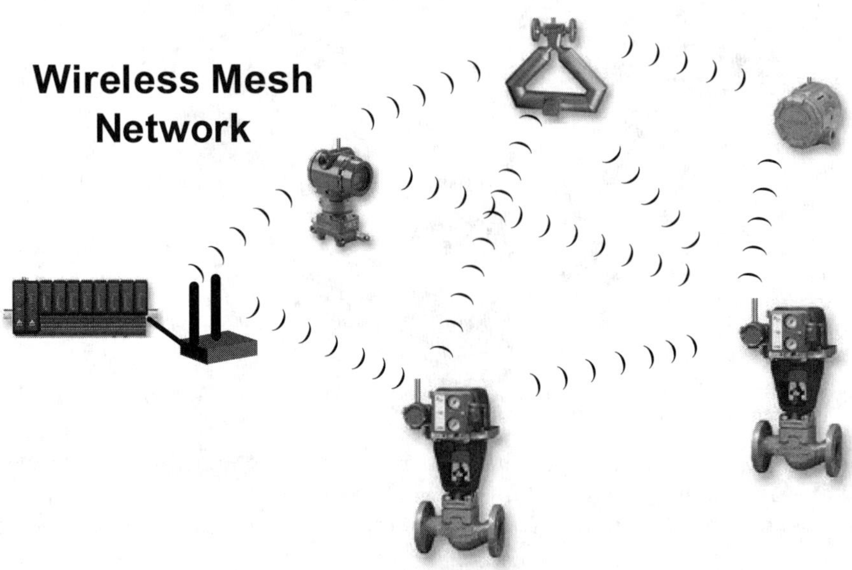


*Figure 23-8: Wireless Mesh Network*

## 23.5 Wireless Networks

Few question the future role of wireless networks in industrial automation, but in 2005 there are few installations or standards for using wireless in this industry. Already mentioned in context of network topologies has been the wireless mesh network as differentiated from wired mesh networks. Clearly, the cost reductions made possible by eliminating wiring is beneficial, especially in the industrial environment where the cost of wiring through plant areas densely packed with equipment and power wiring may be prohibitive.

There are several areas in which wireless networks and radio technology are very useful in data communications within the manufacturing environment. In many cases, wireless replaces wired communications with a resulting installation and operational cost savings. There is also one area in which wireless technology has no equivalent in wired communications: radio-frequency tagging.

### 23.5.1 Wireless LAN

Several IEEE, ISO, and ITU wireless standards have the potential to substitute for wired networks in industrial automation. As of mid-2005 the only available substitutions are for some field instrumentation links typically used where wired costs have been prohibitive. There are some situations in which wireless LANs have been substituted for wired Ethernet network segments for a variety of cost savings or feasibility reasons, but no industrial automation supplier has standardized on the use of wireless LANs. ISA has formed the SP100 standards committee to study the subject and to develop guidelines, recommended practices, and standards.

In general, wireless LANs may be substituted for the lowest layers (physical and media) of the ISO/OSI communications stack. However, the characteristics of the wireless LAN will be quite different from the wired LAN segment displaced. As long as those differences are acceptable for the application, there will be few performance effects. Radio communications may not be as reliable as wired communications due to interference effects and fading that can come from many sources. As long as the overall protocol provides for error detection and recovery, few actual differences from wired LANs may actually be noted during use.

### 23.5.2 Radio-frequency Tagging

Radio-frequency tagging is part of the Automatic Identification field that also includes bar coding. For this reason radio-frequency tags are often called RFID for radio frequency identification, but there are more applications than just ID.

RFID tags may be simple "license plates" such as barcode labels. In these ID applications, the RFID tag has a 64- to 128-bit field filled with a unique number. A reader may obtain this value as the item with the tag passes within its reading range. Data on the tagged item is then retrieved from a database and processed. RFID tags do not contain data other than their ID value, and so may be as secure as the database containing the item data.

Reading range for RFID tags may vary from a few centimeters to a few meters, depending upon tag and reader technology. Passive tags are powered by an electromagnetic field from the reader, and send a short message containing the ID value. Using a high gain antenna, it is possible to read such tags at distances up to about 3 meters. Active tags contain a battery and can usually be read from distances up to about 10 meters.

A variety of agencies issue tag encoding and protocol standards, but the worldwide authority is EPC Global. There are a series of standards currently being considered as ISO/IEC Publicly Available Specifications (PAS), a step in the process of becoming full international standards. Standards have been issued for both Type 1 and Type 2 EPC Global tags. Generally, EPC Global Type 1 standard tags are for passive, rewritable, RFID tags. Type 2 tags are active and have both a rewritable ID field as well as a read/write data field. EPC Global also controls the standards for Electrical Article Numbering (EAN) that is an extension of the older UPC (Universal Product Code) now used for barcode labels in North America.

Type 2 tags may contain data that can be written as part of the manufacturing and inspection process. Currently, there are no standards for the format of this data, except when it is a database to be passed between a supplier and customer. This type of data is generally called a transaction that has been the subject of another set of standards called Electronic Data Interchange (EDI). Standards for EDI have existed for several industries for transmission on private value-added wired data communications networks. Recent work makes these data formats available for writing to Type 2 tags. The work has been done by UN/EDIFACT (United Nations rules for Electronic Data Interchange for Administration, Commerce and Transport), which has been defining the format for any electronic data interchange independent of the network or medium.

## 23.6 Bibliography

ANSI/ISA-50.02, Parts 2-6 – *Fieldbus Standard for Use in Industrial Control Systems.*

Caro, Richard (Dick). *Wireless Networks for Industrial Automation.* Second Edition. ISA, 2005.

IEC 61158-SER Ed. 1.0 b:2005 – *Digital data communications for measurement and control – Fieldbus for use in industrial control systems.*

## About the Authors

**Richard "Dick" Caro** is CEO of CMC Associates, a business strategy and professional services firm in Acton, Mass. Before working at CMC, he was Vice President of ARC Advisory Group in Dedham, Mass. He is the Chairman of ISA SP50 and formerly of IEC (International Electrotechnical Committee) Fieldbus Standards Committees. Before joining ARC, Dick was Senior Manager with Arthur D. Little, Inc., in Cambridge, Mass., a founder of Autech Data Systems, and Director of Marketing at ModComp. In the 1970s, The Foxboro Company employed Dick in both development and marketing positions. He holds a BS and MS in chemical engineering and an MBA.

**Tom Phinney** is a member of the Honeywell ACS Advanced Technology Lab in Phoenix, Ariz. He has more than 35 years of experience designing software and hardware for real-time systems, primarily as an architect and system designer with General Electric and Honeywell. He has specialized in industrial communications since the late 1970s. An ISA Fellow, he has served in a leadership role in ISA/IEC fieldbus activities. In 2002 he was the recipient of ISA's annual Standards & Practices award for outstanding service. He is the U.S. Technical Advisor to ANSI for digital communications within the industrial process measurement and control industries (IEC/SC 65C), chairs three IEC working groups in that area, and is active in the ISA SP99 industrial cybersecurity and SP100 industrial wireless networking efforts. His primary work focus today is on advanced biometric identification systems and the cybersecurity of industrial real-time control networks. He has five patents awarded and seven pending, five publications in the area of industrial time-critical communications, and has been the primary author or editor of more than 5,000 pages of international computer communications standards.

# 24 Industrial Networks

*By Tom Phinney and Dick Caro*

## Topic Highlights

*Network Classifications*
*Industrial Network Standards*
- *Actuator/Sensor Interface (AS-i)*
- *Control Area Network (CAN)*
- *ControlNet*
- *EtherNet/IP*
- *Foundation Fieldbus H1*
- *Foundation Fieldbus HSE*
- *Interbus*
- *LonWorks*
- *Modbus*
- *Profibus-DP*
- *Profibus-PA*
- *Profinet*
- *WorldFIP*

*Bibliography*

## 24.1 Network Classifications

There are several ways to classify industrial networks. Each form of classification has utility for certain industries, applications, or standards bodies.

Networks used in industrial automation may be classified by the type of application:

- process control
- materials handling
- motion control
- discrete automation

They can also be classified by the type of data they transport:

- ON/OFF data expressed in bits or bit strings
- scalar data representing engineering units values
- general database exchanges

Finally, they can be classified by level of automation:

- sensor level
- field level
- control level
- information level

With the convergence of network protocols and multiple uses for networks, none of these classifications are very productive. However, there are networks designed for efficiently transporting certain types of industrial automation data. This discussion will use these network classifications, with the following titles:

- discrete sensor data
- process control fieldbus
- control-level networks
- information technology (IT) networks

### 24.1.1 Discrete Sensor Data

Discrete sensors such as limit switches, proximity switches, level switches, and relays of all types—as well as simple actuators such as solenoid valves and motor contactors—are represented by one binary bit of data. Networks designed for the distribution and collection of data to and from these types of devices usually have nodes that concentrate data from several such devices at one location: the network connection hub. Wiring for these networks focuses on reducing the cost of connection to control I/O equipment. Typically, little or no signal conditioning or data processing is possible at the network connection interface. In many cases, network connection is further simplified with connectors for prefabricated wiring on both the devices and the network connection hubs.

### 24.1.2 Process Control Fieldbus

Process control data is usually the digital representation of several scalar items, and is a packet of several data items packaged with discrete status data. When this data is used for feedback regulatory control purposes, synchronization of sample time with control loop execution is necessary. Traditionally, process control field instruments have been powered from the communications wiring; therefore the process control fieldbus provides power for field instruments, and, often, the electronic portion of field actuators. Furthermore, because many (but not all) data measurement points in the process industries are in explosive hazardous areas, there must be provision for current-limiting technology such as intrinsic safety or nonincendive classifications. Finally, wiring in the process plant is often subject to electrical interference from a wide variety of sources, including welding machines, radios, and weather. Fieldbus wiring designs and protocols are usually required to operate with a high degree of noise rejection in these environments.

### 24.1.3 Control-level Networks

Control-level networks are usually adaptations of IT networks to the industrial environment. Usually, the factory floor or process plant is a dirty and hazardous environment with lots of electrical noise and equipment vibration. The trend has been to run the control-level network into the factory, and even the process plant, where special considerations are necessary to adapt the wiring, and especially the connectors, to the needs of this harsh environment.

A typical wiring solution, when the network must be exposed to high degrees of electrical noise, is to use shielded cable and to ground the shield at one point close to the network power source. Although control-level networks are not usually required for traditional reasons to deliver power to process devices, using commercially available options to electrically power network equipment is now possible.

Control-level networks often use the same connectors as IT networks, but in plant or shop areas where there may be high vibration, moisture, oil mist, and dust that would make such connectors unreliable, connectors specially designed for the factory floor environment are usually advised.

### 24.1.4 IT Networks

IT networks have standardized world-over on the use of Ethernet at the lowest layers of the communications stack, and the TCP/IP suite for the upper communications layers. The only caution when industrial automation networks are interconnected with IT networks is to make sure the industrial automation network is isolated by a router designed to keep higher-level network messages off the control system networks. Firewall protection to allow only authorized access to the industrial automation network is usually needed as well.

## 24.2 Industrial Network Standards

It has already been said there are too many standards for industrial networks. In this section, many of the most popular industrial "standard" networks are profiled. The word "standard" is being used to mean either *de jure* standards, established by a consensus voting process within a formally recognized standards-making body, or *de facto*, established by the marketplace through popular use, and supported by public documentation. There are many more industrial networks that are proprietary to suppliers, not yet publicly documented, or designed for a narrow industry focus, that are not covered in this book.

The following are brief overviews of the most common standard networks used in industrial automation. For a more complete description, refer to the bibliography.

### 24.2.1 Actuator/Sensor Interface (AS-i)

AS-interface is the most popular sensor network for connecting discrete I/O to controllers, remote I/O, and even to some fieldbuses. Although the industry usually refers to this technology as "AS-i," the membership of the AS-International Association prefers to spell it out as "AS-interface." AS-interface is a low-cost multidrop bus topology, using either a two-conductor round cable or a unique two-conductor flat cable. Both cables deliver DC power to the node and the device using the same wires for both power and data. AS-interface taps are wired into the round cable and clamped onto the flat cable with vampire taps penetrating the insulation and the tap forming a vapor-tight seal to protect the connection. Each tap provides an interface either to a device or to an I/O connection module that may connect up to four discrete inputs and four discrete outputs. The AS-interface bus is wired close to clusters of I/O devices. A single AS-interface network can connect up to 124 inputs and 124 outputs and can be about 100 meters long. Many other options exist.

Installation of AS-interface is very simple, especially when you use the flat cable and purchase sensors and actuators with M12 connectors. The flat cable is routed close to the sensors and actuators on the production line. A four- to eight-port interface module is clamped to the flat cable at some convenient location. Each sensor and actuator is connected to the AS-interface module using pre-formed jumper cables with M12 connectors at each end. Care must be made to wire inputs to the input connections and actuators to output connections on the AS-interface modules. Custom wiring to AS-interface bare-wire terminations may cost less for materials, but the labor and maintenance cost is much higher. Bare-wire AS-interface terminations should only be used for connections made inside a wiring cabinet and not those exposed to factory floor or plant environments.

Finding suppliers of sensors and actuators equipped with M12 connectors meeting the AS-interface wiring specifications may be difficult, limiting the choice of devices. For example, Rockwell's Allen-Bradley brand and Cutler Hammer's brand of limit switches, photocells, and proximity switches are all available with M12 connectors wired to DeviceNet specifications, but not AS-interface.

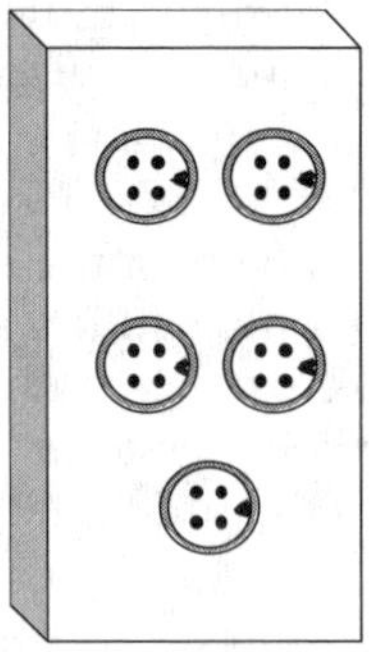

*Figure 24-1: AS-interface Module, Round Cable*

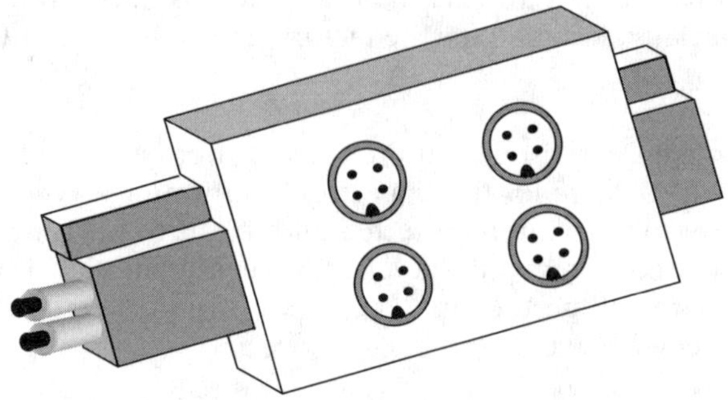

*Figure 24-2: AS-interface Modules, Flat Cable*

*Figure 24-3: AS-interface Bare-wire Interface Module*

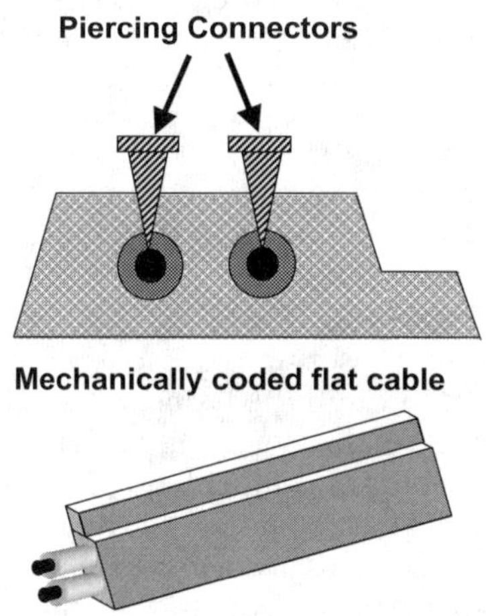


*Figure 24-4: AS-interface Flat Cable*

### 24.2.2 Control Area Network (CAN)

Control Area Network (CAN) was originally developed to replace wiring harnesses for distributed I/O in automobiles and trucks. It has been used for this purpose to some degree, but the major automobile manufacturers have yet to deploy it broadly. CAN has been used as the technical basis of DeviceNet, SDS, CAN Kingdom, and CAN Open networks, but these have not concentrated on cost reduction of the wiring. The CAN chip is usually coupled with a small microprocessor, and some of these networks have concentrated upon providing a highly functional user layer for discrete I/O, much like the fieldbus networks have done for analog I/O. For this reason, these networks are often called "fieldbuses." It is expected that the number of major automotive manufacturers deploying CAN to reduce the cost and lighten the automobile by creating an under-the-hood sensor/actuator network will greatly expand over the next few years and reduce the cost of the CAN chip underlying each of these protocols.

One of the features of all CAN-based networks is the use of producer/consumer messaging, in which a device "producing" or originating the data may send it on the network as a broadcast message. All units receiving the message may choose to use the data or not as the application demands. Producer/consumer messaging is usually more efficient for control networks where more than one station needs access to data from the originating point.

#### 24.2.2.1 CAN Open

CAN Open uses any physical wiring developed for the CAN bus at the hardware level, including all the wiring developed for both SDS and DeviceNet. CAN Open is an object-oriented application layer for use on top of the physical and data link layers of CAN. The producer/consumer version of CAN messaging is organized by the CAN Open application layer to provide event-based control, remote requesting for control action, and sync-triggered control action.

#### 24.2.2.2 DeviceNet

DeviceNet is offered as a sensor/actuator network. I/O modules similar in function to those of AS-interface are available to terminate conventional sensors and actuators and communicate on the DeviceNet network. Although the higher-level application layer, called CIP, can be used with DeviceNet as a fieldbus, it need not be used, making DeviceNet no more complex as a sensor network than AS-interface. The cost of a DeviceNet node may be a factor for any application, but using DeviceNet as a sensor network will yield the same wiring savings as AS-interface.

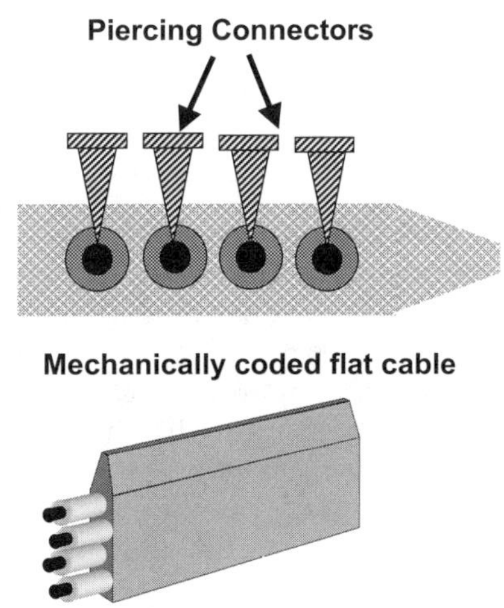


*Figure 24-5: DeviceNet Flat Cable*

Rockwell developed DeviceNet, but gave the specifications to the open DeviceNet vendor association (ODVA), where the membership now controls any revisions. DeviceNet specifications are published on the ODVA Web site.

DeviceNet is offered with four-conductor round and flat cables, illustrated in Figure 24-5, delivering electrical power to the nodes and devices. Power is delivered separately from data in both types of cable. The DeviceNet specification does not call for a wiring standard between the device and the DeviceNet interface module. Most DeviceNet interface modules use compression screw terminals for bare wire, but the M12 round connector is often used to connect devices to the DeviceNet interface module. The flat cable with DeviceNet interface modules, as illustrated in Figure 24-6, provides a low-cost installation very close to AS-interface in cost. DeviceNet is also available with higher density interfaces using bare-wire terminations. The installed cost of a DeviceNet network can rival the installed cost of AS-interface for similar applications.

The CIP application layer used with DeviceNet can use the producer/consumer capability of CAN to organize control by exception and other services of CIP through its electronic data sheet specifications.

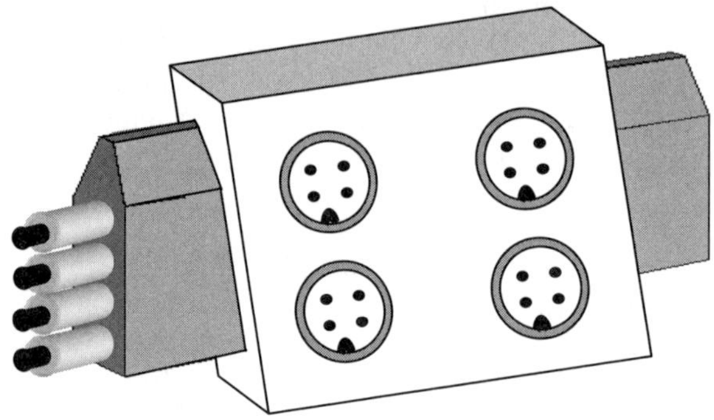

*Figure 24-6: DeviceNet Flat Cable Interface Module*

#### 24.2.2.3 SDS

SDS has been embedded into sensors and actuators implementing many signal processing features using the programmability of the microprocessor usually coupled with the CAN interface chip. SDS has remained somewhat proprietary to its inventor, Honeywell Sensing and Control, although the specification is fully open and available at the Honeywell Web site: http://content.honeywell.com/sensing/prodinfo/sds/sdspec.stm.

Honeywell has implemented SDS throughout its line of sensors and actuators as *smart devices*, and offers an SDS termination block similar to AS-interface or DeviceNet. SDS devices provide the user with lots of functionality not provided with other industrial communications networks for discrete sensors and actuators, but usually implemented with PLC programming. For example, simple discrete contact sensors may have contact bounce suppression configured for the switch itself. Additionally, the following features may be implemented through software configuration in switches or output relays: batch counting, tag-name addressing, motion/jam detection, normally open/normally closed, on delay/off delay, and timed response input. One of the advantages of SDS is that these high-speed functions can be assigned to the sensor or actuator, allowing a lower-cost PC-based controller to be used.

### 24.2.3 ControlNet

ControlNet was developed by Allen-Bradley in early 1995. This real-time, deterministic, peer-to-peer network links PLCs and I/O subsystems at 5 Mbps. Relative to DeviceNet, ControlNet is faster, supports a larger number of nodes, and can directly accommodate longer data values such as floating point process variables. Another application for ControlNet is the peer-to-peer communications between PCs and PLCs. The ControlNet system architecture operates with multiple processors and has the capability to install up to 99 addressable nodes anywhere along the coaxial RG-6 and RG-59 cable of the network.

ControlNet is positioned between EtherNet/IP and DeviceNet in the automation hierarchy, but the emphasis on highly distributed I/O-level communications results in an overlap with both. ControlNet, DeviceNet, and EtherNet/IP all use the same network model called producer/consumer in which each node can be a producer (sender) of data, consumer (receiver) of data, or both. ControlNet also offers multicast capability. This is the ability to send the same data to all network nodes at the same time. Producer/consumer, or a similar service on Foundation Fieldbus, publisher/subscriber, uses the multicast protocol with a data identifier field, allowing nodes interested in the data to quickly identify it for local use. This is more efficient than token passing or master-slave models. Time-critical data on ControlNet is deterministically transferred during reserved time slots, whereas non–time critical data is sent during the time available after the reserved time slots.

In October 1996, Allen-Bradley placed the specifications for ControlNet in the hands of ControlNet International, a membership-supported nonprofit organization similar to ODVA, which controls DeviceNet. The ControlNet specification is available to all automation suppliers in the form of a developer's guide, including a description of the ControlNet protocol, instructions on how to develop a product, and guidelines for installing and implementing a ControlNet system. In addition to the specification, development tools such as example software, a developer's starter kit, ControlNet firmware, and ASICs are also available.

At present, the future of ControlNet appears somewhat in doubt due to the broad availability, higher speed, and generally lower cost of EtherNet/IP on all of the same products. Although there is no doubt about the value of ControlNet's highly synchronous messaging, the lower cost of EtherNet/IP and its higher speed allow EtherNet/IP to effectively replace ControlNet in the Rockwell product line. The application and user layers, called Control and Information Protocol (CIP), developed for ControlNet and DeviceNet are available on EtherNet/IP, as well.

### 24.2.4 EtherNet/IP

Ethernet Industrial Protocol (EtherNet/IP) was created through the combined efforts of ODVA, ControlNet International, and Rockwell Automation. EtherNet/IP was designed to efficiently implement data transfer using CIP at the application layer. It operates on commercial Ethernet, but spawned one of the ODVA special-interest groups to investigate alternative physical wiring and connectors more suitable to industrial automation.

Through the efforts of ODVA and similar organizations, the Electronic Industries Association/Telecommunications Industries Association (EIA/TIA) 42.9 committee formed a standardization committee to prepare specifications for industrial Ethernet cabling and connectors. Through this effort, a new standard has been established. As of mid-2005, the work was not yet completed, but the committee has already selected the use of Category 5E, Category 6, and Category 7 cable. Bulkhead connectors with an RJ-45 form factor have been specified as well as a round M-12 (12 mm) four-pin connector. The objectives for this work are to establish standards for constructing industrial cable plants according to the environmental concerns for the industrial area.

The objectives of EtherNet/IP are to provide a full industrial-grade data communications service using as much commercial off-the-shelf Ethernet hardware and cabling as possible. The benefits are to obtain the speed of Ethernet at the low cost possible from broad usage in the commercial market.

EtherNet/IP uses TCP/IP to establish all network connections and for routine network applications. However, any cyclic high-speed data traffic uses UDP/IP protocol frames for efficiency and determinism.

### 24.2.5 Foundation Fieldbus H1

Foundation Fieldbus was created to supply a bidirectional all-digital data transmission network technology for process control. From the beginning it was to replace the 4–20 mA DC transmissions previously used for analog control instrumentation. It was also to use the same type of wire typically used for analog transmission, to supply power to field instruments, and to fully conform to intrinsic safety requirements. The ANSI/ISA 50.02 standard met all of these requirements and was the basis for Foundation Fieldbus.

The Fieldbus Foundation was formed in 1994, at the urging of many users, from two competing organizations, WorldFIP North America and Interoperable Systems Project (ISP). The users emphasized the importance of a single fieldbus protocol and the imperative of basing it upon a recognized standard. The control systems industry had previously been divided between the prior two bus proposals. With the energy now concentrated upon a single specification, the Fieldbus Foundation rapidly completed its implementation specification based on the ANSI/ISA 50.02 documents. Almost immediately, the Fieldbus Foundation began to create a testing suite to validate that field devices conformed to their specifications. The validation program was called Foundation Fieldbus Registration. Devices passing the validation testing would then be allowed to carry the Foundation's registration symbol, illustrated in Figure 24-7.

*Figure 24-7: Fieldbus Foundation Registration Mark*

**24.2.5.1 Wiring and Signaling**

The initial Foundation Fieldbus specification was for Hunk 1 (H1) for the targeted instrumentation connection 4–20 mA replacement application. H1 operates at 31,250 bps, a very low speed for a communications bus, but necessary because of the need to reject noise, deliver DC power, and provide intrinsic safety. Noise rejection is enhanced by using a trapezoidal waveform, rather than the traditional digital square wave. Encoding is Manchester Bi-phase, which requires both a high and low state for each bit. The signal is differential between the two wires rather than the more noise-prone single-ended signals of RS-232. The wire is a shielded single twisted-pair in which the impedance is not specified to allow conventional analog instrument cable to be used.

The topology of Foundation Fieldbus H1 is called trunk and spur. In development of the 50.02 standard, it was called a *chicken foot,* because of the way it was always illustrated, as in Figure 24-8. There are very few wiring restrictions for H1, allowing the cable to be installed in the most economical way. The illustrated chicken foot wiring is used most often, because it is easy to maintain, but daisy-chain wiring from instrument to instrument is also allowed. Wiring from each instrument may be joined into a single segment within the junction box, and then routed to a field marshalling junction box. The wiring from the field marshalling junction box is usually shielded multicore cable with a shield for each pair and an overall shield. This continues the wiring from each H1 segment to the H1 termination I/O card. However, since H1 instruments are addressed by a bus address, all of the field devices of a single segment use only one pair of the multicore cable. This is illustrated in Figure 24-9.

**24.2.5.2 Intrinsic Safety and Power Delivery**

Intrinsic safety is supported by H1 but not required. The basic requirement for intrinsic safety is a barrier located where the wiring passes from a safe zone into the hazardous zone. The barrier ensures that no electrical current with enough energy to ignite a flammable gas mixture can cross the barrier but is instead shunted to an earth ground. Because Foundation Fieldbus H1 is a differential signal, the intrinsic safety barrier is bipolar. This is different from an analog signal, where one side is usually referenced to an earth ground. The barrier limits the voltage that can be transmitted on the H1 link and consequently also the power available to the connected field instruments. The original H1 specification only limits the number of units connected with a single fieldbus segment to the number of available addresses (240), but it does limit the maximum current draw for intrinsic safety. As field device power requirements are decreased with low-powered electronics, the number of units per segment can be increased while maintaining intrinsic safety. The revised H1 specification is based on the European-developed Fieldbus Intrinsically Safe Concept (FISCO) specification, which considers the electrical wiring parameters to be distributed along the entire segment, thus reducing the energy available at a fault and permitting more devices on an intrinsically safe Foundation Fieldbus H1 segment than the original "Entity Concept" does. A later specification, Fieldbus Non-Incendive Concept (FNICO), was

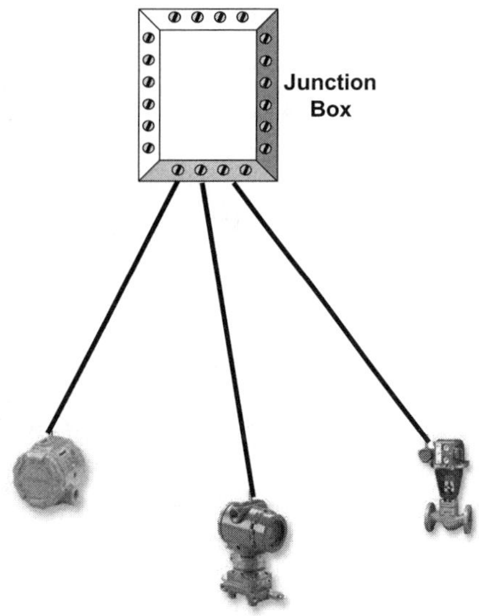


*Figure 24-8: Trunk and Spur (Chicken Foot)*

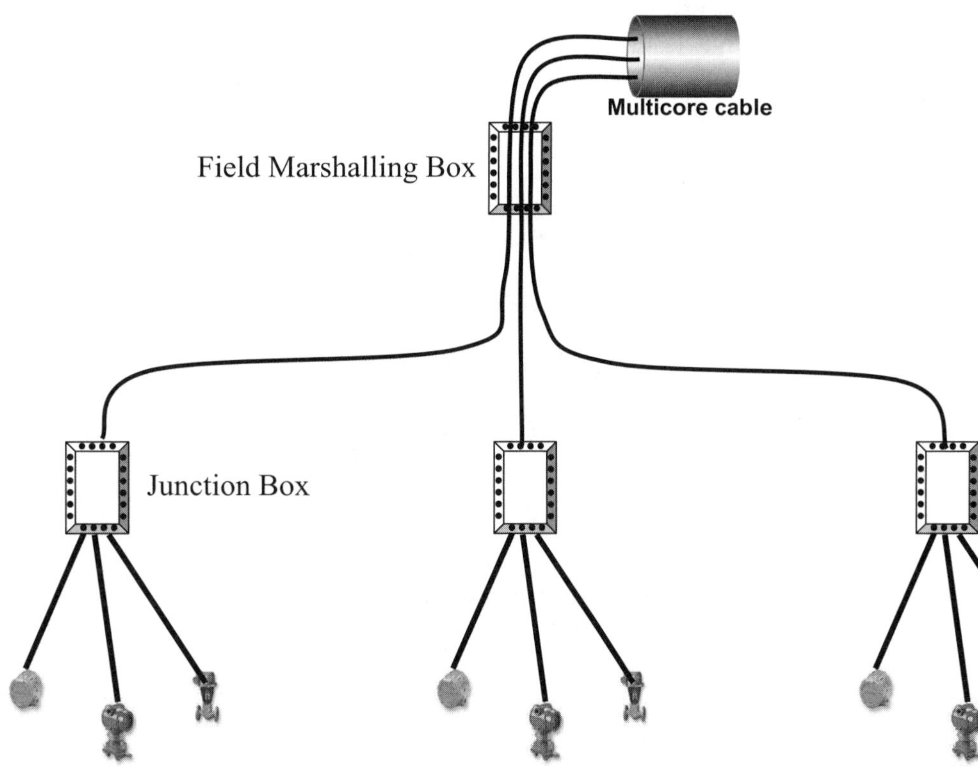


*Figure 24-9: H1 Fieldbus*

established recognizing that most process plants do not have explosive vapors present at all times. It allows another increment in the number of devices per Foundation Fieldbus H1 segment.

While DC power for the field device can be extracted from a Foundation Fieldbus H1 segment, it is not required. Devices may be self-powered much as they were in analog control systems. Devices powered from the fieldbus usually operate with 9–32 VDC. Most commercial fieldbus power supplies operate at approximately 24 VDC. The maximum number of devices that can be powered from the H1 bus depends upon the use of intrinsic safety and nonincendive classifications. Typically, about nine devices can be powered from an intrinsically safe FISCO-conforming fieldbus. FNICO increases this to about 12 devices. Nominally, about 30 devices can be powered from a non-intrinsically safe fieldbus. However, good engineering practice limits the number of devices per H1 fieldbus segment to fewer than any of these limits.

### 24.2.6 Foundation Fieldbus HSE

HSE is the backbone or control-level bus for process control using Foundation Fieldbus H1 for transmitters and actuators. Originally, as defined in IEC 61158 Part 2 and the ANSI/ISA 50.02-2 standard, there was a Hunk 2 (H2) for the higher-level backbone bus functions. However, the AC-current mode bus at 1 Mbps, the voltage mode bus at 1 Mbps, and the fiber-optic buses at 1 and 2.5 Mbps defined in that standard were too expensive and too slow. Instead, Foundation Fieldbus HSE was created to use commercial off-the-shelf Fast Ethernet and Internet software standard protocols. This created a very fast bus (100 Mbps) as well as providing the economies of scale from use of Ethernet components. Foundation Fieldbus HSE is standardized as Type 5 of the IEC 61158 fieldbus standard.

Process control networks based on only Foundation Fieldbus H1 must have each bus segment terminate at a process controller or a similar device. This does not allow a device on one segment to communicate directly with a device on another segment. Many process controllers offer a communications service to allow these "bridging" communications, but they cannot be guaranteed to perform the data transfers fast enough for use in cascade loop control. The plan for Foundation Fieldbus was always to have an H2-level bus to join segments directly; HSE is that bus. Part of the Foundation Fieldbus architecture includes linking devices to join several H1 bus segments to one or more HSE networks. This architecture is illustrated in Figure 24-10.

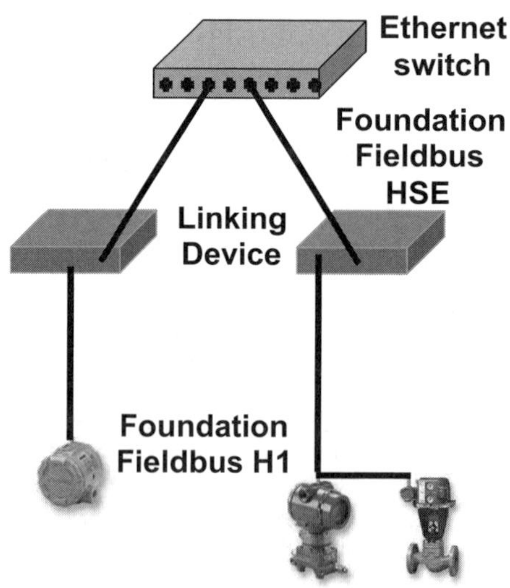


*Figure 24-10: Foundation Fieldbus Wiring*

Foundation Fieldbus HSE uses the same application and user layers as H1 and so completely interoperates with it. The linking device performs the spanning tree bridge functions specified in the data link layer of the ISA and IEC fieldbus standards and so fulfills all of the requirements for the H2 bus. However, instead of mapping the application layer functions to the H1 data link layer, HSE maps the fieldbus functions to User Datagram Protocol/Internet Protocol (UDP/IP) data frames carried on standard 100BaseT Ethernet. This means that you use commercial everyday Ethernet wiring, accessories, and terminations where environmental conditions permit. In other locations, use industrial Ethernet wiring and components.

HSE is designed to provide fault tolerance using cable redundancy. Because HSE is to be used as part of the closed-loop control network, real-time message delivery is critical. Where the control data message passes over HSE, that cable segment should be redundant. The HSE redundancy scheme does not wait for cable failure and switchover, because that methodology cannot guarantee real-time message delivery. HSE redundancy requires that the same message be sent on all active HSE network segments at the same time with the same message identity. Only one such message is used; the others are intercepted with a redundancy manager. Failure to receive a redundant message on a redundant segment indicates a cable failure. More than dual redundancy is supported.

HSE supports exactly the same software interface as Foundation Fieldbus H1 and therefore has all of the same features as H1, except for intrinsic safety and power delivery to field equipment. Although there are currently no commercial products supporting HSE in field devices, in the future this could be done for non-intrinsically safe environments. It may also become possible to distribute DC electrical power on the HSE network by adapting the 48V IEEE 802.3af Power Over Ethernet standard to 24V, which is more acceptable in process automation. It may also be possible to develop intrinsic safety barriers for HSE in the future, but there are no commercial products at this writing.

### 24.2.7 Interbus

Interbus is a highly efficient fieldbus and an integrated sensor network called the local loop. Figure 24-11 illustrates Interbus ring topology. Field devices (sensors and actuators) are typically connected to a local loop I/O module. The I/O modules are connected to each other in a ring, receiving data from previous I/O modules and sending data to the next I/O module in the local loop until the ring ends in a loop termination module.

The Interbus fieldbus interconnects all remote bus modules, including the local loop termination modules, together into a ring. The last I/O module in the Interbus remote bus closes the ring, returning the signal to the master.

The maximum I/O count is 4,096 points per Interbus network, which is made up of both local loops and I/O terminated on remote nodes. There is a maximum number of local and remote modules of 512, of which there can be a maximum number of local loop modules of 192. The data rate is 500 Kbps, and the maximum bus length between any two remote bus modules is 400 m. Because each remote bus module includes its own repeater, very long networks, up to 13 km, can be configured. Local loop modules can be a maximum of 20 m apart.

Interbus modules are available for a variety of functions such as variable-speed drives, motor contactors, motion controllers, encoders, barcode readers, as well as for analog and digital discrete I/O. The Interbus Club Web site lists products implementing Interbus at http://www.interbusclub.com.

Phoenix Contact first developed Interbus to reduce the wiring and installation cost for factory networks of discrete I/O. Later, the inclusion of AS-interface ports into the overall Interbus architecture reduced cost even more. Interbus is integrated with Profinet as the preferred control-level network.

### 24.2.8 LonWorks

The LonWorks system was originally developed by Echelon Corp. in the late 1980s to be a low-cost and moderate-performance network for residential, building, commercial, and industrial automation. It has been applied in all of these markets but dominates the building automation market. Although originally developed for a simple two-wire twisted-pair network, alternate media such as power line modulation, fiber optics, radio, and infrared have always been offered. LonWorks power line modula-

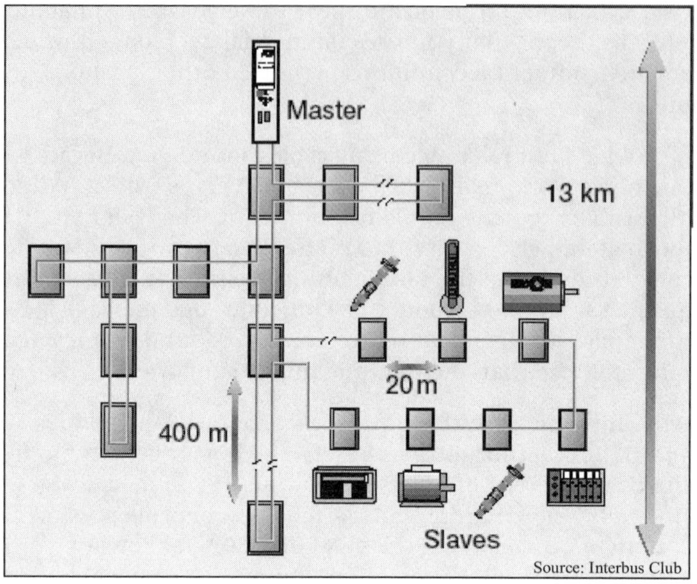


*Figure 24-11: Interbus Topology*

tion is probably the most popular alternative media in actual use. Progress in wireless LonWorks has been demonstrated with constantly reduced cost.

The protocol for LonWorks is called LonTalk. Echelon originally held it as a trade secret, but it now has been standardized as ANSI/EIA 709.1. The entire protocol, all seven ISO layers, is implemented in silicon on *neuron* chips functionally designed by Echelon but produced and sold by Toshiba and Cypress Semiconductor. Each neuron chip has three microprocessors to handle the protocol, the media modulation, and the application. Simple applications such as I/O processing can be accomplished using only the microprocessor power of the neuron chip. LonTalk is also the basis for IEEE 1473-L, a standard for rail transportation communications.

LonWorks is a peer-to-peer network intended for linking clusters of I/O to a controller. Although this objective is similar to CAN, the applications for automation are more demanding and therefore require much greater microprocessor capacity on the neuron chip. The cost of the neuron chip is about triple the cost of a CAN chip, reflecting the greater capability, but it often eliminates the need for a local microprocessor at the I/O node.

With the standardization of the LonTalk protocol, it is possible to port the protocol to chips other than the neuron. Although the potential exists, there are currently no other implementations of chips supporting the ANSI/EIA 709.1 protocol.

Interoperability of LonWorks devices is the responsibility of the LonMark Interoperability Association, which offers a battery of tests for interoperability. The LonMark Interoperability Association Web site (http://www.LonMark.com) lists thousands of products that are certified to carry the LonMark logo, the symbol of LonWorks interoperability.

LonWorks networks can be connected to the Internet and other TCP/IP networks by means of the i.LON™ 1000 Internet Server, a Cisco product. The i.LON is a LonTalk/IP router enabling devices on a LAN to communicate directly with devices on a LonWorks network.

### 24.2.9 Modbus

Modbus in all of its forms is the most popular control-level bus. Modicon, now a brand name of Schneider Electric, originally created it as a means for computers to gather information and control the operation of their PLCs. The data model for all PLCs is a set of addressable registers organized into sets of I/O, control relay, analog inputs, analog outputs, and variables. PLC I/O is organized so each digital discrete input device appears as a single bit in the I/O registers according to its location in the I/O hardware. This usually means that digital discrete output is often mixed together in some of the registers, requiring a register mask to define the outputs.

Modbus commands provide ways to transfer the content of one or many registers from the PLC to the host device, which may be a computer or another PLC. The Modbus command set was so popular when it was first created in 1979 that it has often been copied for other PLCs. The most popular Modbus spin-off is J-Bus, which Siemens, Telemechanique, and many other smaller PLC suppliers used as a secondary access protocol on many PLCs not originating from Modicon roots. The idea behind Modbus, a command set operating on 16-bit registers, has been used by all PLC suppliers and by the ISO 9506 Manufacturing Message Specification (MMS). Additionally, Modbus continues to be the most common protocol for use in supervisory control and data acquisition (SCADA). The same structure is used in OPC/DA.

The identical command set for Modbus has been implemented using several different physical layers. Table 24-1 shows the physical layers used for Modbus, also illustrated in Figure 24-12.

*Table 24-1: Modbus Physical Layers*

| Physical Layer | Standard | Speed | Comments |
|---|---|---|---|
| Modbus | EIA-232 | 19.2 kbps | Original serial Modbus |
| | EIA-485 | Up to 1 Mbps | High-speed serial Modbus |
| Modbus+ (Plus) | HDLC (ISO 3309) on EIA-485 | Up to 1 Mbps | Token passing multipeer bus. Proprietary protocol. |
| Modbus/TCP | IEEE 802.3 | 10/100/1000 Mbps | Modbus on Ethernet |

Modbus was originally developed for operation on an asynchronous serial line, now defined by the standard ANSI/EIA/TIA-232F. This made it compatible with serial port modems without further definition. When remote termination units (RTUs) were developed for SCADA systems, Modbus was a natural selection for the transport protocol across serial dial-up communications lines. This version of Modbus has become known as Modbus/ASCII, and it is still very popular for many applications, not just RTUs.

In local plants, where many PLCs need to be connected to a single computer, Modbus/ASCII is considered too slow, and it does not support multidrop communications. For these applications, Modbus has been implemented on an ANSI/EIA/TIA-485 multidropped serial bus as Modbus/RTU. The 485 bus uses a balanced differential line enabling much longer distances, much higher speeds, and better noise rejection than 232F on a twisted-pair cable. This continues to be a very popular choice for implementation of Modbus.

Schneider/Modicon also supported a version of Modbus called Modbus Plus or, usually, *Modbus+*. The specification for Modbus+ has never been formally released as an open network, but it is still used on Modicon-brand products as well as those external products developed with Schneider agreements. Modbus+ uses a serial line with the High-level Data Link Control (HDLC) protocol, which enables multidrop communications.

Modbus/TCP was developed in 1998 and was declared "open." The Modicon/Schneider Web site has published the specification, but ownership has now been transferred to the independent Modbus-IDA

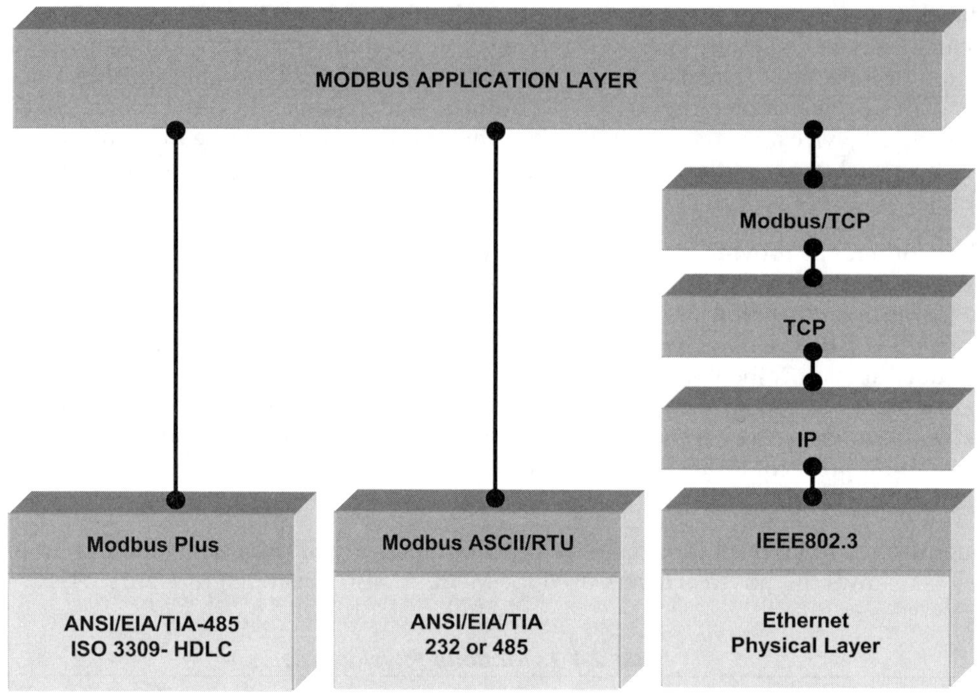


*Figure 24-12: Modbus Structure*

association. The specification is published on the association's Web site (http://www.modbus-ida.org). Modbus/TCP enables several improvements to Modbus communications: it lowers cost by using commercial Ethernet components, enables remote operation via the corporate LAN or the Internet, and increases operational speed to the LAN choices of 10/100/1000 Mbps of Ethernet. It also exposes the PLC to the usual Internet security problems, requiring security and privacy protection with known methods. Schneider divisions have implemented more than 75 products with Modbus/TCP as part of their Transparent Factory initiative. Many other companies have also chosen to use Modbus/TCP as the primary application layer for their Ethernet interfaces.

The Modbus application layer commands are given in Table 24-2. Modicon first introduced the terms used in Table 24-2 in the creation of relay ladder logic (RLL) for its PLC, but they are now in common use. The following are the definitions used in RLL:

| | |
|---|---|
| Coil | A single output bit |
| Input | A single input bit |
| Register | A 16-bit assembly of bits or a value |
| Holding | Internal register storage, typically in the 40,000 range |
| Force | Sets the actual state of an output bit or multiple bits |
| Preset | Sets a value into a holding register |
| Mask | XOR (logical exclusive OR) with a mask register before output |

*Table 24-2: Modbus Command Set*

| Command | Function | Command | Function |
|---|---|---|---|
| 01 | Read Coil Status | 13 | Program Controller |
| 02 | Read Input Status | 14 | Poll Controller |
| 03 | Read Holding Registers | 15 | Force Multiple Coils |
| 04 | Read Input Registers | 16 | Preset Multiple Registers |
| 05 | Force Single Coil | 17 | Report Slave ID |
| 06 | Preset Single Register | 18 | Reserved for Programming |
| 07 | Read Exception Status | 19 | Reset Communication Link |
| 08 | Diagnostics | 20 | Read General Reference |
| 09 | Reserved for Programming | 21 | Write General Reference |
| 10 | Poll | 22 | Mask Write 4X Registers |
| 11 | Fetch Communication Event Counter | 23 | Read/Write 4X Registers |
| 12 | Fetch Communication Event Log | 24 | Read FIFO Queue |

## 24.2.10 Profibus-DP

Although Profibus was created to be a standard communications link between PLCs and host systems such as HMI, the original Profibus-FMS was too slow to support HMI update. When a standard connection with PLC RTUs or remote multiplexers became a requirement, Profibus-DP was created to solve both problems. The high speed of Profibus-DP, up to 12 Mbps for short distances, became its most attractive asset. This makes Profibus-DP both a control-level bus and a fieldbus. Profibus International prefers the term Profibus rather than any of its modifiers such as FMS, DP, or PA, but industry continues to use these designations.

Many companies support Profibus communications for their products. Although it began as a German national standard, it has now been fully internationalized with inclusion into the IEC 61158 Fieldbus standard. Among the benefits of Profibus are that it is the factory communications standard for Siemens, one of the world's largest industrial automation system integrators and the largest manufacturer of PLCs in the world. Integration with Siemens products becomes much easier when using Profibus.

Rather than using FMS as a programming interface to a network of automation devices, Profibus International has created an object-oriented method using GSD (Gerätestammdaten or equipment master data) files and EDD (Electronic Device Descriptions). Having the GSD for a device allows the user to access all the available data for that device. The EDD is very similar to the DD of both HART and Foundation Fieldbus, and now shares a common format with these networks based on IEC 61804 Function blocks (FB) for process control – Part 2: Specification of FB concept and Electronic Device Description Language (EDDL).

## 24.2.11 Profibus-PA

Profibus-PA was created as a two-way all-digital data transmission network technology for use in process control. From the beginning, just like Foundation Fieldbus H1, it was to replace the 4–20 mA DC transmissions previously used for analog control instrumentation. It was also to use the same type of wire typically used for analog transmission, supply power to field instruments, and to fully conform to intrinsic safety requirements.

Profibus-PA is also fully specified as part of the international fieldbus standard, IEC 61158. It has the same physical layer as Foundation Fieldbus, but uses Profibus-DP's data link layer. This makes it easy for manufacturers to build field devices for either bus, but they cannot share the same network due to differences in the data link layer.

Profibus-PA is intended for traditional process control applications where delivery of DC power to the field instrument and support of intrinsic safety is necessary. Unlike Foundation Fieldbus H1, Profibus-PA is a master/slave network that is an extension of Profibus-DP.

Normally, the field instruments are wired to a field junction box where they are terminated in a Profibus DP/PA coupler. Profibus-DP is used as the higher-level control-level fieldbus to connect PA segments to the control system master. Field instrument power is often supplied from the junction box. Because intrinsic safety barriers do not exist for Profibus-DP, intrinsically safe systems require the junction box with the DP/PA coupler to be in a safe area and the intrinsic safety barrier to be placed on each Profibus-PA segment.

#### 24.2.11.1 Wiring and Signaling

Profibus-PA targets instrumentation connection in a 4–20 mA replacement application. Like Foundation Fieldbus H1, Profibus-PA operates at 31,250 bps, a very low speed for a communications bus, but necessary because of the need to reject noise, deliver DC power, and provide intrinsic safety. Noise rejection is enhanced by using a trapezoidal waveform, rather than the traditional digital square wave. Encoding is Manchester Bi-phase, which requires both a high and low state for each bit. The signal is differential between the two wires rather than the more noise-prone single-ended signals of RS-232. The wire is a shielded single twisted-pair in which the impedance is not specified to allow conventional analog instrument cable to be used.

The topology of Profibus-PA is called trunk and spur, as illustrated in Figure 24-13. There are very few wiring restrictions for Profibus-PA, allowing the cable to be installed in the most economical way. The illustrated wiring is used most often, because it is easy to maintain, but daisy-chain wiring from instrument to instrument is also allowed. The junction box usually contains the Profibus PD/PA Coupler and power supplies for each wiring segment. In cases where intrinsic safety must be provided, the segment may be joined in the junction box, and a single twisted-shielded wire is routed to intrinsic safety barriers in a safe area.

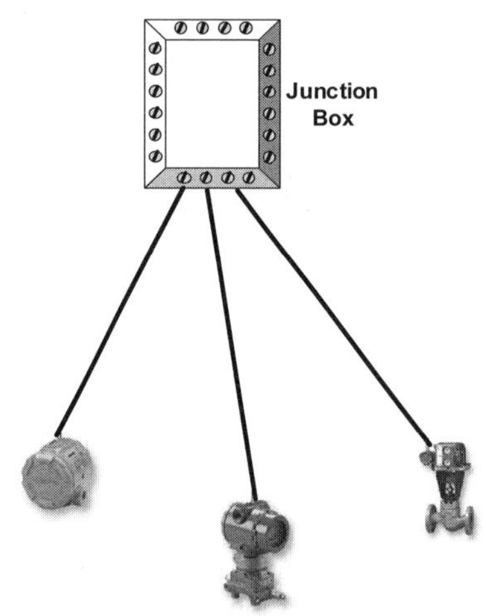


*Figure 24-13: Trunk and Spur*

#### 24.2.11.2 Intrinsic Safety and Power Delivery

Intrinsic safety is supported by Profibus-PA but not required. The basic requirement for intrinsic safety is a barrier located where the wiring passes from a safe zone into the hazardous zone. The barrier ensures that no electrical current with enough energy to ignite a flammable gas mixture can cross the barrier. Instead, any such currents are shunted to an earth ground. Because Profibus-PA is a differential signal, the intrinsic safety barrier is bipolar, unlike an analog signal where one side is usually referenced to an earth ground. The barrier limits the voltage that can be transmitted on the Profibus-PA link and consequently also limits the power available to the connected field instruments. The original Profibus-PA specification only limits the number of units connected with a single fieldbus segment to the number of unreserved addresses (125), but does limit the maximum current draw for intrinsic safety. As field device power requirements are decreased with low-powered electronics, the number of units per segment can be increased to maintain intrinsic safety. The revised Profibus-PA specification is based on the FISCO specification, which refers to the *actual* power consumption of the field devices being used. This is often significantly lower than the maximum power requirements and allows many more field devices per intrinsically safe Profibus-PA segment. A later specification, FNICO, recognizes that most process plants do not have explosive vapors present at all times and allows another increment in the number of devices per Profibus-PA segment.

Although DC power for the field device can be extracted from a Profibus-PA segment, it is not required. Devices may be self-powered much as they were in analog control systems. Devices powered from the fieldbus usually operate with 9–32 VDC. Most commercial fieldbus power supplies operate at approximately 24 VDC. The maximum number of devices that can be powered from the Profibus-PA segment depends upon the use of intrinsic safety and nonincendive classifications. Typically, about nine devices can be powered from an intrinsically safe FISCO-conforming fieldbus. FNICO increases this to about 12 devices. Nominally, about 30 devices can be powered from a non-intrinsically safe fieldbus segment. However, good engineering practice usually limits the number of devices per Profibus-PA segment to fewer than these limits. Generally, because Profibus-PA does not support control in field devices, more devices may be connected to each segment than for Foundation Fieldbus H1.

### 24.2.12 Profinet

Profinet is the next step beyond Profibus. Like Foundation Fieldbus, Profinet is an architecture for a control system that includes a multilevel communications structure. The basic concept is that communications exist between component objects in the network. The Profinet object model expands the GSD and EDD concepts of Profibus into full XML (extensible markup language) based object descriptions. As with all object-based systems, the external or exposed attributes (also called parameters) of each object are the items that can be communicated using well-defined rules. The data format for all Profinet object attributes is an XML data structure. Configuration of Profinet is the action of linking the Profinet objects through their attributes.

Profinet uses a series of international standards to define all aspects of its protocol. The underlying technology is based on ISO/IEC 8802-3 standard Ethernet with TCP/IP and the Microsoft COM/DCOM object model and protocol. It is designed to interoperate with other networks but supports Profibus networks fully. The eventual scope of Profinet is to be used in all operations where Profibus is used today, as long as the physical environment is suitable. Many parts of Profinet use commercial Ethernet wiring components or the more rugged versions now available. Although the ability to interface with Profibus networks is fully supported, Profinet is much more than Profibus on Ethernet.

Profinet uses a software concept called a *proxy* to model legacy bus devices to make them an integral part of the Profinet data structure. The proxy is implemented on the fieldbus master device for Profibus systems, although it can be used to interface any other industrial automation network as well. The role of the Profinet proxy is to make the data of the control system accessible on the fieldbus transparently available to Profinet. While the proxy software may be added to existing controllers, it is most often implemented with a coprocessor on the Ethernet interface to an existing controller.

Profinet was created for all types of control systems to make the task of I/O addressing much simpler and more error-free than traditional PLC hardware-oriented addressing. I/O points are named as objects with attributes appropriate for their device type by the end device supporting Profinet or in a Profinet proxy. This allows object-oriented engineering tools to create self-documenting programs in IEC 61131-3 compatible languages referring to I/O by its object name.

Profibus International makes available a library of Profinet object classes defined in XML in the same way that it has made text-oriented GSD and EDD files available. Most of the GSDs and EDDs are expected to be made available as Profinet object classes.

### 24.2.13 WorldFIP

WorldFIP is the protocol from which ANSI/ISA Fieldbus was developed. Although this work became Type 1 of IEC 61158 in 2000, WorldFIP is also included as originally specified as Type 7 of the IEC fieldbus standard. The original model for WorldFIP was as a data acquisition network for the thriving French nuclear power industry. The objective was to collect large volumes of analog and discrete data very rapidly, using minimal electronics at the device, and using only one hole through the nuclear reactor containment wall. Although it has not been used for this purpose, WorldFIP has been used widely in real-time segments of the rest of the electric power industries and the transportation sector in many countries.

The WorldFIP protocol, like Foundation Fieldbus, is based on distributed arbitration, rather than collision detection, master/slave, or token passing. Also, they both provide data distribution using publish/subscribe logic, which allows tight time synchronization for distributed control action.

The versions of WorldFIP are called *profiles* that relate to the applications. Table 24-3, reproduced from the WorldFIP Web site, describes the WorldFIP profiles. Profile 1 is intended to be used only for very simple devices and lacks features supported by higher-level profiles. The lowest numbered levels are often called *MicroFIP or WorldFIP-I/O,* while the higher levels are often called *FullFIP.*

*Table 24-3: WorldFIP Profile Descriptions*

| Profile Level | Description |
|---|---|
| Profile 1: | For devices with few configuration options, which start on receipt of a simple command (e.g., basic sensors). |
| Profile 2: | For configurable devices that handle small amounts of data for configuration and parameter setting (e.g., I/O, more complex sensors). Key data is of cyclic nature. The device may also handle events. |
| Profile 3: | For devices that handle a lot of data for configuration and parameter setting (e.g., most AC or DC drives). Event handling is essential. |
| Profile 4: | For devices with total configuration flexibility (e.g., PLCs). |

WorldFIP specifies several physical layers for implementing different applications. The same low-speed (31,250 bps) shielded twisted-pair cabling system used by both Foundation Fieldbus H1 and Profibus-PA is also specified for WorldFIP for intrinsically safe, field-powered process control instrumentation. Additionally, higher speeds (1.0 and 2.5 Mbps) are specified for copper-wired networks. A speed of 5.0 Mbps is specified for fiber-optic networks. All of these specifications are also included in the physical layer of IEC 61158, the international fieldbus standard.

## 24.3 Bibliography

1. Caro, Richard H. (Dick). *Automation Network Selection*. ISA, 2004.
2. ANSI/ISA-50.02, Part 2-1992 – *Fieldbus Standard for Use in Industrial Control Systems Part 2: Physical Layer Specification and Service Definition.*
3. Web site: Profibus International http://www.profibus.com
4. Web site: WorldFIP http://www.worldfip.org
5. Web site: Interbus http://www.interbusclub.com

6. Web site: SDS http://content.honeywell.com/sensing/prodinfo/sds/

7. IEC 61158-2003-04 – *Digital data communications for measurement and control – Fieldbus for use in industrial control systems.*

8. IEC 61804-2-2004 – *Function blocks (FB) for process control – Part 2: Specification of FB concept and Electronic Device Description Language (EDDL).*

## About the Authors

**Tom Phinney** is a member of the Honeywell ACS Advanced Technology Lab in Phoenix, Ariz. He has more than 35 years of experience designing software and hardware for real-time systems, primarily as an architect and system designer with General Electric and Honeywell. He has specialized in industrial communications since the late 1970s. An ISA Fellow, he has served in a leadership role in ISA/IEC fieldbus activities. In 2002 he was the recipient of ISA's annual Standards & Practices award for outstanding service. He is the U.S. Technical Advisor to ANSI for digital communications within the industrial process measurement and control industries (IEC/SC 65C), chairs three IEC working groups in that area, and is active in the ISA SP99 industrial cybersecurity and SP100 industrial wireless networking efforts. His primary work focus today is on advanced biometric identification systems and the cybersecurity of industrial real-time control networks. He has five patents awarded and seven pending, five publications in the area of industrial time-critical communications, and has been the primary author or editor of more than 5,000 pages of international computer communications standards.

**Richard "Dick" Caro** is CEO of CMC Associates, a business strategy and professional services firm in Acton, Mass. Before working at CMC, he was Vice President of ARC Advisory Group in Dedham, Mass. He is the Chairman of ISA SP50 and formerly of IEC (International Electrotechnical Committee) Fieldbus Standards Committees. Before joining ARC, Dick held the position of Senior Manager with Arthur D. Little, Inc. in Cambridge, Mass., and he was a founder of Autech Data Systems, and Director of Marketing at ModComp. In the 1970s, The Foxboro Company employed Dick in both development and marketing positions. He holds a BS and MS in chemical engineering and an MBA.

# 25 Manufacturing Execution Systems & Business Integration

*By Dennis Brandl*

## Topic Highlights

*MES Integration with Business Planning and Logistics*
*Level 3 Equipment Hierarchy*
*MES and Production Operations Management*
*Detailed Production Scheduling*
- *Production Dispatching*
- *Production Execution Management*
- *Production Data Collection*
- *Production Tracking*
- *Production Resource Management*
- *Product Definition Management*
- *Production Performance Analysis*

*Other Manufacturing Activities*
*Level 3-4 Boundary*

## 25.1 Introduction

Automation does not end with equipment control; it also includes higher levels of control that manage personnel, equipment, and materials across production areas. Effectiveness in manufacturing companies is not based solely on equipment control capability. Manufacturing companies must also be efficient at coordinating and controlling personnel, materials, and equipment across different control systems in order to reach their maximum potential. This is usually accomplished using software systems and documented procedures that are collectively called the "manufacturing execution system (MES)" layer. MES defines a diverse set of functions that operate above automation control systems, reside below the level of enterprise business systems, and are local to a site or area. This chapter explains the functions of the MES layer and how these functions integrate with other corporate business systems.

The ANSI/ISA-95.00.03-2005 - *Enterprise-Control System Integration, Part 3: Models of Manufacturing Operations Management* standard defines 5 levels of activities in a manufacturing organization. Automation and control supports one level and MES supports a higher level, as shown in Figure 25-1.

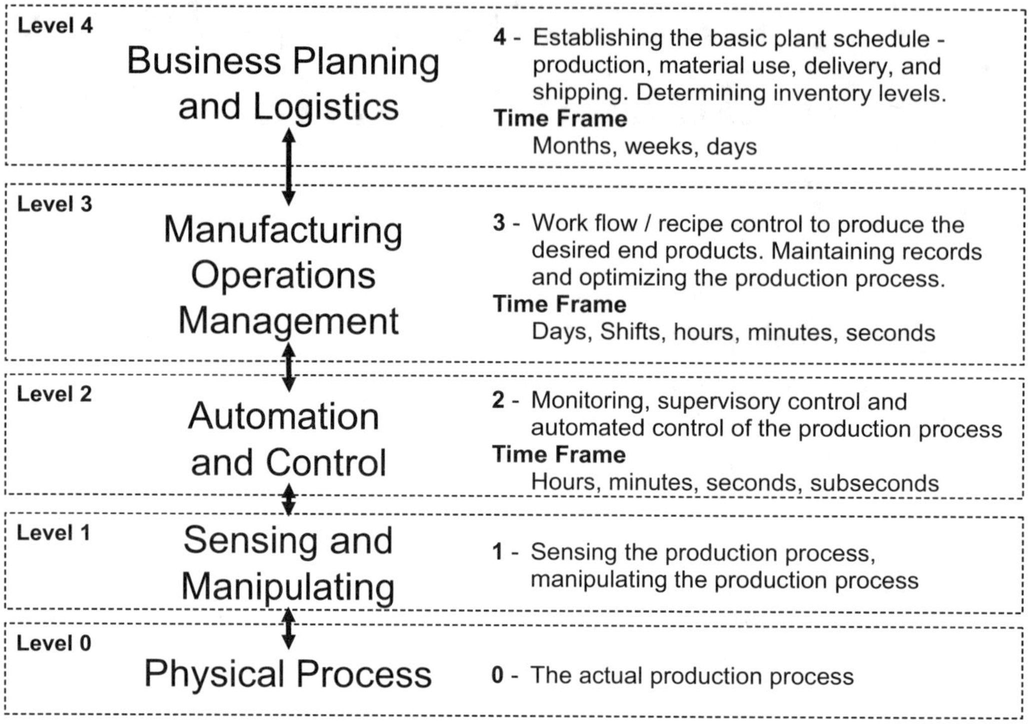


*Figure 25-1: Activity Hierarchy in a Manufacturing Company*

- ***Level 0*** defines the actual physical processes.
- ***Level 1*** defines the activities involved in sensing and manipulating the physical processes. Level 1 elements are the sensors and actuators attached to the control functions in automation systems.
- ***Level 2*** defines the activities of monitoring and controlling the physical processes and in automated systems this includes equipment control and equipment monitoring. Level 2 automation and control systems have real-time responses measured in subseconds and are typically implemented on programmable logic controllers (PLC), distributed control systems (DCS), and open control systems (OCS).
- ***Level 3*** defines the activities that coordinate production resources to produce the desired end products. It includes, work-flow "control" and procedural "control" through recipe execution. Level 3 typically operates on time frames of days, shifts, hours, minutes, and seconds. Level 3 functions also include maintenance functions, quality assurance and laboratory functions, and inventory movement functions, and are collectively called Manufacturing Operations Management. Level 3 functions directly related to production are usually automated using manufacturing execution systems (MES).
- ***Level 4*** defines business-related activities that manage a manufacturing organization. Manufacturing-related activities include establishing the basic plant schedule (such as material use, delivery, and shipping), determining inventory levels, logistics "control," and material inventory "control" (making sure materials are delivered on time to the right place for production). Level 4 is called Business Planning and Logistics. Level 4 typically operates on time

frames of months, weeks, and days. Enterprise resource planning (ERP) logistics systems are used to automate Level 4 functions.

It is important to remember that each level has some form of control and each level has its own definition for real-time. Level 3 systems consider *real-time* to mean information available a few seconds after shop floor events occur. Level 4 systems consider *real-time* to mean logistics and material information is available daily or within a few hours after the end of a shift.

## 25.2 MES Integration with Business Planning and Logistics

The ANSI/ISA-95.00.01-2000 - *Enterprise-Control System Integration Part 1: Models and Terminology* and ANSI/ISA-95.00.02-2001 - *Enterprise-Control System Integration Part 2: Object Model Attributes* standards define terminology to be used for interfaces between Level 3 systems and Level 4 systems. This information is used to direct production activities and to report on production.

Formal data models for exchanged information include:

***Personnel Class, Person, and Qualification Test Information*** – This is the definition of the persons and personnel classes (roles) involved in production. This information may be used to associate production with specific persons as part of a production record, or with personnel classes to allocate production costs.

***Equipment Class, Equipment, and Capability Test Information*** – This is the definition of the equipment and equipment classes involved in production. This information may be used to associate production with specific equipment as part of a production record, or with equipment classes to schedule production and allocate costs.

***Material Class, Material Definition, Material Lot, Material Sublot, and QA Test Information*** – This is the definition of the lots, sublots, material definitions, and material classes involved in production. This information allows Level 3 and Level 4 systems to unambiguously identify material specified in production schedules and consumed or produced in actual production.

***Process Segment Information*** – This is the definition of the business views of production, based on Level 4 business processes that must send information to production, or receive information from production. Examples include: setup segments, inspection segments, production segments, and cleanup segments.

***Product Definition Information*** – This is the definition of the materials, equipment, personnel, and instructions it takes to make a product. This includes the *Manufacturing Bill* (a subset of the Bill of Material [BOM] that contains the quantity and type of material required for producing a product). It also includes product segments, which define the routing and specific resources required at each segment of production.

***Production Capability Information*** – This is the definition of the capability and capacities available from production for current and future periods of time. Capability and capacity information is required for both Level 4 scheduling and Level 3 detailed production scheduling.

***Production Schedule Information*** – This specifies what products are to be made. It may include the definition of the specific personnel or roles to be used, equipment or equipment classes to be used, material lots or material classes to be produced, and material lots or material classes to be consumed for each segment of production.

***Production Performance Information*** – This specifies what was actually produced. It may include the definition of the actual personnel or personnel classes used, the actual equipment or equipment classes used, the actual material lots and quantities consumed, and the actual material lots and quantities produced for each segment of production.

## 25.3 Level 3 Equipment Hierarchy

Figure 25-2 shows the equipment and organizational hierarchy defined in the ANSI/ISA-95.00.03-2005 - *Enterprise-Control System Integration, Part 3: Models of Manufacturing Operations Management* standard. Level 4 ERP Logistics systems will typically coordinate and manage the entire enterprise and sites within the enterprise, but it may also schedule to the area or work center level. Level 3 MES systems will typically coordinate and schedule areas, work centers, and work units.

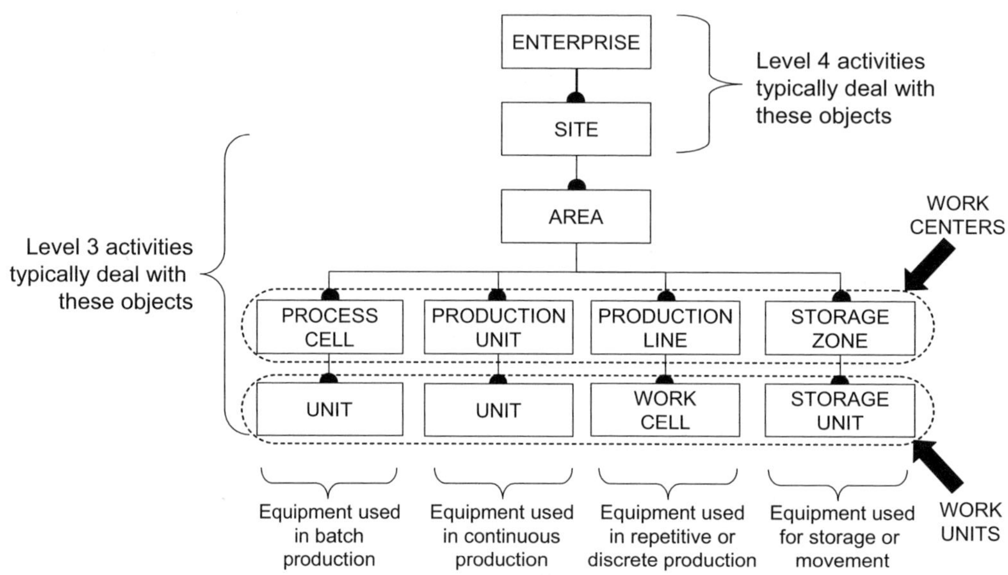


*Figure 25-2: Equipment Hierarchy for Level 3 and Level 4 Functions*

The equipment hierarchy is an expansion of the equipment hierarchy defined in the ANSI/ISA-88.01-1995 batch control standard to include equipment types used in continuous production, discrete production, and inventory storage and movement. The equipment hierarchy provides a standard naming convention for the organization of equipment, automation control, and manual control.

## 25.4 MES and Production Operations Management

Figure 25-3 illustrates the different Level 3 production-oriented functions that take place in sites and areas. Each bubble in the figure represents a collection of activities that occur in a production facility as a production schedule is converted into actual production. It illustrates how production requirements from the business are used to coordinate and control plant floor activity. The top four arrows identify previously defined information that is exchanged with business logistics systems.

The production model is driven by *production schedules* developed by the business and sent to production. The *production schedules* are used by detailed production scheduling activities that define *detailed production schedules* containing *production work orders*. The *production work orders* are dispatched to work centers and work units based on time and events, the *production work order* is executed and data is collected in a production data collection activity. (**Note**: In batch systems a *control recipe* is the equivalent of a *production work order.*)

The collected data is used in production tracking activities that relate the time-series information to the work order information to generate a report on *production performance* and tracing and tracking

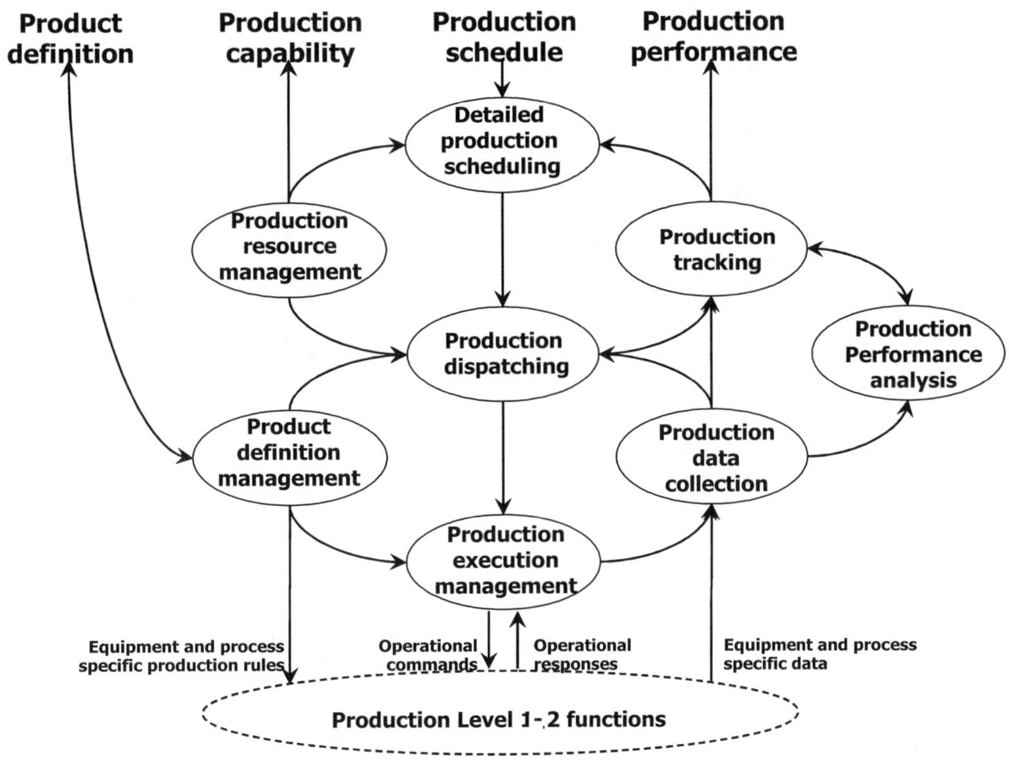


*Figure 25-3: Manufacturing Operations Management Functions*

information. The collected data and the data from tracing and tracking is used in production analysis functions to generate reports and KPIs (Key Performance Indicators). *Production capability* information about the current and future availability is provided to business scheduling systems by production resource management activities. *Product definition* information about the recipe, procedures, Bill of Material (BOM), and work routing needed for production is managed by product definition management activities.

## 25.5 Detailed Production Scheduling

These are the activities in a facility that take a production schedule and use information about local resources to generate a detailed production schedule. This can be an automated process, but in many plants scheduling is done manually by expert production planners or production planning staff. Automated systems are sometime referred to as plant level advanced planning and optimization systems. The key element of this activity is detailed scheduling of work assignments and material flows to a finer level of granularity than the business schedule. While Level 4 schedules may schedule work assignments to areas and work centers, detailed production scheduling will schedule work assignments to work centers and work units.

### 25.5.1 Production Dispatching

Once a detailed production schedule is available, that schedule is dispatched to production lines, process cells, production units, and storage zones. This can take the form of supervisors receiving daily schedules and dispatching work to technicians, or automated systems performing campaign management of batches and production runs. Production dispatching includes handling conditions not anticipated in the detailed production schedule. This may involve judgment in managing workflow and buffers. Unanticipated conditions may have to be communicated to maintenance operations management, quality operations management, and/or inventory operations management. This is one of the core functions of an MES.

### 25.5.2 Production Execution Management

Production execution management activities receive the dispatched work requests and, using paper based systems, MES systems, or recipe execution systems, coordinate and control the actual work execution. This may include the execution of procedural logic in recipes and display of work flow instructions to operators. The activities include selecting, starting, and moving units of work (such as a batch or production run) through the appropriate sequence of operations to physically produce the product. The actual equipment control is part of the Level 2 functions. Production execution management is one of the core functions of an MES system, but it may also be performed by recipe or manual work flow instruction systems in DCS systems or batch execution systems. The standards for information flows from Level 3 to Level 2 are defined in the ANSI/ISA-88.01-1995, OPC, and Fieldbus standards.

### 25.5.3 Production Data Collection

Production data collection are the activities that gather, compile, and manage production data for specific units of work (batches or production runs). Manufacturing control systems generally deal with process information such as quantities (weight, units, etc.), properties (rates, temperatures, etc.), and equipment information such as controller, sensor, and actuator statuses. Collected production data includes sensor readings, equipment states, event data, operator-entered data, transaction data, operator actions, messages, calculation results from models, and other data of importance in the making of a product. The collected data is inherently time or event based, with time or event data added to give context to the collected information. This information is usually made available to various analysis activities, including product analysis, production analysis, and process analysis. Real-time data historians and automated batch record logging systems support this activity.

### 25.5.4 Production Tracking

The production tracking activities convert sensor and equipment data into information related to assigned work (batches and production runs), and into tracking information about equipment, material, and personnel used in production. Production tracking also merges and summarizes information that is reported back to the business activities. This is one of the core functions of an MES. When automated systems are used they usually link to data historians and batch record logging systems.

### 25.5.5 Production Resource Management

The resource management activities monitor the availability of personnel, material, and equipment production resources. This information is used by detailed production scheduling and business logistics planning. These activities take into account the current and future predicted availability, using information such as planned maintenance and vacation schedules, in addition to material order status and delivery dates. This activity may also include material reordering functions, such as Kanban. Kanban is a material management system used as part of just-in-time production operations where components and sub-assemblies are produced, based upon notification of demand from a subsequent operation. A Japanese word for "sign," Kanbans are typically a re-order card or other method of triggering new production of material based on actual usage.

Resource management is usually a mixed operation, with manual work, automation, and database management. Management of the resources may include local resource reservation systems, and there may be separate reservation systems for each type of managed resource (personnel, equipment, and material). This is one of the core functions of an MES.

### 25.5.6 Product Definition Management

Product definition management includes activities associated with the management of product definitions. These may be recipes, work instructions, assembly instructions, standard operating procedures, and other information used by production to make or assemble products. This is one of the core functions of an MES.

### 25.5.7 Production Performance Analysis

The activities associated with the analysis of production, process, and product are defined as production performance analysis. These are usually off-line activities that look for ways to improve processes through chemical or physical simulation, analysis of good and bad production runs, and analysis of delays and bottlenecks in production. Production performance analysis also includes calculating performance indicators, leading, and trailing predictors of behavior. These activities generally are major users of information collected in plant data historians. There are often separate tools for production, process, and product analysis, and the tool sets vary based on the type of production (continuous, discrete, or batch).

## 25.6 Other Manufacturing Activities

The above list does not define all of the activities of a production facility. There are also maintenance operations management activities, quality operations management activities, and inventory operations management activities.

***Maintenance Operations Management*** – The activities that coordinate, direct, and track the functions that maintain the equipment, tools and related assets to ensure their availability for manufacturing.

***Quality Operations Management*** – The activities that coordinate, direct, and track the functions that measure and report on quality. The broad scope of quality operations management includes both quality operations and the management of those operations to ensure the quality of intermediate and final products.

***Inventory Operations Management*** – The activities that coordinate, direct, and track the functions that transfer of materials between and within work centers and manage information about material locations and statuses.

Manufacturing operations also require infrastructure activities that may be specific to manufacturing, but which are often elements also required by other parts of a manufacturing company. The infrastructure activities include:

a) Managing security within manufacturing operations

b) Managing information within manufacturing operations

c) Managing configurations within manufacturing operations

d) Managing documents within manufacturing operations

e) Managing regulatory compliance within manufacturing operations

f) Managing incidents and deviations

## 25.7 Level 3-4 Boundary

There are four rules which can be applied to determine if an activity should be managed as part of Level 4 or as part of Levels 3, 2, or 1. An activity should be managed at a Level 3 or below if the activity is directly involved in manufacturing, includes information about personnel, equipment, or material, and meets any of the following conditions:

a) The activity is critical to plant safety.

b) The activity is critical to plant reliability.

c) The activity is critical to plant efficiency.

d) The activity is critical to product quality.

e) The activity is critical to maintaining product or environmental regulatory compliance.

**Note**: This includes such factors as safety, environmental, and cGMP (current good manufacturing practices) compliance.

This means, in some cases, the Level 3 activities defined above may be performed as part of logistics instead of operations. Typically, this involves detailed production scheduling and production dispatching. The scope of an MES system is determined by applying the above rules to each site or area within a site.

## 25.8 References


1. ANSI/ISA-95.00.01-2000. *Enterprise/Control System Integration – Part 1: Models and Terminology.*
2. ANSI/ISA-95.00.02-2000. *Enterprise/Control System Integration – Part 2: Object Model Attributes.*
3. ANSI/ISA-95.00.03-2005. *Enterprise/Control System Integration – Part 3: Models of Manufacturing Operations Management.*
4. Cox III, James F. and John H. Blackstone, Jr. *APICS Dictionary.* Ninth Edition. APICS, 1998.
5. *MES Functionalities and MRP to MES Data Flow Possibilities.* White Paper Number 2. MESA International, 1994.
6. Williams, Theodore J. *The Purdue Enterprise Reference Architecture: A Technical Guide for CIM Planning and Implementation.* ISA, 1992.


### 25.8.1 Practical References


1. Brown, Mark Graham. *Baldrige Award Winning Quality: How to Interpret the Malcolm Baldrige Award Criteria.* ASQC Quality Press, 1995 and 1996.
2. Goldratt, Eliyahu M. and Jeff Cox. *The Goal: A Process of Ongoing Improvement.* North River Press, 1992.


## About the Author

**Dennis Brandl** is the chief consultant for BR&L Consulting, specializing in Manufacturing IT applications, including business-to-manufacturing integration, MES solutions, general, and site recipe implementations, and automation system security. He has been involved in automation system design and implementation in a wide range of applications over the past 25 years. They have included biotech, pharmaceutical, chemical plants and oil refineries, food manufacturing, consumer packaged goods, PLC-based systems, and batch control systems. He is an active member of ISA's SP95 Enterprise/Control System Integration Committee and is editor of the set of SP95 standards. He has a BS in Physics and an MS in Measurement and Control from Carnegie-Mellon University, and an MS in Computer Science from California State University.

# 26 System and Network Security

*By Robert C. Webb*

## Topic Highlights

*Essential Concepts*
*Security Programs, Plans, and Policies*
*Basic System and Network Security Techniques*
*Conclusions on Automation System and Network Security*

## 26.1 Essential Concepts

### 26.1.1 What Is System and Network Security?

There are many definitions of security; many of them could apply to automation and control systems. Security is often thought of as the robustness of a system and its ability to deliver a service regardless of upsets, faults, failures, or other problems. In the electrical transmission and distribution industry, the term "operational security"[1] is defined as "the ability of a power system to withstand or limit the adverse effects of any credible contingency to the system including overloads beyond emergency ratings, excessive or inadequate voltage, loss of stability or abnormal frequency deviations."

However, a number of factors have mandated a new, equally important meaning. These factors include:

- the advent of ubiquitous, easy to use, remote and network access to almost every automation and control system component;
- the trend to connect these components to phone lines, control networks, business systems, and even the Internet; and
- the risk of intentional or even inadvertent network induced system failures.

In this context we need to re-define "system and network security."

System and network security includes the use of physical protection and electronic identification, authentication, authorization, filtering, blocking, access control, encryption, validation, detection, measurement, audit, monitoring, logging, and other technologies, with the objective of precluding unauthorized or unintended use, modification, disclosure, or destruction of automation and control systems, or associated informational assets. These activities are undertaken in an effort to reduce the risk of personal injury or possibility of endangering public health, loss of public or consumer confidence, disclosure of sensitive assets, and protection of business assets. These concepts are applied to any system in the production process and include both stand-alone and networked components. Com-

1. Duke Energy Web site. Energy 101. Glossary of Terms. 2005. http://www.duke-energy.com/company/energy101/glossary/O.asp

munications between systems may be either through internal messaging or by any human or machine interfaces that authenticate, operate, control, or exchange data with any of these automation or control systems[2].

There are several aspects of this definition that should be highlighted:

- It should be applied in the broadest practical sense, encompassing all types of plants, facilities, and systems, in all industries. Automation and control systems include, but are not limited to:
    - hardware and software systems such as distributed control systems (DCSs), programmable logic controllers (PLCs), supervisory control and data acquisition (SCADA) systems, networked electronic sensing, and monitoring and diagnostic systems;
    - associated local or remote internal, human, network, or machine interfaces used to provide control, safety, and manufacturing operations functionality to continuous, batch, discrete, and other processes;
    - basic process control systems (BPCS), safety instrumented systems (SIS), and associated systems such as advanced or multivariable controls, online optimizers, dedicated equipment monitors, and graphical interfaces.

Hereinafter, we will use the singular term "automation systems" to mean all of these types of systems and equipment, in any similar or related application, in all industries and sciences.

It is a relatively new consideration, because the ubiquitous connectivity which creates the risk is relatively new, and hence the need for this "discipline" is new. It is also an immature discipline. Like any other new consideration that is part of an overall process of concept, design, construct, startup, operate, maintain, and decommission, it is not well understood. Most installed legacy systems today do not have the inherent design features to provide adequate security. They are typically at great risk when connected or connectable to networks or communications outside the specific automation systems themselves. Thus, special treatment and programs to assure they are adequately protected are essential for these legacy systems, and careful consideration, configuration, and design is equally essential for new systems.

With time, automation system and network security will become routine, as standards are developed, accepted, and implemented by vendors and users. Then, our vision of automation networks will routinely include all of the necessary aspects and features to assure adequate security. However, until that time, we must take steps to assure that the systems we specify, design, install, operate, maintain, and decommission, will not fail or otherwise compromise the owners because of security we failed to apply. This chapter will help you to be certain you have taken the appropriate steps.

Automation systems security is not the same thing as typical business system or "information technology" (IT) security. In fact, there are many critical differences. They share many similarities, and we apply appropriate IT practices, where and when appropriate, to automation systems and networks. If that was all we needed to do, there would be no need for this chapter in this book. We would include a sentence which says: "Follow Microsoft's recommendations for secure computing," and we would be done. Well, perhaps a bit of an exaggeration, but close. There is a well established domain of IT security, and rather than ignore it, or use it as it is, we need to combine that expertise with automation system domain expertise to assure the functionality of our systems is unhindered by the security technology and practices we apply. A few quick examples will highlight these differences, and the need to work with all of the expert domains to accomplish the objectives, without harming the systems more than the electronic intrusions we are attempting to prevent:

---

2. Adapted from ANSI/ISA-TR99.00.02-2004 - *Integrating Electronic Security into the Manufacturing and Control Systems Environment.* www.isa.org

- *Learning your system* – In order to properly secure your system, you need to know the system—services, connections, ports used, and so forth. There are a lot of automated software tools which can be run on a network to help inventory and define all the networked equipment and details of the network, including connections, ports, etc. So, you set off to "scan" your network with one of these tools, and your control system fails! A number of automation systems have been shut down or halted by typical IT network scans.

  One should never use a network scanning tool on an automation network without first assuring it will work and not stop the controllers or other key parts of the system. How can you be sure? Test it in a non-production environment that is really representative of your system, or talk to an expert such as your system vendor who has done so. It may be OK to reboot a user computer in a business environment. It is not OK to reboot or reset a controller that is performing critical 24/7 safety or control functions, because someone thought it would be nice to find out which ports it is using.

- *Virus protection, access control, including authentication, authorization, and other technologies* – These security measures simply cannot be applied to many legacy systems, nor to many of the systems still on the market today. However, those systems may be vulnerable. They do not have the processing power or other necessary features to handle the otherwise "essential" features or practices like antivirus programs. So do you give up? No, you work with IT and other experts to find alternatives which will provide adequate security commensurate with the risk,

- *Software updates and patches* – Microsoft has done an outstanding job developing a process to provide users with "critical" security updates as new vulnerabilities are discovered. If your operator workstation is running Windows XP, do you install the latest update? Not without full testing and vetting somewhere else, on a non-production system. Again, just like network scans, many automation and control systems failures have been caused by applying seemingly innocent patches or updates. Don't "just do it." Find out if it is OK, and then and only then, do it. Are your systems too important to stand the risk until then? Then disconnect all external access in the interim.

- *Passwords* – Just think how your operators will react if you require them to type in a "strong" password to respond to a safety alarm and trip a system or component. In many companies, they would throw you out of the control room and tell you not to come back. Adding time for certain otherwise "good" security steps is not OK when operator response time is critical and practiced. The issues need to be addressed, but in different ways.

Enough you say! Stop it! What do I do? The rest of this chapter provides a simple roadmap to address these kinds of issues, and more, in a comprehensive manner. All of the detail and required action is not spelled out here. Rather, the elements of a process you should follow are laid out, and references to detailed information on the process and technologies which you need to apply are provided. Adherence to good processes and procedures, and application-appropriate technologies, will not guarantee adequate security, but they will reduce risk and minimize the likelihood of system failures from electronic intrusions. Correctly implemented and applied, they will assure that you have taken reasonable steps to deal with a real threat.

### 26.1.2 Different Systems Require Different Approaches

An additional point is critical in this overview—recognition there is no single "silver bullet" that will provide adequate security for your automation systems. There is no single "security appliance," or "security procedure," that you can apply, and then rest assured your networks and systems are secure.

Automation systems networks are extraordinarily diverse. Technologies needed to protect them also are diverse (just look at the definition above). The systems are often mixtures of old and new technologies and hardware and software. And even when they are relatively homogenous, they involve people—people to design, install, operate and maintain them. Thus, having a securely designed hardware and software system is of little value if the operator posts his password on a little yellow sticky. Having excellent access controls, encryption, a complete understanding of your network, and personnel who understand the security risks, and are immune to social engineering[3], will not overcome a weakness introduced by the addition of a typical modem left connected so your vendor can help you troubleshoot and resolve problems.

What all of this is saying is that, today, automation network and system security is necessarily an engineered solution, because each network and system is different—they have different vulnerabilities, and different abilities to protect themselves.

Equally important, and perhaps obvious from these few simple examples, automation network and systems security is only as good as the weakest link in the chain. So, to have adequate security, you have to understand the chain and insist that security be applied to the entire chain! This means using a considered, comprehensive process—comprehensively applied to all aspects of system design, procurement, fabrication, installation, startup, operation and maintenance—from initial concept through operation and maintenance. It must be applied to your entire system—from hardware and software, (sensor or Intelligent Electronic Device[4] [IED], controllers, HMIs, and other components), to operator, maintenance, and engineering procedures, training, and actions. These procedures and training comprehensively applied to your personnel, your contractor's personnel, your system and network vendors and their personnel, and your design, operation and maintenance activities, to name a few areas that need to be covered. We will discuss essential elements of this process below.

Once you understand your networks and their vulnerabilities, you can apply the technologies and measures, where appropriate, to reduce risks of intrusion to acceptable levels.

Yes, there are a number of simple rules of thumb you can use to improve security in your systems. But they will be a waste of time and money, and may even reduce the reliability of your systems without benefit, unless you look at the entire picture with a comprehensive process. This chapter will help you to do that.

## 26.2 Security Programs, Plans, and Policies

### 26.2.1 Know Your Systems' Weaknesses—Then Fix Them

There are two fundamental steps to securing automation networks and systems. The first is to understand your system and know its vulnerabilities and weaknesses. The second is to address the weaknesses which are unacceptable.

As discussed in the last section, security of a system is only as good as the security of the "weakest link" in that system—be it human or electronic. Thus, a comprehensive approach is needed to identify all of the "links" or vulnerabilities, so they may be appropriately prioritized and addressed. This normally takes the form of a "program" developed and applied by the entity responsible for the automation system and network.

---

3. A technique used to unwittingly extract secret information, such as passwords or other network access information, from knowledgeable personnel. For example, an intruder may call and pose as an IT network support person, asking for an operator's login ID and password to solve a network problem. Sound stupid? It works more frequently than you would expect.

4. IEDs are the term used in the utility industry to describe microprocessor or computer based field devices like time distance and other system protection relays and other similar devices. They represent a growing class of "smart" field devices used in all industries which not only perform the sensor and control functions of old transmitters, switches, final control elements, and other similar devices, but also communicate over control and other networks and provide more control, safety, and diagnostic information to those networks than ever before.

The exact form or owner of such a program is not critical. It could be a part of an overall corporate IT security program, or a separate operating program, or a separate design process program. It could be owned by Operations, IT or Engineering. It could be part of an integrated risk management program or just focused on automation systems security. It doesn't matter, as long as it actively involves all the affected stakeholders and uses expertise from IT, Operations, and other perspectives. And, it *must* be focused on the unique aspects of automation systems and networks. It is not OK to simply say, "Oh, it is covered by our corporate IT security program—not unless it really is—and that means lots of "cautions," qualifiers, and additions that deal with the unique aspects of automation systems and network security.

In a similar manner, there is no single "right answer" to the magnitude of such a program. It needs to comprehensively cover the risks faced by the entity. This could result in a very simple program, or a very complex one, depending on the vulnerabilities and risks faced by a given entity.

Having said that, the following activities and steps have been found to be important to developing and employing an effective program, which addresses automation system and network security[5].

### 26.2.2 Developing a Program

A program will cover all of the elements discussed in Section 26.2. As such, it will normally be more than a few evenings of work for some otherwise already busy manager or engineer. As such, the nature of the risk and importance of minimizing it needs to be understood and agreed upon by the organization. The program must be supported by top management and have sufficient priority to obtain the resources needed to identify and prioritize the high risk vulnerabilities—and to do something about them. Everyone has to do their part, even if it is only to understand the risk and immunize themselves against "social engineering." All of this will typically require the following activities:

- *The business case* – Any effective automation systems and network security program needs management support at all levels. In order to obtain that, management will normally want to understand the risks and costs to minimize those risks at a macro level. They will want to see the "business case" for developing such a program. Industry organizations can help put this into a context that is meaningful for your industry; in some cases they have established guidelines and requirements to have security programs. Regardless, all entities should be able to quantify the costs of failure to operate, or of mis-operation due to automation systems failures. Indeed, many have already done so, in justifying current automation systems and automation system integration with business systems. Cyber security vulnerabilities are a new failure mechanism that needs to be considered, and the business case must be built to reduce this failure mechanism to acceptable levels.

- *Leadership commitment, support, and funds* – All stakeholders need to know this is an important activity, and they will be held accountable for its success. The scope of the effort should be proportional to the risk and funding needs to be commensurate with the risk and the scope.

- *Charter and scope of program* – A written charter and scope of the program is needed. In this area, as in many other parts of the "program" discussion, results can be faster and more cost effective if this activity is integrated into existing corporate programs. At the same time, the unique aspects of automation systems need to be made clear to the stakeholders, and addressed specifically.

- *Building the team* – This is key to the success of any program. Traditionally, there are organizational turf wars over computerized automation systems between IT organizations and operations or engineering organizations. This is no place to waste time and energy on such

5. These steps or elements to developing a program are based on the program elements developed by the ISA's Standards and Practices SP99 Committee – Manufacturing and Control Systems Security, as published in ANSI/ISA-TR99.00.02-2004 - *Integrating Electronic Security into the Manufacturing and Control Systems Environment.* www.isa.org

nonsense. Both disciplines' expertise is needed to identify vulnerabilities, develop countermeasures that will not cause more problems than the vulnerabilities they are attempting to resolve, and implement effective responses that meet the needs of the operating facilities at all levels of the organization. There are typically a number of others who could be involved as well—construction, vendors, contractors, and so forth. The program needs to address all of these areas. To be effective, participation and buy-in is needed from all of those affected. Make them part of your team.

- *Training* – As noted previously, social engineering and not using little sticky tags with passwords on them are two of many details on which all personnel need to be trained. Training is needed at several levels—training for all personnel at an awareness level and to avoid social engineering traps, and training for programmers, engineers, and operators on how to minimize risks in all steps—from system design to operation in the control room, and so forth. The main point is not to forget that people will be the weakest link, unless you address their lack of understanding and provide them with the knowledge to do the right thing.

- *Corporate policy* – There are several kinds of corporate "policy." What we are talking about here is not the detailed policy that deals with things like how many characters (and of what type) are required in passwords, but rather the broad overall corporate view on a subject and how it should be handled by the organization. Having a policy statement that speaks specifically of automation system and network security is essential. It is the evidence of management commitment and support, and the recognition that automation system and network security is unique and needs to be treated carefully.

- *Organization* – Organizational assignments as a result of the policy and other factors are another important part of program development. They need to define who/which parts of the organization has responsibility, and how they will coordination and work with the other stakeholders. Most entities will find they need a singular leader to develop and run the program effectively.

### 26.2.3 Defining Risk Goals

- *What are the risks and what is acceptable?* – The program should identify the risks from lack of security and have objectives and requirements which define the level of risk the entity is willing to accept. They may be stated in terms of production or impact, or any other measure that allows the entity's security team and management to make decisions as to what is needed. They should be commensurate with the treatment of other similar risks. They may be dynamic—changing as knowledge is gained and the vulnerabilities and potential countermeasures are better understood. In any case, they are needed as a framework to judge the scope and extent of the control systems and networks security program.

### 26.2.4 Assessing and Defining Existing System

- *Inventory and mapping* – This is the first extensive step involving the actual hardware and software used for automation systems and networks. There is significant information on developing and comprehensive inventory and mapping of equipment, networks and software in the referenced ISA Technical Reports. Making sure the maps are comprehensive, and defining the physical, logical, and electronic boundaries of the systems and networks is an essential base to further work, including identification of all systems and network connections—human interfaces as well as electronic interfaces.

- *Identification of all connectivity* – Identifying all connectivity is essential. This includes defining the manner in which system boundaries can be compromised, and the points where countermeasures should be taken if the risks are too high. Extra care needs to be exercised to find all connectivity, and define its nature. It includes not only the obvious, such as Ethernet ports, but also the operator and engineering human interfaces, floppy drives, USB ports, modem connections. A great deal of time can be spent categorizing such connections, but it may be more productive to simply treat them all as potential points of compromise and make sure the system security can deal with potential attacks from that vector. A key to remember is that the intended purpose of a connection is often irrelevant. The size or significance of the connection is only relevant from the viewpoint of how it may be used to attack the system or network. Any network path to a critical system component needs to be identified and addressed.

### 26.2.5 Conducting Risk Assessment and Gap Analysis

- *Identify threats, associated risks, and unacceptable risks.* Once the systems, system boundaries, and connectivity or access points have been identified, the vulnerabilities of these access points (based on the type and nature of the security provided for each) can be compared to the risk goals or objectives established by the program. Where the risks exceed the goals, countermeasures are required.

### 26.2.6 Developing Countermeasures

Once unacceptable risks are defined, countermeasures are required to address each and reduce the risk to an acceptable level. These can range from improving passwords to adding a layer of security around a legacy automation system that cannot do the fundamental things required (like provide access automation), to disconnecting the system from the external world.

A number of technologies are suggested in Section 26.3, and can be considered, along with others. Some may simply require configuration. Others will require new equipment and/or systems. In any case, steps should be taken in all cases where risks exceed acceptable levels to reduce them.

### 26.2.7 Define Component, Integration, Post Installation Test Plans

Once countermeasures have been designed or procured, they need to be tested thoroughly to assure they will not cause systems problems that are worse than the risk they are intended to mitigate. In general, this should not be done on an operating or production automation system or network. Manufacturers may be able provide such testing and or certification. Some entities have training or test systems they can use. Some users groups may be able to work with the vendors to accomplish such testing. National labs may be able to support such testing. The point is, don't apply a "fix" to a production system—whether it is a virus update, an operating system patch or configuration change, or a new firewall—without first testing it somewhere else. When considered in the big picture, this is no different than any other system integration activity. After developing a change, it is thoroughly reviewed and tested before applying it to an operating system. This is just good practice.

### 26.2.8 Finalize & Implement Operational Security Measures

Once countermeasures are developed and tested, they are applied where necessary. This is done in the same manner the entity would apply any other change, with appropriate coordination, controls, training for personnel, documentation, and all the other elements normally applied.

### 26.2.9 Routine Reporting and Analysis

Reviewing system logs, network traffic, intrusion detection results, and other information, as well as reviewing new threats identified by suppliers and other users, is an ongoing process. Monitoring system performance considers the possibility of external or internal intrusions, and indications of such problems are analyzed to determine if additional countermeasures are required.

### 26.2.10 Periodic Audit and Compliance

The program is audited or reviewed to assure it is being carried out, and it is effectively dealing with the potential risks of automation system and network intrusions.

### 26.2.11 Reevaluation

Whenever system or network design is changed; whenever routine review of reports and logs or analysis, or audits, indicate the existing program is not controlling risks to acceptable levels; or new threats are identified that increase risks above acceptable levels, actions are taken to identify and apply new or revised countermeasures to bring risks back to acceptable levels.

### 26.2.12 Work with Suppliers and Consultants

Suppliers of automation system and networking components and systems are one essential source of information on steps to take to minimize vulnerabilities and apply countermeasures. They may not know the answers to your questions, but they should be consulted and often can add valuable expertise to your team. They should be part of your program when they have shown capability to improve it. As noted previously, automation systems and network security is a relatively new discipline. As such, there are not a large number of corporate experts with the combination of automation systems and IT security expertise. There are consultants that can provide that expertise when needed. They are another valuable resource. In selecting support from such personnel, the users should avoid those without expertise in both the automation systems and IT security domain, and those associated with one particular solution. The industry is not mature enough yet to have a single vendor or supplier who has the final answer.

### 26.2.13 Work with Industry Groups

As noted, industry groups can provide support, and should be used where appropriate. Some, such as the Chemical Process Industry's CIDX, offer extensive guidance and are supported by industry leaders. Beware of those industry associations which say this is not a significant issue. The only case where this is not a significant issue is for those who have essentially no connectivity to the outside world. There are very few systems today, even older legacy systems, which meet this criterion.

## 26.3 Basic System and Network Security Techniques

The following topics provide a list of most tools and technologies used to secure automation systems and networks today. All are appropriate when the circumstances dictate. These are the countermeasures that can be applied when your program indicates a greater level of security is needed. Each tool or set of tools has advantages and disadvantages, and many have special considerations when applied to automation systems and networks. Discussion of each of these technologies and those limitations is well beyond the scope of this chapter. However, it is the focus of an ISA Technical Report, where 80 pages are dedicated to just that discussion[6]. There are several principals that should be considered and employed, as these technologies are selected and deployed.

### 26.3.1 Using Technologies to Improve Automation Systems and Network Security

- Don't accept the argument the system or network is incapable to doing what is needed. If security is important, as shown by risk analysis, and the legacy systems you are using cannot support the features you need, find a way to encapsulate or augment your system to provide the necessary features. Or unplug it, and let it run without the connectivity that brings the risk. Don't leave it at risk.

- Apply the concept of "defense in depth." Do not assume that one barrier will preclude intrusion into your automation systems and networks. This must be balanced against the cost and other practical considerations. However, the more significant the automation system (typically controlling the more significant systems for production or safety), the more important this becomes. And within such systems, there are usually more important functions or data that can be provided with additional protection. As an example, important process parame-

6. ISA's ANSI/ISA-TR99.00.01-2004 - *Security Technologies for Manufacturing and Control Systems.* www.isa.org

ters exchanged across large dedicated automation networks might require encryption for a variety of reasons. For critical data, one could take additional steps to re-calculate the data using diverse measurements, further assuring its validity. Capability to change key parameters in an automation system or network could be limited to a smaller set of individuals or system components than, say, normal operating parameters. Defense in depth is not just good for nuclear plants or important IT networks (as suggested by Microsoft). It is a good practice to employ here, but with the extra caution to avoid unwittingly making the system less responsive or reliable.

### 26.3.2 Authentication and Authorization

- Role Based Authorization Tools
- Password Authentication
- Challenge Response Authentication
- Physical/Token Authentication
- Smart Card Authentication
- Biometric Authentication
- Location-Based Authentication
- Password Distribution and Management Technologies
- Device-to-Device Authentication

### 26.3.3 Filtering/Blocking/Access Control

- Dedicated Firewalls (Hardware Based)
- Host-based Firewalls (Software Based)
- Virtual Local Area Networks (VLANs)

### 26.3.4 Encryption and Data Validation

- Symmetric (Private) Key Encryption
- Public Key Encryption and Key Distribution
- Virtual Private Networks (VPNs)
- Digital Certificates

### 26.3.5 Audit, Measurement, Monitoring and Detection Tools

- Log Auditing Utilities
- Virus/Malicious Code Detection
- Intrusion Detection Systems
- Network Vulnerability Scanners
- Network Forensics and Analysis Tools
- Host Configuration Management Tools
- Automated Software Management Tools

### 26.3.6 Operating Systems

- Server and Workstation Operating Systems
- Real-time and Embedded Operating Systems
- Web and Internet Technologies

### 26.3.7 Physical Security

- Physical Protection
- Personnel Security

## 26.4 Conclusions on Automation System and Network Security

In the future, every automation system object will have the appropriate security attributes and capabilities to address security transparently to the user. The system components will take steps, commensurate with the vulnerability and significance of the information they are acting on and the importance of their actions, to assure the validity of that information and to protect themselves from other forms of cyber attack. However, until that time, the automation system and network engineers, IT specialists, vendors, maintenance personnel, operators, and others who work on automation systems and networks will need to consider security carefully throughout the lifecycle of the systems with which they are associated.

Because of the many considerations necessary, manufacturers will need effective programs to assure they are doing a comprehensive job, and they will need to carefully review and understand the automation systems' centric limitations of the security technologies they apply. Such a program is outlined in this chapter, details are provided in ISA SP99's Technical Reports and Standards.

## 26.5 References

ANSI/ISA-TR99.00.01-2004 - *Security Technologies for Manufacturing and Control Systems.*

ANSI/ISA-TR99.00.02-2004 - *Integrating Electronic Security into the Manufacturing and Control Systems Environment.*

## About the Author

**Robert C. "Bob" Webb** is an automation expert with more than 30 years of in-depth experience in many fossil, geothermal, and nuclear power plants, where he has performed and led design, construction and operations support activities. His technical experience ranges from control loop stability and measurement error analyses to selection of valves and flow measuring equipment, to SCADA systems built from Ethernet communications among networks of disparate DCSs and PLCs. He holds BS degrees in physics and electronics engineering from California State Polytechnic University, San Luis Obispo, and is a registered electrical, mechanical, and control systems engineer. He is a past president of ISA's Northern California Section, managing director of ISA SP99, ISA's Manufacturing and Control Systems Security Standards Committee, and past vice president of ISA's Standards and Practices Department.

# 27 Operator Interface

*By Jonas Berge*

## Topic Highlights

*Graphics, Components, & Controls*
*Trend*
*Alarms*
*Reports*
*Scripts*
*Human Engineering*

## 27.1 Introduction

The human-machine interface (HMI) is the software application running in the operator consoles that permits operators to visualize the process. Other common names for this type of software are "process visualization" or supervisory control and data acquisition (SCADA) software. The primary aspects of HMI configuration are graphics, historical trend, alarms, reports, and scripts. These capabilities may either be lumped into a single monolithic software application, or made available as individual components in a suite.

Technologies such as the various flavors of OPC—as well as OLE_DB, VBA, and ActiveX—have made it very easy to configure and to use HMI software.

## 27.2 Graphics, Components & Controls

OPC-DA is the primary technology used to get live data from all the underlying control and device networks through OPC-DA servers. Older HMI software may use DDE (Dynamic Data Exchange) instead.

### 27.2.1 Live Data Access

At the time graphic displays are created, designers using HMI software locate data in the OPC server by browsing the name space. Communication aspects are already configured in the server and need not be set in the HMI software.

### 27.2.2 Direct OPC vs. Intermediate Database

Different implementations of OPC-DA in HMI software access data either directly from the OPC server or through a separate OPC client application that maps the data to an intermediate database, such as a traditional device driver. When access is direct to the OPC server, parameters are simply chosen by pointing and clicking. Only parameters displayed or currently used are polled from networks and devices.

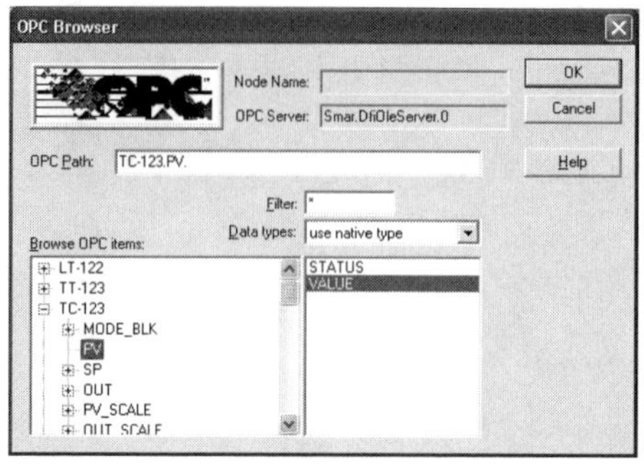


*Figure 27-1: Hierarchical Namespace (Screenshot: Wonderware InTouch)*

The use of an intermediate database requires tag renaming and additional configuration work. Drawbacks of an intermediate database include performance bottlenecks due to DDE and the fact that networks and devices may be polled even if the data is not displayed or used.

### 27.2.3 OPC Communication

Deep down in the "plumbing" of OPC, data exchange can be done in three different ways; some HMI software allows the designer to choose. Asynchronous is the most common scheme, since it results in much better application performance because other tasks are not idle waiting to get the data back from the server. The synchronous scheme may result in sluggish application response, but may be necessary in conjunction with some VBA scripts. Subscription is a third option that reduces OPC transactions by transmitting data when changes exceed a specified deadband. Most applications use asynchronous mode and leave the user no option.

OPC servers have mandatory and optional features (interfaces). Therefore, features available in one server may not be possible in another.

### 27.2.4 Graphics Configuration

HMI packages permit creating totally customizable and dynamically updated screens such as process flow mimic graphics, but often also generate standard displays such as overview, faceplate groups, point detail, tuning, etc. Templates and ActiveX components can be used to simplify consistent display creation. Variables are picked simply by browsing the server. Font, size, color, and other appearance aspects can be customized.

Mimic graphics is application-specific and is customized for every project. Ready made symbols for pumps, tanks, etc., make the design of true-to-life animated process visualization easy. Bar graph indication and dynamic level depiction will require the range to be entered.

### 27.2.5 Semantics and Conversion

OPC does not specify the semantics, or meaning, of data. It just transfers values from server to client. Almost every industrial protocol and device has different representation. For example, in one protocol, manual mode may be indicated by the value 16 in a specific parameter in a block, while in another protocol it may be indicated as a bit in a device-specific register. Some protocols use floating point measurement values while others use scaled integers. It is necessary to configure the graphics to show data coming in through different OPC servers in a consistent way if protocols are mixed. Frequently, it is necessary to use scaling and mathematical or logical expressions to convert the data. Enumerated parameters may have to be converted into text.

### 27.2.6 OPC Status

With every OPC value there is a status that indicates the validity of the value from an OPC perspective. Usually the status is "Good" but if a client cannot get the data from the server, it is flagged as "Bad."

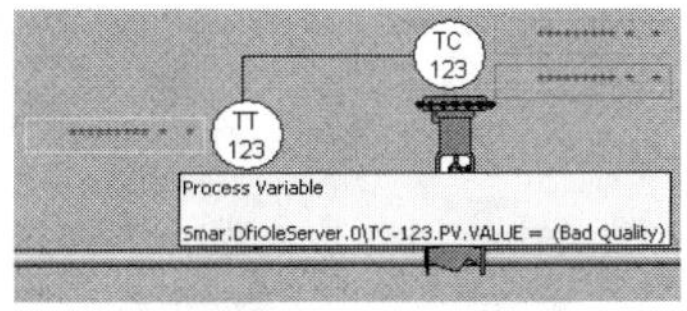


*Figure 27-2: The OPC Client Shows Server Error Messages (Screenshot: SMAR SYSTEM302)*

### 27.2.7 Operation

Additional range checking can be set in the HMI to validate operator input before it is written to the device.

### 27.2.8 Screen Management

Divide the screen space into panels with one pane dedicated to alarms, which is always on top and cannot be hidden behind others to ensure it is always visible. Typically, the screens have top and bottom fields that remain the same in all displays.

### 27.2.9 DDE Client

DDE was used in many HMI applications before OPC was introduced. Some HMI still use it, and some devices only have servers for DDE. Business applications like Excel also rely on DDE. Gateway applications allow data to be transferred between OPC and DDE. The link to the data must be typed manually, as DDE does not support browsing.

### 27.2.10 OPC-XML-DA Client and Communication

The URL for the OPC-XML-DA machine and server must be typed in manually. Data polling may be either basic or advanced. Basic means all values are reported every time. Advanced means the response only contains values which have changed.

### 27.2.11 ActiveX

ActiveX components and controls is a simple way of providing animated graphics, but these are also available in the form of document viewers, reporting tools, communication drivers, etc.—easily added as part of the HMI, provided it is an OLE container.

## 27.3 Trend

OPC-HDA is the primary technology used to access historical trend data logged in a database.

### 27.3.1 Historical Data Collection

HMI software includes a trend logger, which is an OPC-HDA server that samples and logs data in a database and also responds with historical records when requested by a display client. Sophisticated servers can filter and process data for clients.

### 27.3.2 Trend Logger Configuration

The trend logger must first establish connection with the underlying database, such as a SQL database engine. This is typically done through ODBC or OLE_DB and requires login and password to be provided.

Groups of parameters are created for different sampling rates. The OPC-HDA server may be an OPC-DA client, as well, to get the live data it logs. Picking parameters is a matter of pointing and clicking. To reduce the size of the database, it is possible to configure deadband and "aggregates." Aggregates

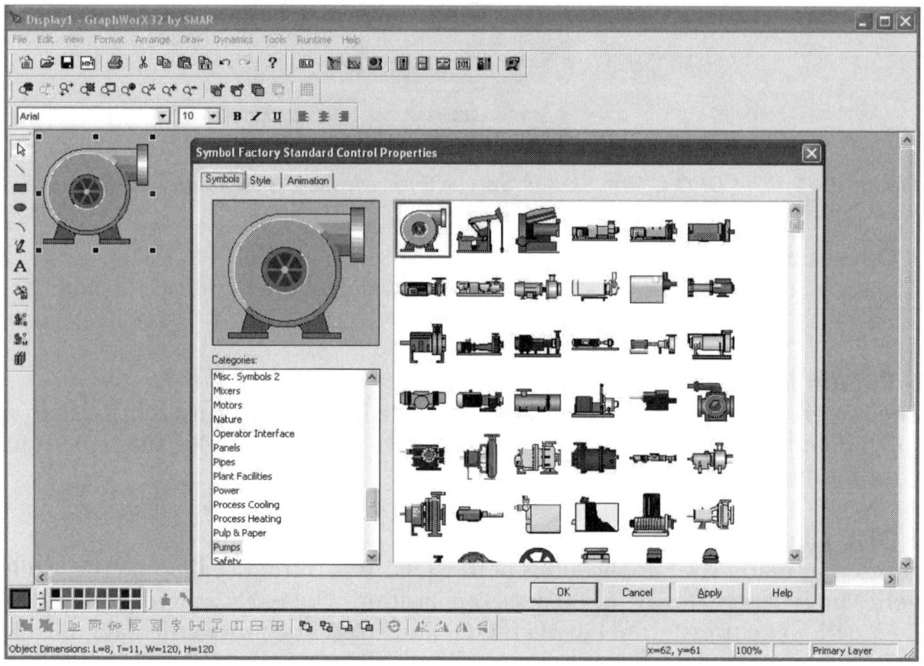


*Figure 27-3: Third-Party Library of ActiveX Components (Screenshot: Software Toolbox Symbol Factory)*

means preprocessing of multiple samples to obtain average, minimum, maximum, total, etc. Deadband means only significant changes exceeding this value are logged.

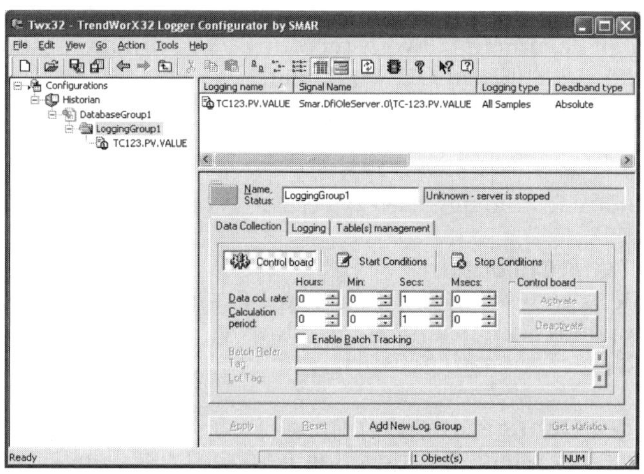


*Figure 27-4: Logger Configuration (Screenshot: SMAR SYSTEM302)*

### 27.3.3 Trend Display

A trend chart display embedded in the HMI graphics uses OPC-HDA to access data from the server. It is possible to "playback" data, zoom, analyze, and compare. A client can display data from many servers.

### 27.3.4 Trend Chart Viewer Configuration

Different display types such as strip chart, circular chart, and X-Y plot can be selected, depending on operator preference. Set the time span, or period, covered in the display. The pens to be displayed are picked by simply browsing the OPC-HDA server. Customized ranges can be applied. Again, it is possible to select aggregates such as average, minimum, maximum, etc. Colors, grids, size, and other appearance aspects can also be set.

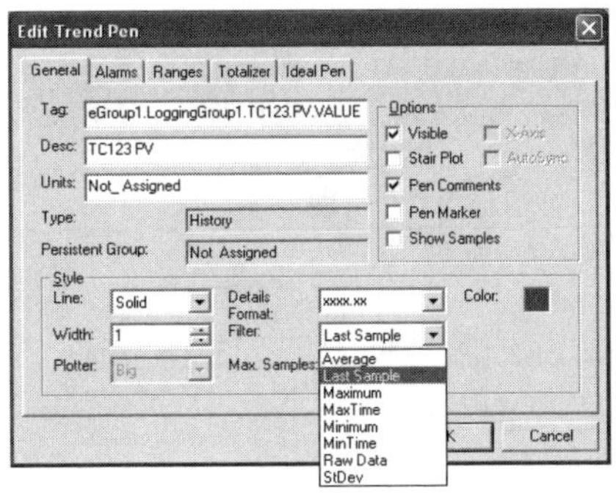


*Figure 27-5: Configuration of a "Pen" in a Trend Client (Screenshot: SMAR SYSTEM302)*

### 27.3.5 Live Data Trend

It is possible to display live trend data direct from an OPC-DA server together with historical trend from an OPC-HDA server.

### 27.3.6 Working with Live and Historical Trends

It is possible to "go back in time" by scrolling back and forth, playing back the values as they were logged. It is possible to zoom, and to see minimum, maximum, and other statistics. Pens can be added by dragging values into the trend chart. A current batch can be compared against an "ideal run" from the past. Annotations can also be included.

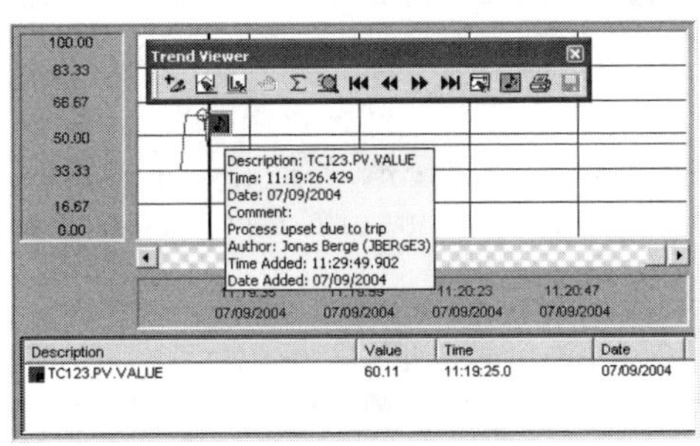


*Figure 27-6: Annotations (Screenshot: SMAR SYSTEM302)*

## 27.4 Alarms

OPC-A&E is the primary technology used to propagate alarms and events. An OPC-A&E server can either generate alarms on monitored values or capture alarms detected in the devices and communicated over the network. OPC-A&E clients include the alarm display and alarm logger.

### 27.4.1 Alarm Generator

HMI software includes an alarm generator that monitors variables to detect level and deviation conditions such as hi-hi, hi, lo, and lo-lo. Configuration is a matter of creating an event area, a hierarchical organization of the alarms, set limits, and priorities. The OPC-A&E alarm generator server is also an OPC-DA client. Variables are picked simply by browsing the OPC-DA server. Alarm condition, trip levels, and priority are then chosen. Other options may include "first-out" disarm, suppression, and hysteresis.

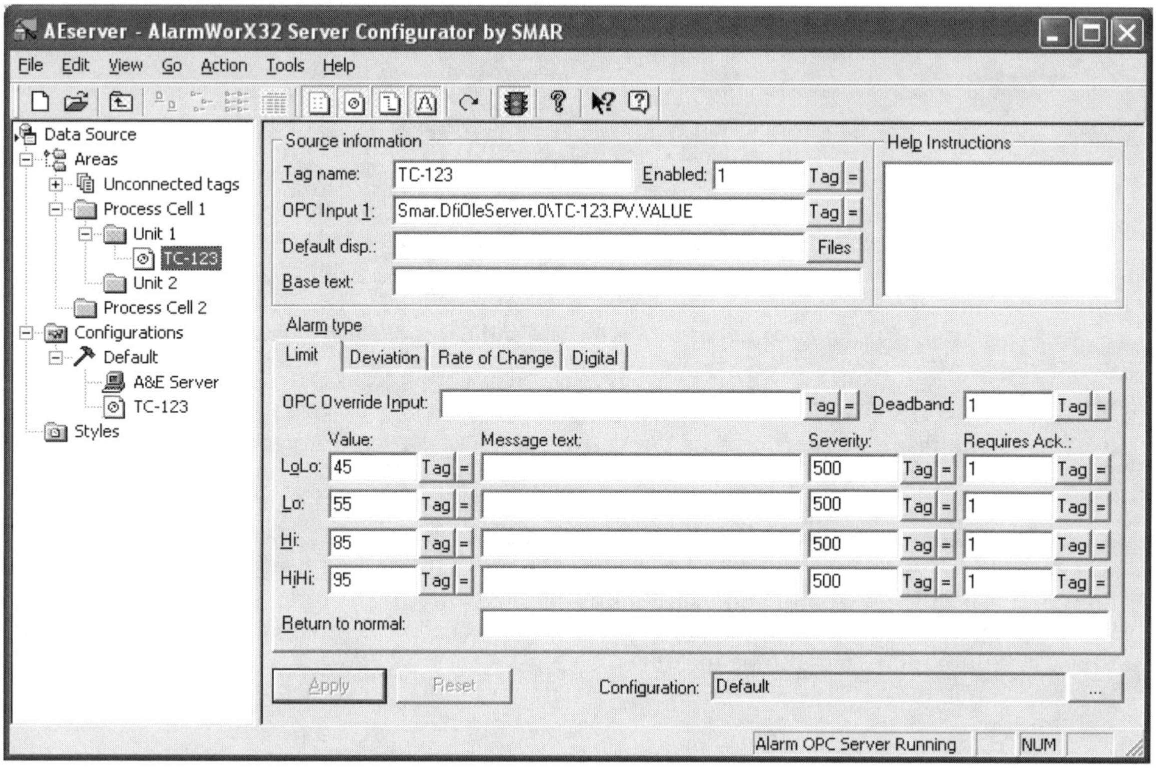


*Figure 27-7: OPC-A&E Alarm Generator (Screenshot: SMAR SYSTEM302)*

There is a tendency to create too many alarms, resulting in nuisance alarms and alarm flooding. Therefore, create alarms sparingly and assign priorities systematically.

### 27.4.2 Alarm Display

An alarm summary display may be embedded in the HMI graphics, which displays alarms as a list and permits the operator to acknowledge them. An OPC-A&E display client can subscribe to alarms from many servers. Communication aspects are already configured in the server and need not be set in the HMI software.

### 27.4.3 Server Filtering

When alarm displays and logger clients subscribe to alarms from servers, they specify filters for the server to sift out alarms each client is interested in. Filtering is based on area, type, priority, etc.

### 27.4.4 Alarm & Event Data Field

Not all information from the A&E server may be of interest for display, database log, or printing. Select the data fields of interest when the OPC-A&E client is configured.

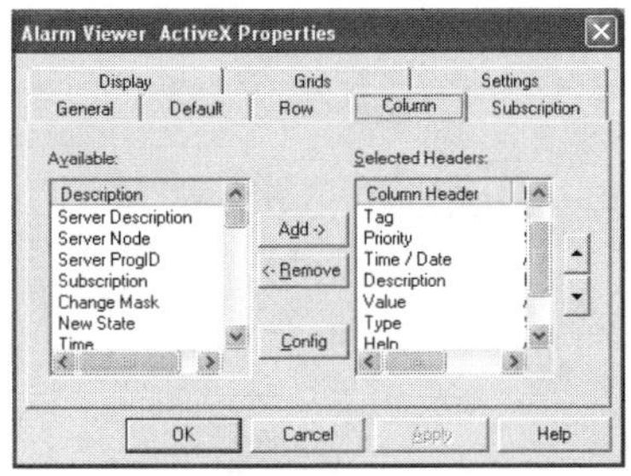

*Figure 27-8: Selecting Columns for the Data Fields (Screenshot: SMAR SYSTEM302)*

### 27.4.5 Alarm Summary Display

An alarm summary display is an OPC-A&E client. Configuring the display includes setting up the subscription filters, select data fields, and configuring appearance such as background color. Priorities are distinguished using different colors, blinking, and sound.

### 27.4.6 Working with Alarms and Events

In the alarm summary display, alarms can be sorted chronologically, or by other criteria such as priority. Alarms can be acknowledged and comments entered.

### 27.4.7 Alarm Logger Configuration

An alarm logger is an OPC-A&E client. Configuring the logger includes setting up the subscription filters, select data fields, and configuring if logging shall be done to the database or printer.

The database logger must first establish connection with the underlying database, such as a SQL database engine. This is typically done through ODBC or OLE_DB and requires login and password to be provided. There is no OPC flavor for historical A&E.

## 27.5 Reports

Historical trend as well as alarm and event data is stored in a database. ODBC and OLE_DB are the primary open technologies used to extract data from databases to generate reports based on some filter criteria. The report display may be embedded into graphics or may be a stand-alone tool. Usually the logging application and reporting tool have a semi-proprietary relationship. Without open standards, extracting data from a database is nearly impossible.

### 27.5.1 Configuration

Historical trend loggers and alarm and event loggers usually have an accompanying reporting tool preconfigured to extract information from the database and present it according to predefined layouts.

HMI report configuration may include designing the report format and selecting when the report shall be generated, such as daily, weekly, monthly, etc.

### 27.5.2 Alarm Report

An alarm report is configured by selecting the database and table, defining appropriate filters, and selecting on-screen or print presentation.

### 27.5.3 Historical Report

A historical report is a subset of logged data, based on a particular tag or a time window. The data is exported to a file. Configuring is done by selecting the database, tags, and time span.

### 27.5.4 Advanced and generic reporting

Free format reporting tools provide greater flexibility for form design. Database fields can be dragged and dropped into the report design.

### 27.5.5 Distributing reports

Reports can be presented on screen, printed on paper, faxed, saved as .PDF files, or rendered as a Web page or e-mail.

## 27.6 Scripts

The language VBA familiar from Microsoft Office and other applications is the primary language used for HMI scripts. Elements of VBA are functions, statements, methods, properties, objects, and events. VBA functions permit sophisticated scripts to perform all kinds of tasks including program flow control, data conversion, file management, file access and printing, arithmetic, comparison, logic, and text string manipulation. Familiar keywords include IF...THEN...ELSE, FOR...NEXT, GO TO, etc. HMI software comes with libraries that include OPC support. VBA is most powerful when used with ActiveX components and applications that support OLE automation.

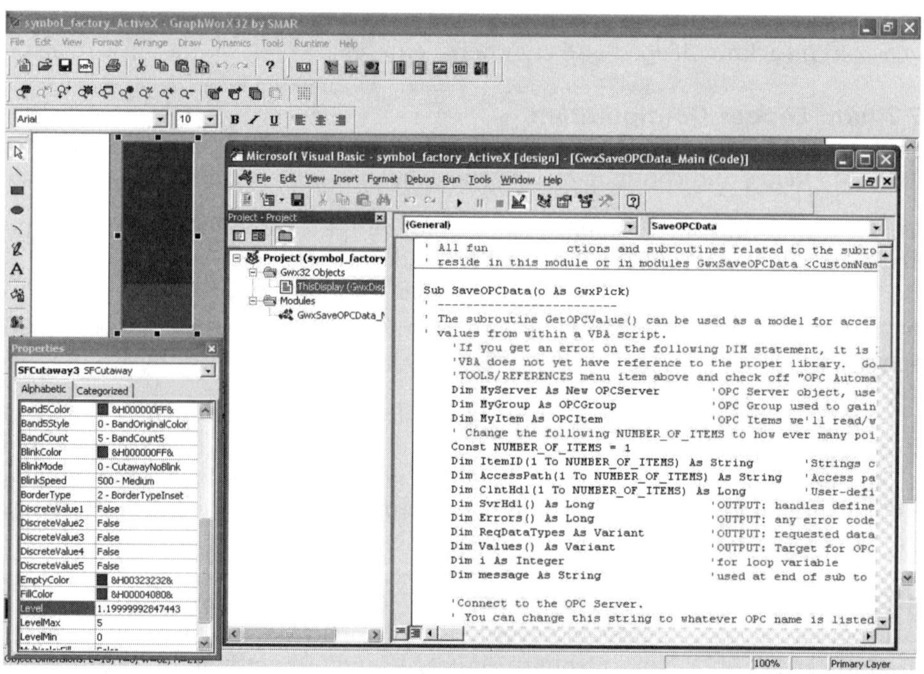


*Figure 27-9: VBA Script for Third-Party ActiveX Component Accesses OPC Data (Screenshot: SMAR SYSTEM302)*

## 27.7 Human Engineering

The introduction of computers in the control room has changed the way operators work. Workers spend more time sitting while performing repetitive movements with mouse and keyboard. This can result in pain in their backs and hands. An awkward posture such as twisting the neck, extended reaching, or maintaining the same posture for prolonged periods of time leads to pain and poor blood circulation. Eye strain may be caused by glare or incorrect lighting. Proper ergonomics results in fewer injuries, subsequent compensation claims, and fewer lost working days. Human engineering is an important aspect in control center design and selection.

### 27.7.1 Ergonomic Operator Consoles

Ergonomics has to be an integral part of the design of a control center. To minimize risks, it is possible to use control center furniture designed to ergonomic standards, which allows the operator to change posture throughout the day. These consoles are adjustable to meet the diverse build of operators. Less discomfort results in reduced fatigue, higher performance, increased productivity, elevated attention level, and better work quality.

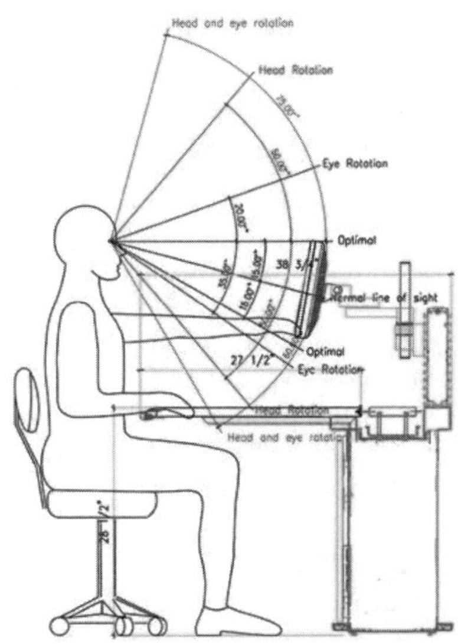


*Figure 27-10: Ergonomics Improve Operator Performance (Courtesy: Evans Console)*

Adjustment possibilities may, for example, include variable work surfaces to accommodate a wide range of users, as well as sitting or standing positions. The monitor may be mounted on a swivel arm to ensure proper line-of-sight, optimum viewing angle, and minimum glare. Anti-glare monitor shields reduce eye strain. Sufficient leg room below the console is also important. Wrist comfort is enhanced by eliminating sharp edges. Some consoles can even transform by adjusting the work surface from sitting to standing levels at the touch of a button.

### 27.7.2 Lighting

The illumination level at the computer monitor shall be 500-1,000 lux (a luminance equal to one lumen per square meter). Be sure to install lighting so as to cause a minimum of glare. Adjustable task lighting for each work surface may also be used. Glare on the surface of the display device may be reduced by using antiglare glass or materials for the monitors. Make use of shading hoods that extend over the monitor tops, particularly for cathode ray tube (CRT) monitors. Monitors with adjustable angles also help to reduce glare and reflections.

### 27.7.3 Information Presentation

The almost limitless graphical capabilities that HMI software inherits from the Windows operating system makes it easy to get overly artistic in creating graphics and ambitious when adding information. In the past, displays were limited to simple symbols in 16 colors. Today, the process flow mimic display can be done with 3D rendering, or photos and live streaming video. Networked instrumentation makes it possible to display hundreds of parameters per instrument. True-to-life graphics are helpful to some point but, when over done, result in cluttered displays that are harder to use. Adding too much information makes the essential parameters harder to locate. "DCS-style" displays are arranged in a hierarchical manner where it is possible, from process flow mimic screens and simple overview displays, to drill down to faceplate group screens and further down to tuning detail screens with ever-increasing details. Advanced HMI software also permit graphics to be layered, permitting operators to "declutter" the display by "zooming" out from fully populated layers to layers with only the basic information, hiding what is less essential.

The alarm system needs to be designed carefully to help, not hinder, operation. Taking measures at design time is important in order to eliminate alarm flooding during a process upset or fault situation. As a rule of thumb, the alarm system should be designed to generate no more than 10 alarms the first 10 minutes after a major upset. Other basic rules include alarms only on abnormal and unexpected events—not normal or expected. Assign a maximum 10% of alarms as "high" priority and a maximum 20% of alarms as "medium." Assign only one pre-alarm per alarm.

Alarms are useful to get the operator's attention to a beginning process upset at an early stage—especially considering the process can, during normal operation, be quite uneventful, almost boring. However, during a fault or process upset, many alarms may subsequently trip as a result of the ripple effect of the real problem. The operators are not able to deal with the ensuing flood of alarms, nor do the alarms reveal the real problem. To avoid alarm flooding, do not configure nuisance alarms—alarms that will not provide the operator with any new or useful information. Nuisance alarms include those that trigger as a direct result of an earlier problem already reported, or situations the operator can do nothing about, or which require no action. The Engineering Equipment Materials and Users Association (EEMUA) Specification 191 gives some useful hints for alarm management practices.

As per the EEMUA Specification 191 alarms shall be:

| | |
|---|---|
| Relevant | Justified and pertinent to the operator's priorities |
| Unique | Not a repetition of what the operator knew from previous alarms |
| Timely | Not too early, and not too late |
| Prioritized | Indicating the urgency or severity |
| Understandable | A clear, easy-to-understand message |
| Diagnostic | Helps find the problem |
| Advisory | Helps identify the correct action |
| Focusing | Direct the operator's attention in the right direction |

## 27.8 References


1. Berge, Jonas. *Software for Automation: Architecture, Integration, and Security.* ISA, 2005.
2. Tryan, Jana Lee. "Ergonomics in the Control Room." *Version Control: EC-MK-WP-ECC–V1.0.* March 30, 2005.
3. Tryan, Jana Lee. "Ergonomic Overview Paper." *Version Control: EC-MK-WP-EP-1.2.* January 5, 2005.
4. ISA–RP60.1–1990. *Control Center Facilities.*
5. ISA–RP60.3–1985. *Human Engineering for Control Centers.*
6. Engineering Equipment and Materials Users Association (EEMUA). *Alarm Systems: A Guide to Design, Management and Procurement.* Publication No.191. EEMUA, 1999.


## About the Author

**Jonas Berge** was educated in Sweden. He is General Manager at Smar's Asia-Pacific headquarters in Singapore and has been working with development and application in the field of instrumentation since 1987. One of the architects of Fieldbus, he has been instrumental in the development of the FOUNDATION Fieldbus specification and also participated in development of Smar's Fieldbus products and system. He is a Senior Member of ISA, VP of the Fieldbus Foundation Marketing Society in Singapore, and co-founder of the Fieldbus Foundation End-User Council in Singapore. He is the author of the books *Fieldbuses for Process Control: Engineering, Operation and Maintenance* and *Software for Automation: Architecture, Integration, and Security.* He received the 1999 ISA Excellence in Documentation Award and the 2001 Raymond D. Molloy Award.

# 28 Data Management

*By Diana C. Bouchard*

## Topic Highlights

*Data Relationships, Storage and Retrieval, Quality Issues*
*Database Structure, Types, Operation, Software and Maintenance*
*Basics of Database Design*
*Queries and Reports*
*Special Requirements of Real-Time Process Databases*
*Data Documentation and Security*

## 28.1 Introduction

Data are the lifeblood of industrial process operations. The levels of efficiency, quality, flexibility, and cost reductions needed in today's competitive environment cannot be achieved without a continuous flow of accurate, reliable information. Good data management ensures the right information is available at the right time to answer the needs of the organization. Databases store this information in a structured repository and provide for easy retrieval and presentation in various formats.

## 28.2 Database Structure

The basic structure of a typical database consists of *records* and *fields*. A *field* contains a specific type of information—for example, the readings from a particular instrument or the values of a particular laboratory test. A *record* contains a set of related field values, typically taken at one time or associated with one location in the plant. In a spreadsheet, the fields would usually be the columns (variables) and the records would be the rows (sets of readings).

In order to keep track of the information in the database as it is manipulated in various ways, it is desirable to choose a *key field* to identify each record, much as it is useful for people to have names so we can address them. Figure 28-1 shows the structure of a portion of a typical process database, with the date and time stamp as the key field.

## 28.3 Data Relationships

Databases describe relationships among *entities*, which can be just about anything: people, products, machines, measurements, payments, shipments, and so forth. The simplest kind of data relationship is *one-to-one*, meaning that any one of entity *a* is associated with one and only one of entity *b*. An example would be customer name and business address.

In some cases, however, entities have a *one-to-many* relationship. A given customer has probably made multiple purchases from your company, so customer name and purchase order number would have a

| DateTime | Impeller Speed rpm | Additive Flowrate, L/min | Additive Concentration ppm | ... |
|---|---|---|---|---|
| 2005-05-20 02:00 | 70.1 | 24.0 | 545 | ... |
| 2005-05-20 03:00 | 70.5 | 25.5 | 520 | ... |
| 2005-05-20 04:00 | 71.1 | 25.8 | 495 | ... |
| 2005-05-20 05:00 | 69.5 | 23.9 | 560 | ... |
| 2005-05-20 06:00 | 69.8 | 24.2 | 552 | ... |
| ... | ... | ... | ... | ... |

*Figure 28-1: Process Database Structure*

one-to-many relationship. In other cases, *many-to-many* relationships exist. A supplier may provide you with multiple products, and a given product may be obtained from multiple suppliers.

Database designers frequently use *entity-relationship diagrams* (Figure 28-2) to illustrate linkages among data entities.

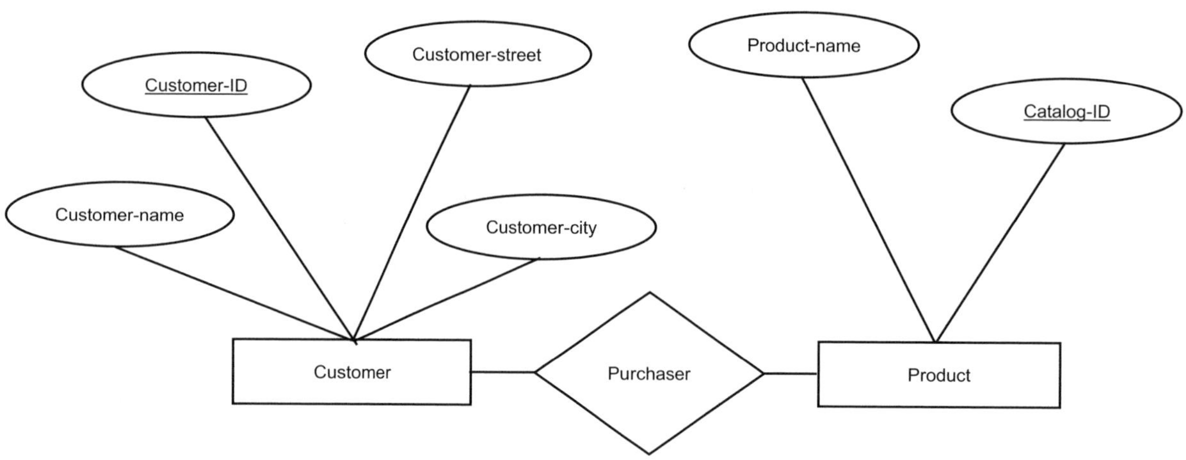


*Figure 28-2: Typical Entity-Relationship Diagram*

## 28.4 Database Types

The simplest database type is called a *flat file,* which is an electronic analogue of a file drawer, with one record per folder, and no internal structure beyond the two-dimensional (row and column) tabular structure of a spreadsheet. Flat-file databases are adequate for many small applications of low complexity.

However, if the data contain one-to-many or many-to-many relationships, the flat file structure cannot adequately represent these linkages. The temptation is to reproduce information in multiple locations, wherever it is needed. However, if you do this, and you need to update the information afterwards, it is easy to do so in some places and forget to do it in others. Then, your databases contain inconsistent and inaccurate information, leading to problems such as out-of-stock situations, wrong customer contact information, and obsolete product descriptions.

A better solution is to use a *relational database.* The essential concept of a relational database is that ALL information is stored as tables, both the data themselves and the relations between them. Each table contains a key field which is used to link it with other tables. Figure 28-3 illustrates a relational database containing data on customers, products and orders for a particular industrial plant.

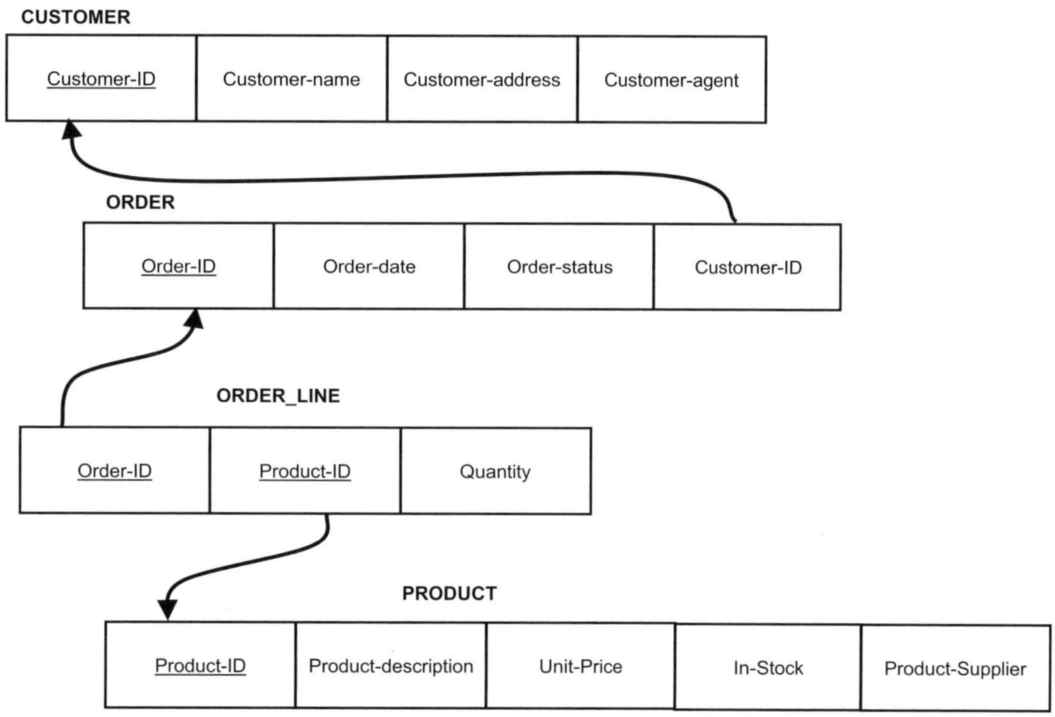


*Figure 28-3: Relational Database Structure*

Additional specifications describe how the tables in a relational database should be structured so the database will be reliable in use and robust against data corruption. The degree of conformity of a database to these specifications is described in terms of degrees of *normal form.*

## 28.5 Basics of Database Design

The fundamental principle of good database design is to create a database that will support the desired uses of the information it contains. Factors such as database size, volatility (frequency of changes), type of interaction desired with the database, and the knowledge and experience of database users will influence the final design.

Key fields must be unique to each record. If two records end up with the same key value, the likely result is misdirected searches and loss of access to valuable information.

Definition of the other fields is also important. Anything you might want to search or sort on should be kept in its own field. For example, if you put first name and last name together in a personnel database, you can never sort by last name.

## 28.6 Queries and Reports

A query is a request to a database to return information matching specified criteria. The criteria are usually stated as a logical expression using operators such as equal, greater than, less than, AND and OR. Only the records for which the criterion evaluates as TRUE are returned. Queries may be performed via interactive screens, or using query languages such as SQL (Standard Query Language) which have been developed to aid in the formulation of complex queries and their storage for re-use (as well as more broadly for creating and maintaining databases). Figure 28-4 shows a typical SQL query.

```
SELECT PRODUCT_NAME, PRODUCT_CATEGORY,
PRODUCT_SERVICERATING, UNIT_PRICE
            FROM PRODUCT_FLOWMETER
            WHERE (PRODUCT_CATEGORY LIKE "%Coriolis"
            AND PRODUCT_SERVICERATING = "%Acid"
            AND UNIT_PRICE < 10000;
```

*Figure 28-4: Typical SQL Query*

Reports pull selected information out of a database and present it in a predefined format as desired by a particular group of end users. The formatting and data requirements of a particular report can be stored and used to regenerate the report as many times as desired using up-to-date data.

Interactive screens or a report definition language can be used to generate reports. Figure 28-5 illustrates a report generation screen.

## 28.7 Data Storage and Retrieval

How much disk storage a database requires depends on several factors: the number of records in the database, the number of fields in each record, the amount and type of information in each field, and how long information is retained in the database. Although computer mass storage has rapidly expanded in size and decreased in cost over the last few decades, human ingenuity in finding new uses for large quantities of data has steadily kept pace. Very large databases such as those used by retail giant Wal-Mart to track customer buying trends now occupy many terabytes (trillions of bytes) of storage space.

Managing such large databases poses a number of challenges. The simple act of querying a multi-terabyte database can become annoyingly slow. Important data relationships can be concealed by the sheer volume of data. As a response to these problems, data mining techniques have been developed to explore these large masses of information and retrieve information of interest. Assuring consistent and error-free data in a database which may experience millions of modifications per day is another problem.

Another set of challenges arises when two or more databases that were developed separately are interconnected or merged. For example, the merger of two companies often results in the need to combine their databases. Even within a single company, as awareness grows of the opportunities that can be seized by leveraging their data assets, management may undertake to integrate all the company's data into a vast and powerful data warehouse. Such integration projects are almost always long and costly, and the failure rate is high. But, when successful, they provide the company with a powerful data resource.

To reduce data storage needs, especially with process or other numerical data, data sampling, filtering and compression techniques are often used. If a reading is taken on a process variable every 10 min-

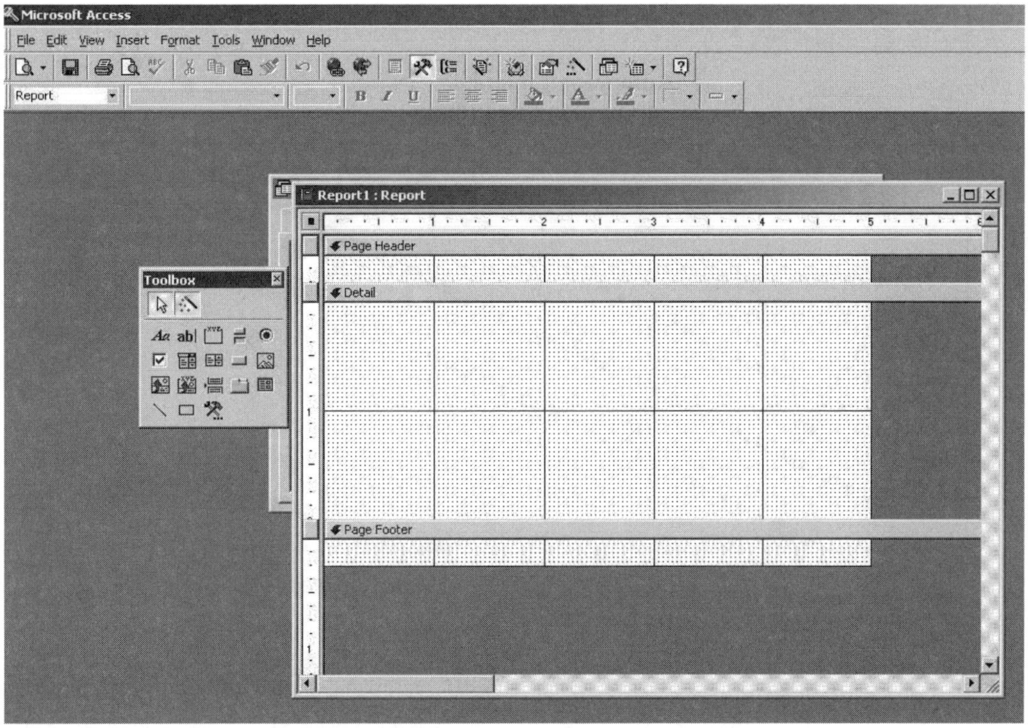


*Figure 28-5: Report Generation Screen*

utes as opposed to every minute, simple arithmetic will tell you that only 10% of the original data volume will need to be stored. However, a price is paid for this reduction: loss of any record of process variability on a timescale shorter than 10 minutes, and possible generation of spurious frequencies (aliasing) by certain data analytic methods. Data filtering is often used to eliminate certain values, or certain kinds of variability, that are judged to be noise. For example, values outside a predefined range, or changes occurring faster than a certain rate, may be removed.

Data compression algorithms define a band of variation around the most recent values of a variable and record a change in that variable only when its value moves outside the band (see Figure 28-6). Essentially the algorithm defines a "dead band" around the last few values and considers any change within that band to be insignificant. Once a new value is recorded, it is used to redefine the compression dead band, so it will follow longer-term trends in the variable. Detail variations in this family of techniques ensure a value is recorded from time to time even if no significant change is taking place, or adjust the width and sensitivity of the dead band during times of rapid change in variable values.

## 28.8 Database Operations

The classic way of operating on a database, such as a customer database in a purchasing department, is to maintain a master file containing all the information entered so far, and then periodically update the database using a transaction file containing new information. The key field in each transaction record is tested against the key field of each record in the master file to identify the record that needs to be modified. Then the new information from the transaction record is written into the master file, overwriting the old information (see Figure 28-7). This approach is well suited to situations where the information changes relatively slowly and the penalties for not having up-to-the-minute information are not severe. Transactions are typically run in batches one to several times a day.

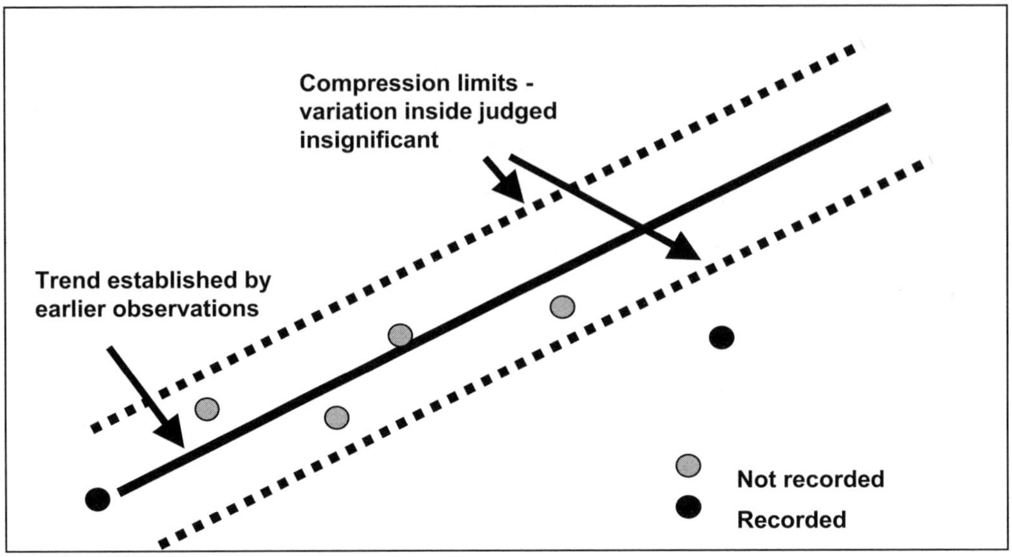


*Figure 28-6: How a Data Compression Deadband Works*

| TRANSACTION FILE | | | |
|---|---|---|---|
| 29177 | company-29177 | **agent-name-29177 (CHANGED)** | agent-phone-29177 |
| 30064 | **company-30064** | **agent-name-30064** | **agent-phone-30064** |
| 30195 | company-30195 | agent-name-30195 | **agent-phone-30195 (CHANGED)** |

New agent name - replace data

NEW RECORD - insert

New phone number - replace data

| MASTER FILE | | | |
|---|---|---|---|
| 28295 | company-28295 | agent-name-28295 | agent-phone-28295 |
| 29003 | company-29903 | agent-name-29903 | agent-phone-29903 |
| 29177 | company-29177 | agent-name-29177 | agent-phone-29177 |
| 29804 | company-29804 | agent-name-29804 | agent-phone-29804 |
| 30018 | company-30018 | agent-name-30018 | agent-phone-30018 |
| 30122 | company-30122 | agent-name-30122 | agent-phone-30122 |
| 30147 | company-30147 | agent-name-30147 | agent-phone-30147 |
| 30195 | company-30195 | agent-name-30195 | agent-phone-30195 |
| 31110 | company-31110 | agent-name-31110 | agent-phone-31110 |

*Figure 28-7: Interaction Between Transaction File and Master File*

As available computer power increased and user interfaces improved, interactively updated databases became more common. In this case, a data entry worker types transactions into an on-screen form, directly modifying the underlying master file. Built-in range and consistency checks on each field minimize the chances of entering incorrect data. With the advent of fast, reliable computer networks and intelligent remote devices, transaction entries may come from other software packages, other computers, or portable electronic devices, often without human intervention. Databases can now be kept literally up-to-the-minute, as in airline reservation systems.

Since an update request can now arrive for any record at any moment (as opposed to the old batch environment where a computer administrator controlled when updates happened), the risk of two people or devices trying to update the same information at the same time has to be guarded against. File and record locking schemes were developed to block access to a file or record under modification, preventing other users from operating on it until the first user's changes were complete.

Other database operations include searching for records meeting certain criteria (e.g., with values for a certain variable greater than a threshold) or sorting the database (putting the records in a different order). Searching is done via queries, as already discussed. A sort can be in ascending order (e.g., A to Z) or descending order (Z to A). You can also do a sort within a sort (e.g., charge number within department) (see Figure 28-8).

**A-Z Sort by Lastname**

| Lastname | PO Number |
|---|---|
| Anderson | 38192844 |
| Anderson | 28691877 |
| Anderson | 31243896 |
| Harris | 31219925 |
| LeMoyne | 36645119 |
| LeMoyne | 30042894 |
| Parrish | 38456712 |
| Williams | 29943851 |

**Ascending Sort by PO Number Within A-Z Sort by Lastname**

| Lastname | PO Number |
|---|---|
| Anderson | 28691877 |
| Anderson | 31243896 |
| Anderson | 38192844 |
| Harris | 31219925 |
| LeMoyne | 30042894 |
| LeMoyne | 36645119 |
| Parrish | 38456712 |
| Williams | 29943851 |

**Z-A Sort by Lastname**

| Lastname | PO Number |
|---|---|
| Williams | 29943851 |
| Parrish | 38456712 |
| LeMoyne | 30042894 |
| LeMoyne | 36645119 |
| Harris | 31219925 |
| Anderson | 38192844 |
| Anderson | 31243896 |
| Anderson | 28691877 |

**Ascending Sort by Purchase Order Number**

| Lastname | PO Number |
|---|---|
| Anderson | 28691877 |
| Williams | 29943851 |
| LeMoyne | 30042894 |
| Harris | 31219925 |
| Anderson | 31243896 |
| LeMoyne | 36645119 |
| Anderson | 38192844 |
| Parrish | 38456712 |

*Figure 28-8: Results of Different Sorting Operations*

## 28.9 Special Requirements of Real-Time Process Databases

When the data source is a real-time industrial process, a number of new concerns arise. Every piece of data in a real-time process database is now associated with a timestamp and a location in the plant, and that information must be retained with the data. A real-time process reading also has an "expiry date" and applications that use that reading must verify that it is still good before using it. Data also come in many cases from measuring instruments which introduce concerns about accuracy and reliability.

In the case of a continuous process, the values in the database represent samples of a constantly changing process variable. Any changes that occur in the variable between sample times will be lost. The decision on sampling frequency is a trade-off between more information (higher sampling rate) and compact data storage (lower sampling rate). Many process databases allow you to compress the data, as discussed earlier, to store more in a given amount of disk space.

Another critically important feature of a real-time process database is the ability to recover from computer and process upsets and continue to provide at least a basic set of process information to support a safe fallback operating condition, or else an orderly shutdown. A process plant does not have the luxury of taking hours or days to rebuild a corrupted database.

Most plants with real-time process databases archive the data as a history of past process operation. Recent data may be retained in disk storage in the plant's operating and control computers; older data may be written onto an offline disk drive or archival storage media such as CDs. With today's low costs for mass storage, there is little excuse not to retain process data for many years.

## 28.10 Data Quality Issues

Data quality is a matter of fitness for intended use. The data you need to prepare a water quality report for a governmental body will be different from the data required for fast-response control of a paper-machine wet end. In the broadest sense, data quality includes not only attributes of the numbers themselves, but how accessible, understandable and usable they are in their database environment. Figure 28-9 shows some of the dimensions, or aspects, of data quality.

| Quality Category | Quality Dimensions |
|---|---|
| Intrinsic | Accuracy, Objectivity, Believability, Reputation |
| Accessibility | Access, Security |
| Contextual | Relevancy, Value-Added, Timeliness, Completeness, Amount of Data |
| Representational | Interpretability, Ease of understanding, Concise representation, Consistent representation |

*Figure 28-9: Aspects of Data Quality*

Data from industrial plants is often of poor quality. Malfunctioning instruments or communication links may create ranges of missing values for a particular variable. Outliers (values which are grossly out-of-range) may result from transcription errors, communication glitches, or sensor malfunctions. An intermittently unreliable sensor or link may generate a data series with excessive noise variability. Data from within a closed control loop may reflect the impact of control actions rather than intrinsic process variability. Figure 28-10 illustrates some of the problems that may exist in process data. All these factors mean that data must often be extensively preprocessed before statistical or other analysis. In some cases, the worst data problems must be corrected and a second series of readings taken before analysis can begin.

## 28.11 Database Software

Many useful databases are built using off-the-shelf software such as MS Excel and MS Access. As long as query and report requirements are modest and real-time interaction with other computers or devices is not needed, this can be a viable and low-cost approach.

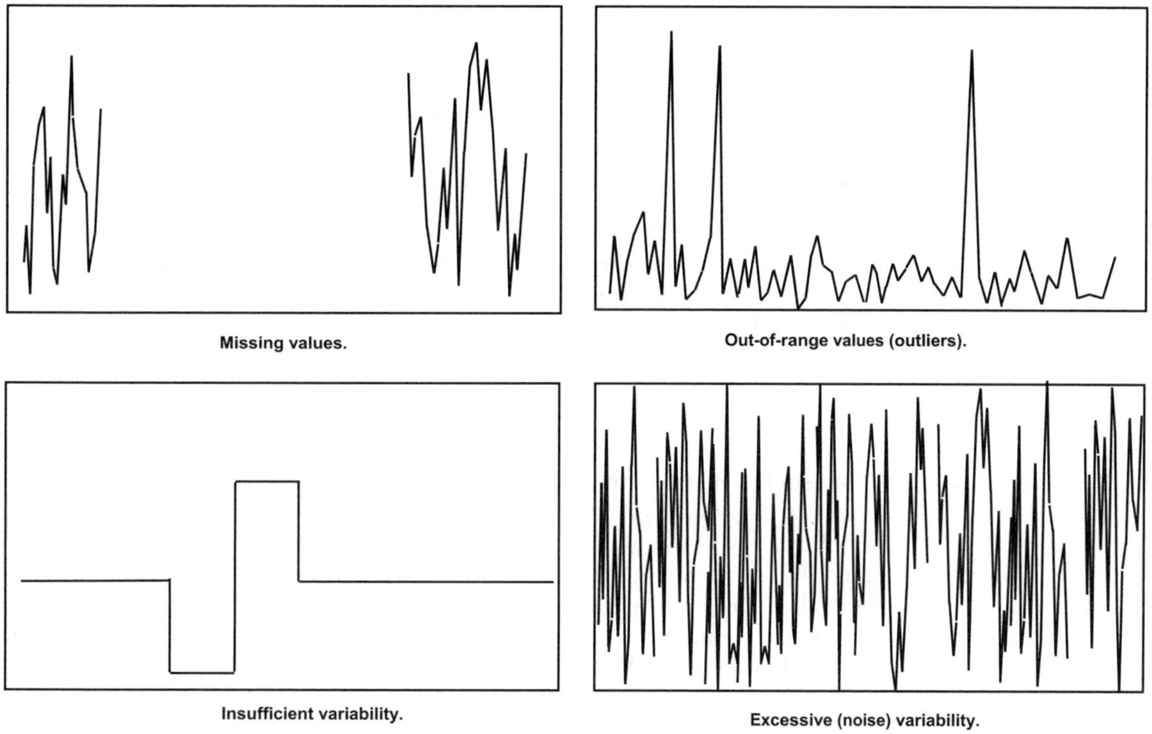

Missing values. Out-of-range values (outliers).

Insufficient variability. Excessive (noise) variability.

*Figure 28-10: Common Problems with Process Data Quality*

The next step up in sophistication is general-purpose business databases such are Oracle. If you choose a database that is a corporate standard, your database can work seamlessly with the rest of the enterprise data environment and use the full power of its query and reporting features.

However, business databases still do not provide many of the features required in a real-time process environment. A number of real-time process information system software packages exist, either general in scope or designed for particular industries. They may operate offline or else be fully integrated with the millwide process control and reporting system. Of course each level of sophistication tends to entail a corresponding increase in cost and complexity.

## 28.12 Data Documentation

Adequate data documentation is a frequently neglected part of database design. A database is a meaningless mass of numbers if its contents cannot be linked to the processes and products in your plant or office. Good documentation is especially important for numerical fields such as process variable values. At a minimum, the following information should be available: location and frequency of the measurement; tag number if available; how the value is obtained (sensor, lab test, panel readout, ...); typical operating value and normal range; accuracy and reliability of the measurement; and any controllers whose operation may affect the measurement. Process time delays are useful information, since they allow you to lag values and detect correlations which include a time offset. A process diagram with measurement locations marked is also a helpful adjunct to the database.

## 28.13 Database Maintenance

Basic ongoing maintenance involves regular checks of the data for out-of-range data and other anomalies which may have crept in. Often the first warning of a sensor malfunction or dropout is a change in the characteristics of the data it generates. In addition, changing user needs are certain to result in a stream of requests for modifications to the database itself or to the reports and views it generates. A good understanding of database structure and functioning are needed to implement these changes while maintaining database integrity and fast, smooth data access.

Version upgrades in the database software pose an ongoing maintenance challenge. All queries and reports must be tested with the new version to make sure they still work, and any problems with users' hardware or software configurations or the interactions with other plant hardware and software must be detected and corrected. Additional training may be needed to enable users to benefit from new software features or understand a change in approach to some of their accustomed tasks.

## 28.14 Data Security

Data have become an indispensable resource for today's businesses and production plants. Like any other corporate asset, they are vulnerable to theft, corruption or destruction. The first line of defense is to educate users to view data as worthy of the same care and respect as other, more visible corporate assets. Protective measures such as passwords, firewalls, and physical isolation of the database servers and storage units are simply good practice. Software routines that could change access privileges, make major modifications to the database, or extract database contents to another medium must be accessible only to authorized individuals. Regular database backups, with at least one copy kept offsite, will minimize the loss of information and operating capability in case of an incident.

## 28.15 References

Date, C. J. *An Introduction to Database Systems*. Seventh Edition. Addison Wesley Longman, 1999.

Gray, J. "Evolution of Data Management." *IEEE Computer,* October 1999. pp. 38-46.

Harrington, J. L. *Relational Database Design Clearly Explained.* Second Edition. Morgan Kaufmann, 2002.

Litwin, P. *Fundamentals of Relational Database Design*. 2003. http://r937.com/relational.html.

Stankovic, J.A., S. H. Son, J. Hansson. "Misconceptions About Real-Time Data Bases." *IEEE Computer,* June 1999. pp. 29-36.

Strong, D. M., Y. W. Lee, R. Y. Wang. "Data Quality in Context." *Communications of the ACM*. Vol. 40, no. 5 (May 1997). pp. 103-110.

Wang, R. Y., V. C. Storey, C. P. Firth. "A Framework for Analysis of Data Quality Research." *IEEE Transactions on Knowledge and Data Engineering.* Vol. 7 (1995), no. 4. pp. 623-640.

## About the Author

**Diana C. Bouchard** (Varanal Data Analysis) offers statistical data analysis services on a consulting basis, as well as scientific and technical writing, editing and translation. She holds an M.Sc. (Computer Science) degree from McGill University in Montreal and worked for 26 years as a scientist in the Process Control Group at the Pulp and Paper Research Institute of Canada (Paprican). Her activities at Paprican included modeling and simulation of kraft and newsprint mills, expert system development, and multivariate statistical data analysis. In the context of the Process Integration Chair at Ecole Polytechnique, she has lectured on steady state and dynamic simulation and multivariate data analysis.

# 29 Software

*By Jonas Berge*

## Topic Highlights

*Benefits, Savings & Doubts*
*Setup*
*Configuration*
*System Integration & Migration*
*Troubleshooting*
*Operation & Applications*
*Availability & Compliance*

A modern automation system needs more than just configuration and monitoring capability. Software infrastructure has therefore become an increasingly important criterion in recent years for selecting an automation system. Because the applications have to work together, an open software architecture is even more important.

## 29.1 Introduction & Overview

Software is the operator display, but software also handles advanced control, data logging, reports, etc.

### 29.1.1 Automation Software

Different plant people need different information. Software is a key part in the information architecture, displaying process variable readings to operators and computing key performance indexes to managers. Standard software interfaces permit applications to exchange data.

Several technologies are involved in open software architecture (see Figure 29-1). In a client-server scheme, a client application displays data from a server application such as a database or hardware source. A basic client-server interface is in the Windows operating system provided by Microsoft's Component Object Model (COM), extended by Distributed COM (DCOM) between networked computers. To provide simpler browsing and exchange of live automation system data, such as process variables, the OLE for Process Control-Data Access (OPC-DA) technology was developed by the OPC Foundation. Some software, particularly non-automation software, still rely on an old Windows technology called Dynamic Data Exchange (DDE) for live data exchange, but it is not as fast or easy to use as OPC. Alarms and events are propagated between applications using the OPC-A&E subscription and filter technology. Database engines are used to store configurations and historical plant data.

At a low level, database interface is based on Structured Query Language (SQL), but a higher interface based on Open DataBase Connectivity (ODBC), OLE for DataBase (OLE_DB), or ActiveX Data Objects (ADO) is used by most applications that store and retrieve data. To make access to historical data easier, process historians use OPC-Historical Data Access (OPC-HDA) that permits browsing and aggregation of data. When incompatible subsystems must be integrated, such as a main control system and a package unit using Ethernet control networks based on different application protocols, the OPC-Data eXchange (DX) technology can be used for one system to write data in the other. DCOM is not firewall-friendly. Therefore Web technologies are used to bring information beyond the automation system. HyperText Transfer Protocol (HTTP) is a firewall-friendly transport protocol for the Internet, ideal for this task. Common basic formats to exchange data on the Internet are HyperText Markup Language (HTML) and eXtensible Markup Language (XML). Based on this, the OPC Foundation developed the Web services based OPC-XML-DA to communicate live data through a firewall, such as from a control system to the rest of the enterprise, albeit with lower performance than DCOM. The upcoming OPC-Unified Architecture (UA) will provide fast and firewall-friendly interface for live and historical data, as well as alarms and events. Complex sequences such as start-up, shutdown, and batch operations are best programmed as scripts. Visual Basic for Applications (VBA) is a scripting language used in many automation applications. ActiveX is a component technology that includes Windows controls, such as a button, as well as animated symbols such as pumps that can be included in graphics that are OLE containers. Applications that support OLE Automation make it possible to include other applications in the graphics, such as an Excel spreadsheet.

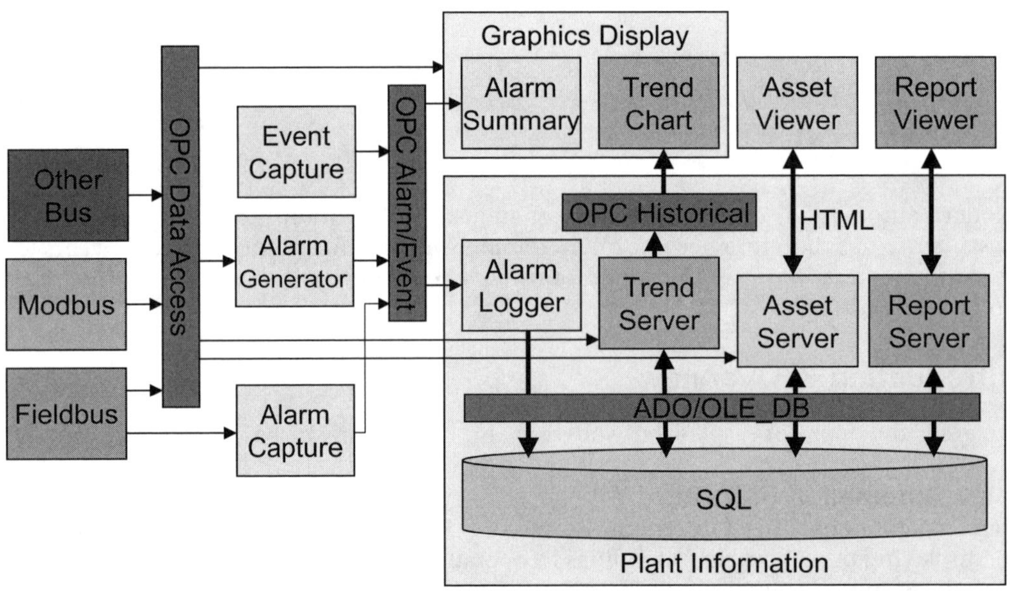


*Figure 29-1: Multiple Software Technologies Make Up an Open Information Architecture*

Software ensures data integrity by storing digitally, rather than recording on paper charts.

Open interfaces like OPC are supported to a different degree. Open systems are built entirely around OPC, while proprietary systems may have OPC only as an external horizontal integration interface to otherwise monolithic software.

### 29.1.2 Software Hierarchy

The software at the automation-level can be characterized as tightly integrated, low volumes of data near real-time, while the software at the enterprise-level is loosely coupled, high volumes of transaction-based data.

### 29.1.3 History of Automation Software

In the past, custom drivers had to be developed to make different applications exchange data, if at all possible. Users were frustrated at not being able to freely select any given combination of hardware and different software applications. Manufacturers were frustrated with endless development of new drivers. As a result, the OPC Foundation was formed to develop a technology to solve this problem. The ubiquitous Windows platform and the OPC interface have now made application connectivity much easier.

### 29.1.4 Evolution of Software Architecture

Legacy distributed control systems (DCSs) had proprietary monolithic software, where third-party applications could not take the place of existing software. "PLC+HMI" solutions were open, but needed laborious configuration work. The OPC interfaces now provide connectivity that is both open and easy to use.

## 29.2 Benefits, Savings & Doubts

An open software architecture enables both capital expenditure (CAPEX) and operational expenditure (OPEX) savings.

### 29.2.1 Realizing Benefits from Software Standards

The OPC interfaces have sparked the creation of a vast array of different automation applications that can work together with all kinds of hardware (see Figure 29-2). All the necessary applications can be combined together, regardless of supplier, to form a complete solution. More can also be added later to meet future needs. Bridging software enables horizontal integration of subsystems and package units. Quality connectivity is ensured by test and certification of client and server products.

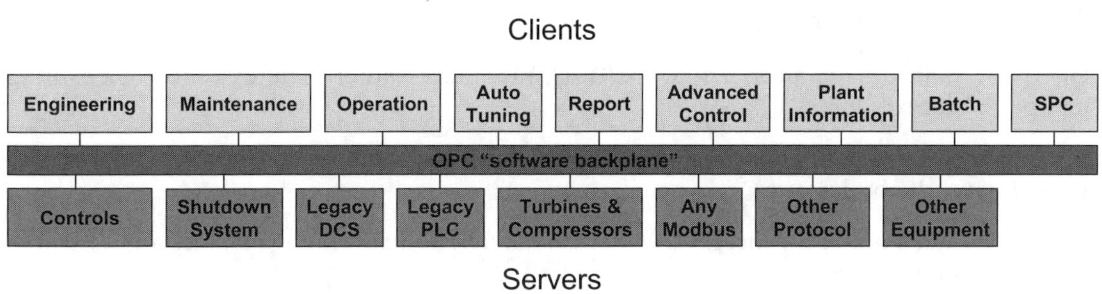


*Figure 29-2: Open Process Automation System Software Architecture*

The ease and performance of OPC makes access to all relevant data feasible, enabling systems to do more. Untimely manual data collection is eliminated. Web technologies enable secure connection to the enterprise level through firewalls. All client applications access the tags in the OPC servers by the same name as in a single integrated database. OPC clients and servers "plug-in," sharing data without using drivers or having to map registers to tags. From the client you point and click to tags in the server, all clients see the same data organized the same way.

Software interfaces eliminate the need for human involvement in data transfer, thus eliminating data reading and re-entry errors.

### 29.2.2 Reducing Capital Expenditure (CAPEX)

OPC has lowered cost of software, as scores of different drivers need not be developed. Eliminating proprietary interfaces ensures market-based software pricing. Tightly integrated applications no longer mean a monolithic DCS; this can also be achieved by software based on OPC. Rather than paying more for something you don't need in order to achieve compatibility, OPC permits you to select the right fit for the job from a wide selection of applications. A Manufacturing Intelligence System (MIS) based on a SQL server can be achieved at low cost and can in many cases be used in place of MES or ERP.

OPC reduces integration effort and cost and eliminates the expense of custom drivers. Training costs are reduced, as OPC is very easy to use and works the same across a wide range of different applications. OPC troubleshooting tools are also available from multiple suppliers at market-based prices, or even for free.

The transparency of OPC makes it easier for a new project team to pick up after the original team at the time of expansion. Since OPC eliminates monolithic software, it becomes possible to exchange applications one-by-one as capacity requirements grow and capability needs change.

### 29.2.3 Reducing Operational Expenditure (OPEX)

Open interfaces allow data from different hardware and databases to be pulled together. New software can then be added to improve control and execution while some operators can be reallocated.

Based on actual condition data, predictive maintenance can reduce cost, and technicians can spend more time fixing actual problems. Reduced vendor dependence ensures market-based service pricing. As legacy systems become hard and expensive to maintain, OPC permits a step-by-step migration, replacing operator workstations, printers, etc.

### 29.2.4 Software Concerns Addressed

Security may be achieved by several means. First, connect the control system to the enterprise network only if really necessary. Implement network security using De-Militarized Zone (DMZ) firewall solutions and software security by employing different authorization levels and password authentication.

Applying a new system based on OPC need not be disruptive. OPC servers for legacy DCS make it possible for new systems to coexist with old, smoothing the migration.

OPC performance is not a concern. COM/DCOM is binary, rather than text, thereby ensuring excellent performance.

Technology getting out of date need not be a worry. Stick to open interfaces and use the latest products to stave off obsolescence.

Availability need not be a concern. OPC can be made redundant for availability and software can run on fault-tolerant servers.

Integration with other operating systems is possible. OPC-XML-DA and the upcoming OPC-UA enable integration with non-Windows platforms.

## 29.3 Setup

Although software is now easy to install, attention must be paid to licensing, as well as networking across firewalls and domains.

### 29.3.1 Application Execution

OPC servers typically should always be running and are often executed as an NT service or set to automatically start and login at power on.

### 29.3.2 System Networking

DCOM-based OPC is used within the automation system. Connection to the enterprise system or Internet/intranet is done through a firewall, interfacing using Web technologies instead. External connection is also possible dialing in to a remote access server (RAS) on the network. Yet another multi-user option is to use a terminal server scheme, meaning that simple computers, or "thin clients," are used as dumb terminals, while all processing is done by a powerful server. The terminal server scheme is centralized and, thus, has availability and performance issues.

### 29.3.3 DCOM Setup

Windows network access security can either be based on a workgroup or a domain scheme. Workgroup is convenient only for systems involving very few computers. Authentication and authorization is best configured by creating groups for different user roles such as operators and administrators.

When OPC clients and servers are located in different computers, Windows DCOM and firewall settings must be modified in both client and server to enable access. First, the default configuration for the computer as a whole is done, followed by custom configuration of properties, security, and protocols for the relevant components.

Typically DCOM connections are not made outside the automation system. However, when it is necessary to bring OPC through a firewall or across media with latency, this can be done using tunneling software. Options include tuned timeout and retry times, use of a single port, and wrapping in Web technologies.

### 29.3.4 License

Most automation software is license controlled using soft key or hard key. A soft key is a character string entered into the software. It is obtained from the manufacturer based on a machine specific code obtained through the software at the time of installation. A hard key is a piece of hardware that plugs into a computer port. The license key controls what features can be used, how many tags it may have, etc. Since the soft key is machine-specific, it may not be trivial to change from a failed machine to a new one.

## 29.4 Configuration

OPC is easy to use from the client end. The server has to be configured to provide the application specific tags.

### 29.4.1 Live Data Interface (OPC-DA and DDE)

Ideally an OPC-DA server is native to the specific hardware and all the tags are automatically configured in the server from the device configuration tool. For example, a specific PLC control strategy builder configures the list of tags organized hierarchically, just as in the control strategy. In lieu of a native server, a generic protocol server can also be used, but requires manual configuration. For example, manual port and register configuration is required for a Modbus server. Universal servers can be configured for rare but simple protocols such as those employed in weighing scales. Older software use DDE instead of OPC-DA and, at best, perform semi-automatic configuration.

### 29.4.2 Historical Database Interface (OPC-HDA)

An open system trend logger application may use OPC-DA to access live data and log it to a database, while at the same time providing access for clients through OPC-HDA. OPC-HDA may also be used as an open interface to a proprietary historian application.

### 29.4.3 Live Alarm & Event (OPC-A&E)

OPC-A&E has some terminology quirks, such as priority being called severity. An alarm is called a condition, and an event is called an occurrence.

An open system alarm generator may use OPC-DA to access live data and compare it to trip levels, while at the same time using OPC-A&E to propagate alarms and events to the clients according to set

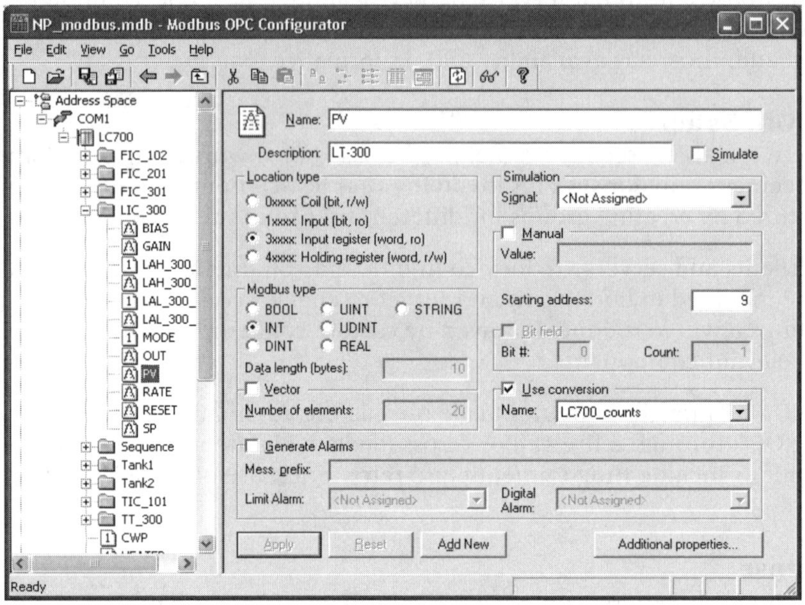


*Figure 29-3: Manually Configured Generic Modbus OPC Server (Screenshot: ICONICS Modbus OPC server)*

filter criteria based on source and priority. An OPC-A&E server may also capture alarms and events from networks supporting such reporting.

There is no OPC interface for historical alarms and events. Historical A&E interface may instead be provided through, for example, OLE_DB.

### 29.4.4 OPC Marshalling

Clever OPC software exists to meet the diverse needs of applications. For horizontal integration between subsystems and package units with OPC-DA servers, a bridge application, through which one server passes data to the other, is used. Another means to achieve this, which has not yet become popular, is the OPC-DX technology that permits one server to write the other directly. Similarly, the software provides for global variables shared by many clients. The application can perform mathematical functions, such as scaling and arithmetic based on multiple tags. Lastly, the application can act as a funnel, aggregating data from multiple servers for clients supporting limited number of servers.

## 29.5 System Integration & Migration

Existing control systems found in plants are typically proprietary with little or no connectivity. Although the long-term objective is to replace legacy DCSs, it may not be feasible short-term.

### 29.5.1 Automation System Coexistence & Migration

By the time a plant expands, the original control system model is not compatible with the new system. OPC servers are available for many legacy DCSs, enabling coexistence with new systems as well as a first phase migration replacing the workstations. Using OPC, data from the old and new hardware can be shown in the same client graphics.

### 29.5.2 Programming

In the past, system integrators were often required to write special software in order to integrate a system. This may, for example, have included writing a device driver for a PLC, or weighing scale to talk to a DCS, or for advanced process control software to talk to a DCS. Other examples may include writing software for batch control or statistical process control. These days, thanks to OPC, special drivers are no longer required since most applications are compatible with each other right out of the box. Similarly, all the different kinds of applications you may need are available with OPC from some supplier. Moreover, most operator visualization software has embedded VBA scripting that permits relatively sophisticated functions and has access to all the OPC data.

In the rare cases that a system integrator actually has to write software, for best connectivity, it should support key software technologies used in off-the-shelf software. This includes: applicable flavors of OPC, as well as ODBC; the software should also be an ActiveX component or an OLE automation container. This inevitably leads to the use of the C++ or C# programming languages and the Visual Studio development environment.

## 29.6 Troubleshooting

Software has its own set of problems.

### 29.6.1 DCOM Troubleshooting

DCOM configuration will almost certainly be wrong unless instructions are followed methodically. Problems that may occur include OPC servers that don't start, don't supply data, or launch in multiple instances. For security reasons, no error messages are shown. Tools are available to assist in DCOM setup and troubleshooting.

### 29.6.2 OPC Troubleshooting

Every OPC tag includes a status that makes OPC troubleshooting simple once DCOM is set up. Problems such as incorrect server configuration are easy to detect. A stopped OPC server automatically restarts. If no client uses the OPC server, it will shut down automatically.

OPC includes mandatory and optional features. Therefore, simple clients and servers may not behave like their more sophisticated counterparts, but they all work with a guaranteed minimum of functionality.

Tools are available to assist in OPC troubleshooting (see Figure 29-4).

### 29.6.3 Other Troubles

Automation systems have very long mission times; they run for years without stopping. Installing new software may replace critical files with a later version, resulting in failures. Poorly designed application may not be releasing unused memory resulting in an eventual failure.

## 29.7 Operation & Applications

Lots of software and hardware specifically designed for automation is available. These applications and platforms can bring about savings, thanks to lower costs of operation and increased efficiencies.

### 29.7.1 Automation Level Software

OPC has made it easy to add a vast selection of software applications to a control system. A key application is operator visualization software that includes graphics, alarms, trends, reporting, etc. This application is often called supervisory control and data acquisition (SCADA), or human-machine interface (HMI), software. Large-scale data collection, beyond the trending in the HMI, is done by a process information management system (PIMS), a.k.a. historian software. Advanced process control (APC) software controls multiple dependent variables considering multiple constraints. Real-time optimization (RTO) software determines optimum operating points for APC. Inferential measurement software estimates values, based on simple measurements, which cannot be measured directly. Opera-

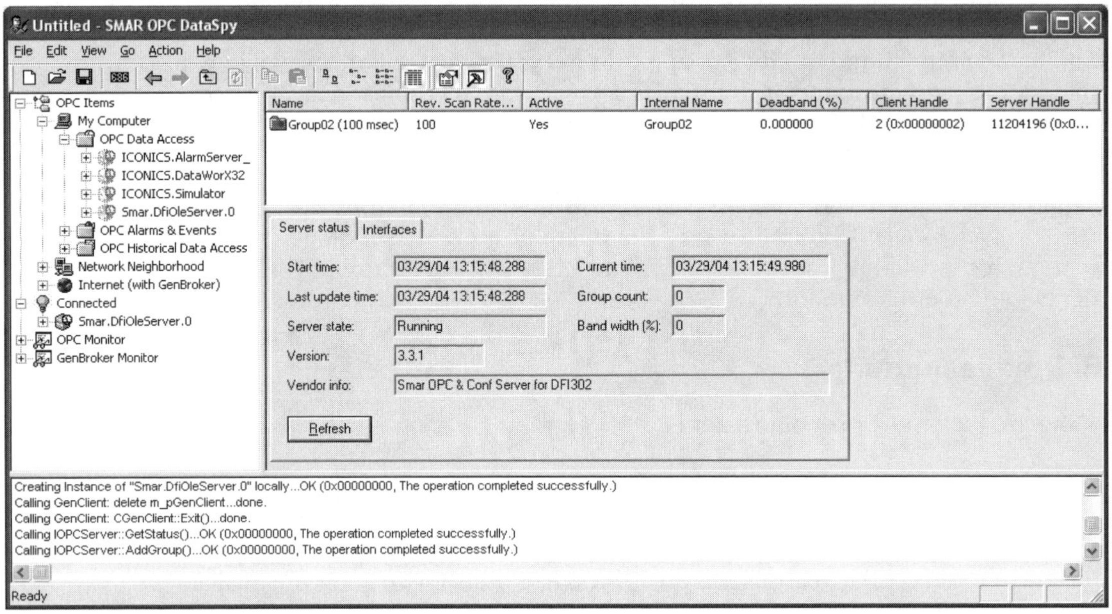


*Figure 29-4: OPC Server Diagnostics (Screenshot: SMAR SYSTEM302)*

tor training software simulates the process enabling training, without risk. Alarm management software prioritizes and filters out nuisance alarms that would otherwise distract operators. Multimedia alarming software announces alarms through mobile messaging and the Internet, etc. Auto-tuning software tune PID controllers (see Figure 29-5). Batch execution software handles multiple streams, recipes, and processing equipment. Statistical process control (SPC) software detects special causes for process fluctuations. Many other specialized applications are available for tank farm management, pipeline leak detection, control, computing, etc.

### 29.7.2 Data Servers

Several third-party companies specialize in making OPC servers for legacy DCSs, making it possible for old systems to coexist with new. OPC servers are available for all major industrial networks and interfaces including FOUNDATION™ Fieldbus, PROFIBUS, Modbus, etc. Special OPC servers exist for Ethernet switch management and Windows task manager. OPC gateways to DDE servers and clients also exist. Universal servers can be configured for simple protocols used in barcode scanners and weighing scales. There are also OPC server simulators generating different waveforms for test.

### 29.7.3 Industrial Computer Peripherals

A vast array of industrially hardened displays, touch screens, keyboards, and pointing devices such as mouse and touch pads are available for use in the harsh industrial environment. Available solutions also include panel-mounted displays, multiple screens, hazardous area screens, and display for sanitary conditions. Control center consoles and display walls are available for the control room.

## 29.8 Availability & Compliance

High plant availability is a must in most industries. Regulatory compliance for electronic records is a requirement in the pharmaceutical industries.

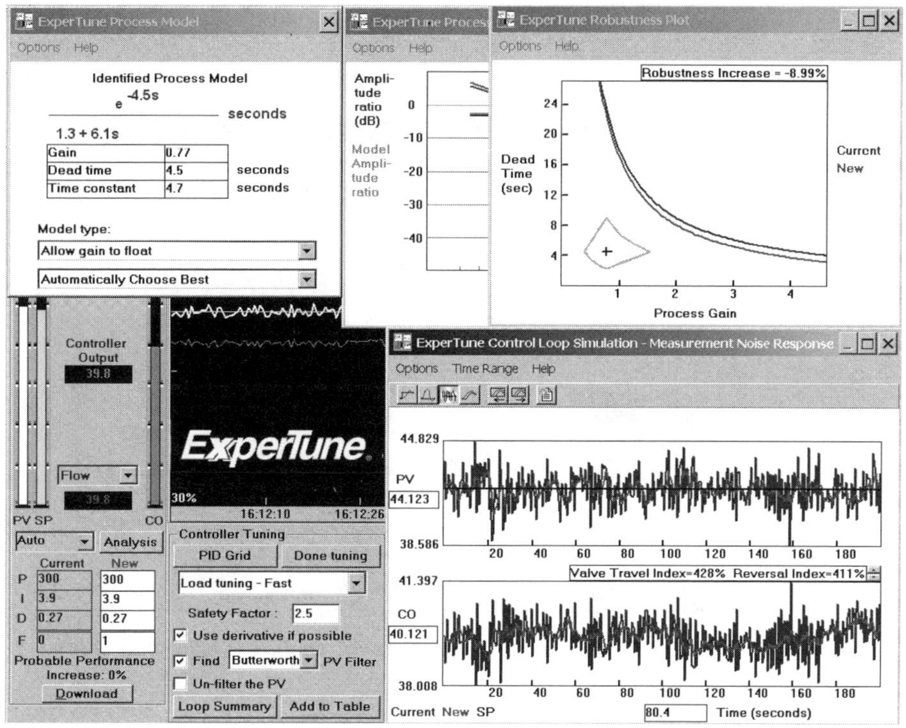


*Figure 29-5: OPC-Based Auto-Tuning Software (Courtesy ExperTune)*

### 29.8.1 Availability

Medium-to-large control systems will have multiple workstations and servers that are networked together using an Ethernet LAN. Greater availability can, to a certain extent, be achieved using a ring topology based on special switches, rather than the regular star topology. However, ring topology still has several single points of failure. The highest availability can be achieved using full LAN redundancy where all switches are duplicated and all devices have dual Ethernet ports. Availability can be further increased using hard disks based on redundant array of inexpensive disks (RAID), redundant power sources and an uninterruptible power supply (UPS), as well as industrial computers hardened for the plant floor.

When redundant OPC servers are used, redundancy management software installed in each workstations access data from the primary and secondary servers and act as a single point from which OPC clients in the machine get their data.

Certain applications, such as batch execution, are not easily implemented in a redundant fashion. In such cases it may be advantageous to instead use a fault tolerant server which has a redundant architecture internally.

### 29.8.2 21CFR11 Electronic Records and Electronic Signatures

One of the main reasons for using software is to put electronic data storage in place of manual records. To comply with the U.S. Food and Drug Administration (FDA) regulation 21CFR11, the system must, among other things, provide electronic signatures consisting of name and password or biometric identifier for these records (see Figure 29-6). Site compliance can be verified by third parties, and is greatly simplified if the software used has built-in features to meet 21CFR11 requirements. Security prevents unauthorized use, and an audit trail logs all activities. Other requirements include recording alarm acknowledgement, enforcement of manual sequences, and revision controls. Many 21CFR11 requirements cannot be met purely by software.

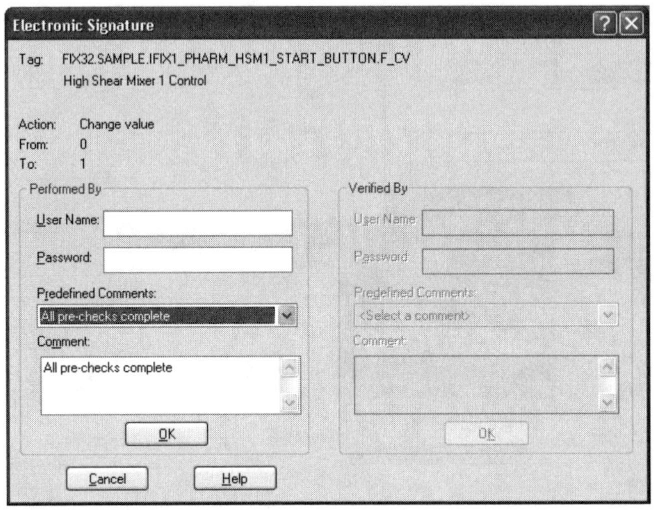


*Figure 29-6: Electronic Signature (Screenshot: GE Fanuc iFix)*

## 29.9 References

Berge, Jonas. *Fieldbuses for Process Control: Engineering, Operation, and Maintenance.* ISA, 2002.

Berge, Jonas. *Software for Automation: Architecture, Integration, and Security.* ISA, 2005.

## About the Author

**Jonas Berge** was educated in Sweden. He is General Manager at Smar's Asia-Pacific headquarters in Singapore and has been working with development and application in the field of instrumentation since 1987. One of the architects of Fieldbus, he has been instrumental in the development of the Foundation Fieldbus specification and also participated in development of Smar's Fieldbus products and system. He is a Senior Member of ISA, VP of the Fieldbus Foundation marketing society in Singapore, and co-founder of the Fieldbus Foundation End-User Council in Singapore. He is the author of the books *Fieldbuses for Process Control: Engineering, Operation and Maintenance* and *Software for Automation: Architecture, Integration, and Security.* He received the 1999 ISA award for excellence in documentation and the 2001 Raymond D. Molloy award.

# 30 Custom Software

*By Dick Caro*

## Topic Highlights

*Specification*
*Programming*
*Revision Control*
*Sourcing*
*Testing*

## 30.1 Introduction

Many processes, particularly those for batch chemical manufacturing, food processing, discrete parts manufacturing, and mechanical assembly operations, must be custom programmed rather than configured from a set of predefined functions offered by the automation system supplier. Programming custom software is a business process requiring organization and management to achieve performance success, on-time completion, and successful long-term support.

## 30.2 Specification

Before writing custom programming, an organization must make several decisions. The primary decision concerns the functions to be performed. You must create a document called the manufacturing specification to describe exactly the desired outcome and methods used to perform the automation functions. Unsuccessful projects are usually traced to incomplete definitions of what is to be done and the ways in which the automation system is to accomplish the desired actions. Successful projects spell out in detail every step of the manufacturing operation, what is to be measured, and the exact flow of the manufactured item. In fact, the manufacturing specification is the base document for testing the completed manufacturing sequence to determine if the system has been programmed correctly.

One highly successful form of the manufacturing specification is a stepwise set of instructions, described as though a human would be performing each step. The benefits of this type of document are that it is more easily diagrammed using Sequential Function Chart (SFC) programming methods and it can be more easily understood by people who will later access the document to maintain, expand, and improve the automation system. This form of document also provides a clear integration path, with the human process/machine operators detailing any manual steps they must take, approvals they must provide, and observations they must make in supervising the automated process. This document then provides the source material for people creating the operator's manual for the process.

In complex projects, you must design the software itself. Software is often composed of a master scheduling program, individual modules that do the work, and a database structured for the real-time and computational data needed and produced by the software. Organization into this form allows the project to be delegated to many programmers. Each program module must itself be specified in a software design specification (SDS) that describes *what* the module is to do, the *results* expected, *how* it will do the work, and how the results can be *tested*. This technique is called *modular programming*.

The modern version of modular programming is called *object-oriented programming*. When object programming languages such as C++ or Java are used, each object is a complete programmed module with both methods and attributes. Some of the attributes are local to the object, are not visible outside the object, and are persistent, while other attributes are made visible to the instantiating program. Object programming relies less on the database for communications of information than conventional programming. Objects must be individually specified and tested.

## 30.3 Programming

The programming language chosen to implement automation applications will usually be selected from one of the IEC 61131-3 languages reviewed in Chapter 9. If this type of programming is appropriate for the application, then it is strongly recommended that the manufacturing specification be encoded into a sequential function chart (SFC) as the first step in programming. Each function block on the SFC can then be expanded into more detailed SFCs to the lowest functional level. Each block of the lowest-level detailed SFCs can be coded in any of the supplier-supported programming languages LD, FBD, IL, or ST as described in Chapter 9.

Although it is not common for automation applications to be programmed in computer programming languages, it may be necessary to write programs to support computer-driven equipment that does not support any of the IEC 61131-3 languages. It is still important to document and plan the program overview using SFC procedures where custom programming is implemented for one of the low-level SFCs when an IEC 61131-3 language is not applicable.

When custom programming is required and the software is more than trivial, the first step is to determine which programming language to use. The choice of programming languages depends strongly on the languages supported by the system on which the programs will be written, and also by the system on which the completed programs will run. They may not be the same. The purpose of this chapter is to provide guidelines to use if programming is required, but the details of each programming language mentioned are beyond the scope of this book.

Table 30-1 is a list of programming languages sometimes used for automation system software and for graphic user interfaces to be used by a process/machine operator. This is not a comprehensive list, because many more programming languages are available, but is provided as guidance.

*Table 30-1: Modern Programming Languages for Automation Systems*

| Programming Language | Platform | Comments |
|---|---|---|
| C | Windows, Linux, Embedded | Detailed system programming and drivers |
| C++ | Windows, Linux, Embedded | Object-oriented programming for systems and drivers |
| Java | Windows, Sun, Linux | Simplified form of C++; requires Java Virtual Machine (JVM) |
| Java Script | Windows, Linux | Web page scripting language |
| C# (C-Sharp) | Windows | Microsoft version of Java, requires C# class library or the .NET engine, supports the ActiveX environment |
| Visual Basic.net | Windows | A visual programming environment for Windows based on use of the .NET engine, supports ActiveX environment. |
| HTML, XHTML | W3C standard web browser | Universal graphical display construction language |
| XML/DTD | Any web browser or database | Universal system-independent tagged data language |

Software projects planned for one of the above programming languages are normally outside the scope and experience of automation engineers. Use one of these languages only for equipment that cannot be integrated using one of the IEC 61131-3 application programming languages or to accomplish some system function unsuited to these languages. Programs written in these programming languages are not easily modified when processes, machines, or user needs change—and such changes always occur.

With those cautions, we can discuss the merits of choosing a programming language. First, certain operating systems (the software controlling the resources of the computer on which your software must operate) favor the choice of certain languages. For example, if Windows is the operating system, then the language should be C# (pronounced "see-sharp") or Visual Basic.NET. If the operating system is Linux or some other version of Unix, then C++ or Java should be the programming language. The programming language C is generally considered obsolete, because it is not object-oriented and is considered more difficult to maintain. C++ is a version of C with added object orientation. Java is a popular choice because it offers the flexibility of C++ but has a few of the more difficult and error-prone constructs removed. If the operating environment is to be a web page (independent of operating systems), then the programming language of choice is usually JavaScript, HTML (Hypertext Markup Language), or XHTML (extended HTML.)

C and C++ will generally result in very efficient code, which is sometimes required for device drivers or software controlling fast hardware or communications functions. Java, C#, and Visual Basic.NET all require the presence of their respective virtual machines, which interpret the language statements at run time. The virtual machines are often part of the resident software because many applications are coded in these languages, but the virtual machines do occupy significant memory for embedded environments such as machine control. Code interpretation for interpreted languages is usually an order of magnitude slower than for programs compiled to machine language such as C or C++. There is no optimal choice of programming language. Your choice should be to use the highest level language (toward the bottom of Table 30-1) possible for the application, suitable for the speed of execution and memory requirements.

The most modern programming trend is to present the results of most program execution to the process/machine operator using web-page techniques. The Internet standards are especially good for the creation, documentation, and validation of web pages to make them independent of the source of the operating system and the web browser chosen for the user. The web standard for data constructs independent of the operating environment is XML. Web pages may be encoded in HTML or XHTML to also be independent of the operating environment. Active web pages require the execution of programs either on the server or in the browser with JavaScript (no relation to Java, except also owned by Sun Microsystems) or VBScript. JavaScript will run on most platforms, while VBScript is primarily used on Microsoft Window's platforms. Web server–side programming is generally done with code generators such as Microsoft Visual Studio or Macromedia DreamWeaver.

## 30.4 Revision Control

Programming has greatly improved in the past several years, but it is still a complex task. Writing perfect programs is very difficult, and even perfect programs must be changed as processes keep changing. This demands that programs be revised frequently to match actual requirements, and to correct errors in both coding and in design resulting from either incorrect specifications or in the implementation of the programs. To make sure you use the most recent revision of a program, you must use proper revision control procedures. In major projects where many programmers are busy building and testing software modules, revision control software should be used to track and document revisions to each module and allow project management to report on progress. However, it is not common in automation systems to manage projects with this level of complexity.

The most important aspect of revision control for any program is to create it with enough internal documentation so that it can easily be understood and maintained later. It is highly likely that software maintenance and enhancement will not be done by the same person who originally creates the program module. Although the software design specification should always present the algorithms and methods to be used by the program, internal program comments must link the program flow to that specification. A successful technique is to actually embed paragraphs from the software design specification into comments of the source program.

One of the reasons for using SFC in the design of batch and discrete manufacturing programs is that the resulting diagram documents the logic and flow of control. When you construct LDs or other program modules into the overall block structure of the SFC, the logic and flow of control remains well documented in this easy-to-read graphical format. It is strongly recommended that the choice of automation system be biased toward systems that make SFC graphical diagramming tools available to the system programmers. Most continuous process control systems (DCS) already support graphical function block diagram strategy builders.

Programmers have many choices to make in coding programs. Some choices make programs run faster or more efficiently, while other choices make the code clearer to the maintainer. Efficient execution is rarely required on today's fast microprocessors. Whenever possible, the programmer should choose programming constructs that make the implementation very clear to the person doing maintenance and enhancement in the future.

The revision control procedures should allow each program module to carry a revision code in both the first line of comments and as a nonexecutable text string appearing as the second line of code in the program. There are three elements to the revision number: a top-level revision that changes only with a specification function change, a second number that changes for each formal release, and a working letter code reflecting each new version being tested. Thus "5.25.aa" will reflect a code module implementing the fifth version of the specification, the twenty-fifth release of that module, and the twenty-seventh revision being tested.

## 30.5 Sourcing

Who should write the specifications and do the actual configuration and programming work? Today, very few user companies have sufficient staff to do the original design specifications, configuration, or programming for automation systems. During the implementation of new construction projects, it is common to assign both control loop configuration and system programming to the suppliers of the major automation systems. This is now called "outsourcing." It is very important that the specifications against which these implementers work reflect the need to maintain and enhance both the control loops and automation programs by the end user in the future using the techniques mentioned in the previous paragraphs.

Peer review is a necessary and frequently overlooked step in outsourcing any project. At each major step of the design and implementation, people knowledgeable about the process and the control system should review the specifications, detailed plans, and even the program code prepared by the outsource suppliers. Persons doing the peer review represent the end users, and work directly for them, although they are frequently consultants hired for this purpose in these days of minimal staffing. It is vital that peer review *not* be undertaken by the outsource supplier, even if assigned to an "independent" unit.

When implementation is assigned to offshore groups, it is even more vital to use strict documentation specifications to assure the future maintainability of the resulting strategies and programs. It is also even more vital to follow formal peer review procedures to assure conformance to the specifications.

## 30.6 Testing

There are two criteria for evaluating an automation system:

1. Does it produce the manufactured product with minimal human involvement?
2. Does it work fast enough that the automation system itself is not a process bottleneck?

The basis for testing is the manufacturing specification described in paragraph 30.2.1. Large custom systems are often tested in a process simulation environment, but most manufacturing automation systems are tested online. In the fluid process industries many of the automation functions are tested with the equipment filled with water rather than the process fluids for safety purposes.

The programmer usually tests custom software as part of the release procedures for software acceptance. Each software module or object must be individually tested prior to integration with other software modules or objects in the automation system. The SDS is the basis for the testing of each software module or object. After integrating all software modules and objects, the entire software project must be tested using the test specifications contained in the SDS.

## 30.7 References

W3C Web site: http://www.w3.org/TR/xhtml1/

## About the Author

**Richard "Dick" Caro** has a BS ChE from University of Florida and a MS ChE from Louisiana State University. He is an independent consultant working for CMC Associates in Acton, Mass. He has worked on automation systems in the chemical and paper industries as well as for Foxboro, Modcomp, Arthur D. Little, and ARC Advisory Group. He was a founding member of the ISA SP50 standards committee in 1985 and managed this committee and the IEC Fieldbus standards committee to complete the ISA/ANSI-50.02 and IEC 61158 Fieldbus standard from 1993 to 2000. He has received three ISA awards for this standards work including the 2000 ISA Standards Award. He is a Life-Fellow of the ISA and an ISA Certified Automation Professional. He has authored three books: *Automation Network Selection, Wireless Networks for Industrial Automation,* and *Consumers Guide to Fieldbus Network Equipment for Process Control.* Caro has also contributed chapters in the book *Process Software and Digital Networks,* as well as other chapters in this *Guide to the Automation Body of Knowledge.*

# VI

# Deployment and Maintenance

Operator Training *continues to increase in importance as systems become more complex, and the operator is expected to do more and more. It sometimes seems that, the more we automate, the more important it is for the operator to understand what to do when that automation system does not function as designed. This topic ties closely with the Modeling topic because simulated plants allow more rigorous operator training.*

*Automation professionals that only work on engineering projects in the office and leave the field work to others may not realize the tremendous amount of work required to get a system operating. Construction staff and plant technicians are doing more and more of the* Checkout, System Testing, and Startup *work today, but that only makes it more important that automation professionals understand these topics.*

Maintenance, Long-Term Support, *and* System Management *takes a lot of work to do well. The difference in cost and effectiveness between a good maintenance operation and a poor one is easily a factor of two and may be much more. Automation professionals need to understand this area more so that their designs can effectively deal with life-cycle cost.*

# 31 Operator Training

*By Joseph A. Ruder*

## Topic Highlights

*Training Process*
*Preparation*
*Training Styles*
*Evaluation and Improvement*

## 31.1 Introduction

A competent operator is a critical element in any manufacturing facility. Even though facilities operate at varying levels of automation, the operators are a key part of that system. The operators' knowledge of how the process is intended to function is crucial in providing a safe, efficient, and cost-effective operation.

It is not sufficient to provide only that training which is mandated by Federal, state or local regulations. The purpose should be to build a knowledge base within the operating personnel. Operators need to be able to respond appropriately to any and all circumstances associated with their work environment.

A recent survey performed by the ARC Advisory Group revealed the trend towards less formal operator training. The exception to this trend was in "Startup-New Processes," which preferred Scheduled Classroom Training. (See Table 31-1).

The favored OJT and use of Operating Manuals are less directed methods and rely on self-training or experience of co-workers. This approach may be adequate to achieve overall performance effectiveness, but it is probably not going to lead to a competitive advantage compared to best in class manufacturers.

*Table 31-1: (Reproduced from ARC Insight # 2005 –21MPH, April 28, 2005—Reprinted with Permission. Courtesy: ARC Advisory Group)*

| Operating Mode | None | OJT | CBT | Scheduled CR | Operating Manual |
|---|---|---|---|---|---|
| Normal Operation | 0.0 % | 82.1 % | 35.7 % | 28.6 % | 50.0 % |
| Startup – Existing Process | 0.0 % | 53.6 % | 25.0 % | 25.0 % | 57.1 % |
| Startup – New Process | 3.6 % | 50.0 % | 32.1 % | 67.9 % | 60.7 % |
| Shutdown - Normal | 10.7 % | 57.1 % | 35.7 % | 14.3 % | 50.0 % |
| Shutdown - Emergency | 10.7 % | 57.1 % | 32.1 % | 25.0 % | 50.0 % |
| Product Transitions | 10.7 % | 60.7 % | 21.4 % | 21.4 % | 39.3 % |
| Abnormal Conditions | 7.1 % | 57.1 % | 39.3 % | 25.0 % | 35.7 % |

OJT – On-the-Job Training

CBT – Computer-Based Training

CR – Scheduled Classroom Training

### 31.1.1 Why We Train

That same ARC survey found the top three objectives for training were to improve safety, increase process knowledge, and improve plant profitability. Running the operation safely was at the top of the list by a two-to-one margin. Operators are considered an important company asset.

## 31.2 Training Process

A successful training program requires planning, execution, and evaluation. As with any project, up-front planning is key. Fitting the training into the big picture is an important consideration.

Even the most informal program requires clear objectives to be effective. Training on a process change or system upgrade must certainly have advantages to the company. Make those benefits clear to the operators. Gaining operator buy-in will make them part of the process and give them reason for the change.

Provide participants with a comfortable non-threatening environment. Where possible, hold training away from outside influences. Distractions during training affect everyone involved.

Evaluation is needed to complete the training cycle. There are two parts to the evaluation. The first is to adequately determine if the operators transferred the knowledge presented to use. This can be formal or informal. It can be performed internally or by a third party. The proper method depends on the critical nature of the subject. The second part is to evaluate the training process itself. This requires feedback from the participants. Both can to be used to adjust and improve the process.

### 31.2.1 Learning Process

There are four elements that must be addressed to ensure learning:

#### Motivation

The instructor needs to establish a rapport with participants to prepare them for learning. Students must not feel offended or intimidated. Motivation requires a friendly open atmosphere. There should be an appropriate level of concern established. This should match the level of importance of the material presented. The level of difficulty should require them to think without frustration.

#### Reinforcement

Positive reinforcement rewards correct behavior. Negative reinforcement may be required to change incorrect behavior or to help "unlearn" an incorrect activity or an old practice. Both may be required to help students retained what they have learned.

#### Retention

If the student sees meaning for information, he or she will attempt to apply it. The ability to interpret and apply correctly means the learning process was successful. This retention is directly related to the amount of practice the student gets during the training process.

#### Transference

This is the final step of the learning process. It is the ability of the student to carry the knowledge outside to a new setting and apply it properly. As with reinforcement, this can take two forms. Positive transference is using the information taught to achieve the desired results. Negative transference occurs when behavior is incorrect, but results in the desired outcome. Negative transference should be evaluated to determine if it is an acceptable approach. If not, it presents an opportunity for retraining and possibly modification to the training process.

Adult learning is a different process than that of children or teenagers. Adults have special needs as learners. There are several different theories on adult learning, but they all focus on some key fundamentals. Adult learning should be based on experiences, be meaningful, and be presented in a positive environment. This type of instruction should focus on the process as well as the content. The instructor needs to take on more the role of a facilitator, or resource, than a lecturer.

Adults have a base foundation of experience and knowledge greater than that of children. This requires them to connect learning to this foundation. In fact, some theories feel it is required for an adult to have prior knowledge to acquire new knowledge.

Adults must see a reason for learning something. They need to see the material has relevance to their situation. They will focus on the aspects that will be useful to them in their work. They are problem-solvers and learn by doing.

Adults need to be shown respect. Instructors must acknowledge the experience base the participants bring to the session. The atmosphere should be non-threatening, using positive reinforcement techniques.

### 31.2.2 Trainers

It is often important to provide instructor training. This is especially important if the body of knowledge about a process is concentrated outside the facility. This "train the trainer" process is normally more intense and requires more time than operator training. It can often help establish a long-term relationship between students and the instructor. If possible, the instructor being trained should have the opportunity to develop training materials for the operators.

#### Know Your Audience

In order to structure the class properly it is important to understand the skill levels of participants. If your new control system is PC-based, and none of the operators have used a mouse or pointing device, terms and techniques such as "Click and Drag" are going to put them at a disadvantage. It would be important to include that basic skills training early in the training with hands-on workshops. It is even more important if the operators have a mixed skill set to adjust training sessions or external training to get everyone at the same level before the main subject information is presented. This goes back to the building-on-knowledge theory of adult learning.

#### Additional Training

If the process is new or significantly different, it is a good idea to offer a modified or overview training to other plant disciplines. For instance, training maintenance personnel on operations can improve their ability to support the system. It is much easier to troubleshoot something when you understand how it is supposed to function.

## 31.3 Preparation

It is hard to put hard numbers to preparation time for a training session. For technical material, it can easily take several hours to a day of development to prepare one hour's worth of training material. Regardless of the time spent preparing, it is important to state the objectives of the session clearly. It is also important to revisit the objectives to keep everyone on track. Allow sufficient time in the schedule to develop visual aids and manuals.

### 31.3.1 Pre-Training Activities

It is important for the students to be prepared, as well as the instructor. Suggestions for pre-training activities or programs to ensure all participants are at similar skill levels can be crucial.

It can also be important to consider who will be trained together. Mixing supervisors or managers with operators can create an atmosphere that suppresses important questions or sensitive issues. Remember the best environment for learning is a non-threatening one.

Communicate what is going to happen in advance. A lot of misinformation can spread quickly when plans are not communicated. This can create "unlearning" requirements before the training even begins.

### 31.3.2 Course Design

As discussed earlier, there are two processes involved: a training process and a learning process. The goal is for the first to facilitate the second.

It is important to limit each session to a few key points that build on each other, or can be related to students' existing knowledge. Allow time for questions and dialogue to draw the class into the discussion. If possible, arrange harder concepts for early in the session. Visual aids help improve retention.

In a classroom setting, the process works well with a balance of lecture, hands-on, discussion, and frequent breaks. Adjust session content if operators are being trained on overtime. Sitting through hours of lecture immediately following an eight-hour shift is not going to enhance the learning process.

On-the-job training (OJT) may benefit from discussion time away from the plant floor environment. It can be very effective in conjunction with classroom instruction.

Self-directed learning programs may work well for some. Not everyone is able to absorb material in that manner. It is important to provide supplemental support for those who need it. This can be in the form of follow-up discussion or assigning a mentor.

## 31.4 Training Styles

Training styles are often dictated by content, budgets, time constraints and a host of other factors. Many of these factors may not have anything to do with the requirements to do a quality job.

Adults generally remember things in the following ways:

- 10% of what they read
- 20% of what they hear
- 30% of what they see
- 50% of what they see and hear
- 70% of what they say and write
- 90% of what they say as they do

These are important keys when determining the mix of training methods to be used. The following methods describe various methods and where they might apply.

### 31.4.1 Operating Manuals

This is more of a distribution of information and self-study approach, which can be useful for more theoretical content. The format could be hard copy or electronic. It may still need to include interaction with the instructor or others in the group that perform the same function.

Electronic or Web technology lends itself to interactive modules. This does not replace the "human touch" factor and instructor reinforcement that is an important aspect in learning.

Retention can be a problem if the only interaction and reinforcement is reading the information.

### 31.4.2 On-the-Job Training

OJT works well for simple changes in operations. It can also be a powerful tool as an adjunct to a more structured training program. It may require follow up time, and also may need group interaction.

This style can cause problems when trying to reinforce abnormal conditions training. Under normal operating conditions, it may be impossible to generate abnormal or emergency conditions for the operator response.

### 31.4.3 Classroom Style

A classroom environment can be very useful setting the tone for the training. Remember the tone of the training helps set expectations and improves retention of participants.

Pure lecture-based training can get boring if it is not tied directly to application-specific activities. Incorporating hands-on activities, both in the field or on a simulator, can really reinforce a point and help retention. A simulator is especially helpful to provide abnormal-condition training. A good practice is to have a lecture, followed by hands-on, followed up by discussion for a few related items at a time.

Keep the atmosphere open and friendly. Engage students. Challenge them. Stimulation breeds learning. Questions are important. Keep a list of questions raised in front of the class. Use a white board or flip chart. Review these questions with other classes if training is performed in shifts. Follow up on unanswered questions.

Have frequent breaks. Make sure refreshments are provided if training is for an extended period. Keep the area comfortable. The focus needs to be on the material.

### 31.4.4 Computer-Based Training

Computer-based training can take two forms. The first form can be as basic as electronic distribution of training manuals. It could also include instructional material, lessons, and question-and-answers sessions. For certain skill sets, this may be an appropriate method.

The second form is coupling instruction with a real-time simulation system, or "virtual plant" environment. This can be extremely useful beyond operator training. Keeping this in mind can help with cost justification of a high fidelity dynamic simulation. Early development and introduction of a high fidelity process simulation into a project can enable faster, safer plant startups—all of which can save money in the long run.

The best all around process simulation is achieved by using the actual control system equipment coupled to a dynamic process model. This gives a much more realistic and functional system. It avoids duplication of effort, since the control system does not have to be emulated. It also provides a testing environment for the control system. Operators training on these systems can also uncover hidden bugs which might manifest themselves as the result of daily use, and not the test-bed environment. It is also the best way to create abnormal or emergency conditions in a safe environment. Hands-on workshops should include these scenarios, as well as normal operations.

Retention is directly affected by the amount of practice during learning. Since they are problem-solvers, hands-on learning with a simulator generates higher levels of retention.

These simulations are not only useful for pre-startup training, they also can be used for refresher training, new operator training, or "what if" scenarios in the future.

### 31.4.5 Dynamic Simulation

It is important to understand what levels of fidelity are available, and their uses.

A simple "tie back" loops outputs back to inputs with some time delay, or filtering, to achieve the simplest form of simulation. This can be useful for system checkout and provide some reasonable level of operator hands-on experience, especially if the process has very simple dynamics. Often operators soon recognize the limits of this type of simulation and lose confidence in its ability to respond as the real process would.

Higher fidelity models that include both mass and energy balances and process reaction dynamics can give the operators a better look and feel of the real system. It can be used for upset conditions and emergency response training that gives operators confidence in both their abilities and the functions of the control system under those conditions. This improves operator retention and allows for transferring their knowledge to the real world. Again these models may already exist in a modeling package—either "steady state" or dynamic. The investment to make them useful for training may be small, compared to the benefits.

## 31.5 Evaluation and Improvement

What is the operator's ability to transfer new knowledge to the workplace? This is the question that needs to be answered to evaluate the overall effectiveness of the training. Often, the only measure of this skill is supervisor observation. It may be necessary to measure this transference more objectively.

Training service companies are available to provide guidance in these areas. Relying solely on internal operator manuals and supervisor evaluations could lead to less than desirable performance in critical situations.

Training is a process. Continued improvement can only occur if feedback is used to refine and adjust. Feedback from students on training relevance, style, level of presentation and retention is important. It can be gathered in the way of anonymous post-training questionnaires or follow-up discussion. Whichever the method, use the information to improve the process.

## 31.6 References

ARC Advisory Group. *ARC Insight* #2005-21MPH, April 2005.

Blevins, Terry, et al. *Advanced Control Unleashed: Plant Performance Management for Optimum Benefit.* ISA, 2003. pages 383-385 and 389.

Dale, Edgar. *Audiovisual Methods in Teaching.* Revised edition. Dryden Press, 1954.

Lieb, Stephen. "Principles of Adult Learning." *VISION*, 1991

## About the Author

**Joseph A. Ruder** is a Principal Control engineer for Nestlé Purina PetCare, St. Louis, Mo. He previously was co-owner of a systems integration company and worked as a systems engineer for a local Fisher Controls representative. He began his career as an E&I engineer at the Monsanto-John F. Queeny Plant. He has worked extensively with batch and continuous controls systems around the world, and has also worked to promote the use of dynamic simulations in control system development and operator training. He has written papers for ISA, Fisher-Provox User Groups, and advance control workshops held by Purdue University. Ruder is a Professional Engineer registered with the state of Missouri. He was recently involved in the development of the ISA Certified Automation Professional program.

# 32 Checkout, System Testing, and Startup

*By Michael Cable*

## Topic Highlights

*Instrumentation Commissioning*
*Software Testing*
*System Level Testing*
*Factory Acceptance Testing (FAT)*
*Site Acceptance Testing (SAT)*
*Safety Considerations*

## 32.1 Introduction

Many automation professionals are involved in planning, specifying, designing, programming, and integrating instrumentation and controls required for process automation. In this section we will describe the various methods of testing the installation, integration, and operation at the component and system level. The goal in the end is to ensure the integrated system functions the way it was intended when everyone got together in the beginning and specified what they wanted the systems to do.

In an ideal world, we could wait until the entire system is ready for operation and perform all testing at the end. This would allow a much more efficient means of testing, with everything connected, communicating, and operational. However, we all know several problems would be uncovered that would lead to long startup delays. Uncovering the majority of these problems at the earliest opportunity eliminates many of these delays and can be corrected at a much lower cost.

An efficient means of testing the system can be developed, resulting in limited duplication of effort by properly planning, communication, and using a standardized approach to instrumentation and control system commissioning. Instrumentation and control system commissioning can be defined as a planned process by which instrumentation and control loops are methodically placed into service. Instrument commissioning can be thought of as building a case to prove the instrumentation and controls will perform as specified.

This chapter does not cover testing performed during software development, as this should be covered in the developer's software quality assurance procedures. However, a formal testing of the function blocks or program code, which will be described later in this section, should be performed and documented. This chapter does not cover documentation and testing required for equipment (for example, pumps, tanks, heat exchangers, filters, and air handling units). The scope for this section begins at the point when instruments are received, panels are installed, point-to-point wiring is completed, and systems have been turned over from construction.

The plan for testing described in this section must consider where the system is being built. There may be several suppliers building skid systems at their facility for delivery to the end user. There may be one main supplier that receives a majority of the components, some of which are used for building panels and skid systems, while others will be installed at the end user's facility. For another project, all components are delivered directly to the end user's facility. Depending on the logistics, some testing may be performed at the supplier's location, even by the supplier, if properly trained. Other testing will be performed at the end user's facility.

The flowchart below illustrates instrument commissioning activities covered in topics 32-1 and 32-2.

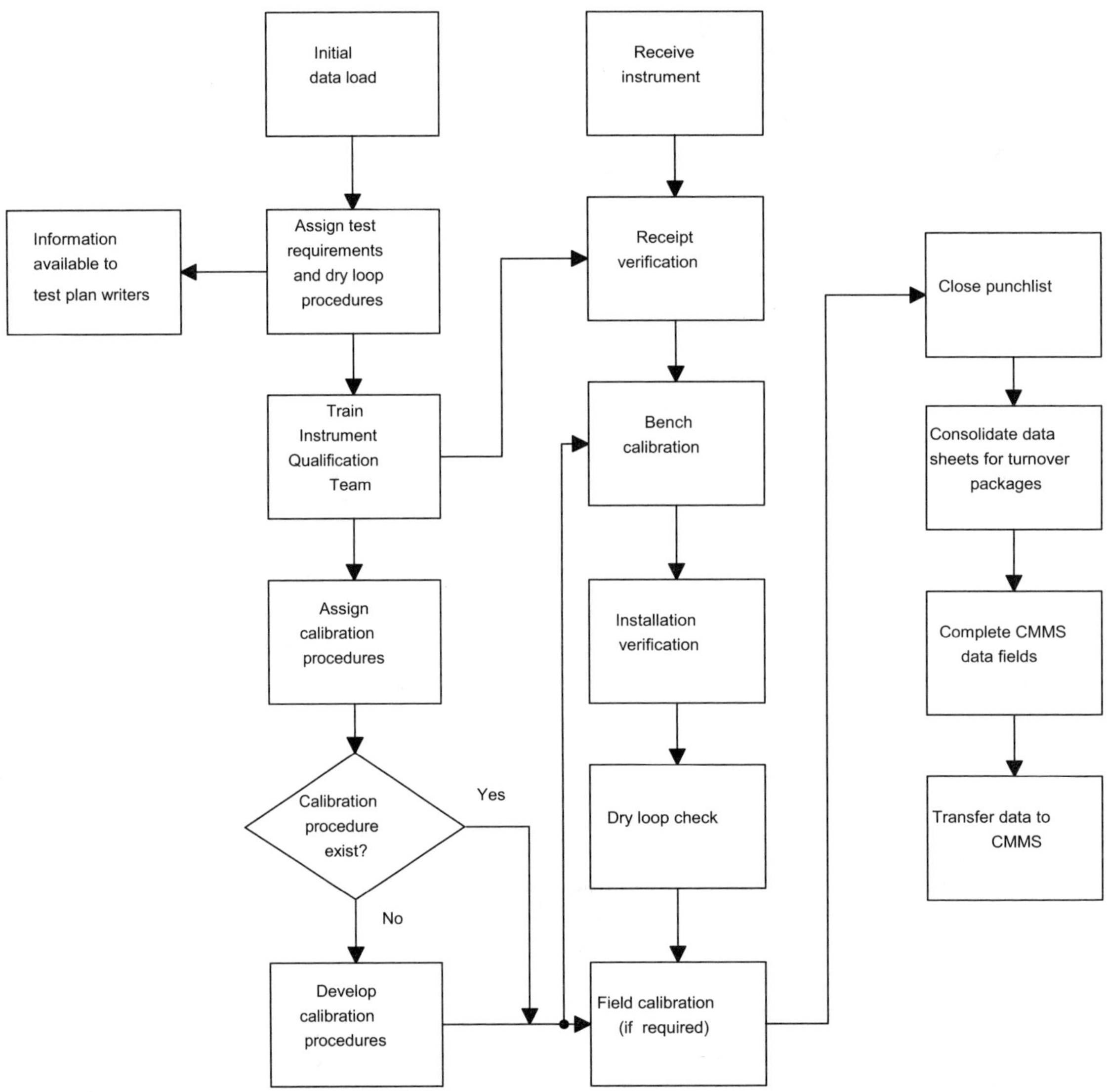


*Figure 32-1: Instrument Commissioning Flowchart*

## 32.2 Instrumentation Commissioning

To begin to build a case, we need to gather evidence that the components will work once they are installed and integrated into the system. At the component level, the first opportunity to gather this evidence occurs when the component is received. Once receipt verification is completed, a bench calibration of calibrated devices can be performed prior to installation. Once installed, instruments should be verified for proper installation. Once all wiring and system connections have been completed and the system has been powered up, loop checks and field calibrations can begin. Preferably, the control system is installed and the control programs are loaded when loop checks are performed. If, for example, the PLC code is not loaded, all testing to verify the proper indications at the human-machine interface (HMI) would need to be duplicated later.

Not all components require all testing. For example, it might make sense to perform all tests for a pressure transmitter, but none of the testing for an alarm light mounted on a panel. You might skip receipt verification and only perform installation verification for a solenoid valve because it is an off-the-shelf common item that would be simple to replace later, if defective. A testing matrix to identify the instrument commissioning activities for each element of the instrumentation and control system should be developed. All instruments and I/O should be accounted for in the testing matrix (i.e., all the diamonds and bubbles for each piping and instrumentation drawing [P&ID]). The I/O listing should be used to verify all control system inputs and outputs are accounted for. Organize the project in a way that makes sense. For example, organize projects with more than one system by P&ID. Use a database program, preferably interfaced with the overall project database, to provide an efficient means of entering the instrument information, required commissioning activities, print test forms, and generate status reports.

A simple example of an instrument commissioning testing matrix is illustrated below.

*Table 32-1: Example Instrument Commissioning Test Matrix*

| Tag# | Description | RV (Yes/No) | IV (Yes/No) | Loop Check (Proc#) | Cal (Proc#) | Vendor, Bench, Field | Loop Tuning (Yes/No) |
|---|---|---|---|---|---|---|---|
| AIT-1002 | Tank 1000 Conductivity Transmitter | Y | Y | AI | CAL-08 | F | N |
| AIC-3453 | Purified Water RO unit pH Controller | Y | Y | AI | CAL-07 | F | Y |
| AY-3453 | Purified Water RO unit pH Control I/P | Y | Y | AO | CAL-02 | F | w/AIC |
| FT-2505 | FRX 2500 Oxygen Flow Transmitter | Y | Y | AI | Vendor | V | N |
| HS-4800 | BRZ-4800 Agitator Hand-Off-Auto Switch | N | N | DI | N/A | N/A | N |
| LIT-4001 | WFI Tank Level Transmitter | Y | Y | AI | CAL-05 | B, F | Y |
| LY-4001 | WFI Tank Level Control I/P | Y | Y | AO | CAL-02 | F | w/LIT |
| LCV-4001 | WFI Tank Level Control Valve | N | Y | w/LY | w/LY | w/LY | w/LY |
| MIT-5108 | Tablet Coater Dew point Indicator/Transmitter | Y | Y | AI | CAL-12 | F | Y |
| MY-5108 | Tablet Coater DP Control Damper Actuator | N | Y | AO | CAL-18 | F | w/MIT |
| PI-1004 | Tank 1000 Pressure Gauge | Y | Y | N/A | CAL-04 | B, F | N |
| PIT-1007 | Tank 1000 Return Loop Pressure Transmitter | Y | Y | AI | CAL-03 | B, F | Y |

| PY-1007 | Tank 1000 Return Loop Pressure Control I/P | Y | Y | AO | CAL-02 | F | w/PIT |
|---|---|---|---|---|---|---|---|
| PCV-1007 | Tank 1000 Return Loop Pressure Control Valve | N | Y | w/PY | w/PY | w/PY | w/PIT |
| PSE-1010 | Tank 1000 Rupture Disk | N | Y | N/A (unless alarmed) | N/A | N/A | N |
| PSV-8111 | Compressed Air Relief valve | N | Y | N/A | Vendor | V | N |
| SIC-4809 | BRX-4800 Agitator Speed | Y | Y | AI | CAL-20 | F | N |
| SV-3080 | Clean Steam to Bioreactor Suite Control Valve Solenoid | N | Y | DO | N/A | N/A | N |
| TE-4003 | WFI Tank Temperature RTD | Y | Y | w/TT | w/TT | w/TT | w/TRC |
| TT-4003 | WFI Tank Temperature Transmitter | Y | Y | AI | CAL-01 | B, F | w/TRC |
| TRC-4003 | WFI Tank Temperature Recorder Controller | Y | Y | w/TT | CAL-10 | B, F | Y |
| TY-4003 | WFI Tank Temperature Control I/P | Y | Y | AO | CAL-02 | F | w/TRC |
| TCV-4003 | WFI Tank Temperature Control Valve | N | Y | w/TY | w/TY | w/TY | w/TRC |
| WIT-200 | Weigh/Dispense Weight Transmitter | Y | Y | AI | CAL-14 | F | N |
| XV-3080 | Clean Steam to Bioreactor Suite Control Valve | N | Y | w/SV | N/A | N/A | N |
| ZSO-3080 | Cln Stm to Brx Suite Control Vlv Open Limit Switch | N | N | DI | N/A | N/A | N |
| ZSC-3080 | Cln Stm to Brx Suite Control Vlv Closed Limit Switch | N | N | DI | N/A | N/A | N |

### 32.2.1 Receipt Verification (RV)

The main objectives of performing receipt verification (RV) are to verify that the device received is the device that was ordered, the device meets the specification for that instrument, and the correct vendor manuals are received for the device. For most projects, instrument specifications are developed and approved prior to purchase. Instrument specifications are developed using ISA-TR20.00.01-2001 - *Specification Forms for Process Measurement and Control Instruments, Part 1: General Considerations – Updated with 20 New Specification Forms in 2004*, or equivalent. The purchase order should also be referenced in case any additional requirements are listed there.

Receipt verification is performed upon receipt of the instrument at the end user's site, supplier's site, or off-site storage location. RV is performed per an approved RV procedure and documented on an approved data sheet printed out as a report from the inputted database information, if applicable. The following minimum requirements should be performed during RV:

- Instrument matches Purchase Order and Instrument Specification
- Manufacturer and model number verified
- Serial number and other nameplate data recorded
- Permanent tag is verified correct and properly applied
- Correct quantity of manufacturers manuals received and logged in

- Any deficiencies noted on RV Data Sheet. Any deficiencies not corrected are added to punchlist

The technical manuals should be organized for turnover to the end user. A good way to organize manuals is by manufacturer and model number. Once the RV is complete, properly disposition the device to bench calibration, designated storage location, or installation.

That being said, RV is optional. It is very convenient to perform the activities listed above at receipt. Even if RV is not performed for most devices, it should be performed for long-lead-time devices to prevent significant delays later. If RV is not performed, all the RV requirements listed above should be performed with the installation verification

### 32.2.2 Installation Verification (IV)

Installation verification (IV) is performed to verify the instrument or device is installed in accordance with project drawings, customer specifications, and manufacturer's instructions. Project drawings may include instrument installation details and P&IDs. For example, it is very important to verify proper orientation for some sensors, such as flow and pressure instrumentation. Pneumatic valves must have air connected to the correct port for proper operation and fail position.

To minimize duplication of effort, IV will typically be performed once all instruments in a system have been installed and authorization to begin has been received from the project manager. Once IV has started, all installation changes must be communicated to the commissioning team. If the start of IV is not communicated, undocumented changes may continue to occur during construction even after the IV has been completed. The following minimum requirements should be performed during IV:

- Instrument is installed per project drawings (i.e., installation detail, P&ID) and manufacturer's instructions.
- Instrument is properly tagged.
- Instrument wiring is properly terminated.
- Instrument air is properly connected, if applicable.
- Instrument is installed in proper location.
- Instrument can be removed for periodic maintenance and calibration (i.e., slack in flexible conduit, isolation valves, RTDs installed in thermowell) Note any discrepancies and whether the discrepancy is a deviation from specification or an observation.
- Any deficiencies should be noted on IV Data Sheet. Any deficiencies not corrected are added to punch list.

### 32.2.3 Loop Checks

An instrument loop is a combination of interconnected instruments that measure and/or control a process variable. An instrument loop diagram is a composite representation of instrument loop information containing all associated electrical and piping connections. Instrument loop diagrams are developed in accordance with standard ISA-5.4-1991 - *Instrument Loop Diagrams*. Minimum content requirements include:

- Identification of the loop and loop components
- Point-to-point interconnections with identifying numbers or colors of electrical and/or pneumatic wires and tubing, including junction boxes, terminals, bulkheads, ports, and grounding connections
- General location of devices such as field, panel, I/O cabinet, control room, etc.

- Energy sources of devices
- Control action or fail-safe conditions

Loop checks are performed for every I/O point. Loop checks are performed to verify each instrument loop is connected properly, indications are properly scaled, alarms are functioning properly, and fail positions are properly configured. A formalized loop check should be documented prior to placing any loop in service. This formalized program should include verifying installation against the loop diagram and simulation of signals to verify output responses and indications throughout the range.

Why is this important? A significant percentage of instrument loops have some problem, some of which would result in hidden failures. As an example, a temperature transmitter output wired to a PLC analog input provides a temperature display on an operator interface. The transmitter is calibrated from 0-100°C to provide a proportional 4-20 mA output. If the PLC programmer writes the code for this input as 0-150°C for a 4-20 mA input and a loop check is not performed, an inaccurate displayed value will result (and possibly improper control action).

Other typical problems found and corrected by performing loop checks include wiring connected to the wrong points, ground loops, and broken wires. Whenever possible, loop checks should be performed at the same time as the field calibration for devices in the loop. The same test equipment will be utilized and some of the same test connections will be used. This makes more efficient use of time and resources.

There are a number of ways to perform loop checks. To perform loop checks with maximum effectiveness, they should be coordinated with the field calibration requirements and should be performed with the control system program, such as programmable logic controller (PLC) code or distributed control system (DCS) program, completed and loaded.

Why is this important? Consider the following.

Let's say a field calibration of a temperature transmitter was performed with the RTD connected and placed in a temperature block. The 4-20 mA transmitter output was checked over the calibrated input range of 0-100°C. No remote indications or alarms were checked during the calibration. That is perfectly normal. There are a few options for the loop check. The loop check could have been performed at the same time as the calibration with the remote indications at the HMI verified and alarms checked. Or, if the field calibration was already completed, the loop check could be performed by using a milliamp simulator connected from the transmitter output to verify remote indications and alarms.

However, if no field calibration is required because the bench calibration was the only calibration requirement, you would have to start the loop check at the RTD to verify all loop components are working together. Of course, it is very important to properly reconnect the loop once the loop check is completed, or the whole thing is null and void.

Let's take the other issue where the control system program is not loaded when performing loop checks. This has been common in my experience, so it takes excellent planning to make sure the program is ready when it's time for loop checks. If the program is not ready and loop check is performed by verifying the correct PLC input displays the correct bits, we have not checked the whole loop. In too many cases I've had to do loop checks multiple times, once to the control system input (just so we could show progress) and again later from the input to verify HMI indications. A little planning would have saved time and money (and a lot of complaining).

Loop checks can be divided into four main categories: analog input, analog output, discrete input, and discrete output. Analog refers to devices that accept or deliver signals that change proportionately (includes analog signals that have been digitalized). Discrete refers to on-off signals. Examples of each signal type include:

*Discrete Input (DI)* - pushbuttons, switches, valve position feedback

*Discrete Output (DO)* - solenoids, alarms

*Analog Input (AI)* - process parameters such as temperature, pressure, level, flow

*Analog Output (AO)* - control output to an I/P or proportional valve, retransmitted analog input

Other loop checks may include RTD input and block valve. The RTD input is an analog input but may require a different test procedure and form. The block valve loop check *can be used for efficiency to test the solenoid valve and valve position inputs all together.*

*RTD Input (RTD)* - RTD input direct to RTD input card without the use of a transmitter

*Block Valve (BV)* - includes testing of solenoid, valve, and valve position feedback switch(es) in one loop check, if desired. (Otherwise, it's OK to do a discrete output check for the solenoid and valve with discrete input checks for the valve position switch(es).

For additional information on loop checks, refer to *Loop Checking: A Technicians Guide*, by Harley M. Jeffery, a part of ISA's Technician Series (2005).

### 32.2.4 Calibration

The ISA definition of calibration is "a test during which known values of measurand are applied to the transducer and corresponding output readings are recorded under specified conditions." To perform a calibration, a test standard is connected to the device input and the input is varied through the calibration range of the instrument while the device output is measured with an appropriate test standard. Once the as-found readings have been recorded, the device is adjusted until all readings are within the required tolerance and the procedure is repeated to obtain as-left readings.

There are a few different times during the project when calibration can be performed. First of all, the instrument vendor can calibrate the instrument to the required specification and provide calibration documentation. In this case, a certificate simply stating the device has been calibrated is not sufficient. A calibration report including calibration data, test standards used, procedure reference, technician signature and date must be included as a minimum. A bench calibration can be performed upon receipt or just prior to installation. A field calibration can be performed once installed and integrated with the instrument loop. A field calibration can be performed at the same time as a loop check, described previously, for increased efficiency. Depending on the instrument and end user requirements, a vendor calibration, bench calibration, and field calibration may be performed for some instruments with only a vendor calibration for other instruments. Here are some suggestions for when to calibrate for commissioning of a new project:

*Vendor Calibration*: The calibration parameters are specified prior to order and bench calibration will typically not be performed for this project.

*Bench Calibration*: Vendor calibration is not performed and/or it is the end-user's practice to perform bench calibrations.

*Field Calibration*: Always, unless it is impossible to access the instrument safely in the field or it is the end-user's practice to perform bench calibrations.

Refer to ISA's *Calibration: A Technician Guide* (2005) for examples of calibration procedures and additional information on the various elements of calibration.

## 32.3 Software Testing

### 32.3.1 Software Development

Software should be developed by the supplier in accordance with approved software development standards and a Software Quality Assurance program. These documents should detail programming standards, version control, internal testing during development, documentation of bugs and corrective actions taken to debug.

### 32.3.2 Program Code Testing

Prior to deployment, the functional elements of the programming code should be testing using simulations. The simulations can be performed by forcing inputs internal to the program and observing outputs. If practicable, it is better to use an external control system simulator setup that can be used to provide simulated inputs and observe outputs. Using a simulator prevents modifying the program for testing purposes, which may lead to errors if the program is not restored to original.

The testing is significantly less demanding if the programming is developed using functional blocks for each specific control actions. For example the ANSI/ISA-88.01-1995 - *Batch Control* standard uses operations or phases in procedural model and equipment logic which can be duplicated for each similar program function. This way, once the functional block is tested once, it does not need to be repeated for other program code that uses the same functional block.

### 32.3.3 Security Testing

Adequate controls must be placed on process control system programs and configurable devices to prevent unauthorized and undocumented changes. The controls are specified during program development or device design. Elements of system security that should be tested at a minimum are login security and access levels.

Login security ensures the appropriate controls are in place to gain authorized access to the program and to prevent unauthorized access. This can be a combination of a user ID and password or use of biometrics such a fingerprint or retina scan. Examples of login security include passwords that can be configured for a minimum number of characters, use of both alpha and numeric characters, force password change at initial login, and require password change at specified intervals. Access should be locked out for any user that cannot successfully log in after a specified number of attempts. All of the specified configurations should be challenged during testing.

Each user should be assigned an access level based on his or her job function associated with the program or device. Examples of access levels include Read-only, Operator, Supervisor, Engineer, and Administrator. Each access level would have the ability to view, add, edit, delete, print, create reports, etc., as applicable to their job function. All of the access levels would be challenged during testing.

## 32.4 System Level Testing

To continue building a case, we need to gather evidence that the components are working together when integrated with the equipment and control systems. At the system level, the first opportunity to gather this evidence is when the instrumentation, controls, equipment, and utilities have been connected, powered up, and turned over from construction.

### 32.4.1 Alarm and Interlock Testing

Alarm and interlock testing is performed to verify all alarms and interlocks are functioning properly and activate at the proper setpoints/conditions. Many of the alarm and interlock tests could be performed with loop checks, since most alarms and interlocks are activated from some loop component or control system output. If alarm and interlock tests are performed separately from loop checks, they are documented as part of the Factory Acceptance Test, Site Acceptance Test, or operational qualification test.

### 32.4.2 Wet Testing and Loop Tuning

Now that we have evidence the specified components are properly installed and calibrated, the loops components are communicating, and the programming has been developed using appropriate quality procedures and tested, we can make a final case by testing the integrated operation of the system. Loop tuning would also be performed with wet testing.

Loop tuning is performed to optimize loop response to setpoint changes and process upsets. Controller **PID** parameters of **P**roportional band (or gain), **I**ntegral, and **D**erivative are adjusted to optimize the loop response and minimize overshoot. Various methods have been developed to perform loop tuning, including the Ziegler-Nichols and trial-and-error method. ISA has several resources for additional information on loop tuning, such as *Tuning of Industrial Control Systems* (Corripio) and *Maintenance of Instruments and Systems* (Goettsche).

Detailed test procedures must be used so wet testing is performed safely. The system must be in operational condition with all utilities connected, and filled with an appropriate medium, if applicable.

Each system should be tested against the specified Sequence of Operations for startup, shutdown, and operations. As an example, the system should be started up and all aspects of the startup sequence verified. In some cases simply pressing the ON button is all that is required to bring the system up to normal operation. In other cases, operator intervention is required during the startup sequence. In either case, all specified operations must be verified.

Once sequence of operation testing is performed, the system should be tested for normal operation, process setpoint changes, process upsets, and abnormal operations (to verify it recognizes improper operating parameters and places the system in a safe condition). Examples include:

*Normal operation* - Tank level drops to the low fill setpoint and initiates a tank filling operation

*Process setpoint change* - temperature setpoint is changed from 40°C to 80°C

*Process upset or disturbance* – heat exchanger steam pressure decreases

*Abnormal operations* – valve(s) out of position, pump not running, filter is clogged

## 32.5 Factory Acceptance Testing (FAT)

Factory acceptance testing (FAT) is a major milestone in a project when the system has been built, the supplier is ready to deliver, and the end-user has an opportunity to review documentation and witness performance testing. FAT is used to verify the system hardware, configuration, and software has been built, assembled, and programmed as specified.

Whenever appropriate, supplier documentation deliverables should be reviewed and testing conducted at the supplier site prior to system delivery. This will allow for troubleshooting and problem resolution prior to shipment, providing a higher level of assurance that the system will meet specifications and function properly upon delivery. In addition, problems found and corrected at the supplier site can be corrected at less cost and with less impact on project schedule. FAT should be performed with end-user representatives from various departments such as manufacturing, engineering, maintenance, information technology, and quality. During testing, systems should be challenged to best simulate actual production conditions. FAT test plan and procedures should be developed by the supplier and approved by the end user prior to testing. Some of the activities to consider during FAT are described below.

### Documentation

Perform a formal review of all project deliverables such as:

- Design specifications
- P&IDs
- Instrument index
- Instrument specifications
- Instrument location drawings
- Loop diagrams
- Instrument installation details
- Sequence of events
- Logic diagrams
- DCS/PLC program
- Operating instructions
- Process flow diagrams
- Instrument, equipment, component technical manuals

### Software

No matter how much review and testing of the programming is performed, it is impossible to test it all. The most efficient use of resources for is to verify:

- Compliance with Software Development Quality Assurance
- Compliance with software programming standard
- Test against software design specifications
- Security and critical functions should be tested during FAT, SAT, commissioning, and/or qualification testing
- Generate and review a sampling of historical data, trend data, and reports for adequacy

### Operator Interface

The end user's ability to interact with the system depends heavily on the effort taken to ensure ergonomic and practical implementation of the operator interface. The following items should be reviewed for each operator interface:

- HMI screen layouts
- Usability
- Readability
- Responsiveness

### Hardware

Hardware testing is performed to verify the system has been built according to approved hardware design specification. In addition, review completed documentation for any instrument commissioning, software testing and system level testing described in sections 32.1 and 32.2 that has been performed.

If at all possible, and many times it is, startup and operate the system. Many suppliers now have the facilities to connect temporary utilities so the system can be operated as if it were installed at the end-user's facility. Take advantage of this opportunity. Any testing completed at the FAT can significantly reduce the testing requirements at the end-user facility which can reduce the burden on an already tight schedule.

## 32.6 Site Acceptance Testing (SAT)

The Site Acceptance Test demonstrates the system is working in its operational environment and interfaces with instruments and equipment from other suppliers. The SAT normally constitutes a repeat of elements of the FAT in the user's environment plus those tests made possible with all process, field instruments, interfaces, and service connections established. A repeat of all FAT testing is not necessary. After all, one of the purposes of FAT is to minimize the testing required at the end-user's facility. Obviously, for systems built at the end-user facility, a FAT is not performed. All elements mentioned previously in the FAT should therefore be performed during SAT. A Site Acceptance Test Plan should be co-developed by the supplier and end-user prior to testing. Some of the activities to consider during SAT are described below:

- Repeat critical FAT elements possibly compromised by disassembly, shipment, and reassembly
- Perform testing that is now made possible with all process, field instrumentation, interfaces, communications, and service connections established
- Interface testing of critical elements of Level 3 System (e.g., MES), if applicable
- Interface testing with critical elements of Level 4 System (e.g., ERP), if applicable.

## 32.7 Safety Considerations

It is worth mentioning safety considerations during commissioning and system testing. Automation professionals who are not permanently assigned to a manufacturing location are typically involved with startup, commissioning, and system testing. In addition, systems are placed in unusual configurations and construction activities are usually still going on during commissioning. Also, more equipment failures occur when initially placed in service, before they are "broken in." For these reasons, more incidents occur during checkout and commissioning. Of course OSHA regulations must be followed, but let's mention a few common sense safety precautions for those engineers that don't often get out in the field.

*Electrical safety* – Technically, any voltage source over 30 volts can kill you. During construction and startup, electrical panels are routinely left open. You should always assume the wires are live until proven otherwise. Taken adequate precautions when working in and around control panels that are energized. For example, remove all metal objects (watches, jewelry), insulate surrounding areas, wear rubber sole shoes, and never work alone. Always have someone working with you (and not in the panel with you!). Depending on the voltages present and risk involved it may be a good idea to tie a rope around yourself so the other person can pull you out, just in case.

*Pressurized systems* – Precautions must be taken for any system or component that could be under pressure. During startup, systems are turned over from construction in an unknown status. Valve line-up checks should be performed to place the system in a known condition. Even then systems are placed in unusual operating conditions to perform testing. Always know what you're working with and always proceed with caution when manipulating system components and changing system conditions. As mentioned before, if a component is going to fail, it will tend to fail when initially placed in service. I've seen, on more than one occasion, tank rupture disks fail the first time a system is brought up under steam pressure. This is even after the rupture disk was visually inspected for defects during installation verification. When those things blow, you better hope you are not in its discharge path. Check system pressure using any available indications before removing a component that would breech system integrity. However, even if a gauge reads 0 psig, carefully remove the component. The gauge could be faulty. And again, **never work alone**. Make sure somebody knows where you are working.

*Extreme temperature* – High- and low-temperature systems, such as steam, hot water, cryogenic storage, and ultra low temperature freezer systems can be very dangerous even if everything is kept inside the pipes, tanks, and chambers. Although insulation minimizes burn risks, there is exposed piping. Do not grab any exposed piping until you know it is safe to touch. We learned this as a kid the first time we touched a hot stove. We seem to have to relearn this for every new project. Even the smallest steam burn hurts for several days. Exposure to cryogenic temperatures such as liquid nitrogen is just like frostbite and steam burn and hurts just as badly.

There are also dangers with using liquid nitrogen in small, enclosed spaces. The nitrogen will displace the air and cause you to pass out and eventually die. The bottom line is to **know what you're working with, be smart, and never work alone**.

## 32.8 References

Cable, Mike. *Calibration: A Technician Guide*. Technician Series. ISA, 2005.

Coggan, D.A. editor. *Fundamentals of Industrial Control.* Second Edition. Practical Guides for Measurement and Control Series. ISA, 2005.

*GAMP Good Practice Guide - Validation of Process Control Systems*. ISPE, 2003.

*GAMP Guide for Validation of Automated Systems in Pharmaceutical Manufacture*. ISPE, 2001.

Jeffery, Harley M. *Loop Checking: A Technicians Guide*. Technician Series. ISA, 2005.

Whitt, Michael D. *Successful Instrumentation and Control Systems Design*. ISA, 2004.

## About the Author

**Mike Cable** is a Level 3 Certified Control System Technician and author of ISA's *Calibration: A Technicians Guide*. Mike started his career as an Electronics Technician in the Navy Nuclear Power Program, serving as a Reactor Operator and Engineering Watch Supervisor aboard the USS Los Angeles submarine. After the military, Mike spent 11 years as a validation contractor, highlighted by an assignment managing instrument qualification projects for Eli Lilly Corporate Process Automation. He is currently the Validation Manager at Argos Therapeutics in Durham, N.C.

# 33 Troubleshooting

*By William L. Mostia, Jr.*

## Topic Highlights

*Logical/Analytical Troubleshooting Framework*
*The Seven-Step Troubleshooting Procedure*
*Vendor Assistance: Advantages and Pitfalls*
*Other Troubleshooting Methods*

## 33.1 Introduction

Troubleshooting can be defined as the method used to determine why something is not working properly or is not providing an expected result. Troubleshooting methods can be applied to physical as well as non-physical problems. As with many practical skills, it is an art but also has an analytical or scientific basis. As such, basic troubleshooting is a trainable skill, while advanced troubleshooting is based on experience, developed skills, information, and a bit of art. While the discussion here centers on troubleshooting of instrumentation and control systems, the basic principles apply to broader classes of problems.

Troubleshooting normally begins with identifying that a problem exists and needs to be solved. The first steps typically involve applying a logical/analytical framework.

## 33.2 Logical/Analytical Troubleshooting Framework

A framework underlies a structure. Logical frameworks provide the basis for structured methods to troubleshoot problems. However, following a step-by-step method without first thinking through the problem is often ineffective. We also need to couple logical procedures with analytical thinking. To analyze information and determine how to proceed, we must combine logical deduction and induction with knowledge of the system, then sort through the information we have gathered regarding the problem. Often a logical/analytical framework does not produce the solution to a troubleshooting problem in just one pass. We usually have several iterations, which cause us to return to a previous step in the framework and go forward again. We can thus systematically eliminate possible solutions to our problem until we find the true solution.

### 33.2.1 Specific Troubleshooting Frameworks

Specific troubleshooting frameworks have been developed which apply to a particular instrument, class of instruments, system, or problem domain. For example, frameworks might be developed for a particular brand of analyzer, for all types of transmitters, for pressure control systems, or for grounding problems. When these match up with your system, you have a distinct starting point for troubleshooting.

Figure 33-1, for example, illustrates a specific problem domain troubleshooting flowchart, or tree, for transmitters.

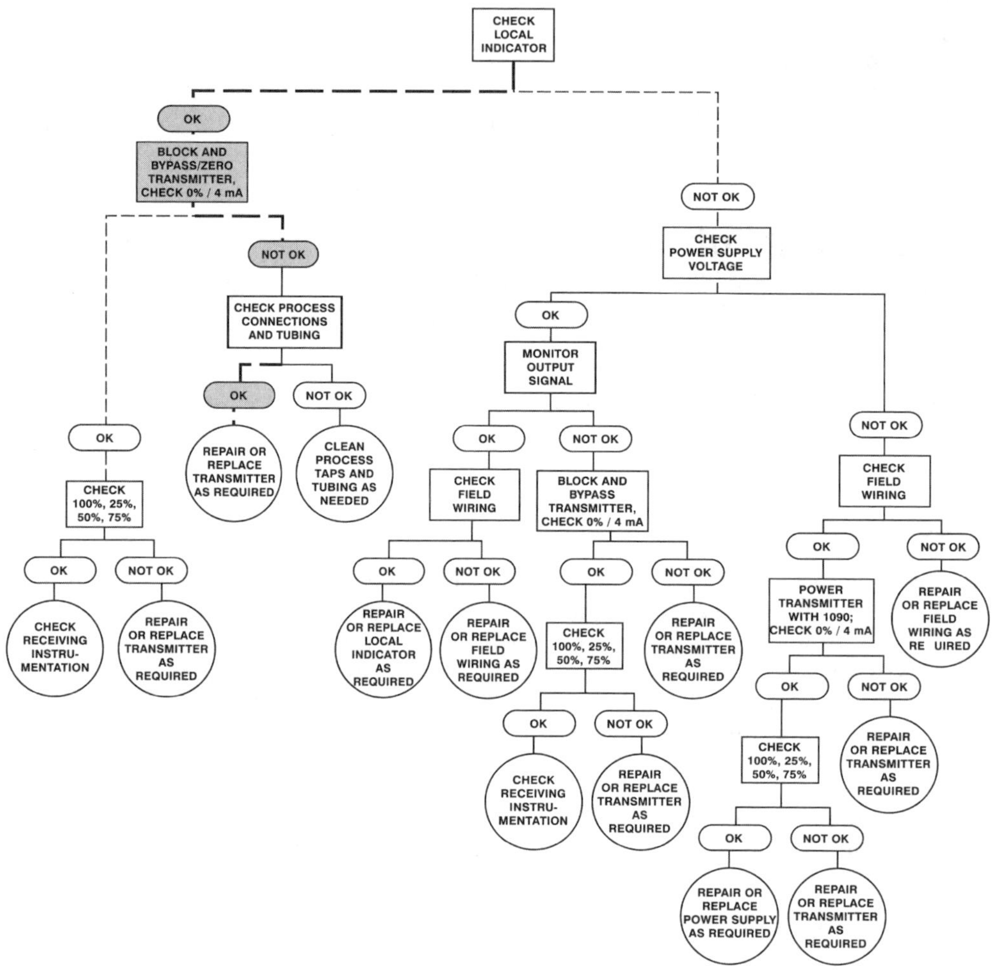


*Figure 33-1: Specific Troubleshooting Framework Example*

### 33.2.2 Generic Logical/Analytical Frameworks

Since we do not always have a specific structured framework available, we need a more general or generic framework that will apply to a broad class of problems. Figure 33-2 depicts this type of framework as a flowchart.

The framework shown in Figure 33-2, while efficient, does leave out some important safety-related tasks and company procedural requirements associated with or related to the troubleshooting process. A few important points should be made here:

- Troubleshooting increases the safety risk to the troubleshooter due to troubleshooting actions that involved on-line systems, energized systems, and moving parts.
- Always make sure what you are doing is safe for you, your fellow workers, and your facility.
- Follow company procedures.

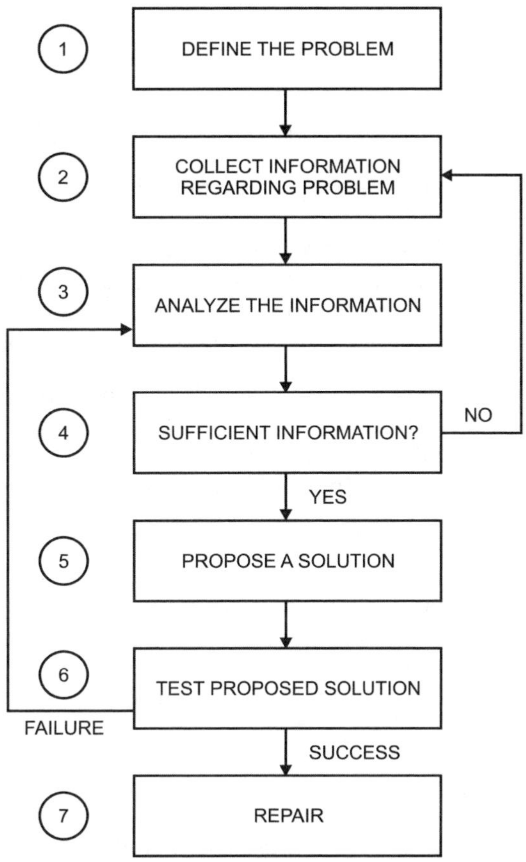


*Figure 33-2: General Troubleshooting Framework Flowchart*

- Get the proper permits, and follow their requirements.
- Always communicate your actions with the operator in charge and other involved people.
- Never put any part of your body anywhere that you do not know exactly what is there.
- Never take unnecessary risks. The life you save may be your own!

## 33.3 The Seven-Step Troubleshooting Procedure

The following seven-step procedure as illustrated in Figure 33-2 provides a generic, structured approach to troubleshooting.

### 33.3.1 Step 1: Define the Problem

You cannot solve a problem if you do not know what the problem is. The problem definition is the starting point. Get it wrong, and you will stray off of the path to the solution to the problem.

#### Communication

When defining the problem, listen carefully and allow the person(s) reporting the problem to you to provide a complete report of the problem as he or she sees it. The art of listening is a key element of troubleshooting. After listening carefully, ask clear and concise questions. All information has a subjective aspect to it. When trying to identify a problem, you have to strip away subjective elements and get to the meat of the situation.

Avoid high-level technical terms or "technobabble." A troubleshooter must be able speak the "language" of the person reporting the problem. This means understanding the process; the plant physical layout; instrument locations; process functions as they are known in the plant; and the "dialect" of abbreviations, slang, and technical words commonly used in the plant. Some of this is generic to process plants in general, while some is specific to the plant in question.

### 33.3.2 Step 2: Collect Additional Information Regarding the Problem

Once a problem has been defined, you should then collect additional information. This step may overlap with Step 1, and, for simple problems, these two steps may even be the same. For complex or sophisticated problems, however, collecting information is a more distinct stage.

Develop a strategy or plan of action for collecting information. This plan should include determining where in the system you will begin to collect information, what sources will be used, and how the information will be organized. Information gathering typically moves from general to specific, though there may be several iterations of this. In other words, you are continually working to narrow down the problem domain.

#### Symptoms

The information you gather typically consists of symptoms (what is wrong with the system) as well as what is working properly. Primary symptoms are directly related to the cause of the problem at hand. Secondary symptoms are downstream effects—that is, not directly resulting from what is causing the problem. Differentiation between primary and secondary symptoms can be the key to localizing the cause of the problem.

#### Interviews and Collecting Information

Typically, a large part of your information gathering will be in the form of interviews with the person(s) who reported the problem and with any other people who may have relevant information. People skills are important here; good communication skills, the use of tact, and non-judgmental and objective approaches can be key to getting useful information.

Then, you review the instrument or system's performance from the control system's faceplates, trend recorders, summaries, operating logs, alarm logs, recorders, and system self-diagnostics. System drawings and documentation can also provide additional information.

#### Inspection

Next, you may inspect the instrument(s) that are suspected of being faulty, or other local instruments (such as pressure gauges, temperature gauges, sight glasses, and local indicators), to see if there are any other indications that might shed light on the matter.

#### History

Historical records can also provide useful information. Your facility's maintenance management system (MMS) may contain information regarding the failed system or ones like it. Also, check with others who have worked on the instrument or system.

#### Beyond the Obvious

If there are no obvious answers, testing may be in order. Plan your testing to ensure it is done safely and gets you the information you need with a minimum of intrusion. When you test by manipulating the system, plan to test or manipulate only one variable at a time. If you alter more than one, you might solve the problem, but be unable to identify what fixed the problem. Always make the minimum manipulation necessary to obtain the desired information. This minimizes the potential upset to the process.

### 33.3.3 STEP 3: Analyze the Information

Once you have collected information, you must start analyzing it to see if you have enough to propose a solution. Begin by organizing what you have collected.

You then can analyze the problem by reviewing what you already know, plus the new information you have gathered—connecting causes and effects, exploring causal chains, applying "If/Then" and "If/Then Not" logic, applying the process of elimination, and applying other relevant analytical or logical methods.

#### Case-Based Reasoning

Probably the first analytical technique that you will use is past experience—you have seen the same problem before. If you have seen this situation or case before, then you know a possible solution. Note that we say "a possible solution" because similar symptoms sometimes have different causes and hence different solutions.

#### "Similar To" Analysis

Compare the system you are working on to similar systems you have worked on in the past. For example, a pressure transmitter, a differential pressure transmitter, and a differential pressure level transmitter are similar instruments. Different PLC brands often have considerable similarities. RS-485 communication links are similar even on very different source and destination instruments. Similar instruments and systems operate on the same basic principles and have potentially similar problems and solutions.

#### "What, Where, When" Analysis

This type of analysis resembles the "Twenty Questions" game. You ask questions about what the gathered information may tell you. These are questions such as:

- What is working?
- What is not working?
- What is a cause of an effect (symptom) and what is not?
- What has changed?
- What has not changed?
- Where does the problem occur?
- Where does it not occur?
- When did the problem occur?
- When did it not occur?

#### Patterns

Symptoms can sometimes be complex and can be distributed over time. Looking for patterns either in symptom actions or lack of action or in time of occurrence can sometimes help in the analysis of symptoms.

#### Basic Principles

Apply basic scientific principles to analyze data—such as electrical current can only flow certain ways. Ohm's and Kirchhoff's Laws always work, mass and energy always balance, physical properties, etc.

### The Manual

When in doubt, read the manual! It may have information on circuits, system analysis, or trouble-shooting that can lead to a solution. It may also provide voltage, current, or indicator readings, test points, and analytical procedures. Often manuals have troubleshooting tables or charts to assist you.

### Logical Methods

Now you will need a logical approach to make this iterative procedure successful. Several approaches, such the linear approach and the "Divide and Conquer" method, may apply.

First, try the linear or walk-through approach. This is a step-by-step process (illustrated in Figure 33-3) that you follow through a system. The first step is to decide on an entry point. If the entry point tests correctly, then you test the next point downstream in a linear signal path. If this test point is all right, then you choose the next point downstream of the previous test point, and so on. Conversely, if the entry point is found to be bad, choose the next entry point upstream and begin the process again. As you move downstream or upstream, each step narrows down the possibilities. Any branches must be tested at the first likely point downstream of the branch.

IN LINEAR ORDER:

STEP #1 - CHECK PROCESS CONNECTION
STEP #2 - CHECK PRESSURE TRANSMITTER
STEP #3 - CHECK DCS INPUT
STEP #4 - CHECK DCS
STEP #5 - CHECK DCS OUTPUT
STEP #6 - CHECK VALVE I/P
STEP #7 - CHECK VALVE

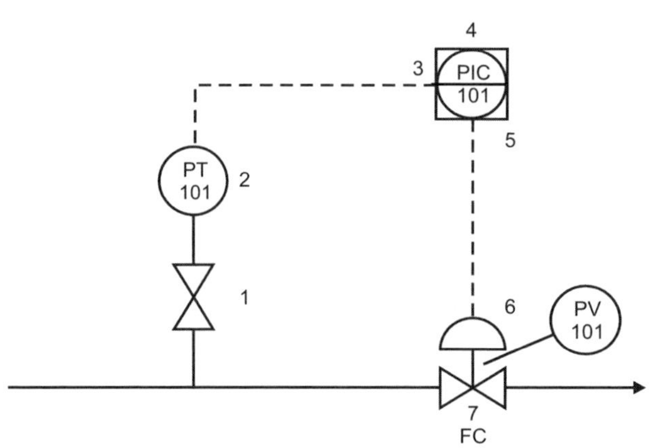


*Figure 33-3: Linear Troubleshooting Approach*

Second, "Divide and Conquer" is a general approach that is illustrated in Figure 33-4. You choose a likely point, or commonly the midpoint of the system, and test it. If it tests bad, then the upstream section of the system contains the faulty part. The upstream section is then divided in two parts and the system is tested at the dividing point. If, on the other hand, the test is good, the downstream section contains the bad part. Divided that in two, and so on, until the cause of the problem has been found.

## 33.3.4 Step 4: Determine Sufficiency of Information

When you are gathering information, how do you know that you have enough? Can you determine a cause and propose a solution to solve the problem? This is a decision point for moving on to the next step of proposing a solution.

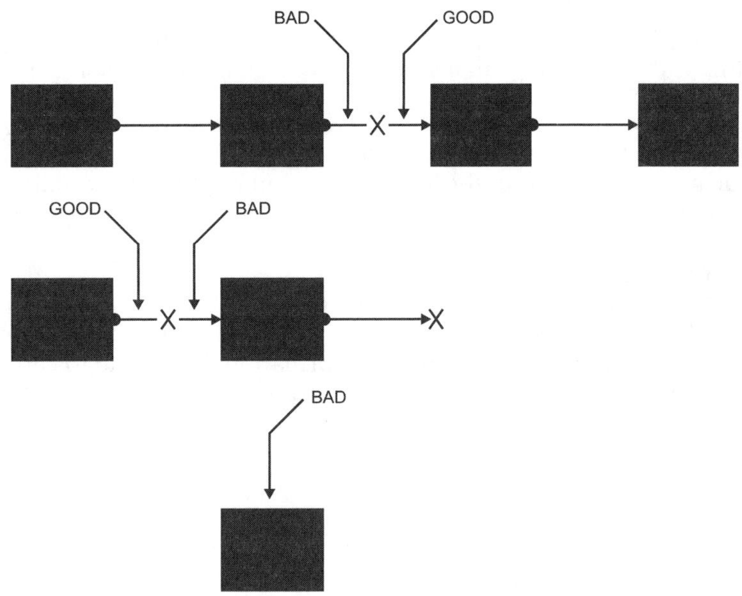


*Figure 33-4: "Divide and Conquer" Troubleshooting Technique (Courtesy of Control Magazine)*

### 33.3.5 Step 5: Propose a Solution

When you believe you have determined the cause of the problem, propose a solution. In fact, you may propose several solutions, based on your analysis. Usually the proposed solution will be to remove and replace (or repair) a bad part. In some cases, however, your proposal may not offer complete certainty of solving the problem and will have to be tested. If you have several possible solutions, propose them in the order of their probability of success. If this is roughly equal, or other operational limitations come into play, you can use other criteria. You might propose solutions in the order of the easiest to the most difficult. In other cases, there may be cost penalties (labor costs, cost of consumable parts, lost production, etc.) associated with trying various solutions; you may propose to try the least costly but viable option.

Do not try several solutions at once. This is called the "Shotgun Approach" and will typically only confuse the issue. Management will sometimes push for a shotgun approach due to time or operational constraints, but you should resist it. With a little analytical work, you may be able to solve the problem and meet management constraints at a lower cost. With the shotgun approach, you may find you do not know what fixed the problem, and it will be more costly both immediately and in the long term. If you do not know what fixed the problem, you may be doomed to repeat it.

Do not rush to a compromise solution proposed by a "committee," either. Consider the well-known "Trip to Abilene" story, illustrating the "group think" concept that is the opposite of synergy. In the story, some people are considering going to Abilene, though none of them really wants to go. They end up in Abilene, however, because everyone in the group figures everyone else wants to go to Abilene. This sometimes occurs in troubleshooting when a committee gets together to "assist" the troubleshooter and the committee gets side tracked by a trip to Abilene.

### 33.3.6 Step 6: Test The Proposed Solution

Once a solution, or combination of solutions, has been proposed, it must be tested to see if your analysis of the problem is correct.

#### Specific Versus General Solutions

However, be careful of specific solutions to more general problems. At this step, you must determine if the solution needed is more general than the specific one proposed. In most cases, a specific solution will be repairing or replacing the bad instrument, but that may not solve the problem.

But what if replacing an instrument only results in the new instrument going bad? Suppose an instrument with very long signal lines sustains damage from lightning transients. The specific solution would be replacing the instrument; the general solution might be to install transient protection on the instrument as well.

#### The Iterative Process

If the proposed and tested solution is not the correct one, then return to Step 3, "Analyze the Information." Where might you have gone astray? If you find your mistake, then move on to propose another solution. If not, move back to Step 2, "Collect Information." It is time to gather more information that will lead you to the real solution.

### 33.3.7 Step 7: The Repair

In the repair step, implement the solution you have proposed. In some cases, testing a solution results in the repair, as in replacing a transmitter, which both tests the solution and repairs the problem. Even in this case, there will generally be some additional work to be done, such as tagging, updating the database, and updating maintenance records, in order to complete the repair. Document the repair so future troubleshooting is made easier.

## 33.4 Vendor Assistance: Advantages and Pitfalls

Sometimes it is necessary to involve vendors in troubleshooting, either directly or by phone. Manufacturer's service personnel can be very helpful (and can provide a learning experience), but some are quick to blame other parts of the system (not their own) when they cannot find anything wrong—in some cases before they have even checked out their own system. Do not let vendors off the hook when trying to solve a problem just because they say it is not their equipment. Ask questions and make them justify their position.

## 33.5 Other Troubleshooting Methods

There are other types of troubleshooting methods available which complement the logical/analytical framework. Some of these are Substitution, Fault Insertion, "Remove and Conquer," "Circle the Wagons," "Complex-to-Simple," "Trapping," Consultation, Intuition, and "Out-of-the-Box" Thinking.

### 33.5.1 The Substitution Method

The substitution method substitutes a known good component for a suspected bad component. For modularized systems or those with easily replaceable components, substitution may reveal the component that is the cause of the problem. First, define the problem and gather information and analyze just as you do in the generic troubleshooting framework. Then, select a likely replacement candidate and substitute a known good component for it. By substituting components until the problem is found, the substitution method may find problems even where there is no likely candidate or only a vague area of suspicion. One potential problem with substitution is a higher-level cause can damage the replacement component as soon as you install it, or a transient (such as lightning) may have caused the failure and your fix may only be temporary. The use of this method can raise the overall maintenance cost due to extra module cost and the cost of inventory of replacement modules.

### 33.5.2 The Fault Insertion Method

Sometimes you can insert a fault instead of a known good signal or value and see how the system responds. For example, when a software communication interface is periodically locking up, you may suspect that the interface is not responding to an I/O timeout properly. You can test this by inserting a fault—an I/O timeout.

### 33.5.3 The "Remove and Conquer" Method

For loosely coupled systems that have multiple independent devices, removing the devices one at a time may help you find certain types of problems. For example, if a communication link with 10 independent devices talking to a computer is not communicating properly, you might remove the boxes one at a time until the offending box is found. Once the problem device has been detected and repaired, the removed devices should be reinstalled one at a time to see if any other problems occur. The "Remove and Conquer" technique is particularly useful when a communication system has been put together incorrectly or exceeds system design specifications. For example, there might be too many boxes on a communication link, cables that are too long, cable mismatches, wrong cable installation, impedance mismatches, or too many repeaters. In these situations, sections of the communication system can be disconnected to see what happens.

A similar technique, "Add Back and Conquer," means removing all the boxes and adding them back one by one until you find the cause of the problem.

### 33.5.4 The "Circle the Wagons" Method

When you believe the cause of a problem is external to the device or system, try the "Circle the Wagons" technique. Draw an imaginary circle or boundary around the device or system; then see what interfaces (such as signals, power, grounding, environmental, and EMI) cross the circle. Then, isolate and test each boundary crossing. Often this is just a mental exercise that helps you think about external influences, which then leads to a solution. Figures 33-5 and 33-6 illustrate this concept.

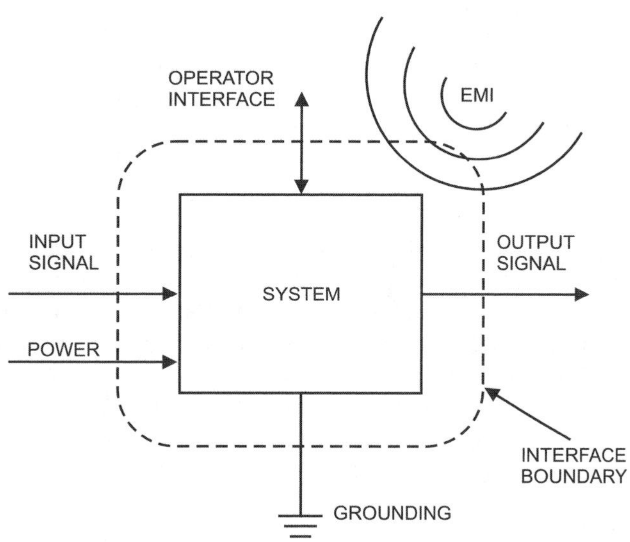


*Figure 33-5: Circle the Wagons—Single Box Example*

### 33.5.5 A Trapping We Shall Go

Sometimes an event that triggers or causes the problem is not alarmed, or is a transient, or happens so fast the system cannot catch it. This is somewhat like akin to having a mouse in your house. You generally cannot see it, but you can see what it has done.

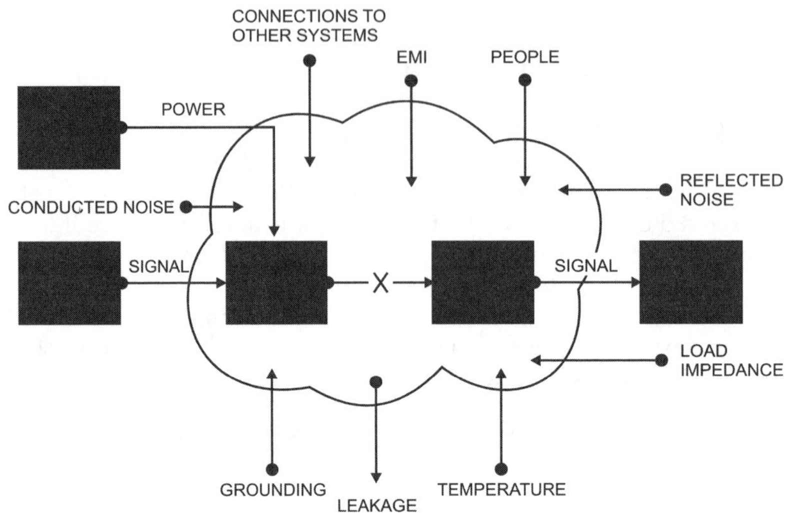


*Figure 33-6: Circle the Wagons—Multiple Box Example (Courtesy of Control Magazine)*

How do you catch the mouse? You set a trap. In sophisticated systems, you may have the ability to set additional alarms or identify trends to help track down the cause of the problem. For less sophisticated systems, you may have to use external test equipment or build a trap. If software is involved, you may have to build software traps that involve additional logic or code to detect the transient or bug.

### 33.5.6 The Complex-to-Simple Method

Many control loops and systems may have different levels of operation or complexity with varying levels of sophistication. One troubleshooting method is to break systems down from complex to simple. This involves finding the simple parts that function to make the whole. Once you find the simplest non-functioning "part," you can evaluate the non-functioning part or, if necessary, you can start at a simple known good part and "rebuild" the system until you can find the problem.

### 33.5.7 Consultation

Consultation, also known as the "third head" technique, means you use a third person, perhaps someone from another department, or an outside consultant, with advanced knowledge about the system or the principles for troubleshooting the problem. This person may not solve the problem but may ask questions that make the cause apparent or that spark fresh ideas for you. This process allows you to stand back during the discussions, which sometimes can help you distinguish trees from forest. The key is to know when you have reached the limitations of your investigation and need some additional help or insight.

### 33.5.8 Intuition

Intuition can be a powerful tool. What many people would call troubleshooting "intuition" certainly improves with experience. During troubleshooting or problem solving, threads of thought in your consciousness or sub-consciousness may develop, one of which may lead you to the solution. The more experience you have, the more threads develop during the troubleshooting process. Can you cultivate intuition? Experience suggests that you can, but success varies from person to person and from technique to technique. Find what works for you!

### 33.5.9 "Out-of-the-Box" Thinking

Difficult problems may require different approaches beyond normal or traditional troubleshooting methods. The term "out-of-the-box thinking" was a buzzword for organizational consultants during the 1990s. "Out-of-the-box" thinking means approaching a problem from a new perspective, not being limited to the usual ways of thinking about it.

The problem in using this approach is that our troubleshooting "perspective" is generally developed along pragmatic lines, i.e., what has worked before; changing can sometimes be difficult.

How can you practice "out-of-the-box" thinking? How can you shift your perspective to find another way to solve the problem? Here are some questions that may help:

- Is there some other way to look at the problem?
- Can the problem be divided up in a different way?
- Can different principles be used to analyze the problem?
- Can analyzing what works rather than what does not work help to solve the problem?
- Can a different starting point be used to analyze the problem?
- Are you looking at too small a piece of the puzzle? Too big?
- Could any of the information on which the analysis is based be in error, misinterpreted, or looked at in a different way?
- Can the problem be conceptualized differently?
- Is there another "box" that that has similarities that might provide a different perspective?

## 33.6 Summary

While troubleshooting is an art, it is also based on scientific principles and is a developed skill that is both trainable and develops with quality experience. An organized, logical approach to troubleshooting is necessary to be successful and can be provided by following a logical framework, supplemented by generalized techniques such as Substitution, "Remove and Conquer," "Circle the Wagons," and "Out-of-the-Box" thinking.

## 33.7 References

Mostia, William L., Jr., PE. "The Art of Troubleshooting." *Control.* (Volume IX, No. 2), February 1996. pp. 65 – 69.

Mostia, William L., Jr., PE. *Troubleshooting: A Technician's Guide*. ISA, 2000.

Goettsche, L.D., editor. *Maintenance of Instruments & Systems.* Second Edition. Practical Guides for Measurement and Control Series. ISA, 2005.

## About the Author

**William L. "Bill" Mostia** has 30+ years of experience in instrumentation, controls, safety, and electrical areas. He is currently a Senior Consultant with SIS-Tech Solutions, a leading consulting firm specializing in hazards & risk analysis, safety instrumented systems, and related engineering. He has worked for Amoco, Texas Eastman, Dow Chemical Co., exida, and as an independent consultant in instrument, electrical, and safety areas. He graduated from Texas A&M University with a BSEE. He is a professional engineer and a senior member of ISA. He is an active member of ISA and serves on a number of ISA standards committees including SP84, SP91, and various SP12 committees. He has published over 50 articles and papers and a book on troubleshooting, as well as been a contributor to several books on instrumentation.

# 34 Maintenance, Long-Term Support and System Management

*By Joseph D. Patton, Jr.*

## Topic Highlights

*Maintenance Is Big Business*
*Service Technicians*
*Big Picture View*
- *No Need Is Best*
- *Evolution of Maintenance*
- *Automatic Analysis of Device Performance*

*Production Losses from Equipment Malfunction*
*Performance Metrics and Benchmarks*

## 34.1 Maintenance Is Big Business

Maintenance is a challenging mix of art and science, where both economics and emotions have roles. Please note that *serviceability* and *supportability* parallel *maintainability*, and *maintenance* and *service* are similar for our purposes. *Maintainability* (i.e., serviceability or supportability) is the discipline of designing and producing equipment so it can be maintained. *Maintenance* and *service* are performing all actions necessary to restore durable equipment to, or keep it in, specified operation condition.

The very word "durable" means the equipment is intended for long life and must therefore be maintained. For the military, a tightening budget coupled with increasing operating and support costs and a drive toward high-technology equipment, means more effort must be invested in maintaining available equipment. A similar situation is occurring in commerce and industry. It is encouraging to note business people in both civilian and government enterprises are paying more attention to life-cycle costs and are at least talking about making the investment required for front-end reliability and maintainability in order to improve system availability and reduce maintenance, repair, operating (MRO) and overall costs.

Organizations that design, produce, and support their own equipment, often on lease, have a vested interest in good maintainability. On the other hand, many companies, especially those with sophisticated high-technology products, have either gone bankrupt or sold out to a larger corporation when they became unable to maintain their creations. Then, of course, there are many organizations such as automobile service centers, TV repair shops, and most factory maintenance departments that have little, if any, say in the design of equipment they will later be called on to support. While the power of these dealers is somewhat limited by their inability to do more than refuse to carry the product line, their complaints generally result in at least modifications and improvements to the next generation of products.

Maintenance is big business. Gartner estimates hardware maintenance and support is $120B per year and growing 5.36% annually. The 10 largest petrochemical producers together spend over $15 billion annually on maintenance, which averages 4.3% of their expected replacement costs. US Bancorp estimates that spending on spare parts is $700B in the U.S.A. alone, which is 8% of gross domestic product.

## 34.2 Service Technicians

Typically, maintenance people once had extensive experience with fixing things and were oriented toward repair instead of preventive maintenance. In the past many technicians were not accustomed to using external information to guide their work. Maintenance mechanics or technicians often focused on specific equipment, usually at a single facility, which limited the broader perspective developed from working with similar situations at many other installations.

Today, service technicians are also called field engineers (FEs), customer engineers (CEs), customer service engineers (CSEs), customer service representatives (CSRs), and similar titles. This document will use the terms "technicians" or "techs." In a sense, service technicians must "fix" both equipment and customer employees. There are many situations today where technicians can solve problems over the telephone by having a cooperative customer download a software patch or perform an adjustment. However, about half of customer calls for service result in a technician traveling to the site, and roughly half of those visits will involve at least one part replacement.

Service can be used both to protect and to promote. Protective service ensures that equipment and all company assets are well maintained and give the best performance of which they are capable. Protective maintenance goals for a technician may include the following:

- Install equipment properly
- Teach the customer how to use the equipment capability effectively
- Provide functions that customers are unable to supply themselves
- Maintain quality on installed equipment
- Gain experience on servicing needs
- Investigate customer problems and rapidly solve them to the customer's satisfaction
- Preserve the end value of the product and extend its useful life
- Observe competitive activity
- Gain technical feedback to correct problems

Service techs are becoming company representatives who emphasize customer contact skills, instead of being solely technical experts. In addition, the business of maintenance service is becoming much more dependent on having the correct part. A concurrent trend is customer demand and service level agreements (SLAs) that require fast restoration of equipment to good operating condition. This is especially true with computer servers, communications equipment, medical scanners, sophisticated manufacturing devices, and similar equipment that affects many people or even threatens lives when it fails.

In focusing on completing a given job, most technicians prefer to take a part right away, get equipment up and running, and enter the related data later. Returning defective or excess parts may be a lower priority, and techs may cache personal supplies of parts if company supply is unreliable. However, there are many situations where on-the-site, real-time data entry and validation are vital to gaining accurate information for future improvement. As a result, a challenge of maintenance management is to develop technology that stimulates and supports maintenance realities.

## 34.3 Big Picture View

Enterprise asset management (EAM) is a current buzzword for the big picture. There are good software applications available to help manage MRO activities. However, most data is concentrated on a single facility and even to single points in time, rather than covering the life cycle of equipment and facilities. As Figure 34-1 illustrates, the initial cost of equipment is probably far exceeded by the cost to keep it operating and productive over its life cycle. Many maintenance events occur so infrequently in a facility that years must pass before enough data is available to determine trends and, by then, the equipment is probably obsolete or at least changed. Looking at a larger group of facilities and equipment leads to more data points and more rapid detection of trends and formation of solutions.

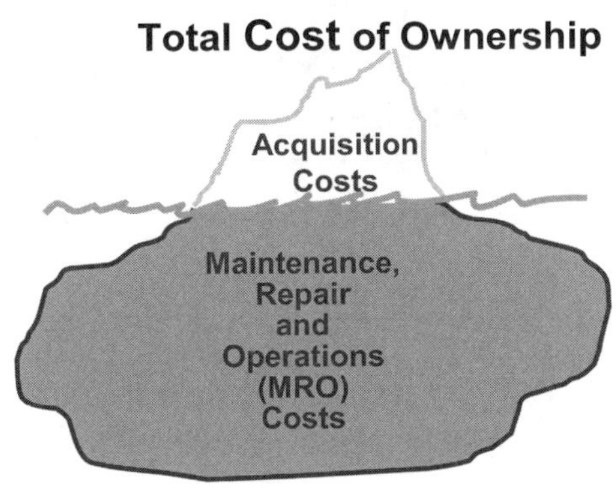


*Figure 34-1: MRO Costs Are Usually Much Larger Than Acquisition Costs*

Interfacing computerized information on failure rates and repair histories with human resources (HR) information on technician skill levels and availability, pending engineering changes, procurement parts availability, production schedules, and financial impacts can greatly improve guidance to maintenance operations. Then, if we can involve all plants of a corporation, or even all similar products used by other companies, the populations become large enough to provide effective, timely information. Optimizing the three major maintenance components of people, parts, and information, shown in Figure 34-2, are all important to achieving that end.

Historically, the two main maintenance costs have been labor and materials (people and parts). Labor costs are increasing. This means organizations must give priority efforts to reducing frequency, time, and skill level and thereby the cost of labor. The costs of parts are also increasing. A specific capability probably costs less, but integrating multiple part capabilities into a single part brings high costs and more critical need for the replaceable costs. A third leg is becoming important to product development and support: information as generally provided by software on computer and communications systems. Digital electronic and optical technologies are measurably increasing equipment capabilities while reducing both costs and failure rates. Achieving that reduction is vital. Results are seen in the fact that a service technician, who a few years ago could support about 100 personal computers, can now support several thousand. Major gains can be made in relating economic improvements to maintainability efforts. Data has been gathered showing a payoff of 50:1; that is a benefit of $50 prevention value for each $1 invested in maintainability.

### 34.3.1 No Need Is Best

Everything will fail sometime—electrical, electronic, hydraulic, mechanical, nuclear, optical, and especially biological systems. People spend considerable effort, money, and time trying to fix things faster.

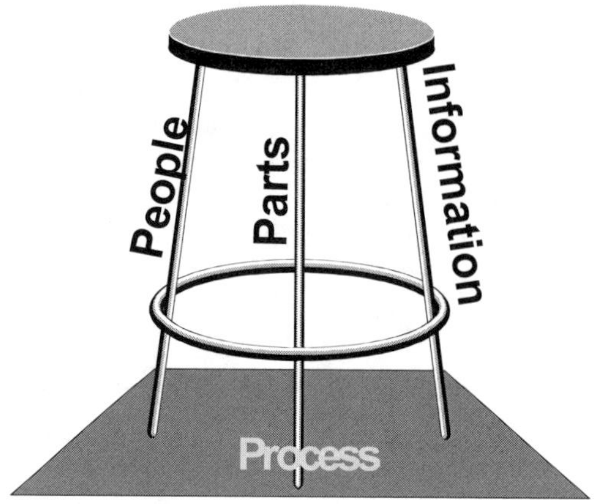


*Figure 34-2: The Three Legs of Support Are People, Parts, and Information*

However, the best answer is to avoid having to make a repair at all. To quote Ben Franklin, "An ounce of prevention is worth a pound of cure." The failure-free item that never wears out has yet to be produced. Perhaps some day it will be, but meanwhile we must replace burned-out light bulbs, repair punctured car tires, overhaul jet engines, and correct elusive electronic discrepancies in computers.

A desirable long-range life-cycle objective is to achieve very low equipment failure rates and require replacement of only consumables and the parts that wear during extended use, which can be replaced on a condition-monitored predictive basis. Reliability (R) and maintainability (M) interact to form availability (A), which may be defined as the probability that equipment will be in operating condition at any point in time. Three main types of availability are inherent availability, achieved availability, and operational availability. Service management is not particularly interested in inherent availability, which assumes an ideal support environment without any preventive maintenance, logistics, or administrative downtime. In other words, inherent availability is the pure laboratory availability as viewed by design engineering. Achieved availability also assumes an ideal support environment with everything available. Operational availability is what counts in the maintenance tech's mind, since it considers a "real world" operating environment.

The most important parameter is failure rate, as a product needs corrective action only if it fails. The main service objective for reliability is mean time between failure (MTBF), with "time" stated in the units most meaningful for the product. Those units could include:

- *Time:* hours, days, weeks, etc.
- *Distance:* miles, kilometers, knots, etc.
- *Events:* cycles, gallons, impressions, landings, etc.

It is important to realize equipment failures caused by customer use should be anticipated in the design. Coffee spilling in a keyboard, a necklace dropping into a printer, and panicked pushing of buttons by frustrated users add more calls for help. Operating concerns by inexperienced users often result in more than half the calls to a service organization. What the customer perceives as a failure may vary from technical definitions, but customer concerns must still be addressed by the business.

For operations where downtime is not critical, the need for a highly responsive maintenance organization is not critical. However, for manufacturing operations where the process performance is directly related to the performance of the automation systems, or any other part of the process, downtime can be directly related to the revenue-generation potential of the plant. Under these conditions, response time represents revenue to the plant itself. Thus, plants that would have revenue generation capacity of $10,000 worth of product per hour, operating in a 24-hour day, seven-day week environment, would be losing approximately $240,000 of revenue for every day that the plant is shut down. A 24-hour response time for plants of this type would be completely unsatisfactory. On the other hand, if a manufacturing plant that operates on a batch basis has no immediate need to complete the batch because of the scheduling of other products, then a 24-hour response time may be acceptable.

A typical automobile, for example, gives more utility at lower relative cost than did cars of even a few years ago; however, it must still be maintained. Cars once required frequent spark plug changes and carburetor adjustments, but fuel injection has replaced carburetion. A simple injector cleaning eliminates the several floats, valves, and gaskets of older carburetors—with fewer failures and superior performance. Computer-related failures that used to occur weekly are now reduced to units of years.

Service level agreements (SLAs) increasingly require that equipment be restored to good operation the same day service is requested, and often specify four hours, two hours, or even faster repair. Essential equipment may cause great hardship physically and financially if it is down for long periods of time.

For example, a production line of an integrated circuit fabrication facility can lose $100,000 per hour of shutdown. A magnetic resonance induction (MRI) scanner that cannot operate costs $4,000 per hour in revenue lost and even more if human life is at risk. Failure of the central computer of a metropolitan telephone system can cause an entire city to grind to a stop until it is fixed. Fortunately, reliability and the availability (uptime) of equipment are improving, which means there are fewer failures. However, when failures do occur, the support solutions are often complex.

### 34.3.2 Evolution of Maintenance

Maintenance technology has also been rapidly changing during recent years. The idea that fixed-interval preventive maintenance is right for all equipment has given way to the reliability-based methods of on-condition and condition monitoring. The process of maintenance is illustrated in Figure 34-3.

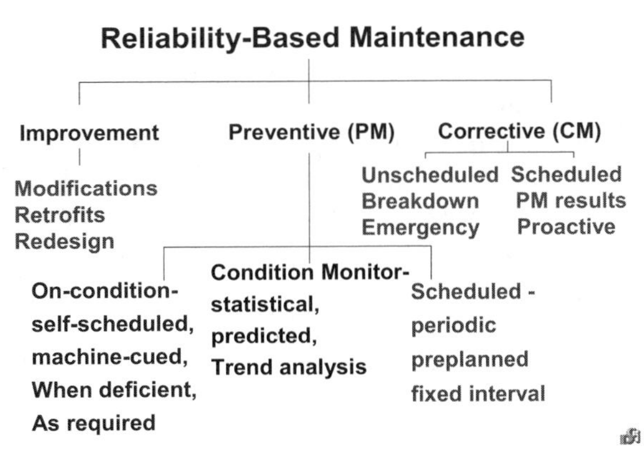


*Figure 34-3: The Branches of Modern Maintenance*

Many parts are now discarded rather than being maintained at organizational or even intermediate levels. The multilevel system of maintenance is evolving into a simplified system of more user participation, local first- and second-level maintenance, and backup direct from a third party or original equipment manufacturer (OEM) service organization. Expert systems and artificial intelligence are being developed to help diagnostics and to predict the need for preventive maintenance. Parts are often supplied directly from vendors at the time of need, so maintenance organizations need not invest hard money in large stocks of parts.

### 34.3.3 Automatic Analysis of Device Performance

There is increased focus and resource deployment to design durable products for serviceability. Durable equipment is designed and built once, but it must be maintained for years. With design cycles of six months to three years and less, and with product lives ranging from about three years for computers through 40+ years for hospital sterilizers, alarm systems, and even some airplanes, the initial investment in maintainability will either bless or haunt an organization for many years. If a company profits by servicing equipment it produced, good design will produce high return on investment in user satisfaction, repeat sales, less burden for the service force, and increased long-term profits. In many corporations, service generates as much revenue as product sales do, and the profit from service is usually greater. Products must be designed right the first time. That is where maintainability that enables condition monitoring and on-condition maintenance becomes effective.

Instruments that measure equipment characteristics are beginning to be directly connected to the maintenance computer. Microprocessors and sensors allow vibration readings, pressure differentials, temperatures, and other nondestructive test (NDT) data to be recorded and analyzed. Presently, these readings primarily activate alarm enunciators or recorders that are individually analyzed. There are, of course, automated control systems in use today that can signal the need for more careful inspection and preventive maintenance. These devices currently are certainly cost effective for high-value equipment such as turbines and compressors. Progress is being made in this area of intelligent device management so all kinds of electrical, electronic, hydraulic, mechanical, and optical equipment can "call home" if they begin to experience deficiencies. Trend analysis for condition monitoring may be assisted by computer records.

Capabilities should also be designed into computer programs to indicate any other active work orders that should be done on equipment at the same time. Modifications, for example, can be held until other work is going to be done and can be accomplished most efficiently at the same time as the equipment is down for other maintenance activities. A variation on the same theme is to ensure emergency work orders will check to see if any preventive maintenance work orders might be done at the same time. Accomplishing all work at one period of downtime is usually more effective than doing smaller tasks on several occasions.

Products that can "call home" and identify the need to replace a degrading part before failure bring major advantages to both the customer and support organization. There are, however, economic trade-offs regarding the effort involved versus the benefit to be derived. For example, the economics may not justify extensive communication connections for such devices as smart refrigerators. However, business devices that affect multiple people need intelligent device management (IDM) with remote monitoring to alert the service function to a pending need, hopefully before equipment becomes inoperable. The ability to "know before you go" is a major help for field technicians, so they have the right part and are prepared with knowledge of what to expect.

It is important to recognize the difference between Response Time and Restore Time. Response Time is the minutes from notification that service is required until a technician arrives on the scene. Restore Time adds the minutes necessary to fix the equipment. Service contracts historically specified only Response Time, but now usually specify Restore Time. Response is action. Restore is results.

The challenge is that, to restore operation, the technician often needs a specific part. Many field replaceable units (FRUs) are expensive and not often required. Therefore, unless good diagnostics identifies the need for a specific part, techs may arrive at the customer location and then determine they need a part they do not have. Diagnostics is the most time consuming portion of a service call. Technicians going to a call with a four-hour restore requirement will often consume an hour or more to complete the present assignment and travel to the new customer. Diagnostics adds even more time, so the techs could easily consume two hours of the four available before even knowing what part is needed. Acquisitioning parts quickly then becomes very important. The value of information is increasing. Information is replacing inventory. Knowing in an accurate, timely way that a part was

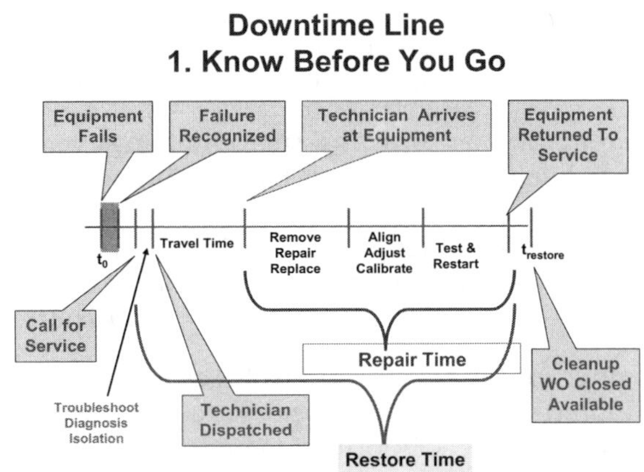


*Figure 34-4: Maintenance Sequence When Tech Knows Needed Resources Before Going to Fix Equipment*

used allows a company to automatically initiate resupply to the authorized stocking site, even to the extent of informing the producer who will supply the warehouse with the next required part.

An organization can fix considerable equipment the next day without undue difficulty. A required part can be delivered overnight from a central warehouse that stocks at least one of every part that may be required. Overnight transportation can be provided at relatively low cost with excellent handling, so orders shipped as late as midnight in Louisville, Ky. or Memphis, Tenn. can be delivered as early as 6:00 a.m. in major metropolitan areas. Obviously those are "best case" conditions. There are many locations around the world where a technician must travel hours in desolate country to get to the broken equipment. That technician must have all necessary parts and, therefore, will take all possible parts or acquire them en route.

Service parts is a confidence business. If technicians have confidence the system will supply the parts they need, then techs will minimize their cache of service parts. If confidence is low, techs will develop their own stock of parts, will order two parts when only one is needed, and will retain intermittent problem parts.

Parts required 24/365 can be shared through either third-party logistics companies (TPLs) or intelligent lockers instead of being carried by the several individual technicians who might provide the same coverage. Handoffs from the stock-keeping facility to the courier to the technician can be facilitated by intelligent lockers. Today, most orders are transmitted to the company warehouse or TPL location that picks and packs the ordered part for shipment and notifies the courier. Then the courier must locate the technician, who often has to drop what he or she is doing, go to meet the courier, and sign for the part. Avoid arrangements that allow the courier to leave parts at a receiving dock or reception desk, because they often disappear before the technician arrives.

Intelligent lockers can facilitate the handoff procedures at both ends of the delivery process. The receiving personnel can put parts in the intelligent locker and immediately notify the courier or technician by page, cell phone, fax, e-mail, or other method that the part is ready for pick up. The receiver can then retrieve parts at his or her convenience, and the access code provides assurance that the correct person gets the part.

A single vendor can manage one-to-many intelligent lockers to provide parts to many users. For example, Granger or The Home Depot could intelligently control sales of expensive, prone-to-shrink tools and accessories by placing these items in intelligent lockers outside their stores where the ordered items can be picked up at any hour. Public mode allows many users to place their items in intelligent lockers for access by designated purchasers. Vendors could arrange space as required so a single courier "milk run" could deliver parts for technicians from several companies to pick up when convenient. This "controlled Automat" use is sure to excite couriers themselves, as well as entrepreneurs who could use the capability around the clock for delivery of airline tickets, laptop computer drop-off and return, equipment rental and return, and many similar activities.

Installation parts for communications networks, smart buildings, security centers, and plant instrumentation are high potential items for intelligent lockers. These cabinets can be mounted on a truck, train, or plane and located at the point of temporary need. Communications can be by wired telephone or data, and wireless cell, dedicated or pager frequencies so there are few limits on locations. Installations tend to be chaotic, without configuration management, and with parts taken but not recorded. Intelligent lockers can improve these and many other control and information shortages.

Physical control is one thing, but information control is as important. Many technicians do not like to be slowed down with administration. Part numbers, usage, transfers, and similar matters may be forgotten in the rush of helping customers. Information provided automatically by the activities involving intelligent lockers should greatly improve parts tracking, reordering, return validation, configuration management, repair planning, pickup efficiency, and invoicing.

## 34.4 Production Losses from Equipment Malfunction

In-plant service performance is primarily directed at supporting the plant operations. As most equipment failures in a plant represent production loss, measuring the amount of loss that results from inaccurate or improper service is a key element to measuring the service operation. Because other parameters can affect production loss, only by noting the relationship of production losses caused by equipment malfunction to production losses caused by other variables, such as operator error, poor engineering, or random failures, can a true performance of the service function be assessed. By maintaining long-term records of such data, companies can visualize the success of the service department by noting the percent of the total production loss that results from inadequate or improper service. The production loss attributable to maintenance also represents a specific performance measure of the generic element of accuracy in problem definition. Effective preventive maintenance (PM) is a fundamental support for high operational availability.

PM means all actions are intended to keep durable equipment in good operating condition and to avoid failures. New technology has improved equipment quality, reliability, and dependability by fault-tolerance, redundant components, self-adjustments, and replacement of hydraulic and mechanical components with more reliable electronic and optical operations. However, many components can still wear out, corrode, become punctured, vibrate excessively, become overheated by friction or dirt, or even be damaged by humans. For these problems, a good PM program will preclude failures, enable improved uptime, and reduce expenses.

Success is often a matter of degree. Costs in terms of money and effort to be invested now must be evaluated against future gains. This means the time-value of money must be considered along with business priorities for short-term versus long-term success. Over time, the computerized maintenance management system must gather data, which must then be analyzed to assist with accurate decisions. The proper balance between preventive and corrective maintenance that will achieve minimal downtime and costs can be tenuous.

PM can prevent failures from happening at a bad time, can sense when a failure is about to occur and fix it before it causes damage, and can often preserve capital investments by keeping equipment operating for years as well as the day it was installed. Predictive maintenance is considered here to be a branch of preventive maintenance.

Inept PM, however, can cause problems. Humans are not perfect. Whenever any equipment is touched, it is exposed to potential damage. Parts costs increase if components are replaced prematurely. Unless the PM function is presented positively, customers may perceive PM activity as, "that machine is broken again." A PM program requires an initial investment of time, materials, people, and money. Payoff comes later. While there is little question that a good PM program will have a high return on investment, many people are reluctant to pay now if the return is not immediate. That challenge is particularly predominant in a poor economy where companies want fast return on their expenditures. PM is the epitome of, "pay me now, or pay me later." The PM advantage is that you will pay less now to do planned work when production is not pushing, versus having very expensive emergency repairs that may be required under disruptive conditions, halting production and losing revenue. Good PM saves money over a product's life cycle.

In addition to economics, emotions play a prominent role in preventive maintenance. It is a human reality that perceptions often receive more attention than do facts. A good computerized information system is necessary to provide the facts and interpretation that guide PM tasks and intervals. PM is a dynamic process. It must support variations in equipment, environment, materials, personnel, production schedules, use, wear, available time, and financial budgets. All these variables impact the how, when, where, and who of PM.

Technology provides the tools for us to use, and management provides the direction for their use. Both are necessary for success. These ideas are equally applicable to equipment and facility maintenance and to field service in commerce, government, military, and industry.

The foundation for preventive maintenance information is equipment records. All equipment and maintenance records should be in electronic databases. The benefits obtained from computerizing maintenance records are much greater than the relatively small cost. There should be a current data file for every significant piece of equipment, both fixed and movable.

The equipment database provides information for many purposes beyond PM and includes considerations for configuration management, documentation, employee skill requirements, energy consumption, financials, new equipment design, parts requirements, procurement, safety, and warranty recovery. Essential data items are shown in Table 34-1.

*Table 34-1: Typical Equipment Data Elements*

| Equipment Identification number |
|---|
| Equipment name |
| Equipment product/family/group/class |
| Supplier(s) |
| OEM and supplier model numbers |
| Geographic location |
| System process location |
| Criticality |
| Responsible user |
| Installation date |
| Warranty end date |
| Original comprehensive cost |
| Current value |
| Safety precautions |
| Use per day |
| Use meter reading (latest plus history) |
| Calibration history and due dates |
| PM interval(s) |
| Last PM date and meter |
| Next PM due date and meter |
| PM average time, personnel, and parts |

The data for new equipment should be entered into the computer database when the equipment is procured. The original purchase order and shipping documents can be the source, with other data elements added as they are fixed. It is important to remember there are three stages of configuration:

1. As designed
2. As built
3. As maintained

The *As Maintained* database is the major challenge to keep continually current. The master equipment data should be updated as an intuitive and real-time element of the maintenance system. If pieces of paper are used, they are often forgotten or damaged, and the data may not get into the single master location on the computer. Part number revisions are especially necessary so the correct part can be rapidly ordered if needed. A characteristic of good information systems is that data should only need to be entered once, and all related data fields will be automatically updated. Many maintenance applications today are Web-based so they can be accessed from anywhere a computer (or even a personal digital assistant [PDA] or enabled cell phone) can connect to the Internet.

Computers are only one component of the information system capability. Electronic PDAs, Blackberry two-way pagers, voice recognition, bar codes, and other technologies are coming to the maintenance teams, often with wireless communications. A relatively small investment in data entry technology can gain immediate reporting, faster response to discovered problems, accurate numbers gathered on the site, less travel, knowledge of what parts are in stock to repair deficiencies, and many other benefits.

It is important that the inspection or PM data be easily changeable. The computer program should accomplish as much as possible automatically. Many systems record the actual odometer reading at every fuel stop, end of shift, and other maintenance, so meter reading can be continually brought up to date. Other equipment viewed less often can have PM scheduled more on predicted dates. Meter information can be divided by the number of days to continually update the use per day, which then updates the next due date. When an inspection or PM is done and the work order closed, these data automatically revise the date last done, which in turn revises the date next due.

Companies can store preventive maintenance procedures in the computer and print them at the same time the work order is dispatched. Most computers have standard software for word processing capability that can be used to enter and revise procedures. While paperwork from a computer system should be kept to a minimum, a printed procedure checklist that the inspector can sign should help assure responsible accomplishment of tasks. Single-point control over procedures is a big help, especially on critical equipment. The risk of pulling an obsolete procedure from someone's file drawer is greatly reduced. If all items on a procedure cannot be accomplished at one shift, the document can be passed to the next shift supervisor or held for completion until the next day. When completed, the work order would be closed out and the related information entered automatically onto history records for later analysis.

Safety inspections and legally required checks can be maintained in computer records for most organizations without any need to retain paper copies. If an organization must maintain those paper records for some legal reason, then they should be microfilmed or kept as electronic images rather than in bulky paper form.

Humans are still more effective than computers at tasks that are complex and are not repeated. Computers are a major aid to humans when tasks require accurate historical information and are frequently repeated. Computer power and intelligent software greatly enhance the ability to accurately plan, schedule, and control maintenance.

## 34.5 Performance Metrics and Benchmarks

The heart of any management system is establishing the objectives that must be met. Once managers determine the objectives, then plans, budgets, and other parts of the management process can be brought into play. Too often service management fails to take the time to establish clear objectives and operates without a plan. As service may contribute a majority of a company's revenues and profits, that can be a very expensive mistake.

Objectives should be:

- Written
- Understandable
- Challenging
- Achievable
- Measurable

Each company must develop its own performance measures. Useful performance measures, often referred to a benchmarks or key performance indicators (KPIs), include the following:

### Asset Measures—Equipment, Parts, and Tools

A1. Support Level $= \dfrac{\text{Total Quantity Issued}}{\text{Total Quantity Demanded}}$

A2. Demand Accommodation $= \dfrac{\text{SKUs (Stock-keeping Units) on Authorized Stock List (ASL)}}{\text{SKUs Demanded}}$

A3. Demand Satisfaction $= \dfrac{\text{Total Quantity of ASL Parts Issued}}{\text{Total Quantity of ASL Parts Demanded}}$

A4. Turnover $= \dfrac{\text{Quantity (or Value) Issued per Year}}{\text{Average Quantity (or Value) on Hand per Year}}$

A5. Emergency Rate $= \dfrac{\text{Quantity (or Value) Expended}}{\text{Total Quantity (or Value) Demanded}}$

A6. Assets % $= \dfrac{\text{\$ Book Value of Assets}}{\text{\$ Value of Work, Revenue, Total Costs, or Profits}}$

A7. Repair Cycle = Days from failure until usable on hand. (Note that this may be divided into a) the technician's days to return and b) the repair time once the decision is made to repair the defective part.)

A8. Parts per Unit Repair $= \dfrac{\text{Sum of All Costs of Parts Used}}{\text{Number of Repairs}}$

A9. Repair Cost Ratio $= \dfrac{\text{Cost to Repair Defective Unit}}{\text{Cost of a New Unit}}$

A10. No Trouble Found (NTF) $= \dfrac{\text{Count of Units with No Defects Found}}{\text{Total Alleged Failures}}$

All. Dead on Arrival (DOA) Rate $= \dfrac{\text{Quantity Defective for All Causes}}{\text{Total Quantity Processed}}$

**Cost Measures**

Cl. Total Maintenance Costs = Sum of Labor \$ + Parts \$ + Travel \$ + - - - + Direct \$ + Indirect \$ + General & Administrative (G&A) \$

C2. Labor Costs = Labor Hours x Loaded Cost per Hour

C3. Parts and Materials Cost = Parts, Expendables, and Consumables Direct + Indirect Costs

C4. Production Loss (Revenue Loss) = \$ Foregone Revenues and/or Cost to Obtain Substitute Capability

C5. Actual versus Estimated $= \dfrac{\text{Actual \$, Time, Events}}{\text{Estimated \$, Time, Events}}$

C6. Revenue per Person $= \dfrac{\text{\$ Total Revenue}}{\text{Number of People}}$

C7. Expense to Revenue Ratio $= \dfrac{\text{\$ Expenses}}{\text{\$ Revenue}}$

C8. Break-Even Quantity: Revenue = Fixed Costs + Variable Costs

C9. Return on Investment (ROI) $= \dfrac{\text{Net Payback}}{\text{\$ Invested}}$

**Equipment Measures**

El. Availability (Uptime):

Ai (Inherent Availability) $= \dfrac{\text{MTBF (Mean Time Between Failures)}}{\text{MTBF + MTTR (Mean Time To Repair)}}$

Aa (Achieved Availability) $= \dfrac{\text{MTBM (Mean Time Between Maintenance)}}{\text{MTBM + M (Mean Maintenance Time)}}$

Ao (Operational Availability) $= \dfrac{\text{Uptime (Operational Time)}}{\text{Total Time}}$

E2. Mean Down Time (MDT) $= \dfrac{\text{Sum of All Down Time}}{\text{Number of Failure Occurrences}}$

E3. Mean Time Between Failures $= \dfrac{\text{Total Time}}{\text{Number of Failure Occurrences}}$

E4. Mean Time Between Maintenance $= \dfrac{\text{Total Time}}{\text{Total of Corrective + Preventive Occurrences}}$

E5. Installation Time = Hours and minutes from installation start until usable. May calculate

Mean Install Time (MIT) $= \dfrac{\text{Total of All Installation Times}}{\text{Number of Installations}}$

### Preventive Measures

P1. PM Rate $= \frac{\text{PM Events, Time}}{\text{Total Events, Time}}$

P2. PM Completion Ratio $= \frac{\text{PM Events Completed}}{\text{PM Events Due}}$

P3. Mean Preventive Time $= \frac{\text{Sum of All PM Times}}{\text{Number of PM Occurrences}}$

P4. Minimize Total Costs = Sum of Preventive Costs + Corrective Costs + Lost Revenue

P5. Defect Detection Rate $= \frac{\text{Total Number of Defects Reported}}{\text{Number of Inspections}}$

### Human Measures

H1. Response Time = Hours and minutes from request for assistance until expected effort is started.

H2. Restore Time = Time from notification of failure until system is operable.

H3. First Call Fix Rate $= \frac{\text{Quantity Satisfied at First Attempt}}{\text{Total Requests}}$

H4. Callback Rate $= \frac{\text{Number of Repeat Attempts}}{\text{Total Attempts}}$

H5. Attempts per Incident $= \frac{\text{Total Attempts}}{\text{Number of Incidents}}$

H6. Maintenance Hour per Operating Hour (MH/OH) $= \frac{\text{Total Support Hours}}{\text{Total Equipment Operating Hours}}$

H7. Administration and Support Ratio $= \frac{\text{Support People Number or Costs}}{\text{Total People Number or Costs}}$

H8. Overtime % $= \frac{\text{Overtime Hours or costs}}{\text{Total Labor Hours or costs}}$

H9. Emergency versus Planned Calls and Time $= \frac{\text{Repair Work Number, Time, Costs}}{\text{Total Work Number, Time, Costs}}$

H11. Backlog Days $= \frac{\text{Demand Total Work Hours}}{\text{Supply Work Hours per Day}}$

H12. Operational Productivity $= \frac{\text{Utilized Time}}{\text{Total (Paid) Time}}$

H13. Achieved Productivity $= \frac{\text{Standard Units Output}}{\text{Total (Paid) Time}}$

H14. Effectiveness $= \frac{\text{Standard Units Output}}{\text{Utilized Time}}$

**Example Calculation:**
The most important measure for production equipment support is operational availability, which we also term "uptime." This is item E1 and definition $A_O$ above. This is the "real world" measure of what percent of time equipment is available for production. In the following example, we evaluate an item of automation equipment for one year, which is 365 days x 24 hours per day = 8,760 total possible "up" hours. Our equipment gets preventive maintenance for one hour every month (12 hours per year) plus additional quarterly PM of another hour each quarter (four (4) more hours per year). There was one failure that resulted in six hours of downtime. Thus, total downtime for all maintenance was 12 + 4 + 6 = 22 hours.

$$Ao = \frac{\text{Uptime (Operational Time)}}{\text{Total Time}} = \frac{(8,760 - 22)}{8,760} = \frac{8,738}{8,760} = 0.9975 = 99.75\% \text{ Uptime}$$

That would be considered acceptable performance in most operations, especially if the PM work can be done at times that will not interfere with production. The maintenance challenge is to avoid failures that adversely affect production operations.

Automation professionals should consider life cycle cost when designing or acquiring an automation system. Design guidelines for supportability include:

1. Minimize the need for maintenance by:
   - Lifetime components
   - High reliability
   - Fault tolerant design
   - Broad wear tolerances
   - Stable designs with clear yes/no indications

2. Access:
   - Openings of adequate size
   - Fasteners few and easy to operate
   - Adequate illumination
   - Work space for large hands
   - Entry without moving heavy equipment
   - Frequent maintenance areas have best access
   - Ability to work on any FRU (field replaceable unit) without disturbing others

3. Adjustments:
   - Positive success indication
   - No interaction effects
   - Factory/warranty adjustments sealed
   - Center zero and increase clockwise
   - Fine adjustments with large movements
   - Protection against accidental movement
   - Automatic compensation for drift and wear
   - Control limits
   - Issued drawings show field adjustments and tolerances
   - Routine adjustment controls and measurement points in one service area

4. Cables:
   - Fabricated in removable sections
   - Each wire identified
   - Avoids pinches, sharp bends and abrasions
   - Adequate clamping
   - Long enough to remove connected components for test

- Spare wires at least 10% of total used
- Wiring provisions for all accessories and proposed changes

5. Connectors:
   - Quick disconnect
   - Keyed alignment
   - Spare pins
   - Plugs cold, receptacles hot
   - No possible misconnection
   - Moisture prevention, if needed
   - Spacing provided for work area and to avoid shorts
   - Labeled; same color marks at related ends

6. Covers and panels:
   - Sealed against foreign objects
   - Independently removable with staggered hinges
   - Practical material finishes and colors
   - Moves considered—castors, handles, and rigidity
   - Related controls together
   - Withstand pushing, sitting, strapping and move stress
   - Easily removed and replaced
   - No protruding handles or knobs except on control panel

7. Consumables:
   - Need detected before completely expended
   - Automatic shutoff to avoid overflow
   - Toxic exposure under thresholds

8. Diagnostics:
   - Every fault detected and isolated
   - Troubleshooting can not damage
   - Self-Tests preferred
   - Go/no go indications
   - Isolation to field replaceable unit
   - Never more than two signals observed simultaneously
   - Condition monitoring on all major inputs and outputs
   - Ability for partial operation of critical assemblies

9. Environment—equipment protected from:
   - Hot and cold temperatures
   - High and low humidity
   - Airborne contaminants
   - Liquids
   - Corrosives
   - Pressure
   - Electrical static, surges and transients

10. Fasteners and hardware:
    - Few in number
    - Single turn
    - Captive
    - Fit multiple common tools
    - Non-clog
    - Common Metric

11. Lubrication:
    - Disassembly not required
    - Need detectable before damage
    - Sealed bearings and motors

12. Operations tasks:
    - Positive feedback
    - Related controls together
    - Decisions logical
    - Self-guiding
    - Check lists built-in
    - Fail-safe

13. Packaging:
    - Stacking components avoided
    - Ease of access guides replacement need
    - Functional groups
    - Hot items high and outside near vents
    - Improper installation impossible
    - Plug-in replaceable components

14. Parts and components:
    - Labeled with part number and revision level
    - Breakable knobs and buttons replaceable separate from switch
    - Delicate parts protected
    - Stored on equipment if user replaceable
    - Standard, common, proven
    - Not vulnerable to excessive heat
    - Mean time between maintenance known
    - Wear-in/wear-out considered

15. Personnel involvement:
    - Weight for portable items 35 lb. (16 kg.) maximum
    - Lowest ability expected to do all tasks
    - Male or female
    - Clothing considered
    - Single-person tasks

16. Refurbish, rejuvenate and rebuild:
    - Materials & labels resist anticipated solvents & water
    - Drain holes
    - Configuration record easy to see and understand
    - Aluminum avoided in cosmetic areas

17. Safety:
    - Interlocks
    - Electrical shut off near equipment
    - Circuit breaker and fuses adequate
    - Protection from high voltages
    - Corners and edges round
    - Protrusions eliminated
    - Electrical grounding or double insulation
    - Warning labels
    - Hot areas shielded and labeled
    - Controls not near hazards

- Bleeder and current limiting resistors on power supplies
- Guards on moving parts
- Hot leads not exposed
- Hazardous substances not emitted
- Radiation given special considerations

18. Test points:
    - Functionally grouped
    - Clearly labeled
    - Accessible with common test equipment
    - Illuminated
    - Protected from physical damage
    - Close to applicable adjustment or control
    - Extender boards or cables

19. Tools and test equipment:
    - Standardized
    - Minimum number
    - Special tools built into equipment
    - Metric compatible

20. Transport and storage:
    - Integrated moving devices, if service needs to move
    - Captive fluids and powders
    - Delivery and removal methods practical
    - Components with short life easily removed
    - Ship ready to use

The preferred rules for modern maintenance are to regard safety as paramount, emphasize predictive prevention, repair any defect or malfunction, and, if the system works well, strive to make it work better.

## 34.6 References


1. Patton, Joseph D., Jr., and William H. Bleuel. *After the Sale: How to Manage Product Service for Customer Satisfaction and Profit*. Solomon Press, 2000.
2. Patton, Joseph D., Jr. *Maintainability & Maintenance Management*. Fourth Edition. ISA, 2005.
3. Patton, Joseph D., Jr. *Preventive Maintenance*. Third Edition. ISA, 2004.
4. Patton, Joseph D., Jr. and Roy J. Steele. *Service Parts Handbook*. Second Edition. ISA, 2003.


**Author's note:** With the Internet available to easily search for publications and professional societies, using a search engine with key words will be more effective than a printed list. Internet search on specific topics will be continually up-to-date, whereas materials in a book can only be current as of the time of printing. Search with words like *maintenance, preventive maintenance, reliability, uptime* (finds better references than does the word *availability*), *maintainability, supportability, service management*, and *maintenance automation* will bring forth considerable information, from which you can select what you want.

## About the Author

**Joseph D. Patton, Jr.** is Chairman of Patton Consultants, Inc. (www.PattonConsultants.com), advisors to management on product service, logistics, and support systems. Before founding Patton Consultants in 1976, Patton was a Regular Army Officer and spent 11 years with Xerox Corp. He is author of over two hundred published articles and eight books. He earned a BS degree from the Pennsylvania State University and an MBA in marketing from the University of Rochester. He is a Registered Professional Engineer (PE) in Quality Engineering and a Fellow of both ASQ – The American Society for Quality and SOLE – The International Society of Logistics. He is a Certified Professional Logistician (CPL), Certified Quality Engineer (CQE), Certified Reliability Engineer (CRE), and a senior member of ISA.

# VII

# Work Structure

*Automation professionals who work for engineering contractors and system integrators and see the project only after it has been given some level of approval by the end-user management may not realize the months or even years of effort that goes into identifying the scope of a project and justifying its cost. In fact, even plant engineers who are often responsible for doing the justifications often do not realize that good processes exist for identifying benefits from automation.*

*In the process of developing the CAP certification program, the first step was to do a job analysis of the work of an automation professional. By focusing on describing the job, it became clear just how big a part project leadership and interpersonal skills are in the work of an automation professional. That effort helped the team realize these topics had to be included in any complete scope of automation.*

*If the Job Analysis Team had jumped immediately into defining skills, they might have failed to recognize the importance of project and interpersonal skills for automation professionals—whether they are functioning in lead roles or not.*

# 35 Automation Benefits and Project Justifications

*By Peter G. Martin*

## Topic Highlights

*Capital Projects*
*Return on Investment*
    *Net Present Value*
    *Internal Rate of Return*
*Lifecycle Costs*
    *Lifecycle Economics*
*Barriers to Success*
*Real-Time Cost Accounting*

## 35.1 Background

When the digital computer was first introduced to process manufacturing in the late 1960s, the promise of that new technology was unbounded. Many manufacturing managers saw computer technology as the key to driving the performance of their plants to new levels, and driving real competitive advantages into their manufacturing lines. For the most part, after over 30 years of using this technology in the process industries, this vision has still not been realized.

Decisions on the initial installations of computer-based automation systems appeared to have been made on a binary basis. That is, many manufacturers just felt it was important to get the new technology installed in order to run their operations. Little consideration seems to have been given to the economic impact the system would provide. A survey of manufacturing managers indicated the primary motivators driving manufacturers to purchase automation systems include their desire to:

- Improve plant quality
- Improve safety
- Increase manufacturing flexibility
- Improve operations reliability
- Improve decision-making
- Improve regulatory compliance
- Increase product yields
- Increase productivity

- Increase production
- Reduce manufacturing costs

Although few would disagree with this list, it appears these criteria are seldom taken into consideration either during the purchase of an automation system or over the system's lifecycle. However, most of the criteria listed have a direct impact on the ongoing economic performance of the manufacturing operation.

## 35.2 Capital Projects

Automation systems are typically purchased from manufacturers' capital budgets. Therefore, any discussion of the economic benefits of automation systems and technologies must be from the perspective of the capital budgeting and project process in manufacturing companies. Capital budgeting is typically a long process, often involving multiple years for each capital project. The process is initiated when a manufacturing operation identifies a need for a capital project and then develops a nomination package for the proposed project that is forwarded to corporate planners. Corporate planners evaluate all the nominated projects against qualifying criteria, as well as available capital and then select a set of nominated projects for implementation, typically for the following fiscal year. At this point, the project moves from planning to execution. A project team is convened and provides a bid package to a set of vendors who can provide the products and services necessary to satisfy the defined project requirements. The vendors are evaluated, one is selected, and the order is negotiated and purchased. The project is executed to install the system, start it up and get it operational. The system is then operated for its lifecycle and, in theory, work is done to get continuous improvement from the system.

It is interesting to note that, in the typical capital project process, automation systems vendors have little-to-no say on what the actual solution is. By the time the request for proposal (RFP) is put out for bid in Step 5 of the typical process shown in Figure 35-1, the solution has already been defined. Vendors must only respond with the lowest possible priced system that meets the solution definition. Over the past decade, as manufacturers have significantly reduced headcount in their engineering departments, this issue has begun to become very important because the vendors now may have a stronger engineering talent base than the manufacturers and may be in a better position to define performance-generating automation solutions.

Since automation systems are purchased from capital budgets, it is very important to understand capital budget economics in order to effectively analyze the economic benefits of automation systems. Figure 35-2 displays a classic lifecycle capital economic profile. The lower bar chart represents the cost of the capital project including hardware, software, engineering, installation, start-up, commissioning, operations, and maintenance.

The costs in an automation project tend to be quite high at the beginning of the lifecycle, due to the purchase of the system, engineering, installation, and startup. The costs tend to level out after startup and are largely comprised of ongoing engineering, operations, and maintenance of the system.

Toward the end of the lifecycle, annual automation costs tend to increase, due to aged equipment, spare parts and repairs, as well as increased training levels. A review of a number of automation projects revealed that most manufacturers have a fairly good understanding of their automation costs, even if they do not have specific programs to capture them over the lifecycle of the equipment. The upper dashed line represents the economic benefit derived by the deployment of the automation system.

Notice that this value begins at start-up and is expected to continuously grow over the useful life of the automation system. The same review of the automation projects that revealed that most manufacturers have a good handle on the cost of their automation systems also revealed that almost none of

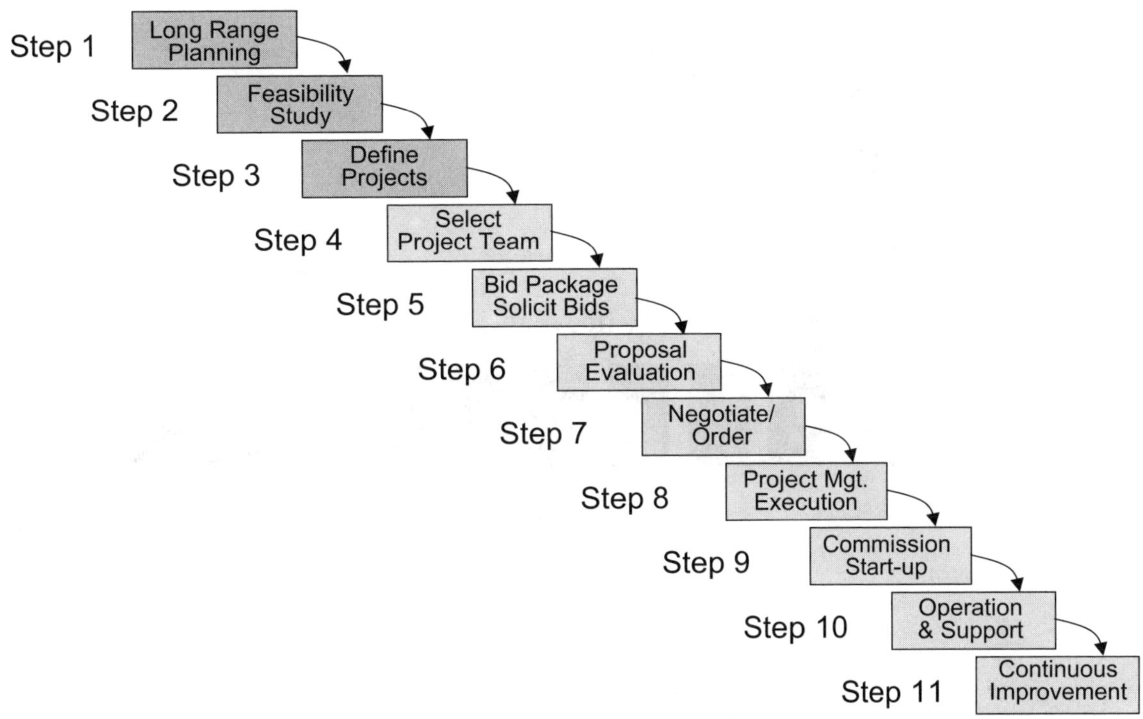


*Figure 35-1: Typical Capital Process Project*

them had any understanding as to the true benefits provided by the automation. This is because the benefit line in Figure 35-2 is seldom, if ever, measured. It was determined that many engineers felt the finance people in their plants were measuring the benefit of each capital investment and were not passing that information back to engineering. The fact is the finance systems in place in most manufacturing operations cannot capture benefit information at this level of specificity. This is a huge problem when trying to assess the true economic benefit provided by automation investments.

## 35.3 Return on Investment

The most common way to discuss the economic benefit for any capital investment is in terms of return on investment (ROI). Basically, ROI is defined to be the cash inflows resulting from a capital investment, such as an automation system, divided by the initial investment made over a given period of time. ROI can be determined in a number of ways.

The simplest and most common approach is to evaluate the purchase price as the initial investment of the automation system against the cash inflows that result from the deployment of the system. Although the price approach is often utilized, a more complete view of ROI would be to evaluate the purchase price and all other initial (project) costs associated with the project against the accumulated cash inflows.

A more complete investment evaluation would be to evaluate the purchase price, project costs, and ongoing operational and maintenance costs against the accumulated cash inflows, but this approach is seldom if ever done. In any case, the basic evaluation approach is the same, when the accumulated cash inflows equal the purchase price or the purchase price plus initial project costs, depending on which method is utilized, 100% return on investment is achieved.

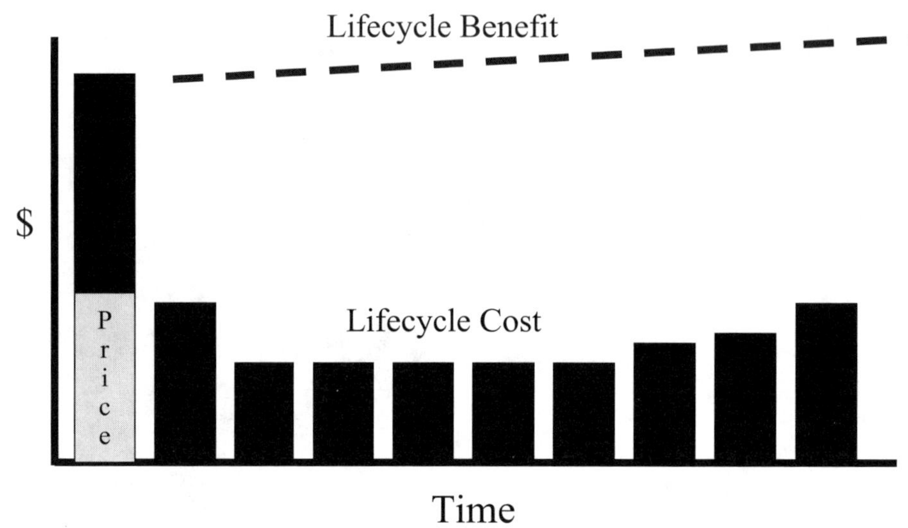


*Figure 35-2: Lifecycle Capital Economic Profile*

ROI is often stated in terms of time, if it takes less than one year to reach 100% return, or in terms of percentage, if it takes greater than one year to reach 100%. The time to reach 100%, ROI is often referred to as the "payback period." Unfortunately, the point in the capital project process at which ROI is addressed is typically in Step 6 (Figure 35-1), when automation vendors are often asked to provide a projected ROI analysis with their proposals. This analysis is constructed based on what the vendor believes the unique features of their offering might be able to provide if effectively used in the manufacturing operations.

The good news for these vendors is that, after the automation systems are installed and operating, manufacturers almost never go back to check to see if the ROI projections were ever really achieved. One of the major contributing factors for not doing this analysis is the aforementioned lack of any effective way to capture the benefit side of the ROI model in an operating plant. The accounting systems in place just do not have the level of resolution necessary to systematically calculate and verify an ROI analysis of this type.

### 35.3.1 Net Present Value

An ROI analysis is quite simple, but it may also be a bit deceptive from the perspective of making a decision as to whether or not to invest in an automation project. The reason for this is, if it takes a fairly long time to get to 100% return, the amount of the investment is based on the value of the money at the time the investment is made, while the inflows will occur at a later time. From a value perspective, an expected cash inflow of a given amount at some future time is not worth the same as if that amount were paid today. Therefore, valuing future cash inflows at an equivalent amount to what they will actually be when the inflow occurs overvalues the return. The larger the amount of time that is expected to pass prior to receiving a particular cash inflow, the less it is worth in today's currency. An ROI analysis ignores this situation. When trying to make a determination on where a company's capital budget will be invested, a more appropriate evaluation of the worth of money over time is required.

For example, if a company has a capital budget of $1,000,000 and there are two potential projects that each would cost $1,000,000, one of which has a return of $2,000,000 over 20 years and the other

$1,500,000 over 2 years, which would be the best capital investment? The answer to this question is not obvious. A more sophisticated approach to ROI is required to effectively answer.

The Net Present Value (NPV) function is designed to address this issue. The NPV of an expected cash flow over a period of time is a function that takes the time value of money into consideration. The inputs to this function are the expected cash inflows over time, the initial investment, and the discount rate. The discount rate represents the value of a future cash inflow in today's terms. The result of this function provides the current value of an expected time series of cash inflows minus the initial investment amount. This NPV can be compared to other potential investments to determine which potential capital investment is better for a company to make.

The equation of NPV is as follows:

$$\text{NPV} = \sum_{t=1}^{n} \frac{(\text{Cash Inflows})^t}{(1+\text{rate})^t} - \text{Initial Investment} \tag{33-1}$$

### 35.3.2 Internal Rate of Return

An alternative approach to conducting an NPV analysis to determine the value of a capital investment that will pay back over an extended period of time is referred to as Internal Rate of Return (IRR). IRR uses that same basic calculation approach as NPV, but the variable in the calculation is the discount rate that will result if the NPV calculation results in $0 over the selected time period. In other words, what discount rate used in an NPV calculation results in a value of $0? The higher the value of the IRR, the better the investment. Solving the following equation for IRR would result in the appropriate calculation.

$$\sum_{t=1}^{n} \frac{(\text{Cash Inflows})^t}{(1+\text{IRR})^t} = \text{Initial Investment} \tag{33-2}$$

Both NPV and IRR are common financial functions in standard spreadsheet packages such as Excel, as well as standard functions of financial calculators. Using a spreadsheet or financial calculator is an easy way to perform either calculation. It should be noted that, although both NPV and IRR provide a more appropriate projection of a capital project's expected value, neither addresses the fundamental issue involved with automation projects. That is: the benefit value or cash inflows associated with an automation system are seldom, if ever, captured in a cost accounting system.

This means ROI, NPV, and IRR are not verifiable for any automation project over time. No matter which of these methods is employed to justify the automation investment during the evaluation process, it is still difficult to verify the actual value of a project after it is implemented. This issue must be addressed for automation projects to have any investment credibility at all.

Interviews with industry executives have revealed an interesting progression in credibility that has resulted from the lack of verifiability of automation investments. Thirty years ago, when management really felt they needed automation technology to compete, they readily approved capital investments; when they later inquired as to whether the ROI, NPV, or IRR were realized, the project team answered in the affirmative, and the managers believed it. Twenty years ago the managers accepted it. Ten years ago they questioned it. Today they reject it. It is becoming more essential than ever to validate the value generated by automation to regain its credibility as a value-adding technology.

## 35.4 Lifecycle Costs

The lack of an effective way to measure the benefit from automation investments has resulted in the selection criteria for automation systems being relegated to either the low price system or the lowest lifecycle cost system. Clearly, taking a full lifecycle cost view (Figure 35-3), as compared to a price-only view, provides a much more comprehensive evaluation. But either view relegates the selection of the automation system to a cost, with no associated measurable benefit. Any offering that can be evaluated only from the perspective of cost sooner or later is categorized as a "necessary evil," and the market moves toward commoditization. This has been the case with automation systems over the past decade.

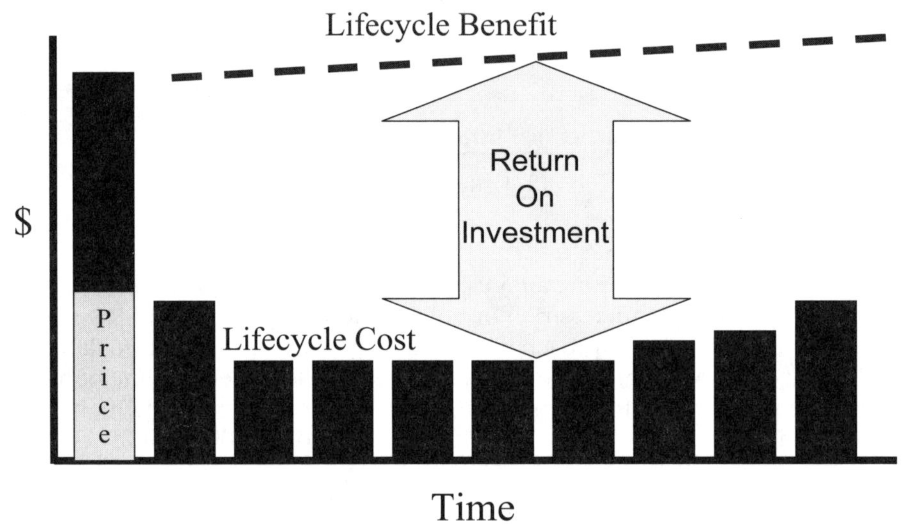


*Figure 35-3: Lifecycle Capital Economic Profile*

There has been fairly significant movement toward a lifecycle cost view of automation systems, rather than a price-only view. The following model was developed to reflect this expanded view. The basic equation used for the analysis of the lifecycle cost is:

LCC = Price + Project Engineering + Installation + NPV(Ongoing Annual Costs)

wherein:

LCC = Lifecycle Costs

Price = Automation system price

Project Engineering = Total cost to engineer the project

Installation = Total cost to install the system (including start-up)

NPV = The net present value function of the ongoing costs

Ongoing Annual Costs = The annual engineering, operations, and maintenance cost of the system (33-3)

The net present value function serves to calculate the time-value of money over extended periods (Figure 35-4).

It is interesting to note that the system price tends to be a reasonably small component of the overall automation system cost. In fact, studies have placed the average price at less than 35% of the total

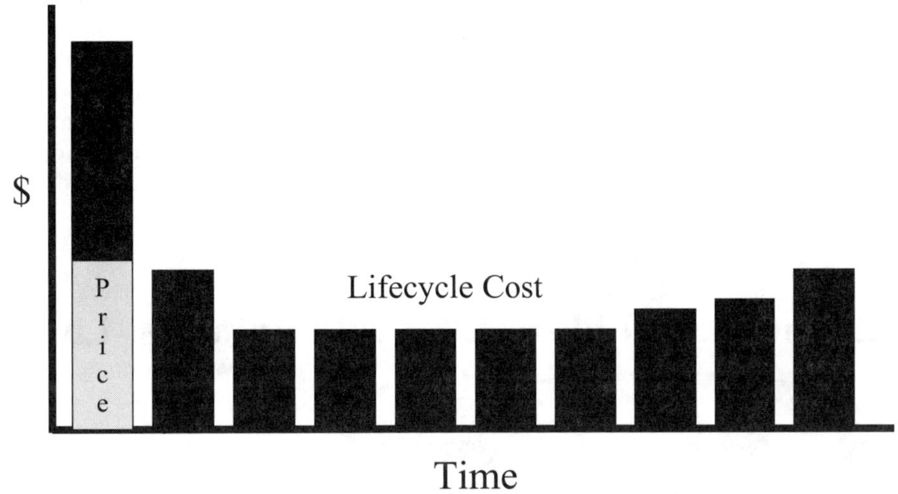


*Figure 35-4: Lifecycle Capital Cost Profile*

project cost, without even taking into consideration the ongoing costs. This tends to demonstrate one of the deficiencies in the price-only approach.

The expansion of economic perspective from price to lifecycle cost was a major step forward for the industry, but was still very limiting. This is the perspective of most of the industrial automation users today, since most of them still use a *cost-only* economic evaluation approach. If there is no economic benefit to the manufacturing operation for putting in an automation system, then automation systems are certainly not meeting management's objectives, and probably nothing but the most rudimentary system should ever be deployed.

### 35.4.1 Lifecycle Economics

Many automation system users have discussed returns on their automation investments. However, calculating returns requires an accurate measure of the lifecycle benefits the manufacturing operation derives from the utilization of the automation system, not just the lifecycle costs. In interviews conducted with hundreds of manufacturing executives, it was learned none was actually measuring the benefits derived from automation. Most admitted they did not know how to get at these metrics in any reasonable way, and their current cost accounting systems did not provide the detailed economic data to be able to infer the benefit value.

A manufacturing operation sees the economic benefits from automation systems essentially in two major areas:

- First, manufacturing cost savings through such things as reduced power consumption, raw material costs, and manpower requirements
- Second, the increase in production that can be gained through better asset utilization

Of these, the only one that was regularly monitored was reduced manpower due to automation, because it is relatively easy to measure. The other elements of the benefit calculation are variables that constantly change as products are produced and are, therefore, very difficult to measure.

As the concept of automation systems' lifecycle economic profiles started to gain recognition during the late 1990s, the author was asked to lead a session on the subject at the ISA Technical Conference of 1996. At that conference, and at subsequent meetings, a number of professionals from companies such as E. I. DuPont, General Foods, Eli Lilly, and Dow Chemical contributed a considerable amount of data from a number of automation projects to help build a lifecycle economic profile. A high-level view of the profile used for this exercise is shown in Figure 35-5. This profile defines the benefit as the net present value (NPV) of the annual manufacturing cost savings and the annual production increases resulting from the automation system. The net present value is a function that calculates the current value of money paid on a regular basis over a number of years. It is appropriate when trying to make an up-front decision of the best economic value among a number of possible choices that will either be generating cost or economic benefit over a period of time. An automation system does both.

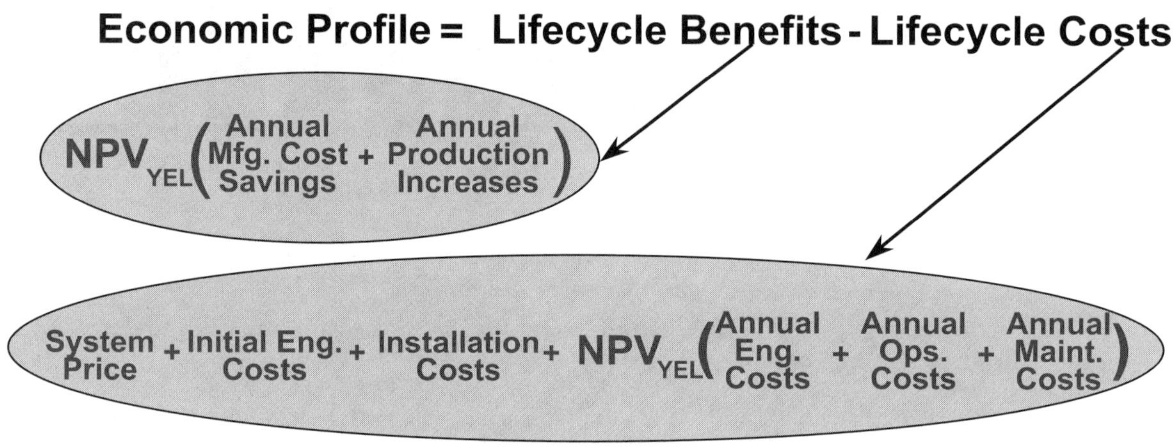


*Figure 35-5: Lifecycle Economic Model*

Years of Expected Life (YEL) is an indicator that the NPV calculation should be done over the automation system's expected lifecycle. This helps to determine the value of automation systems that have different expected installed lifecycles. The lifecycle cost calculation has two basic components, the project costs and the ongoing costs. The project costs can be captured in the three general categories of price, initial engineering costs and installation costs. The net present value function was also utilized for the ongoing aspect of annual engineering, operations and maintenance costs.

As part of this activity, a number of automation projects were analyzed and, from the data collected, an actual economic profile developed. Considerable data was available for the cost side of the profile, but only limited data was available to analyze the benefit side. This is not surprising in light of the lack of focus industry-wide on the benefits component of the equation.

A number of interesting results developed from the analysis of this data. First, as Figure 35-6 depicts, most of the projects for which the benefit side was measured at all showed a continuous decline in the benefit value over the system's lifecycle. This result was shared with a number of the professionals who contributed the data. They were not overly surprised by this result. Many of the professionals attributed the continuous decline to the obsolescence of the system. This conclusion was later demonstrated to be erroneous, at best, since there are many projects that do realize continuous improvement in the economic benefit derived from the automation system. In fact, the decline seems to be more associated with the lack of measurement of the benefit side of the profile than to the reduction due to obsolescence.

One of the basic principles of process control is that if you cannot measure the controlled variable, you cannot control it. The same is true for financial variables such as lifecycle benefits. It also appears that,

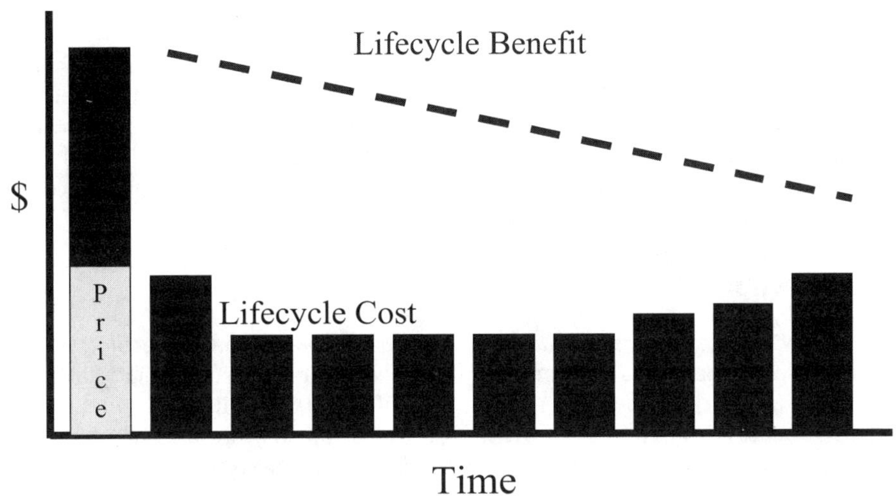


*Figure 35-6: Actual Lifecycle Capital Economic Profile*

if the variable is not measured, it will most probably move in the wrong direction, which is exactly what the data showed.

Another interesting aspect of the data collected was how the lifecycle cost data was actually distributed across the set of data, as shown in Figure 35-7. The price, which had traditionally been the primary economic variable in the automation system decision process, represented less than a quarter of the first five-year cost of the system. Between the initial engineering costs and the five-year lifecycle engineering costs, the engineering of the automation system accounted for 37.8% of the costs. This is considerably greater than the price of the initial system.

| | System Price | Initial Eng. Costs | Installation Costs | Annual Eng. Costs | Annual Ops. Costs | Annual Maint. Costs |
|---|---|---|---|---|---|---|
| Average first 5-year % of cost | 23.2% | 28.5% | 16.1% | 9.3% | 7.6% | 15.3% |

*Figure 35-7: Lifecycle Cost Breakdown*

Perhaps the most interesting result of the analysis was found on the benefit side of the model. The benefit data collected, which was not statistically valid due to the small sample size of projects for which benefit data was available, showed the benefit to cost ratio over the first five years was 3.4:1. This means that the average ROI on automation technology for projects for which the benefit data was measured realized very strong economic returns. The analysis team did not believe the 3.4:1 ratio was representative of the average return on automation realized. In fact, the automation users who were measuring the benefit resulting from automation were considered to be among the best performers. This would mean that the 3.4:1 ratio is a "best practices" result as compared to an average result.

When this data was shared with a larger number of automation users, the consensus seemed to be that these results were not surprising. In fact, most of those interviewed readily accepted the results. This caused us to ask why more users were not focused on the economic benefit resulting from automation. Although most of those asked did not have any solid response, an additional, more detailed analysis uncovered three practices that were significant barriers to a total lifecycle economic approach to automation.

## 35.5 Barriers to Success

The first barrier was the *replacement automation* approach to automation projects that is almost universally used. The replacement automation approach begins at most industrial plants when their capital budget is approved for an automation system upgrade. At that point, a project team is typically established to determine the specifications for the new automation system to be installed. In most of these cases the specification is developed by looking at the functionality of the currently installed system, subsequently building the new system specification around the existing system.

This leads to a request for proposal (RFP) that is provided to a number of automation suppliers for a competitive bid. Unfortunately, this RFP exactly defines the old system that is already in place. The suppliers realize that the one who will win the order is the one who meets the specification at the lowest price. Therefore, even though they may have added significant performance enhancing capabilities to their new computer-based automation systems since the old system had been installed, they propose their lowest-cost system. Typically, this means that all of the advanced capability in which the suppliers have invested is left out of the proposal. The net result is the new system being installed is an exact functional replacement for the system being replaced. Replacing old technology with new technology that does the exact same thing seldom leads to breakthrough improvement.

When this is pointed out to the user's project team, their response is typically "don't worry, once the new system is up and running, we will take full advantage of all of the advanced capabilities." This is when the second barrier starts to take effect. The second barrier is the *project team approach* used on most automation projects.

Project teams of highly qualified engineers are established when capital budgets for automation projects are approved. The project team works on the project throughout the project cycle. Once the project is completed, the project team goes away. Some of the members may go on to other projects. Some may remain in the plant to manage ongoing engineering activities. But the resources required to take advantage of the advanced features of the automation system are no longer available. As a result, the "later on" when everyone was hoping to take advantage of advanced capabilities never happens. The initial benefit, if any, provided by an automation system is typically the best benefit that will be realized through the system over its lifecycle. From that point, there is a continuous degradation of benefit.

## 35.6 Real-Time Cost Accounting

The good news in all of this is that, as this continuous degradation of economic performance happens, nobody knows it. This is because, in most cases, the benefit side of the economic profile is not measured. This is the third barrier to success. Measurement systems are typically only valued if they are providing good news.

The news that might be provided by a measurement system that continuously communicates the economic benefit due to automation would probably not be very positive. But since these measurements are not being made, most manufacturers do not feel the pain of poor performance. In this case, no news is good news. Or is it? This approach was fine when process manufacturers could not help but make significant profits no matter how good—or bad—their operations were. In today's tough global business environment, this approach is just not acceptable.

One promising result from all of this analysis was that, when the professionals in industrial operations paid attention to the economic benefit measures, even if poorly calculated on an infrequent basis, they were able to realize phenomenal results. This implies the potential economic benefit from automation is limited by the ability to measure the benefit side of the ROI model. The implication is that, if the benefit side of the model were made measurable in a systematic and effective manner, it would uncover all kinds of untapped potential benefits from automation technologies.

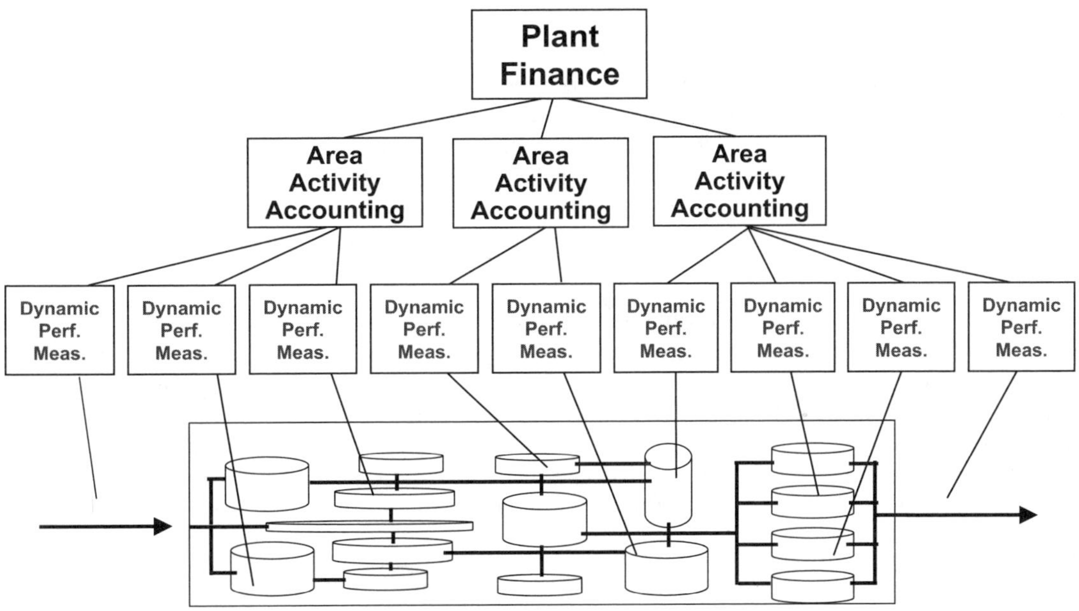


*Figure 35-8: Real Time Accounting*

The good news in this entire discussion is a number of executives in manufacturing companies have been pressing for a fundamental change in the way cost accounting systems work. This change is starting to bring visibility to the benefit of automation systems, as well as other improvement initiatives, and the automation systems are critical to the implementation of this change.

Traditional cost accounting systems have been developed to provide monthly financials for manufacturing operations at the plant level. The common accounting vehicle is the variance report, which presents cost per unit product made for each product line manufactured on a monthly basis. This information is just not sufficient to get the necessary level of visibility into plant operations.

Executives have been pushing accounting to be able to provide higher-resolution financial information from two dimensions: time and space. That is, the executives have been pushing for real-time cost data right down to the process unit level. If such data were available, the visibility into the benefit due to automation expenditures would significantly increase and true ROI calculations for any plant capital investment would become much more visible.

Accountants have been stymied as to how to generate this financial information on a real-time basis. They have not been able to determine a data source that is available at such frequencies and that will enable the calculation of the appropriate cost and profit information. Fortunately, plant engineers have been aware of an available real-time plant database in the form of the plant instrumentation that is already used for process monitoring and control. It has been demonstrated that this plant instrument database can also be effectively used as source data for plant accounting systems enabling the necessary real-time accounting calculations that can be used to evaluate the actual ROIs for the automation system investments.

The appropriate location for these real-time accounting calculations to be made is in the real-time automation systems. Trying to execute them in the IT systems is very difficult to accomplish due to the inherent design of the IT systems. Automation systems are designed to work in real time and are the ideal location for the origination of the real-time accounting models. It is important to note that this approach essentially dissolves the traditional separation between the IT and automation systems. The real-time accounting models are referred to as dynamic performance measures, and although the use of such measures is still in its infancy, the initial results have proven to be very promising.

Every process control engineer realizes that, if a variable is not measurable, it is not controllable. This goes for physical, chemical, and financial variables. It has been nearly impossible for plant personnel to manage the ROI of automation systems because, to this point, it has not been measured. With the advent of dynamic performance measures, the benefit side of the ROI model is finally measurable and available. Now, plant personnel can really work to control the ROI from automation. As this has begun to take place, the resulting returns have been even greater than expected. Automation systems are starting to become vehicles for performance improvement once again, rather than the "necessary evils" they had started to become.

The only true way, therefore, to prove to manufacturing management and executives that an investment they made in automation system technology has realized the economic value projected is to have an accounting system that can measure it. For the most part, today's accounting systems, although providing absolutely necessary financial reporting information, are insufficient in measuring the impact of automation system investments or almost any other investment made to improve plant floor operations. They just do not provide the necessary data. In addition, there is no way to extract the improvement information from the data they do provide.

Real-time cost accounting down to the process unit level, and perhaps even below that level, is required to get the necessary measurements of improvement value. Anything short of implementing and analyzing the real-time accounting models will not provide the necessary level of information to credibly provide automation system payback.

It should also be noted that the development and collection of real-time accounting information is what is required to do accurate projections of ROI and IRR for future automation projects. The projections being provided for the most part today are based on sets of unsubstantiated assumptions that are losing credibility with manufacturing management.

A real-time accounting system provides a history of capital project economics which can be used with a high degree of credibility to project the expected payback of proposed projects. Unfortunately, this is a bit of a *chicken-and-egg* situation in that the real-time accounting systems must be installed and operating for a period of time in order to collect sufficient historical data to make reasonable projections.

The move toward real-time accounting has been very slow, but appears to be accelerating. The good news is that initial experience has shown the economic value improvements that can be realized through the effective application of automation technologies can be much greater than has been projected. Once this fact becomes general knowledge, the focus on value improvement through automation system technology will increase significantly, which should lead to a new era of value-based automation.

An interesting corollary to this analysis is that traditional ROI may be a very poor measure of the value of automation systems. The reason is that, once 100% return is reached, ROI as a measure is typically ignored, but the automation systems keep generating economic value. It is a shame and a huge disservice to ignore the ongoing value generation incurred by automation systems once the initial project cost has been covered. With real-time accounting measures available, a better approach to the measurement of economic benefits of automation systems may be based on cash flow. The cash flow benefit from automation investment can keep accruing and improving over the life of system and the plant assets that are impacted. Viewed in this manner, the economic value of automation, effectively applied and measured, can and should be many times the original price and cost of the system. The capital cost involved in acquiring automation technology in manufacturing operations has been under such limitations in recent years that it may actually prove to be the most significant business investment a manufacturing company can make.

The potential benefits from automation are huge, but, unfortunately, have not been realized or visible to this point. Changes occurring in the way businesses measure performance may finally start to drive visibility into the value that this technology could really provide—if we just manage it and measure it in an appropriate business manner.

## 35.7 References


1. Berliner, Callie, and James A. Brimson (editors). *Cost Management for Today's Advanced Manufacturing*. Harvard Business School Press, 1988.
2. Blevins, et al. *Advanced Control Unleashed: Plant Performance Management for Optimum Benefit.* ISA, 2003.
3. Cooper, Robin, and Robert S. Kaplan. *Cost & Effect: Using Integrated Cost Systems to Drive Profitability and Performance*. Harvard Business School Press, 1998.
4. Friedmann, Paul G. *Economics of Control Improvement*. ISA, 1995.
5. Gitman, Lawrence J. *Principles of Managerial Finance*. Tenth Edition. Addison-Wesley, 2003.
6. Martin, Peter G. *Bottom-Line Automation*. ISA, 2002.
7. Martin, Peter G. *Dynamic Performance Management: The Path to World Class Manufacturing*. Van Nostrand Reinhold, 1993.


## About the Author

**Peter G. Martin**, PhD, has more than 30 years of industry experience and education. After joining the Foxboro Company in the 1970s, Martin worked in a variety of positions in training, engineering, product planning, marketing, and strategic planning. He later became Vice President at Automation Research Corporation before returning in 1996 to the Foxboro Company (now part of Invensys Process Automation) where he currently serves as Vice President of Strategic Initiatives. Martin has BA and MS degrees in Mathematics, an MA degree in Administration and Management, and a PhD in Industrial Engineering.

# 36 Project Management and Execution

*By Michael D. Whitt*

## Topic Highlights

*Contracts*
- *Constraints*
- *Cost-Plus (CP)*
- *Time and Material (T&M)*
- *Time and Material/Not-To-Exceed (T&M/NTE)*
- *Lump-Sum, or Fixed-Price*
- *Hybrid*

*Project Life Cycle*
- *Feasibility Study*
- *Definition*
- *System Design*
- *Software Development*
- *Deployment*
- *Support*

*Project Management Tools*
- *The Scope of Work (SOW)*
- *The Estimate*
- *The Design Schedule*
- *The Status Report*

*Project Management Techniques*
- *Assessing Project Status*
- *Management of Change (MOC)*

## 36.1 Introduction

**Project:** A project is a temporary activity whose purpose is to create a product or service. Temporary projects have a defined beginning and end. Projects usually involve a sequence of tasks with definite starting and ending points. These points are bounded by time, resources, and end results.[1]

An engineering project is a means to an end, with a specific objective. For such a project to exist there must be a perceived need and an expectation that the need can be met with a reasonable investment. The owner must weigh the risks against the rewards and conclude the project is worthwhile. Making that risk/reward assessment is sometimes more of an art than a science. *Every project— particularly an automation project—involves some level of risk.*

1. Cockrell, Gerald W., *Practical Project Management: Learning to Manage the Professional* (ISA, 2001), p. 2.

*All pilots take chances from time to time, but knowing—not guessing—about what you can risk is often the critical difference between getting away with it and drilling a fifty-foot hole in Mother Earth.*

—Chuck Yeager, 1985[2]

More than in most endeavors, the effects of mishandling risk in an automation project can be catastrophic. Beyond the economic ramifications of a poor estimate, which are bad enough, the potential risk to the operators and the public at large can be extensive. Therefore, a well-conceived process of preliminary evaluation, short- and long-range planning, and project control is necessary. This evaluation begins with a thorough understanding of the issues. Like General Yeager, the key is to know, not to guess.

Proper project management starts with the project manager (PM). What are some attributes of a good PM? A good PM will:

- Understand who his real customer is and gain a thorough understanding of the forces driving the customer/owner (risk-taker) to make the investment in the first place.
- Be the customer's advocate when working within his own organization and be an advocate for his own organization when working with the customer. The PM must, more than any other individual on the project team, live with one foot in each camp.
- Have knowledge relating to the techniques of project management and the technological, logistical, and interpersonal challenges facing him.
- Ensure that each member of the project team has a thorough knowledge of the issues.
- Continually monitor the project's progress, measuring against the agreed-upon parameters.

Whether you are a project manager, or a project team member, a thorough understanding of the principles and concepts discussed here will broaden your avenue to success. The following major topics are discussed in this chapter:

- *Contracts* – What are some of the most common project types?
- *Project Life Cycle* – What is the normal order of things?
- *Project Management Tools* – What is in the project manager's toolbar (i.e. project controls)?
- *Project Execution* – What are some of the key techniques for managing an ongoing project?

## 36.2 Contracts

Each member of the project team defines *success* from his or her own unique perspective, as viewed through the prism of the project parameters. From the customer's point of view, success is achieved when the desired end is reached within the time allotted and/or the funds allocated. This is likely to be a broader interpretation than that of the service provider, who is also interested in making a profit.

A truly successful project is one in which both the customer and the service provider are satisfied with the outcome. For this to occur, a zone of success must be created that is as large as possible (see Figure 36-1).

A "success triangle" is formed at the point where the goods delivered meet the customer's cost and quality expectations, while still allowing the service provider to make a fair profit and maintain his reputation for good quality work. Staying within this comfort zone is sometimes a bit tricky, and it

2. Yeager, General Chuck and Leo Janos, *Yeager: An Autobiography* (Bantam Books, 1985), p. 84.

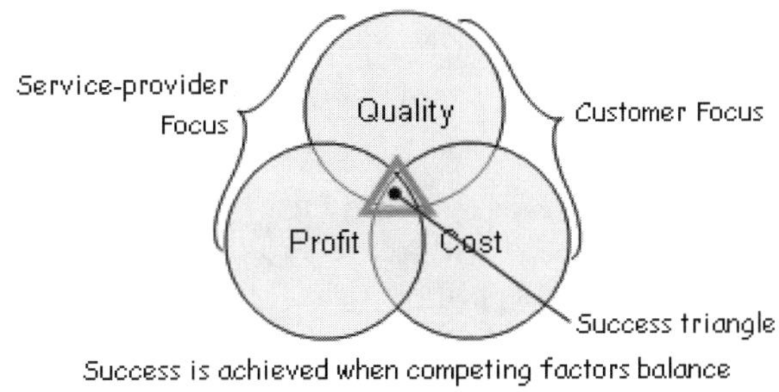


*Figure 36-1: Success Triangle*

really helps to understand the "physics" involved. The physics of a project are defined by its constraints.

### 36.2.1 Constraints

Time and resources are two parameters that impose limits on the design process (see Figure 36-2). Time, as it relates to a project, can mean either duration (calendar time) or intensity (labor hours). The term *time driven* in this context implies the project is constrained by the calendar; the term *cost driven* implies the project is constrained by cost. Cost is calculated by finding material cost and then measuring the level of intensity per unit of project time (in man-hours) required to design and construct.

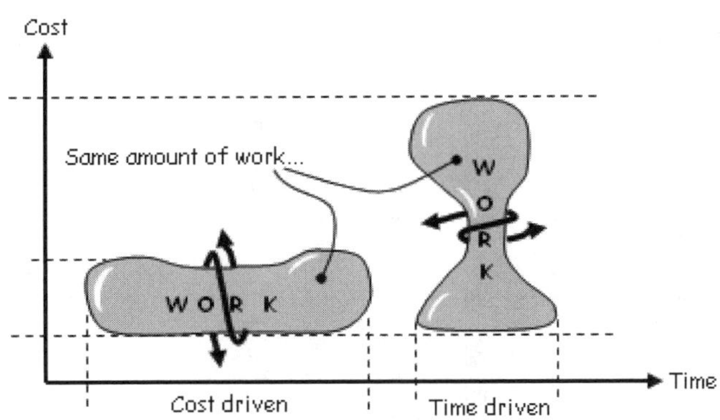


*Figure 36-2: Effects of Constraints on Project Structure*

The relationship between the customer and the engineer or constructor requires clear definition. If a vendor provides a quote, then that vendor is bound by it, and must provide the materials or services for the price offered in the quote. The same concept applies to the automation service provider (seller). As the seller, the act of submitting a proposal constitutes initiating the contracting process. If the customer (buyer) accepts the proposal, the seller is legally bound to execute according to the contract, and the buyer is legally bound to honor it. Following are some of the most common types of contracts:

- "Cost-Plus"
- "Time and Material" (also, "Time and Material, Not-to-Exceed")
- "Lump Sum" or "Fixed-Fee"

- "Turnkey"
- "Hybrid"

Each contract type strikes a different balance of risk/reward for each participant as noted in the following commentary.

Table 36-1 (below) depicts the relative risk/reward factors of several of the most common types of contracts. Each contract type is analyzed with respect to the project constraint, and rated on a scale of 0 – 4 as follows:

0: None

1: Minimal

2: Moderate

3: Maximum

A risk/reward ratio of 1/3, therefore, would indicate the condition has minimal risk, with maximum reward... a very desirable state, indeed, for the concerned party.

*Table 36-1: Risk/Reward Analysis*

| Contract Type | (Seller) | (Shared) | (Buyer) | Project Constraint |
|---|---|---|---|---|
| | **Profit** | **Quality** | **Cost** | |
| Cost-Plus | 0 / 1 | 0 / * | 3 / # | Content |
| Cost-Plus | 0 / 1 | 2 / * | 2 / # | Schedule |
| T&M | 0 / 2 | 0 / * | 3 / # | Content |
| T&M | 0 / 2 | 2 / * | 2 / # | Schedule |
| T&M, NTE | 1 / 2 | 0 / * | 1 / # | Content |
| T&M, NTE | 3 / 2 | 2 / * | 1 / # | Schedule |
| T&M, NTE | 3 / 2 | 2 / * | 1 / # | Cost |
| T&M, NTE | 3 / 2 | 3 / * | 1 / # | Cost & Schedule |
| Lump Sum | 1 / 3 | 0 / * | 3 / # | Content |
| Lump Sum | 3 / 3 | 2 / * | 2 / # | Schedule |
| Lump Sum | 3 / 3 | 2 / * | 0 / # | Cost |
| Lump Sum | 3 / 3 | 3 / * | 0 / # | Cost & Schedule |
| * High quality is the reward for all parties in all cases...<br># Low cost is the reward in all cases for the Buyer...<br>( ) Indicates concerned party: Seller, or Buyer | | | | |

The following commentaries discuss each of these major contract types in detail:

### 36.2.2 Cost-Plus (CP)

A cost-plus contract guarantees a profit for the seller. In return, the buyer may retain control over project content and the seller's method of execution. The parameters defined by the contract are unit rates—not project price. The unit rates can be applied to either the various employees' hourly rates, or to negotiated rates based on employee's classification (e.g., engineer, programmer, designer, clerk, etc.). These unit rates are valid over the life of the contract, which can extend into the future until the scope of work is satisfied. For the seller, this guarantees a minimal, negotiated profit regardless of project constraint (Profit: no risk, minimal reward). If the seller completes the task below budget, the buyer realizes a windfall. But, if the seller exceeds budget, the buyer must pay. However, if the seller is constrained by schedule, then a sort of budgetary control can be established that can mitigate the risk/reward ratio.

Refer to Table 36-1: If the constraint is content, the seller will eventually get what he wants (Quality: 0, no risk), but with a very real risk of high cost (Cost: 3, high risk). In this scenario, the seller may work indefinitely until the buyer is satisfied with the result. But, if schedule is defined as a constraint, then the risk to quality rises to moderate levels with a corresponding drop in the risk of cost overruns from high to moderate. The seller just has no time to spend an infinite amount of resources. Note that Cost is not a constraint in this format, even though the buyer always has a budget that was approved internally. The buyer's internal budget has no bearing on this type of contract. Though it is in the seller's best interest to know the buyer's target and try to work within it, he is not contractually obligated to do so.

### 36.2.3 Time and Material (T&M)

The T&M contract and the Cost-Plus contract are very similar. For Cost-Plus, the service provider is reimbursed for his cost, plus profit and expenses. The T&M contract reimburses for cost, plus profit and expenses, plus material and markup. Again, the parameters defined by the contract are unit rates – not project price. Therefore, the only major difference between the two is a higher potential profit for the seller due to his markup on materials (see Table 36-1). Please note that the T&M contract's risk/reward scenarios for Quality and Cost are the same as for the Cost-Plus format. Please refer to the narrative description for Cost-Plus for more insight into this contract type.

Both the T&M and the CP formats are tried-and-true structures that tend toward stable, long-term business relationships between buyer and seller—provided the seller guards the buyer's interests and manages the relationship. The allure of future project work is the carrot that offsets the seller's tendency to become complacent. Good project management techniques are as important for management of these contracts as for any other—perhaps more so, since the need for it is sometimes not readily apparent. See the section on Management of Change (MOC) later in this chapter.

### 36.2.4 Time and Material/Not-To-Exceed (T&M/NTE)

The CP and T&M scenarios place most of the budgetary risk squarely on the shoulders of the buyer. The "not-to-exceed" stipulation reverses this by adding the constraint of "cost." The seller's potential reward remains moderate, as in the straight T&M format, but his profit risk climbs to very high levels very quickly as constraints are added. In addition, if the seller finishes below budget the buyer receives the windfall—not the seller. High-risk/moderate reward for the seller causes an increase in his intensity and focus, and decreases his desire to work with the buyer to "tweak" the product. Thus, the buyer loses some control over the way the project is executed, and the likelihood of disagreements between the two organizations increases somewhat.

### 36.2.5 Lump-Sum, or Fixed-Price

The terms lump-sum and fixed-price are interchangeable. In the lump-sum contract, the parameters defined by the contract are a fixed project price for a fixed set of tasks or deliverables. Since the price is negotiated before the services are rendered, the fixed price contract minimizes cost risk for the buyer. Further, the buyer usually signs a contract only after a bidding process in which several potential sellers compete for the contract by providing their lowest bids. The buyer can either accept the lowest bid, or he can analyze them to determine the lowest "reasonable" bid that will save him the most money while letting him retain a sense of comfort that the seller can execute to the terms of the contract.

The profit risk to the seller depends on whether the service or product is "deterministic" or "probabilistic." If selling "widgets," in which little or no research and development is needed, the product is deterministic, and presumably the seller knows exactly how much resources and funds are required in their production. If making retrofit modifications to an existing facility in which drawings and/or software listings are out of date, for example, then the project is deemed probabilistic, and his or her risk is maximized. One way for the seller to mitigate risk, and also maximize reward, is to add "contingency" to the bid.

Contingency is a factor used to cover normal design development issues that are hard to quantify, and also to cover those things unknown. If the level of uncertainty is low, then the level of contingency can be low. If not, then contingency should be high. The proper amount of contingency reflects the balance between the level of uncertainty that exists, and the level of risk deemed acceptable. Sometimes the seller is willing to forego profit to retain staff, reducing contingency; sometimes the staff is busy, leading the seller to raise it.

Profit, for the seller, is at risk, depending on the scenarios described above. But, the possibility of reward is also high. If the seller manages to execute his or her task below budget, the buyer must pay the full amount of the contract. High risk/high reward.

A fixed-price, or lump-sum, project offers the buyer several benefits, foremost of which is an enhanced ability to allocate resources. The buyer can set aside project funds (plus a safety buffer) with a high level of confidence that additional funds will not be required. This works to the seller's advantage in planning other work. The buyer gains these benefits, but, in comparison to the cost-plus format, loses much control during the execution of the project. To submit fixed-price bids, bidders work at their own expense to clearly define not only the set of deliverables, but the methods they expect to employ to meet the owner's defined scope of work.

A fixed-cost project, if properly managed, can be the most efficient and effective format for both organizations. However, the seller must be ever vigilant in managing scope-creep.[3] Once the buyer accepts a bid, the seller has no obligation to adjust the deliverables or methods if it can be demonstrated that doing so will negatively affect his ability to turn a profit. If the buyer makes a request that is out of bounds with respect to the scope of work, the seller has the right—and obligation—to refuse the request until the buyer approves an engineering change order (See Management of Change). This defensive posture on the seller's part can lead to a fractious relationship with the customer if a previously agreed upon method of change control is not employed.

### 36.2.6 Hybrid

A particular project may exhibit several of the characteristics in the various scenarios described above. For example, the preliminary engineering (discovery) phase is frequently done on a cost-plus basis because there is simply not enough information available to produce a responsible fixed quote. Subsequent phases such as detail engineering and construction may then be done on a lump-sum basis.

## 36.3 Project Lifecycle

No matter what style of project, whether lump sum or T&M, most automation projects are similar in the way they are developed and executed. Figure 36-3 depicts the ISA Certified Automation Professional (CAP) program's model for this cycle.

### 36.3.1 Feasibility Study

Automation projects frequently begin on the production floor. A problem needs to be fixed or a process needs to be streamlined. For the issue to be addressed, someone must isolate the problem and prepare a written description detailing its properties and effects. Some of the areas to consider when describing the effects are as follows:

3. "Scope-creep" describes the case where the seller agrees to perform additional out-of-scope work without amending the contract.

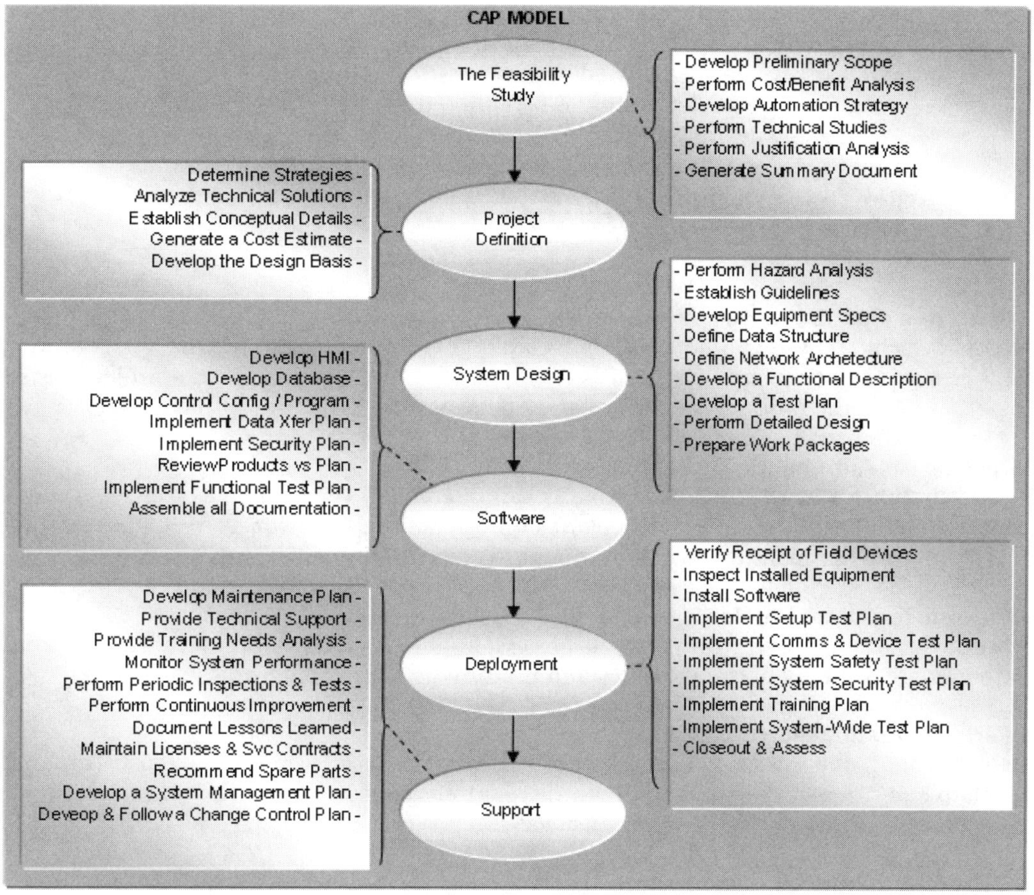


*Figure 36-3: ISA-CAP Model for Automation Project Flow*

**A) Describe the Need**

Issues to consider are:

- Personnel safety
- Production rate
- Product quality
- Equipment reliability
- Maintenance
- Operability

**B) Identify Possible Solutions**

- Analyze the problem
- Estimate the cost

**C) Develop a Preliminary Scope of Work**

Write a project scope of work that clearly describes the project goals. It should provide enough information to let someone prepare a reasonably accurate estimate with minimal expense. The following information should be discussed in the document:

- Document Lists
- Equipment Lists

- Software/hardware performance specifications
- Service provider performance criteria such as:
  - How should documents be transmitted?
  - What media should be used to prepare the documents? CADD? Manual?
  - What design standards should be adhered to? NFPA? NEC? Internal?
  - What is the desired timetable?
  - What specific deliverables will be expected?
- Approved vendors lists
- Safety concerns
- Security

#### D) Perform a Cost/Benefit Analysis

The cost/benefit analysis compares the most likely investment cost to the most likely return. The result of this analysis can be expressed in units of calendar-time. For example, if a project costs $10M, and the net profit on the product is expected to be $2M/yr, then the project has a 5-year payback.

#### E) Develop an Automation Strategy

The automation strategy must take into account operability issues such as plant shutdowns, equipment availability, and the state of the plant's infrastructure. It must accommodate the operations department and the maintenance department's concerns, as well as satisfy technology issues.

#### F) Perform Technical Studies

Technical studies to prove basic concepts are essential to a successful project. The aim of these studies is to eliminate as many of the variables and unknowns as possible.

#### G) Perform Justification Analysis

The justification analysis looks at marketplace effects and risk, in addition to cost. This analysis reviews the relative position of the company with respect to its competitors in the marketplace; it forecasts the effects of a successful project, and an unsuccessful project, in the marketplace.

#### H) Generate a Summary Document

The feasibility study should be published as a document that summarizes the findings of each of the aforementioned activities to support a value assessment of the merits of the project.

### 36.3.2 Definition

The definition phase of the project identifies customer requirements and completes a high-level analysis of the best way to meet those requirements.

#### A) Determine Operational Strategies

Key stakeholders should be identified and interviewed to establish the true impetus of the project. What is the need this project is envisioned to fulfill? The service provider should take the time to satisfy himself or herself that the customer has considered all the issues and will be happy with the results of the project if it is executed well. The Law of Unintended Consequences should be considered in this step.

#### B) Analyze Technical Solutions

Once the true reason for the project is identified, and the operational strategy developed, technical solutions to the problem should be sought. This can entail visits to sites with similar situations, vendor and/or manufacturer interviews, and sometimes test-bed evaluations or pilot projects.

#### C) Establish Conceptual Details

Several sets of documents are produced during the conceptual stages of the project. These documents are referred to as "upper-tier" documents. They describe the conceptual details to develop a design basis.

#### D) Generate a Cost Estimate

The conceptual details developed above provide a baseline to which the design may be applied. A detailed Work Breakdown Structure[4] (WBS)-based estimate can now be produced. Once the "first-draft" of the estimate is done, a review cycle should be used to look for cost-reduction characteristics such as economy of scale, duplication of work, etc.

#### E) Develop the Design Basis

The Design Basis is a detailed plan of execution for the System Design Phase. A complete list of deliverables is developed that is based on the research done in the Definition Phase.

### 36.3.3 System Design

Steps to the "System Design" phase of the project are:

#### A) Perform Hazard Analysis (HAZOP)[5]

After the design basis is defined to a fairly high degree, but before the detail design has begun, the envisioned system should be analyzed for operability and safety.

#### B) Establish Guidelines

Establish standards, templates, and guidelines as applied to the automation system using the information gathered in the definition stage and considering human-factor effects to satisfy customer design criteria and preferences.

#### C) Develop Equipment Specifications and Instrument Data Sheets

Create detailed equipment specifications and instrument data sheets based on vendor selection criteria, characteristics and conditions of the physical environment, regulations, and performance requirements to purchase long-lead-time equipment and support system design and development.

#### D) Define the Data Structure Layout and Data Flow Model

Define the data structure layout and data flow model, considering the volume and type of data involved to provide specifications for hardware selection and software development.

#### E) Select the Physical Communication Media, Network Architecture, and Protocols

Select the physical communication media, network architecture, and protocols based on data requirements to complete system design and support system development.

#### F) Develop a Functional Description of the Automation Solution

Develop a functional description of the automation solution (e.g., control scheme, alarms, HMI, reports) using rules established in the definition stage to guide development and programming.

#### G) Develop a Test Plan

Design the test plan using chosen methodologies to execute appropriate testing relative to functional requirements.

#### H) Perform Detail Design

Perform the detailed design for the project by converting the engineering and system design into purchase requisitions, drawings, panel designs, and installation details consistent with the specification and functional descriptions to provide detailed information for development and deployment.

#### I) Prepare Construction Work Packages

Prepare comprehensive construction work packages by organizing the detailed design information and documents to release project for construction.

---

4. Work Breakdown Structure: A task-based method of organizing a project, usually sub-grouped by process area or physical location.
5. HAZOP: Hazard and Operability Study. See *Process Safety and Safety Instrumented Systems* chapter.

**J) Procurement**

Several procurement activities occur during the latter part of the design phase. Delivery lead times must be considered with respect to timing of orders.

### 36.3.4 Software Development

If the system design phase is properly executed, the software development phase of the project should be simply a matter of executing to the plan. In most cases, the content being developed in the areas described below are defined in the plans produced previously. Following are some of the steps involved in developing software:

**A) Develop the Human-Machine Interface (HMI) System using:**

- The Alarm Grouping and Annunciation Plan
- The Alarm and Alarm Setpoint List
- The *Human-Machine Interface (HMI)* Screen Hierarchy and Navigation Plan
- The HMI Color Scheme and Animation Plan
- The Operational Security Plan

**B) Develop Database and Reporting Functions using:**

- The Data Historian List and Archival Plan
- The Data Backup and Restore Plan
- The Report and Report Scheduling Plan

**C) Develop the Control System Configuration and/or Program using:** [6]

- Logic Diagrams
- Device Control Detail Sheets (DCDS)
- Process Control Detail Sheets (PCDS)
- Sequence Control Detail Sheets (SCDS)
- Recipes
- Interlock Lists and Alarm/Trip Setpoint Lists

**D) Implement the Data Transfer Plan**

Implement data transfer methodology that maximizes throughput and ensures data integrity using communication protocols and specifications to assure efficiency and reliability.

**E) Implement the Operational Security and Data Integrity Plan**

Implement security methodology in accordance with stakeholder requirements to mitigate loss and risk.

**F) Perform a Scope Compliance Review**

Review configuration and programming using defined practices to establish compliance with functional requirements.

**G) Develop and Implement the Functional (or Factory) Acceptance Test Plan (FAT)**

Test the automation system using the test plan to determine compliance with functional requirements.

**H) Assemble Documentation, and Prepare for Turnover**

Assemble all required documentation and user manuals created during the development process to transfer essential knowledge to customers and end users.

6. Many of the deliverables listed in this section are described fully in *Successful Instrumentation and Control Systems Design* by Michael Whitt.

### 36.3.5 Deployment

The "deployment" phase, or construction phase, is the phase in which the design is implemented. The deployment phase frequently overlaps the design phase activities by some margin, beginning construction as WBS area designs are completed. During this time of overlap, the needs of the constructor ascend in importance over the needs of design. Work stoppages should be avoided at all costs, even to the point of interrupting ongoing design activities to concentrate on a phase-three problem.

Upon arrival at the site, the automation professional should begin the processes described below.

#### A) Verify Field Device Status

Perform receipt verification of all field devices by comparing vendor records against design specifications to ensure devices are as specified.

#### B) Inspect Installed Equipment

Perform physical inspection of installed equipment against construction drawings to ensure installation in accordance with design drawings and specifications. Construction activity status should be captured in a report similar to the Startup Readiness Report, shown below.

| Startup Readiness Report | | | | | | | |
|---|---|---|---|---|---|---|---|
| | Installation | | | | Checkout | | |
| Tag Number | Mounted? | Wired? | Tubed? | Rung Out? | Local | Remote | Summary |
| PT-10 | 1 | 1 | 1 | 1 | 1 | 1 | 100.0% |
| PT-11 | 1 | 1 | 1 | 1 | NA | | 80.0% |
| PT-12 | 1 | | 1 | | | | 33.3% |
| LT-15 | | | | | | | 0.0% |
| PCV-10 | 1 | 1 | | | | | 33.3% |
| HV-11 | | | | | | | 0.0% |

Excel readily lends itself to this type of use. The Summary field can be used to help develop a completion percentage if desired. In this case, the summary assumes each category is weighted equally. The numbers are summed and then simply divided by the number of categories. PT-10 has six categories and PT-11 has five. The equation "=SUM(C5:H5)/6" is loaded into the summary cell for PT-10, for example.

*Figure 36-4: Startup Readiness Report*

Using the instrument database and I/O list as a basis, a Startup Readiness Report can be produced that will make this an organized, verifiable process.

#### C) Install Hardware and Software

Install configuration and programs by loading them into the target devices to prepare for testing.

#### D) Perform Preliminary System Checks; Troubleshoot and Correct Faults

Solve unforeseen problems identified during installation using troubleshooting skills to correct deficiencies.

#### E) Perform the Site Acceptance Test for the Control Software

Test configuration and programming in accordance with the design documents by executing the test plan to verify the system operates as specified.

**F) Perform the Site Acceptance Test for Communications and Field Devices**
Test communication systems and field devices in accordance with design specifications to ensure proper operation. This is sometimes referred to as the "checkout," or "bump and stroke" phase, each field device should be exercised to demonstrate proper operation.

**G) Perform the Site Safety Test**
Test all safety elements and systems by executing test plans to ensure safety functions operate as designed.

**H) Perform the Site Security Test**
Test all security features by executing test plans to ensure security functions operate as designed.

**I) Execute the Training Plan**
Provide initial training for facility personnel in system operation and maintenance through classroom and hands-on training to ensure proper use of the system.

**J) Perform the Site System Integrity Test (i.e., Startup & Commissioning)**
Execute system-level tests in accordance with the test plan to ensure the entire system functions as designed. A successful test at this stage constitutes the system startup. The system is generally considered to be "commissioned" upon conclusion of this step. Startup should be well organized and sequential. A formal startup procedure is recommended.

**K) Closeout & Assess**
The final step in the deployment phase is the closeout step. This is the step in which:

- Project documentation is finalized to reflect as-built conditions (thus documenting changes that accrued during the deployment process). Redline markups that occurred due to construction adjustments should be transcribed into the original design package.
- Project documents are turned over to the end user.
- Final billing issues are resolved.
- A post-mortem is held to evaluate and discuss lessons learned, and to recognize high achievement.
- A project celebration is held to commemorate conclusion of the project. Recognizing the staff for their work is of great importance.

### 36.3.6 Support

The "support" phase of the project should be very important to the automation professional. In many cases, this is where the next big project is derived.

**A) Develop and Implement a System Performance Monitoring Plan**
Verify system performance and records periodically using established procedures to ensure compliance with standards, regulations, and best practices.

**B) Develop and Implement a Long-Term Technical Support Plan**
Provide technical support for facility personnel by applying system expertise to maximize system availability.

**C) Develop and Implement a Long-Term Training Plan**
Perform training needs analysis periodically for facility personnel using skill assessments to establish objectives for the training program.

**D) Provide Periodic Training for Facility Personnel**
Provide training for facility personnel by addressing identified objectives to ensure the skill level of personnel is adequate for the technology and products used in the system.

**E) Monitor System Performance**
Monitor performance using software and hardware diagnostic tools to support early detection of potential problems.

**F) Perform Periodic Inspections and Tests to Re-certify the System**
Perform periodic inspections and tests in accordance with written standards and procedures to verify system or component performance against requirements.

**G) Develop and Implement a Continuous Improvement Plan**
Perform continuous improvement by working with facility personnel to increase capacity, reliability, and/or efficiency.

**H) Document Lessons Learned**
Document lessons learned by reviewing the project with all stakeholders to improve future projects.

**I) Develop and Implement a License and Service Contract Maintenance Plan**
Maintain licenses, updates, and service contracts for software and equipment by reviewing both internal and external options to meet expectations for capability and availability.

**J) Provide a Recommended Spare Parts List**
Determine the need for spare parts, based on an assessment of installed base and probability of failure, to maximize system availability and minimize cost.

**K) Develop a System Management Plan**
Provide a system management plan by performing preventive maintenance, implementing backups, and designing recovery plans to avoid and recover from system failures.

**L) Develop and Implement a Change Control Plan**
Follow a process for authorization and implementation of changes in accordance with established standards or practices to safeguard system and documentation integrity.

## 36.4 Project Management Tools

A well-conceived, well-managed project is one that has been given a chance to succeed. A set of tools have been developed to help the project team stay on track. If properly developed and used, these tools will help define the tasks to be performed, facilitate accurate progress reporting, and provide early warning of potential cost overruns and/or scheduling conflicts. These tools include the scope of work, the project estimate, the project schedule, and various project reports.

### 36.4.1 The Scope of Work (SOW)

A scope of work is simply a description of tasks that need to be accomplished to achieve the desired end. It is the result of a cycle of investigation intended to eliminate or reduce the number of assumptions that need to be made later. Thus, an SOW should list each major task for each process area that falls under the scope of the project. Each task should then be evaluated from the perspective of each discipline. All execution statements should begin with verbs to convey the type of work being done.

Each task should have a unique number called a work breakdown structure (WBS) number. An estimate and material cost will be derived for this item and stored as part of the scope of work.

### 36.4.2 The Estimate

The estimate is an under-utilized tool for the project manager. All too often, a project is estimated using a set of criteria that have no bearing on how the project will be executed, yielding a throw-away document that has little or no use beyond generation of a price. The price generated might be accurate, but an opportunity has been lost to provide meaningful structure to the project. A better way is to build an estimate formatted to reflect the execution plan in the scope of work.

The relationship between the estimate and the scope of work results in a success envelope that is either big enough or too tight. Scope and schedule define the "mechanics" of the project. The estimate defines the "physics" of the situation, setting limits that affect the way in which the project will be executed. These physical measures are the primary tools of the project manager when tracking the project on the strategic level and of the design supervisor managing the work on the tactical level. The scope of work defines the work to be done. The material cost estimate and labor cost estimate define the parameters within which the work will be done.

There are three main types of estimates: budget, bid, and definitive. Following is a brief description of each:

- The *budgetary cost and labor estimate* is produced by a group intimate with the site and process. This group may or may not execute the remainder of the project. The purpose of this type of estimate is primarily to obtain initial funding. It is typically "quick and dirty" and is expected to be rather inaccurate. In fact, an error margin of ±30% is acceptable for this type of estimate. Prior to this estimate, a formal scope of work has probably not been done and, quite possibly, the project specification has not been finalized. This type of estimate is generally unfunded by the customer, or at least under-funded.
- The *bid material cost and labor estimate* is produced by the various engineering bidders vying for the work. Again, this estimate is typically unfunded by the customer.
- The *definitive material cost and labor estimate* is prepared by the engineering firm that was awarded the contract based on their bid. The customer typically includes this as a part of the project, so it probably is fully funded. Once the contract has been awarded and any secrecy agreement issues have been settled, the engineering contractor is given full access to the information developed by the customer during the internal evaluation process. The engineering firm then does some research to validate the basic assumptions and develops a finalized scope of work. This information sometimes alters the picture significantly, and the engineering contractor is given an opportunity to adapt the estimate and schedule to the new information. Of course, this re-estimating process is bypassed if full disclosure was made during the bid process. In that case, the bid estimate becomes the definitive estimate.

The estimate becomes the baseline document for project management, providing a yardstick to which performance will eventually be measured. It is based on the scope of work and should reflect the work scope on a task-by-task basis, wherever possible. Many estimating tools are available. Whichever is chosen, you should be able to produce estimates that are:

- *accurate*, by identifying each task relating to the scope of work and then quantifying the labor and expense of its execution.
- *timely*, by being repeatable.
- *verifiable*, by adding references, notes and amplifications, and by making the calculations available for dissection.
- *meaningful*, by relating all the deliverables to hard data.
- *adaptable*, by being easy to modify to accommodate "what-if" scenarios.

### 36.4.3 The Design Schedule

A schedule can be built from the project estimate, provided the estimate has enough detail and is task-oriented. Each task should be given a unique WBS identifier. Linked relationships should be identified and analyzed. In most cases, these relationships may be defined as follows:

- *Start-to-start*: Task B can't start until Task A starts. This relationship is used when an outside trigger initiates a scheduled chain of events. For example, receipt of a specific material shipment might trigger several tasks at the same time.
- *Finish-to-start*: Task B can't start until Task A finishes.
- *Start-to-finish*: Task A can't finish until Task B starts.
- *Finish-to-finish*: Task B can't finish until Task A finishes.

After the relationships are defined and the resources have been assigned to each task, the overall project duration may be estimated. The customer may have a time frame in mind already, which constrains the schedule to a particular set of milestones.[7] When the schedule is compared to this time line, resources may need to be modified to lengthen or shorten the schedule.

### 36.4.4 The Status Report

Periodically, the design team will be asked to provide some feedback as to their execution status. This feedback will answer questions such as the following: Did you start the project on time? How far along are you? Which tasks have you started, and when did you start them? Will you finish the project on time and within budget?

Using the WBS milestone schedule is a way to subdivide the available man hours and/or costs. This effectively breaks the budget into smaller, easier to manage sections that can be reported against individually. Data should be collected for each task (WBS item) and then plowed back into the overall project schedule by the PM. Status reporting typically includes updating the following parameters:

- Start date.
- Finish date.
- Percent complete.
- Man-hours estimate to complete (Manhrs ETC).

Figure 36-5 shows a status data collection form that has 10 WBS groupings. Each grouping is subdivided by six subtasks, which, in this case, equates to the six phases of a project (CAP Model), discussed previously. Of the 10, only WBS-T01 and WBS-T02 have been started. WBS-T01 covers the Railcar Unloading area of the plan, and has 250-manhours allocated. Ideally, the estimate would have been done in this format, with each subtask being estimated on its own merit, and the hours rolling up into the main task equaling 250. The Subtask Weight must equal 100%.

Total manhours, subtask weight, and subtask manhours values were loaded at the beginning of the project, forming the baseline data to which all updates are compared. The data to be updated are the project timeline dates and the Manhrs ETC. The scheduling team plows the date information back into the schedule, and the project manager plows the manhour data back into the budget for analysis (to be discussed later). Notice that the ETC for WBS-T01 shows 188 manhours to complete, with 250 available. From this, it might be inferred the design staff is 188/250 = 75% complete. This is incorrect, as we shall see.

---

7. Milestone – A schedule item pegged to a particular date. These usually represent targeted points in the engineering process, such as "issue drawings."

| Project Status Reporting Structure - by WBS Area | | | | | | | | | |
|---|---|---|---|---|---|---|---|---|---|
| WBS Item# | Description | Project Timeline Start | Finish | Float | Total Mhrs | Subtask Weight | Subtask Manhrs | % Comp | Manhrs ETC |
| T01 | Railcar Unloading | 04/03/05 | 03/01/07 | | 250.0 | | | | 188.0 |
| T01 -1 | Feasibility Study | 04/03/05 | 04/20/05 | 5d | | 10.0% | 25.0 | 100% | - |
| T01 -2 | Definition | 05/01/05 | 06/01/05 | 15d | | 10.0% | 25.0 | 95% | 8.0 |
| T01 -3 | System Design | 06/15/05 | 07/15/05 | 15d | | 25.0% | 62.5 | 25% | 40.0 |
| T01 -4 | Software | 08/01/05 | 09/01/05 | 15d | | 20.0% | 50.0 | 10% | 60.0 |
| T01 -5 | Deployment | 12/15/05 | 02/15/06 | 15d | | 25.0% | 62.5 | 0% | 56.0 |
| T01 -6 | Support | 03/01/06 | 03/01/07 | na | | 10.0% | 25.0 | 0% | 24.0 |
| T02 | Bulk Storage | 04/03/05 | 03/01/07 | | 400.0 | | | | 276.0 |
| T02 -1 | Feasibility Study | 04/03/05 | 04/20/05 | 5d | | 10.0% | 40.0 | 100% | 16.0 |
| T02 -2 | Definition | 05/01/05 | 06/01/05 | 15d | | 10.0% | 40.0 | 95% | 32.0 |
| T02 -3 | System Design | 06/15/05 | 07/15/05 | 15d | | 25.0% | 100.0 | 25% | 16.0 |
| T02 -4 | Software | 08/01/05 | 09/01/05 | 15d | | 20.0% | 80.0 | 10% | 72.0 |
| T02 -5 | Deployment | 12/15/05 | 02/15/06 | 15d | | 25.0% | 100.0 | 0% | 100.0 |
| T02 -6 | Support | 03/01/06 | 03/01/07 | na | | 10.0% | 40.0 | 0% | 40.0 |
| T03 | River Water | 04/03/05 | 03/01/07 | | 80.0 | | | | 80.0 |
| T04 | Fire Protection | 04/03/05 | 03/01/07 | | 400.0 | | | | 376.0 |
| T05 | Material Handling | 04/03/05 | 03/01/07 | | 1,000.0 | | | | 980.0 |
| T06 | Crush & Slurry | 04/03/05 | 03/01/07 | | 1,500.0 | | | | 1,490.0 |
| T07 | Convey | 04/03/05 | 03/01/07 | | 400.0 | | | | 400.0 |
| T08 | Truck Loading | 04/03/05 | 03/01/07 | | 80.0 | | | | 80.0 |
| T09 | System Services | 04/03/05 | 03/01/07 | | 80.0 | | | | 80.0 |
| T10 | Miscellaneous | 04/03/05 | 03/01/07 | | 80.0 | | | | 80.0 |
| | Project Summary: | 04/03/05 | 03/01/07 | | 4,270.0 | | | | 4,030.0 |

*Figure 36-5: Reporting Project Status*

## 36.5 Project Management Techniques

### 36.5.1 Assessing Project Status

The project status update data that was collected in Section 36.4.4 can be used by the project manager to develop additional data that will help forecast the likelihood of success for the project (See Figure 36-6). Some of the additional information needed will be:

- *Earned hours = Estimated percent complete X budgeted man hours.*
- *Actual hours:* The number of hours actually expended to date, as reported by the timesheet system.
- *Estimate to complete (ETC):* Entered by the design team to reflect the amount of work remaining.
- *Estimate at completion (EAC) = Manhrs actual + manhrs ETC;* Indicates the expected final manhrs.
- *Apparent percent complete = Manhrs actual/allocated manhrs; m*easures actual vs. initial expectations.
- *Actual percent complete = Manhrs actual/manhrs EAC; m*easures actual vs. final expectations.
- *Efficiency rating = Actual percent complete/apparent percent complete.*

| Project Status Reporting Structure - by WBS Area | | | | | | | | | | | | | | | |
|---|---|---|---|---|---|---|---|---|---|---|---|---|---|---|---|
| WBS Item# | Description | Project Timeline Start | Finish | Float | Budget Mhrs | Subtask Weight | Allocated Manhrs | Est. %Comp | Manhrs Earned | Manhrs Actual | Manhrs ETC | Manhrs EAC | Apparent % Comp | Actual %Comp | Efficiency Rating |
| T01 | Railcar Unloading | 04/03/05 | 03/01/07 | | 250.0 | | | 42% | 69.4 | 106.0 | 188.0 | 294.0 | 42% | 36% | 0.85 |
| T01 -1 | Feasibility Study | 04/03/05 | 04/20/05 | 5d | | 10.0% | 25.0 | 100% | 25.0 | 32.0 | - | 32.0 | 128% | 100% | 0.78 |
| T01 -2 | Definition | 05/01/05 | 06/01/05 | 15d | | 10.0% | 25.0 | 95% | 23.8 | 40.0 | 8.0 | 48.0 | 160% | 83% | 0.52 |
| T01 -3 | System Design | 06/15/05 | 07/15/05 | 15d | | 25.0% | 62.5 | 25% | 15.6 | 32.0 | 40.0 | 72.0 | 51% | 44% | 0.87 |
| T01 -4 | Software | 08/01/05 | 09/01/05 | 15d | | 20.0% | 50.0 | 10% | 5.0 | 2.0 | 60.0 | 62.0 | 4% | 3% | 0.81 |
| T01 -5 | Deployment | 12/15/05 | 02/15/06 | 15d | | 25.0% | 62.5 | 0% | - | - | 56.0 | 56.0 | 0% | 0% | - |
| T01 -6 | Support | 03/01/06 | 03/01/07 | na | | 10.0% | 25.0 | 0% | - | - | 24.0 | 24.0 | 0% | 0% | - |
| T02 | Bulk Storage | 04/03/05 | 03/01/07 | | 400.0 | | | 18% | 111.0 | 72.0 | 276.0 | 348.0 | 18% | 21% | 1.15 |
| T02 -1 | Feasibility Study | 04/03/05 | 04/20/05 | 5d | | 10.0% | 40.0 | 100% | 40.0 | 16.0 | 16.0 | 32.0 | 40% | 50% | 1.25 |
| T02 -2 | Definition | 05/01/05 | 06/01/05 | 15d | | 10.0% | 40.0 | 95% | 38.0 | 32.0 | 32.0 | 64.0 | 80% | 50% | 0.63 |
| T02 -3 | System Design | 06/15/05 | 07/15/05 | 15d | | 25.0% | 100.0 | 25% | 25.0 | 16.0 | 16.0 | 32.0 | 16% | 50% | 3.13 |
| T02 -4 | Software | 08/01/05 | 09/01/05 | 15d | | 20.0% | 80.0 | 10% | 8.0 | 8.0 | 72.0 | 80.0 | 10% | 10% | 1.00 |
| T02 -5 | Deployment | 12/15/05 | 02/15/06 | 15d | | 25.0% | 100.0 | 0% | - | - | 100.0 | 100.0 | 0% | 0% | - |
| T02 -6 | Support | 03/01/06 | 03/01/07 | na | | 10.0% | 40.0 | 0% | - | - | 40.0 | 40.0 | 0% | 0% | - |
| T03 | River Water | 04/03/05 | 03/01/07 | | 80.0 | | | 0% | - | - | 80.0 | 80.0 | 0% | 0% | - |
| T04 | Fire Protection | 04/03/05 | 03/01/07 | | 400.0 | | | 0% | - | - | 376.0 | 376.0 | 0% | 0% | - |
| T05 | Material Handling | 04/03/05 | 03/01/07 | | 1,000.0 | | | 0% | - | - | 980.0 | 980.0 | 0% | 0% | - |
| T06 | Crush & Slurry | 04/03/05 | 03/01/07 | | 1,500.0 | | | 0% | - | - | 1,490.0 | 1,490.0 | 0% | 0% | - |
| T07 | Convey | 04/03/05 | 03/01/07 | | 400.0 | | | 0% | - | - | 400.0 | 400.0 | 0% | 0% | - |
| T08 | Truck Loading | 04/03/05 | 03/01/07 | | 80.0 | | | 0% | - | - | 80.0 | 80.0 | 0% | 0% | - |
| T09 | System Services | 04/03/05 | 03/01/07 | | 80.0 | | | 0% | - | - | 80.0 | 80.0 | 0% | 0% | - |
| T10 | Miscellaneous | 04/03/05 | 03/01/07 | | 80.0 | | | 0% | - | - | 80.0 | 80.0 | 0% | 0% | - |
| | Project Summary: | 04/03/05 | 03/01/07 | | 4,270.0 | | | | 180.4 | 178.0 | 4,030.0 | 4,208.0 | 4% | 4% | 1.01 |

*Note:* Efficiency rating is calculated by dividing "Apparent % Comp" by "Actual % Comp".

*Figure 36-6: Analyzing Project Status*

Note the differences between the estimated, apparent, and actual completion percentages. In the case of WBS-T01-2, the design staff thought they were 95% complete (estimated). But, they had already exceeded the original budget by 60% (apparent), and should have been done a long time ago. From the standpoint of the EAC, the one that counts, they are really at 83% (actual).

### 36.5.2 Management of Change (MOC)

Every project, regardless of the type of contract, has both implicit and explicit sets of expectations. This is because it is impossible to spell out every permutation in a contract. Project parameters as noted in the contract (explicit expectations) can be met, and yet the project can still fail due to poor relationships, misunderstandings, bad perceptions, and the like (implicit expectations).

Thinking back to the Success Triangle (Figure 36-1), the buyer's focus is on quality and cost; the seller's focus is on quality and profit. Few projects get off the ground if either party knows that the triangle is either too small, or non-existent. However, projects still fail. So, in those cases there must be a disconnect between what is suspected at the beginning, and what is known at the end.

The key to this mystery is both parties' ability to manage change. The type of contract has a bearing on the amount of energy that is put into the MOC process, but MOC needs to be addressed in some fashion across the board. In fact, in the most benign case of the cost-plus contract, the buyer still has an expectation of ultimate price. If the scope of work is allowed to expand unchecked over a long period of time, no budgetary quote is safe and the likelihood of the buyer's remaining happy with the seller, regardless of the seller's abilities, is remote.

Therefore, it is important that the management of change (MOC) topic be discussed openly and frankly at the project's inception. A procedure for processing change notices should be approved by all parties, and documented thoroughly—even to the point of amending the contract if necessary.

To manage change, a good baseline should be established. This baseline is the design basis. The design basis includes not only the list of deliverables, but also the project execution plan (PEP). The PEP incorporates the scope of work, and details methods and assumptions. Every member of the project team should be cognizant of the design basis, and should continually compare the tasks being done to that document. As changes are discovered, a change notice should be generated to be passed through the management team for review and approval.

## 36.6 References

Cockrell, Gerald W. *Practical Project Management: Learning to Manage the Professional.* ISA, 2001.

Courter, Gini, and Annette Marquis. *Mastering Microsoft® Project 2000.* Sybex Books, 2000.

ANSI/PMI 99-001-2004, *A Guide to the Project Management Body of Knowledge.* Third Edition. Project Management Institute, 2004.

Whitt, Michael D. *Successful Instrumentation and Control Systems Design.* ISA, 2004.

Scholtes, Peter R. *The Team Handbook: How to Use Teams to Improve Quality.* Joiner Associates, 1988.

## About the Author

**Michael D. Whitt**'s experience includes 15-years with Raytheon Engineers and Constructors as an I&C Design Supervisor; five years with Mesa Associates, Inc., as Industrial Controls Department Manager responsible for Instrumentation Engineering & Design, Systems Integration, and Panel Fabrication groups. He is the author of *Successful Instrumentation and Control Systems Design* (ISA, 2004). He is a native of Asheville, N.C.

# 37 Interpersonal Skills

*By David Adler*

## Topic Highlights

*Communication*
- *Presentations*
- *Reports*
- *E-Mail Etiquette*
- *Conversations*

*People*
- *Roles and Responsibilities*
- *Social Styles*
- *Interviews*
- *Negotiations*
- *Conflict Resolution*

*Leadership*
- *Motivation*
- *Leading Meetings*
- *Leading Teams*
- *Working with Stakeholders*
- *Building Consensus*

## 37.1 Introduction

This chapter covers the topic of interpersonal skills needed by an automation team member, team leader, or project manager. It covers communicating, writing, interviewing, motivating, negotiating, leading teams, and leading meetings. What an ambitious goal to cover in only a few pages!

No, the author of this topic is not a human resources professional. Rather, I am a 30-year practicing automation professional who has had many leadership opportunities. I also have tuned process control loops, wrote application code for many production facilities, and managed numerous automation projects. I have a passion for the automation discipline and a love for automation professionals. I have found over the years that automation professionals need to 1) know the process and the equipment, 2) have appropriate business processes and project management skills, and 3) have interpersonal skills. A balance of these technical, business, and interpersonal skills are needed on every project or team. All my projects had hardworking and technically competent professionals. The successful projects, however, also had:

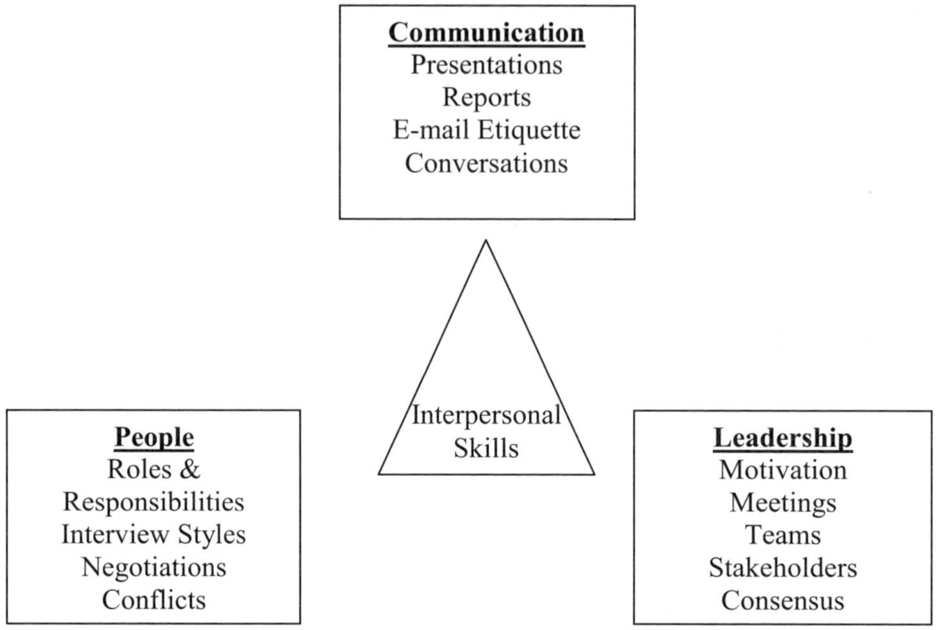


*Figure 37-1: Automation Professional Interpersonal Skill Domain*

1. a strong purpose and plan,
2. genuine respect and caring among co-workers, and
3. high levels of personal integrity and accountability from all participants.

## 37.2 Communication

Automation engineers need to communicate their ideas and accomplishments. Yet many technical professionals do not have any confidence in their ability to describe and sell automation projects and technology to their business leaders. With preparation and practice, communication skills can be developed just like any technical skill.

### 37.2.1 Presentations

What are some general tips for speaking and presenting effectively? (Gruhn 2003)

1. Make sure you have a crisp message. Outline your thoughts ahead of time. Practice.
2. Speak clearly. Manage your pace. Vary your tone. Be enthusiastic. Remember, it is not just what you say but how you say it.
3. Speak with the confidence of an expert, but recognize that the audience may have valuable experiences and insights to share.
4. Look for non-verbals in your listeners (eye contact, posture, facial expression, and hands). If you are losing your audience, ask them a question.

### 37.2.2 Reports

Take the time to summarize your work in a report. With the pressure of day-to-day assignments increasing all the time, writing a report is lost in the shuffle by the automation engineer when a project or assignment is completed. However, don't consider your project complete until the final report is finished. What are some rules for successful reports? (Ulman 1972)

1. Consider your audience. Avoid the overuse of process control system jargon. Hopefully, business leaders will want to read your report, and they will want to understand it.
2. Give the reader a clear idea of what it is you are going to talk about at the beginning of the report. A one-page summary and the purpose on the first page of the report is a great way to start.
3. Document the details in the body of the report. Keep it clear and crisp. Don't make it too long. Don't hide your important work under a mountain of minor details. Make liberal use of headings. Make sure to repeat the most important points.
4. Use active verbs. Be positive. Celebrate your and your co-workers' success.
5. Use data in charts and graphs. It is an effective way to reinforce and emphasize key points.
6. Be sure to include the date on every technical report.
7. Be sure to state your conclusions and recommendations.
8. When you are finished with the report, ask yourself if you clearly articulated a crisp theme. Ask a peer to review.

### 37.2.3 E-mail Etiquette

When you write an e-mail, consider the audience. The ease of sending an electronic message can cause some loss of the quality of your written communications. But the "brain dead" e-mail that was just sent can quickly ripple throughout the whole corporation in a matter of minutes. Take the time to proof it. And if you know the content is controversial, get a trusted peer to read it before you send it out.

### 37.2.4 Conversations

Automation professionals need to influence others in their daily conversations. Communicating between automation professionals requires you to create a dialogue that is open and honest. It requires speaking and listening skills. The conversation needs to stay focused on what you really want, but it has to be safe for others to talk to you. Be open to critical feedback, and don't take it personally. Speak persuasively and not abrasively. A training associate once told me communicating is just 20% what you say and 80% how you say it. Listen to what is said, understand it, and then respond. Credibility and trust is more a function of good listening than being a good speaker. (Patterson 2002)

## 37.3 People

Successful teams of automation professionals have the right personnel with solid skills in all the roles needed and a diverse set of social styles. Selecting the right team of automation professionals with all the needed skills is important because people are the most important factor in achieving the automation team's objectives. (Boehm 1981) Automation professionals must also work together on teams and need to learn to negotiate issues and to resolve their conflicts with their peers.

### 37.3.1 Roles and Responsibilities

As with any major undertaking that involves a group of professionals working together, it is important to define each individual's roles and responsibilities on an automation team. This will ensure each professional knows his area of responsibilities. It will minimize overlap of effort and reduce the possibility of a gap in deliverables. Some examples of roles are automation leader, project manager, automation engineer, testing engineer, and instrument engineer. (Smith 2005)

The *automation leader* is the supervisor of all automation personnel involved. He is accountable for ensuring that design meets operational needs and approves requirement, design, testing plans, and records. He ensures implementation meets corporate standards. He interfaces with site operations and maintenance leaders. He is accountable for testing, acceptance, and start-up.

The *automation project manager* is accountable for meeting the approved scope, schedule, and cost plans. He prepares the project plan. He monitors progress against plan and produces a written report. He coordinates team activities. He interfaces with the overall project leaders and produces project metrics.

The *automation engineer* provides overall technical directions and is responsible for the technical execution of the work. This engineer understands the processing equipment and the control strategy. He defines the automation scope and functional requirements. He designs and implements the automation application. He also interfaces with process engineering and operations.

The *testing engineer* is responsible for the commissioning activities and oversees the documentation process. He ensures the project is delivered to site standards. He drives the completion of automation testing and commissioning activities. He ensures this work is executed in a logical fashion according to plan and cost.

The *instrument engineer* is responsible for the design of the hardware and produces instrument specification, field layout, and design drawings. He ensures the deliverables are listed, understood, and produced on schedule. He coordinates design document review and approval. He oversees procurement and installation. He participates in production start-up.

### 37.3.2 Social Styles

Social styles can help you understand how to manage interpersonal relationships. One must start by understanding one's own social style. Understanding one's style and others with different styles can help us to relate to each other in more positive ways. Social style does not define our knowledge, experience, or work capacity. The four social styles are driver, expressive, amiable, and analytical. Roughly 25% of a normal population will fall into each of these four styles. (Wilson 1999)

The *driver* gets things done now. He can take independent action and has tight control. To some he can seem overpowering and overly demanding. On an automation project, he will get the project completed on schedule, but he might leave a few dead bodies in his wake.

The *expressive* is extremely creative, has an intuitive understanding of the overall project scope, and can create a lot of excitement and energy. He can be seen as impractical by some and lacking a detail-oriented approach to completing the task. On an automation project, he always has the best overall view of how the whole project fits together, but he often needs help to complete his deliverables.

The *amiable* gets broad acceptance of common objectives and gets everyone to work together in a positive environment. To some, he might be slow to make decisions and seem risk adverse. On an automation project, he is the one who keeps the team working together as a cohesive happy team, but he often is not worried about project costs and schedule.

The *analytical* thinks through problems systematically and does a thorough analysis. He is a deep thinker and solves hard problems. On an automation project, he will solve the most difficult control problem, but he will often get lost in the details and lose sight of the big picture.

There is no "best" social style. Automation professionals can be successful with any social style. It is our flexibility in relating to others with different social styles that promotes success. In fact, I have found it is best to have a balance of all social styles on a project team.

### 37.3.3 Interviews

The selection of personnel is the most important decision an automation leader makes. No other decision will have a longer lasting impact and be more difficult to undo. Take the extra time to get the right professional that has technical skills, integrity, and interpersonal skills. It is not uncommon for automation leaders to spend huge blocks of time selecting the process control system vendor and architecture, while spending less time selecting personnel. I have found over the years the right team of automation professionals can make even the worst vendors' hardware and software systems work successfully. Let me select the team of automation professionals and let anyone pick the distributed control system (DCS) or programmable logic controller (PLC); the team will make it work. A key skill in selecting the right personnel is the interview process. (Byham 1979)

Before you start interviewing, make sure you have defined the needed roles and responsibilities for the automation team. Write a job description. Acquire the needed training and guidance to know legally what questions are off limits. Develop the questions you will ask before the interview. Review the applicant's resume. Three key areas of focus for these questions are 1) leadership and problem solving skills, 2) team work and interpersonal skills, and 3) technical proficiency.

At the start of the interview, build rapport with the candidate, as it puts him or her at ease. Describe the position and the selection process. Take written notes. Manage the pace. Keep the candidate focused or draw him or her out, as appropriate. Avoid making a quick decision in the first few minutes of an interview. Focus on all three specific areas. Don't let the candidate talk about what "we" did, but probe to find out in particular what "I" did.

### 37.3.4 Negotiations

Many automation professionals need to negotiate with their peers, customers, and vendors. With a systemic approach and preparation, an automation professional can increase his chances of getting the outcome he wants. Take the time to prepare when negotiating. Some negotiating techniques are: take it or leave it, silence, surprise, good cop/bad cop, and deadlock. What are the keys to successful negotiation? (Benedict 2005)

1. Know your leverage. Aim high. Know what it is you want.
2. The first few minutes are important, so watch your words. Don't use weak words such as "we like" or "we prefer" but rather use "I need" or "It's essential." Take control of the meeting at the beginning of negotiations. Don't be afraid to change the agenda to your advantage. Stay assertive, confident, and non-manipulative.
3. Trade rather then make concessions. Give each concession its highest psychological value. Look for trades of high value to them, but of low value to you.
4. Negotiate with honesty and integrity.

### 37.3.5 Conflict Resolution

Conflict is a normal part of automation projects. The need to balance schedule, cost, and scope, while maintaining high quality instrumentation design, application development, and process control capability, creates tension. This tension can help engineers to think harder and create even better solutions. But if two strongly opinionated automation professionals choose different paths for the technical solution, and then make it personal, it can result in conflict. Then there is a need to manage the disagreement.

Common responses to conflict are avoid, minimize, attack, compromise, and problem-solve. What do you do if you find yourself in a serious conflict? (Davis 1992)

1. Listen. Put yourself in the other person's situation. Look at the other person's point of view. Don't lecture on why you are right. Attack the problem and not the person.
2. Understand the underlying problem. When you think a topic has been discussed, summarize and ask the other person if all the issues are on the table.

3. Allow the other person to vent. If it gets out of hand, stop the discussion and try to solve the conflict another day.

4. Find areas of mutual agreement. Clearly state the area of agreement and then try to resolve any remaining issues.

If this is a recurring issue, or the parties involved can't resolve the conflicts, the leader shouldn't hesitate to get involved. The team leader needs to help the automation professionals focus on facts and not on personalities. Make sure each is listening, and find common ground. Set up a plan to resolve and follow-up. Don't let the conflict fester and impact the whole team. If necessary, get help from a human resources professional.

## 37.4 Leadership

Automation professionals are often asked to take a leadership role in projects, daily assignments, and meetings. The leader of a successful team must have: 1) good communication skills, 2) good people skills, and 3) good decision-making skills. (Scholtes 2003) Leaders need to serve the other team members. (Block 1996) They need to see the value of influencing individuals as an opportunity for each team member to develop his or her unique gifts. The leader motivates and builds consensus to achieve the common objective. She or he needs to inspire the team members and work with the team's many stakeholders.

### 37.4.1 Motivation

Individuals must be able to find their own motivation to perform. As an automation leader or project manager, you can provide an environment in which personal motivation will flourish. The top cause of loss of motivation is lack of control over one's job. (Connors 1994) An automation leader needs to create an environment of accountability and help team members understand they have responsibility over their work. The leader needs to ensure work/life balance and lead by example. The automation team leader needs to make coaching team members a priority. How does a team leader coach? (Davis 1992)

1. Develop a partnership between leader and automation professional. Trust each other. This first step is often the hardest. Ask each other for feedback. Mutually develop a few critical objectives and priorities.

2. Promote persistence and self-discipline. The leader focuses on results and not just activity. The leader holds the team member accountable to agreed commitments and expects results. He or she celebrates the successes and encourages in a positive way the need to improve.

3. The leader removes barriers, eliminating unnecessary permissions and approvals. She or he gives the support needed.

4. The leader encourages lifetime learning. New skills allow engineers to perform at the highest level.

A good leader also recognizes when it is not appropriate to coach—for example, when the employee is at capacity and any more feedback will overwhelm; the setting is public, such as a group meeting; the leader is not prepared, pressed for time, or preoccupied with other problems; and when the feedback is too general or dated to be of value. Feedback needs to be timely and constructive.

### 37.4.2 Leading Meetings

Many automation professionals dislike meetings. They see them as an obstacle or time away from getting the "real" engineering work done. But the need to sell your ideas often are best done in a meeting. The needed group consensus can often be generated by the group discussions and decisions in a good team meeting. By following a few simple rules of pre-work and meeting management, it is possible to achieve the desired outcome from a meeting.

Take the time to plan the meeting. Define the meeting's purpose. Make sure there is business value in having the participants come together. Develop an agenda and include a list of topics to discuss, and expected results. Define the length. Don't schedule the meeting for longer than is needed. Prepare presentation and handouts. Make the necessary logistical arrangements for audio, video, and/or teleconferencing. How do you lead the meeting? (Streibel 2003)

1. Start the meeting on time. Establish the meeting rules.
2. Assign a facilitator to manage time, keep the group on agenda, and drive the meeting to the desired outcomes.
3. Manage the tone and tempo. Make sure all participate. Don't allow just a few to dominate the discussion. Manage the participants' need to digress into side issues. Have fun.
4. Drive the team to the expected results. Make a decision. Finish on time.
5. Take minutes and publish.

The best way to improve meetings is to make sure that every participant attends with a sense of the meeting's purpose. This is accomplished by publishing an agenda before the meeting to give participants ample time to prepare. Another way to improve meetings is to ask the meeting participants for feedback. Take five minutes at the end of the meeting and give each participant a chance to comment on what went well and what needs to be improved. Finally, if the meeting was worth having, make sure it generates action. Publish the minutes and action items. Follow-up and ensure action item completion.

### 37.4.3 Leading Teams

Fostering teamwork requires focus by the automation team leader. In today's complex automation environment, delivering projects, and resolving problems would be impossible without teams. No individual automation professional can know all there is to know about instrumentation, system applications, and automation. Teams go through normal dynamics that can be categorized as forming, storming, norming, and then performing. (Levi 2001)

**The Team Forms**. When a team forms, it can generate a lot of excitement and positive energy. But some team members can also be concerned and uncertain about the new activity. The team builds its relationships. The leader must clearly and concisely articulate the team's purpose and priorities.

**The Team Storms**. Before a team can get to the highest level of performance, it must learn to work together. This requires testing roles, authority, and boundaries. Team members learn to work through differences of opinions. The team leader encourages equal participation, and the team grows from the tension as ideas are discussed.

**The Team Norms**. A team has learned to work through differences of opinions and to solve common problems. They accept the team's norms and each other's roles. They are able to give each other feedback. Each team member feels empowered, works hard, and is able to solve problems. The project is now making progress. The team leader is able to share significant responsibilities.

**The Team Performs**. The individual team members now cooperate and perform as a cohesive team. The team delivers at a higher level than the sum of each individual's efforts. Each team member supports and complements the other team members to achieve the highest level of performance. Individuals not only take pride in their work, they also take satisfaction in the overall contribution of the whole team. The leader celebrates the success of the team.

### 37.4.4 Working with Stakeholders

Automation professionals may have many stakeholders. These stakeholders might include the leadership of operations, process engineering, plant engineering, shops, information technology, technical services, quality control, and numerous manufacturing business leaders. To work effectively with stakeholders on a difficult issue or project requires: 1) understanding their problems, 2) delivering good workable solutions, and 3) anticipating future needs. (Block 2000)

*Understand their problems.* Take the time to meet directly with each stakeholder. Find out each stakeholder's expectations, and state clearly your desire for a balanced relationship. Be willing to say "no" to unreasonable stakeholder requests. Develop a service-level agreement with specific deadlines and get the stakeholders to sign off. Remember the purpose of all this planning is to generate action.

*Deliver solutions.* Implementation is more difficult than planning. Make sure you get your hands dirty in the details of the problem. Solicit feedback throughout from stakeholders. Understand that stakeholder criticism and resistance is not directed at you personally. Persevere in solving any issues. Train operators and teach them to solve problems. At the end of a project, take the time to celebrate your success. And for those few things that did not go well, learn from it and forgive.

*Anticipate future needs.* Discuss with stakeholders your strengths and weaknesses. Don't assume that the stakeholders' understanding is the same as yours. Be committed to never ending improvement. Grow new skills and see learning as an adventure. Engage with the stakeholders in long-range planning.

The automation professional must try to influence stakeholders—often without any direct power to make changes or implement programs. The goal is to develop a collaborative relationship with stakeholders, to solve their problems so they stay solved, and to ensure attention is given to both technical problems and interpersonal relationships. Trust and respect from stakeholders are gained by completing each requirement in a quality manner, on schedule, and within cost.

### 37.4.5 Building Consensus

An automation organization needs to have focus and alignment to achieve results. A common purpose is the best way to build consensus among automation team members. Without a common purpose, the team's effort will be random, and it will take heroic efforts and good luck to achieve results. The four steps are: 1) understand current givens, 2) define the future state, 3) define changes needed, and 4) plan and implement.

*Understand the current givens.* Understand your operation's purpose and trends in automation technology. Benchmark other automation groups. Define your team's strengths and weaknesses. This step defines clearly your current state.

*Define the future state.* The automation team develops its purpose. What does this automation team want to accomplish, and how will we achieve it? Define the new investments needed. This step develops a high level view of the future that automation will bring to support production operations or to implement a project.

*Define changes needed.* How will we align organizational structure, leadership, people, management, work process, and culture to achieve? Define the business value this will bring to our customers. This step defines the new capabilities needed to move to this new future state.

*Plan and implement.* The final step is to develop and execute an operational plan. Develop specific scope, schedules, and costs to achieve. Develop an action plan and track progress.

Automation leaders must have a clear purpose and articulate it. They must build consensus in their organizations and inspire team members and stakeholders with their enthusiasm. Building consensus depends on the support of your stakeholders. Equally important is getting the active engagement and buy-in from every team member. Explain the rationale. Ask each team member or stakeholder for their view of the future and how automation fits into the success of operations. And, finally, be prepared to give this crisp message many times. A wise production manager once told me, if you wish to achieve change and have everyone buy-in, you need to be prepared to give your pep talk 37 different times.

## 37.5 Conclusion

People are the most important factor in an automation team achieving its objectives. Develop your interpersonal skills just like you develop your technical skills. Encourage and support other team members to develop and maximize their communication, people, and leadership skills.

If you find your interest has been sparked by this topic, I hope you will read one of the many excellent books on interpersonal skills referenced below and in this chapter.

## 37.6 References


1. Benedict, Robert. "Real World Negotiating." (Seminar) Benedict Negotiating Seminars, Inc. Valrico, Florida. Web site: www.backdoorselling.com
2. Block, Peter. *Flawless Consulting: A Guide to Getting Your Expertise Used.* Jossey-Bass Pfeiffer, 2000.
3. Block, Peter. *Stewardship: Choosing Service Over Self-Interest.* Berrett-Koehler Publishers, 1996.
4. Boehm, Barry W. *Software Engineering Economics.* Prentice Hall, 1981.
5. Byham, William C. *Targeted Selection®: A Behavioral Approach to Improved Hiring Decisions.* (Monograph) Development Dimensions International, 1979 (revised 2004). Website: www.ddiworld.com
6. Connors, Roger, Tom Smith, and Craig Hickman. *The Oz Principle: Getting Results Through Individual and Organizational Accountability.* Prentice Hall, 1994.
7. Davis, Brian, et al. *Successful Managers Handbook: Development Suggestions for Today's Managers.* Personnel Decisions International, 1992.
8. Gruhn, Paul. *Sell More Through Effective Technical Presentations—A pocket guide.* ISA, 2003.
9. Levi, Daniel. *Group Dynamics for Teams.* Sage Publications, 2001.
10. Patterson, Kerry, et al. *Crucial Conversations: Tools for Talking When Stakes are High.* McGraw-Hill, 2002.
11. Scholtes, Peter E., Brian L. Joiner, and Barbara J. Streibel. *The Team Handbook.* Third Edition. Oriel, 2003.
12. Smith, Graham. *Managing Software Contractors.* Brillig Systems, 2005. Website: www.brilligsys.com
13. Streibel, Barbara J. *The Manager's Guide to Effective Meetings.* McGraw-Hill, 2003.
14. Ulman, Joseph N. and Jay R. Gould. *Technical Reporting.* Third Edition. Holt, Rinehart, and Winston, 1972.
15. Wilson Learning Corporation. *Social Styles as a Global Phenomenon.* (White Paper) Wilson Learning Corporation, 1999. Web site: www.wilsonlearning.com (retrieved 06/21/2005)

## About the Author

**David J. Adler** is a Senior Engineering Consultant and Process Control Engineering group leader in Lilly's Engineering Technical Center. He has been with Lilly for 30 years and has had a wide variety of engineering and leadership assignments in development, tech service, project, process, and automation engineering. He received Lilly's "Engineering Excellence Award" in 2000 for his lifetime achievement in engineering technology. In his spare time, he is involved with Purdue University and local church student organizations. He received the Society of Hispanic Professional Engineers' Platinum Award for exemplary contributions to the Purdue student chapter in 2001.

# APPENDIX A
# Control Equipment Structure

*by Sam Herb*

## Topic Highlights

*Definitions*
*Single-loop Control*
*Large Systems*
*Central Computer Control*
*PLC Control*
*SCADA Systems*
*Distributed Control*
*Hybrid Control*
*PC Control*
*Control in the Field*

## A.1 Introduction

Talking about various controllers is like talking about various vehicles. There is no one best type of transportation; it is all in the application requirements. Sport cars, sedans, and pickup trucks all serve different functions. Tractor trailers offer flexibility, yet they are frequently hauled long distances in railway trains. Yachts and cruise ships have different purposes, as do ocean-going oil tankers, coastal freighters, and ships and barges on large rivers. When it comes to selecting suppliers, it's like automobile buyers—there is a large group who swears by a particular brand, and an equally large group who swears *at* that same brand. Controller and control systems are no different. They all control. There are different applications and different emotions—yes, even for "logical" engineers. For lack of a better way, I will arbitrarily approach this topic from an approximately historical perspective, which may explain the existence and purpose of control equipment.

## A.2 Definitions

Before we start, let's define some terms and acronyms often used for controllers to reduce confusion:

**SLC** – single-loop controller

**DAQ** – data acquisition

**SCADA** – supervisory control and data acquisition (pipelines, power distribution, water systems, etc., beyond the plant itself)

**DAC** – data acquisition and control (some now call this SCADA, but it is within a plant)

**DCS** – distributed control system

**PLC** – programmable logic controller

**PAC** – programmable automation controller (ARC term for the combined functionality of PLCs and PCs; some may call this a version of hybrid controller)

**PC Controller** – controller uses industrial personal computer (PC), maybe through an SBC

**SBC** – single board computer (controller); a printed circuit board that contains a complete computer, including processor, memory, I/O, and clock

**Hybrid controller** – architectures that blend benefits of both PLCs and DCSs; generally "small-scale DCSs"; the term is defined differently by different people.

**CIF** – control in the field; the control loop resides across a fieldbus only, not within a controller

## A.3 Single-loop Control

More than your home furnace control, which is relay on-off control, these modulating controllers were initially for operating valves. They were originally pneumatic/mechanical, then electromechanical, then electric, then electronic (vacuum tube, then transistor), and are now microprocessor-based. This later version makes some networking possible, as well as some more sophisticated control, approaching what is sometimes called "hybrid control."

Because the mechanism of the early controllers provided the motion of the pointer and pen of the indicator and recorder, the "faceplate" was integral to the control, and the many instrument companies called the entire assembly "the controller" (Figure A-1). This later influenced the instrument companies' approach to their respective "distributed control systems."

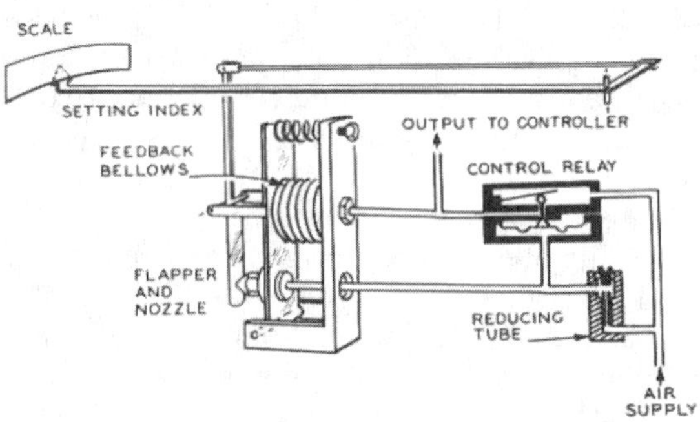


*Figure A-1: Measurement and Control Mechanism was One with the Indicator and Recorder*

Another design circumstance was the three modes of control as covered in the chapter entitled "Continuous Control," were determined by the interaction of the mechanism (Figure A-2). It did not matter if you had a restrictor, a capacity, or a needle valve, or if you later had a resistor, a capacitor, and a vacuum tube (voltage valve)—or transistor (current valve)—the technology was essentially the same.

To change from a proportional (P) controller, to a proportional-integral (PI) or a proportional-integral-derivative (PID) type controller, you had to change controllers, or at least change the chassis. If you wanted to change strategies, or even tune differently, you had to go out and buy more stuff. Later electronic versions still required different circuit boards. There was no way to merely turn off one of the modes. This is one of the reasons why so many automatic controllers were run in manual mode (some say 85–90%).

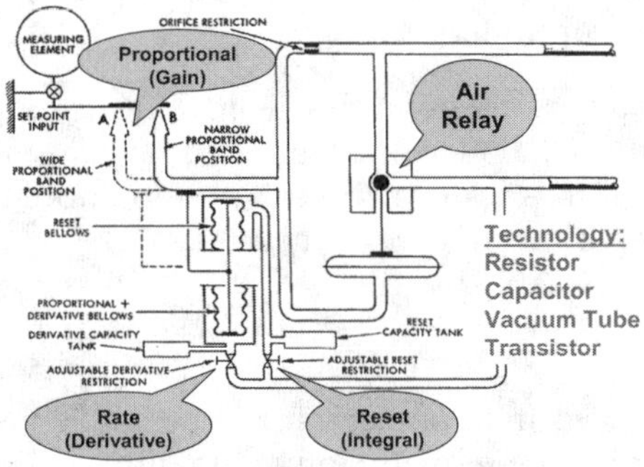


*Figure A-2: Interactive Mechanism, Pneumatic or Its Electronic Counterpart, Restricted Control Action Flexibility*

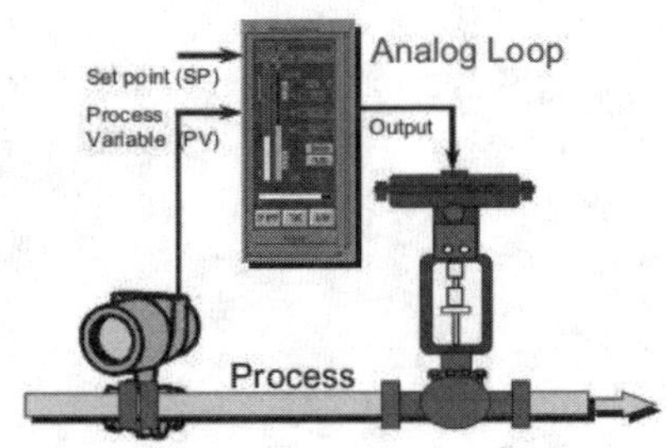


*Figure A-3: Traditional Single-loop Control*

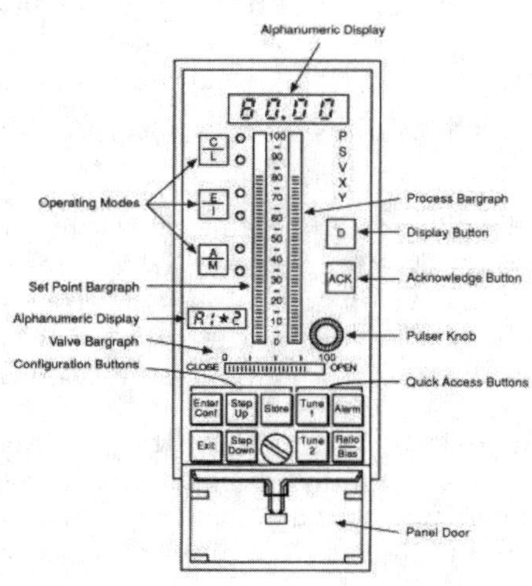


*Figure A-4: Microprocessors Made Powerful SLC Possible*

Until the microprocessor! Unlike the vacuum tube or the transistor, the microprocessor was not a "valve" but a "counter." This allowed a change in how the control algorithm worked, and permitted the so-called parallel algorithm shown in "Continuous Control." It also gave more choices in tuning loops. In addition, the microprocessor made more involved control strategies practical. It allowed plants to operate closer to their potential. Microprocessors led to the flexibilities of the distributed control systems and the hybrid controllers.

Single-loop controllers are a very economical way to control simple stand-alone unit processes such as dryers, small furnaces, and small boilers. Because of advances in microprocessors, they can readily be used in small networks as well as team with PLCs, where more sophisticated continuous control is needed along with larger amounts of discrete actions.

## A.4 Large Systems

When plants became very large and more complex, the challenge was for the operator to keep track of all the operations and coordinate the many individual unit processes. This meant the many individual single control loops required someone, or several people, to "tour the plant" and record all the data on a clipboard to make a decision (Figure A-5). Then another tour was needed to adjust all the valves and drives. This took time and limited effective plant production.

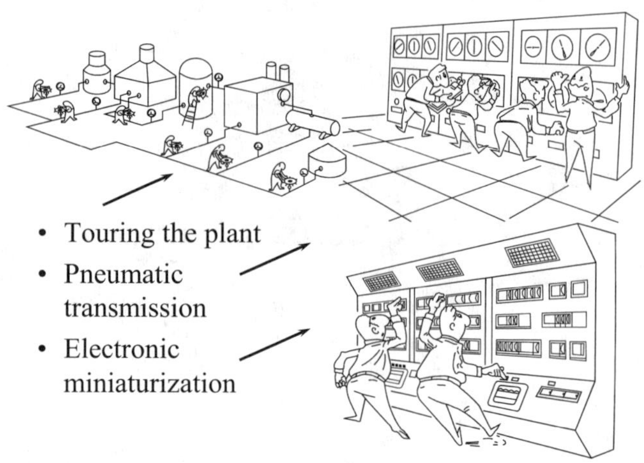


*Figure A-5: Limitations of a Single-Loop Control in Large Plants*

Following World War II, electronic controls became more rugged and practical for industrial environments. More measurements were becoming possible because the cost of sensors was coming down. This also allowed the industry to develop newer types of sensors for measuring parameters not previously measurable. Further, it was becoming possible to continuously measure more parameters online, rather than tediously take laboratory samples and wait for results.

The size of these new electronic controllers was smaller, so more of them could fit on a panel, in a smaller area. All of this led to a more complex control room, as well as the need to bring more wires to that location. This presented information management problems to the operators and the logistics challenge of signal management to the instrument engineer.

As changes in technology decreased the price of computers, their use became more common in large and complex facilities. This also allowed the further development of the single, centralized control room.

## A.5 Central Computer Control

Large central computers were important for large plants and power utilities. While these computers were now able to cope with the new data, most computers were designed for transactional businesses rather than navigational process control. During the 1960s and 1970s, two types of computers emerged for process control:

- Direct digital control
- Digitally directed analog control, more popularly called Supervisory Control

The central control room concept gave a much better picture of the overall plant operation, but when all the distant portions of the plant were connected to this one room, the cost was high because of the following factors:

- Many control cable runs, wire trays, and handling devices
- More complex engineering design
- Craft labor of installing lines and making terminations
- Problems making changes in the control strategy

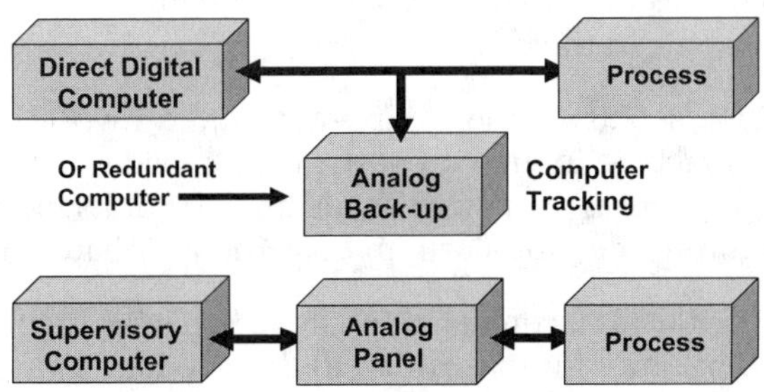


*Figure A-6: Central Computer with Analog Backup*

A serious problem also presented itself—a failure in the computer could shut down the entire plant! To overcome this limitation, backup controllers were often used with the computer system. As a result, companies frequently had to duplicate control systems (which meant buying two sets of controls for everything) to make systems reliable. This redundancy often required analog instruments to keep the plant running and made changes to control strategy more difficult.

Centralized computer control and monitoring had several advantages and disadvantages:

- On the plus side: there was a more organized central view of the operations; control strategies became more flexible; alarms became much more flexible and effective; and there was increased ability to have meaningful history of events.

- On the negative side: there was a large amount of wiring; there was considerable risk to the plant; and it was not very "scalable," meaning you could not scale up a little bit more without reprogramming the whole computer. All of these negatives cost money.

Computer-like control has become more necessary, however, because as each industry matures, it needs to optimize its processing methods. Cost of raw materials, cost of waste, cost of pollution, cost of compliance to government regulations—all are growing factors in the efficiencies of operations.

The technology of the microprocessor arrived and affected the controls industry for both process control with the DCS and factory automation with the PLC.

## A.6 PLC Control

As the acceptance of transistors increased, the "box of relays" grew to be programmable logic controllers. They were primarily used for discrete control in factory automation, and at one time 70% of all PLCs were used in automobile manufacturing. PLCs were designed to be stand alone, and essentially had start-and-stop operator interfaces and little need for communication networks. Each box was configured independently. They provided very fast scanning circuitry for robotic control and numeric control machines for manufacturing parts. Designed to perform in a work cell (workstation for one manufacturing employee), common practice was to have one PLC per function. Work cells in manufacturing are comparable to a unit process in a continuous control plant for such functions as burner management and reactor control.

The chapters "Discrete Input and Output Devices and General Manufacturing Measurements" and "Discrete and Sequence Control" provide more information about PLCs. A large advantage to PLCs has been their low unit price, because they have traditionally been sold as commodities. PLCs are usually installed by local third-party system integrators, and they have usually been configured in Ladder Logic (North America) or Boolean (Europe). Today most have migrated to the IEC 61131-3 "standard," using the five languages that make it easier to configure more sophisticated functions. There is more information on this topic in the chapter.

Another advantage of PLCs, at least in the factory-automation world, is the user's ability to choose an operator station (human-machine interface [HMI]) from a wide selection of different suppliers. For process control, the PLC supplier is likely to provide an HMI of its own design for better compatibility to the greater sophistication needed for alarm management or batch processing.

PLCs became very versatile for many applications such as packaging lines, which led them into food and pharmaceutical applications. This in turn led to the need for more sophisticated control and some single-loop functions. Companies demanded more sophisticated continuous control from PLC suppliers who have headed into the batch markets. The legacy of configuring each controller independently persists (Figure A-7). Most factory-automation applications did not need redundancy, so this has not been a strength of PLCs of the past. Sophisticated redundancy is often a requirement in most process control applications, and those PLC suppliers moving towards those market spaces have begun meeting the challenge.

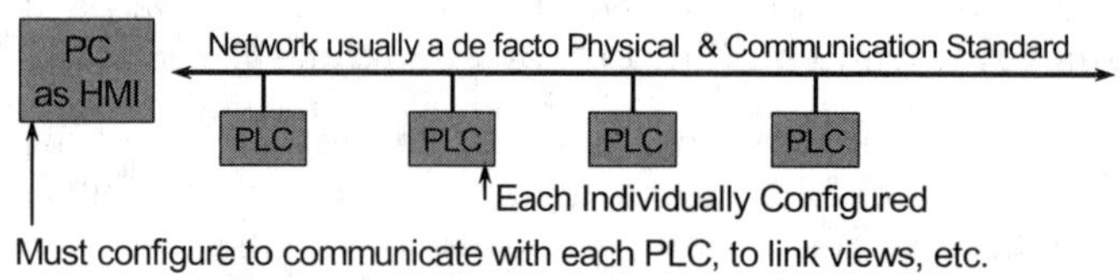


*Figure A-7: Typical PLC System Configuration*

As PLC applications became more complex, their suppliers have been developing appropriate networks and graphic operator interfaces.

## A.7 SCADA Systems

Unfortunately, when suppliers of HMIs began to refer their lash-up of PCs with PLCs as "SCADA" systems, confusion arose in the industry. This is in part because of the simple definition of this acronym for supervisory control and data acquisition. The original reference for systems of PCs with PLCs within a plant was data acquisition and control (DAC).

Traditional SCADA, which has been in existence for more than a half century, has always been used to transmit data over long distances. Common uses include power transmission systems (transmission and distribution), oil and gas pipelines, and water distribution systems. SCADA can be used in any situation where monitoring and control is done over large distances.

A system is made up of a master termination unit (MTU) and one or more remote termination units. The MTU will likely be in the control room, connected to the many functions needed there (Figure A-8). Beyond large control systems, there can be not only operators, maintenance folks, engineers and managers, but also accountants, government regulatory record keeping, and the many functions of custody transfer, depending upon the application.

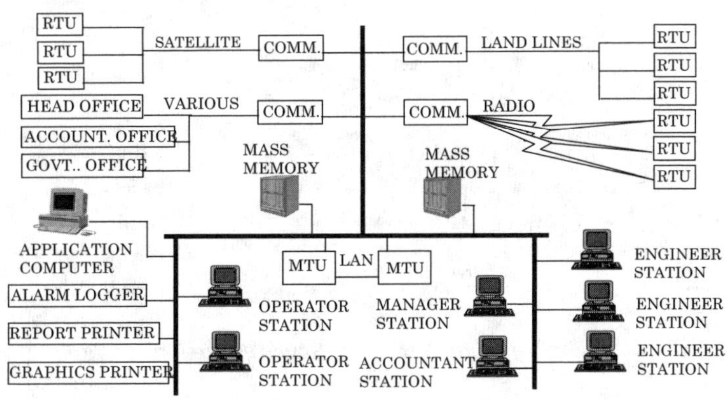


*Figure A-8: SCADA is for Applications Over Long Distances*

Systems transmit over many kinds of media such as landlines, phone systems (wire or wireless), microwave, radio, and satellite. Because of the distances, any process control done remotely must carefully be stand-alone control because of the time lag from distances and sharing of communications (scanning many RTUs). This is an essential feature, requiring specific communication techniques to ensure the integrity of information and direction given for any control action.

Also because of developments in microprocessor technology, RTUs today can come with many more functions than merely remotely reporting values and turning switches on and off as their earlier cousins did. There are different levels of "smart" RTUs, just as there are different and expanding capabilities with "smart" transmitters and end elements. Some very sophisticated remote stand-alone controllers can be added to the RTU functions.

## A.8 Distributed Control

The advent of distributed control came with the useful capabilities of emerging video technologies to display data and to allow an operator to initiate control actions "through the video" as well. The advantage of the central control room was it provided centralized *information* without having all the processing in one vulnerable location, thereby distributing the risk.

Distributed control also saved the cost and complexity of wiring by sending a digital signal through a single cable, which was used as a communication network (data highway) connecting the diverse portions of the plant. The magic of sending many signals through a single wire is an old technology, called the telegraph. The use of Morse code was really digitally communicating analog values (in contrast to analog voice-based information, like radio). There is more on this topic in the chapter "Digital Communications."

This new architecture permitted a *functional* distribution of the tasks among many processors, reducing the risk of everything failing at once. As capabilities for reducing ground loops emerged, it was also possible to allow *physical* distribution as well. These critical features began to open up many new possibilities for tying central information to local control in those plants where this is important.

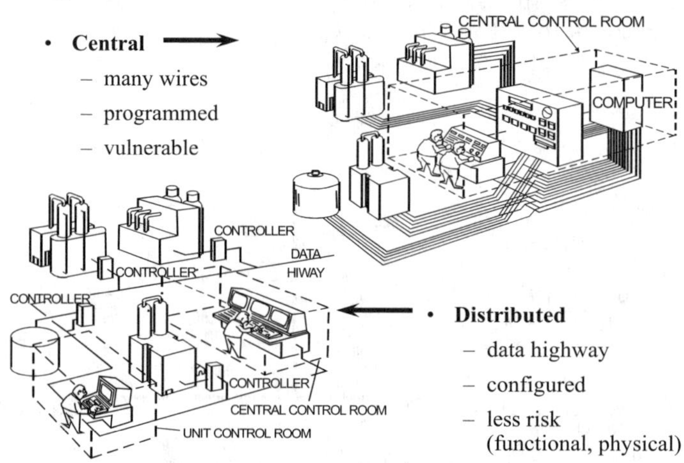


*Figure A-9: DCS Saved More Than Wires*

The central control room view of plant operation gives operators a single window to the process. Now operators no longer have to tour the plant. They can literally "let their fingers do the walking" as they call up each controller or group of controllers on a screen to check the progress of the process. If needed, they can easily make set point and output changes from the keyboard, as well as respond to any alarms if a process is "off normal."

Furthermore, if a plant process requires it, there also can be several operator stations along this network. A local operator station can be positioned in a specific portion of the plant, either working from the same data highway, or wired directly to a cluster of some loops of control.

On the plus side, distributed control and monitoring meant shorter wiring runs, no wires between controllers and control room, less risk of failure, and a more scalable system, if you wished to gradually grow the system without much cost of replacement.

On the negative side, these distributed control systems still must wire sensors and final elements to control cabinets, and interconnection between different vendors' components presented no small difficulty. This is the realm being addressed by digital I/O—called fieldbus, which is covered in "Digital Communications."

Remember the comment about the instrument companies looking at controllers as including the "faceplate"? Unlike PLC suppliers who needed only a "box" to do a specific function, the DCS suppliers were all instrument companies with the philosophy that the operator interface was integral to the function of process control. As a result, the control strategies are configured from the workstation, which was designed integral to the controllers (Figure A-10). This enables a single database and a more integrated operator interface for more sophisticated alarm management and other operator functions. Simplistically, operator actions in factory automation were traditionally "on-off," whereas in process control, the operator turned a valve while watching a gauge—the action was more "adjust until."

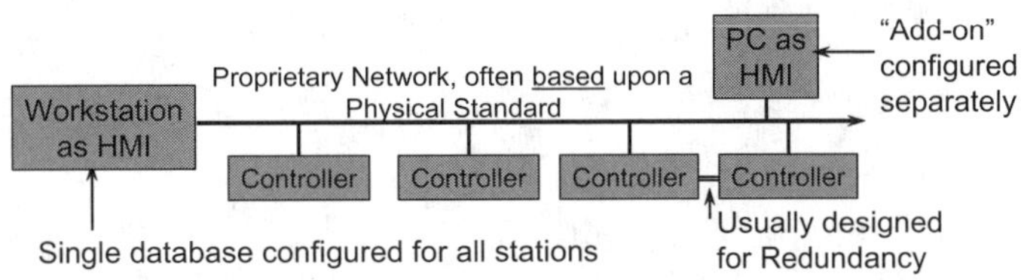


*Figure A-10: Typical DCS Configuration*

Another distinction was that PLCs had traditionally not needed redundancy or sophisticated communication. DCSs, however, usually did need sophisticated redundancy, and so-called "real-time" communication. No computer can do "real-time," but at best has to act in "real-enough-time," because the process is actively "flowing." To achieve this, the instrument suppliers had to design controller, workstations, and communication networks as a single package. No standard communication method was fast enough, so they each modified the standards to achieve the "real-enough-time" networks to match their individual controller and workstation designs.

DCSs originally used UNIX-based workstations, often very proprietary, to allow for the "real-enough-time" seriously needed, as well as for the unique navigational process functions not accounted for in discrete-based transactional "business machines." As companies began using Microsoft throughout their plants, they began demanding PC-based workstations. This presented several significant challenges to the DCS suppliers, which they began to overcome because PCs were also transaction based.

## A.9 Hybrid Control

Wouldn't it be nice to have the cost advantage and simplicity of the PLC combined with the sophistication and scalability of the DCS? Wouldn't it be nice to have the power and sophistication of DCS control on a small scale? Thanks to the microprocessor, these capabilities are emerging. What we haven't mentioned yet was the emerging capabilities of combining a few process control loops and some discrete interlocking that would be helpful for those small-to-midsized industries that have stand-alone unit processes, such as heat treating plants and some food industry functions.

A new market space opened to provide improved production capabilities for these smaller plants, which traditionally had to "kludge together" some sort of control system with a very limited budget. Having a decent answer for these small plants is so new that a perfect definition is wanting. Now for the big question: What is a hybrid system? It depends upon whom you ask. If you look at the many articles in the trade journals and the marketing words by the various vendors, you will begin to realize there are many different understandings of the phrase.

That is why there is some confusion—even to the point where one control magazine suggested the whole idea is really a hoax by the instrument companies to sell more control systems! You must realize that in the early DCS days of the 1970s and 1980s, there were still DCSs that could not do discrete actions, or at least not very well. PLCs at that time were notable for their poor implementation of analog control, especially multiloop control of even low sophistication. Clearly *both* worlds have been moving towards each other, but both typically bring the baggage of their legacies even in their newer designs. That is why there are still several definitions, which *almost* sound alike, but are somewhat different.

Definitions include:

- Industry (ARC Research, Figure A-11)
- Input and output capability
  - Analog + Discrete I/O
- Function (batch capability)
- Architecture
  - Advantages of both PLCs and DCSs
  - Few disadvantages of either (hopefully)

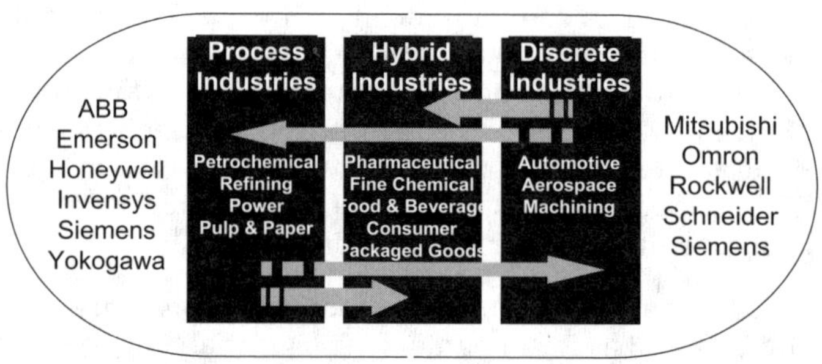


*Figure A-11: Hybrid Defined by Industries (From ARC Advisory Group Strategies, June 2001)*

Hybrid control first appeared (by name) in the early 1990s with a DCS that allowed users to add any supplier's UNIX- or PC-based workstation to the control system. It uniquely allowed the controllers to be configured as a system, and diagnostics were not limited to each of the "boxes," but reported on the system as a whole, like a DCS. By now "out-of-the-box" workstations could be used, especially if a plant had standardized with a particular workstation supplier. Nevertheless, there was still the issue of needing navigational-type process actions, especially for alarm handling requirements and process batch-like functions.

By the turn of the millennium most suppliers offered PC-based workstations, so this has become the standard. It added to the commercial-off-the-shelf (COTS) needs of the industry. Nevertheless, to achieve the needed process functions, most suppliers also offer their own modified software for these PCs and recommend industrial-grade hardware to survive the harsh conditions of the industries served. This is true of the instrument suppliers and the PLC suppliers who have entered the so-called "hybrid industries" markets.

ARC, which defined "hybrids" as an industry category (Figure A-11), began using the term process automation controller (PAC) to describe the combined functionality of PLCs and PCs. A PAC could include equipment from PLC suppliers, but might be stretched to include equipment from suppliers of "small DCSs . . . and perhaps could also include those who didn't exactly fit either supplier category." Also notice that Figures A-7, A-10, and A-12 look nearly alike, except for the words used. The distinctions are becoming much more subtle. Like most things in this business, begin with the application and functions you need rather than the labels when buying a control system. My recommendation? Start with the simplest answer for your needs and build upwards as you discover new requirements.

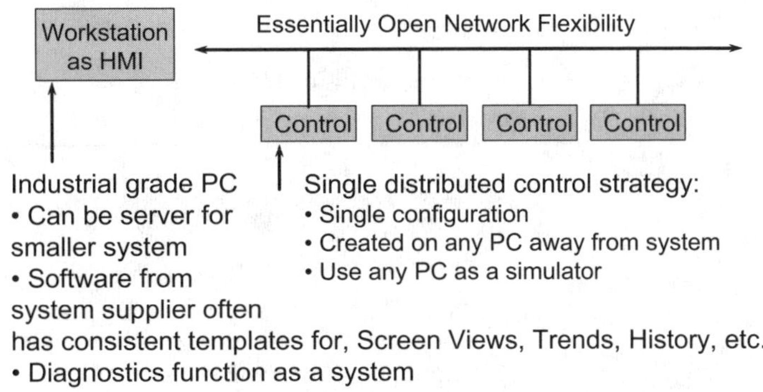


*Figure A-12: Hybrid System Architecture Blends PLC Simplicity with DCS Sophistication*

## A.10 PC Control

Not to be overlooked is the realm of PC controllers. Microsoft technology—and the ability to take advantage of the inexpensive technology—has been hovering over the process control world ever since PCs became so popular. Until recently PC controllers have been limited by the vulnerability of delicate moving components, not practical in many of the harsh environments of the process industries. Nevertheless, this has become a Microsoft world.

The next generation of controllers will be influenced significantly by Microsoft. The trend will be towards off-the-shelf hardware, and away from proprietary systems. Value added by controls vendors will be to the software, which will not only provide control capability, but will still have to overcome the obstacles to "real" real-time. There will also be a significant need for standards in many places for this to be fully realized. This is happening more with PLCs giving way to "soft logic" in factory-automation applications, but some process applications are also suitable. ARC Advisory Group refers to this as Open Control Software (OCS).

Although PC control has the potential to "liberate" the user from being held hostage by a single supplier, there are some remaining considerations about robustness. "Industrial strength" for processes is far more severe than the factory floor where some PC controllers have replaced PLCs. Unlike factory automation, which usually shares the environment with people, process control often occurs in corrosive atmospheres dangerous to people, where vibration is frequently significant. Often, the PC's general-purpose operating system is not stable enough for control. PC-controlled installations are forced to handle system crashes and unplanned rebooting.

As the control and automation market continues to evolve, technologies developed and honed by PC-based control suppliers are finding their way into a diverse range of platforms such as Compact PCI, VME, PC/104, and custom single-board computer (SBC).

## A.11 Control in the Field

We complete the circle, which is made possible through the development of digital fieldbus communications, particularly Foundation Fieldbus. Foundation Fieldbus (see "Industrial Networks") provides control function blocks, allowing a complete control loop to exist among the transmitters and end elements without a specific controller "box" (Figure A-13). With this architecture, a PC can be placed on the fieldbus itself and used as a workstation. The neat part is the field devices can come from different suppliers, because no matter the supplier, the function blocks are written to the same standard so they readily communicate with each other. Nevertheless, the actual content of the function blocks can have unique characteristics, allowing suppliers to differentiate their products.

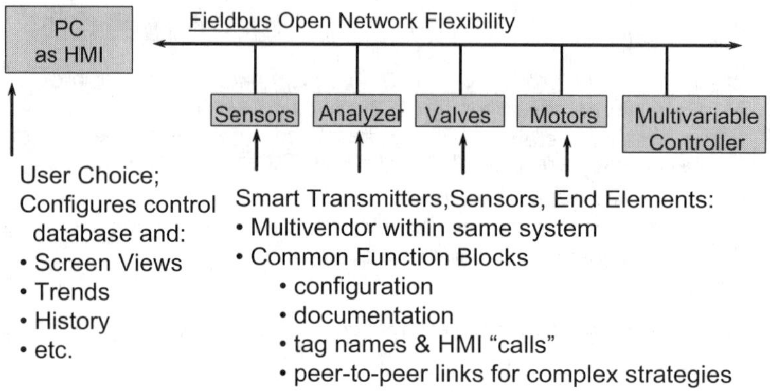


*Figure A-13: Foundation Fieldbus Itself can be the Controller*

This capability can be very useful in small stand-alone applications. It also allows users to incorporate many separate stand-alone systems into a larger system, with access from a central control room. Fragmenting the control hardware may be counterproductive for more coordinated or multivariable control schemes. The arrangement, however, does not preclude the use of multivariable controllers for more complex control strategies. At present, the limitation is that only a few devices (about six) can exist on the same segment of the bus, with no peer-to-peer communication to other segments. The arrangement may not be practical for more intricate batch applications.

Control strategies can be developed among several transmitters and end elements as shown in Figure A-14. Although some have concerns that the vulnerability of "inner signals" of controllers are at the mercy of the field network, the robustness of installed applications seems to be placing those fears to rest.

Some valve companies have released designs that have placed flow measurement across the valve along with temperature measurement, feeding controller chips and providing a signal directly to the valve actuator—all inside the bonnet of the valve.

## A.12 What Next?

Of course new technologies will continually appear to overcome limitations, but I doubt any one of the types covered will fully disappear. Each has some very strong advantages depending upon the application. Remember, it is always the application, and the functions you need, *not* the labels. The good news is that today the *choices* are growing, and generally the prices are coming down, especially when you look at total cost of ownership.

## A.13 References

Herb, Samuel M. *Understanding Distributed Processor Systems for Control.* ISA, 1999.

Boyer, Stuart A. *SCADA: Supervisory Control and Data Acquisition, 3rd edition.* ISA, 2004.

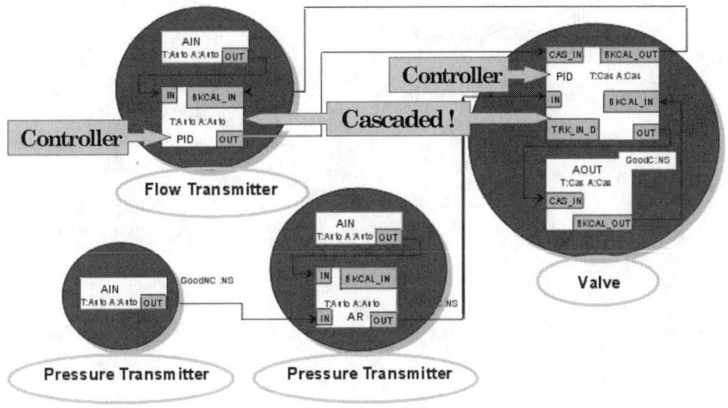


*Figure A-14: Control Strategy Fully "in the field"*

For further information on the evolution of control see:

Feeley, J., et al. "100 Years of Process Automation." *Control Magazine.* (Vol. XII, No. 12), December 1999 (Special Issue).

### Standards

ANSI/ISA-50 Series, Parts 2-6 - *Fieldbus Standard for Use in Industrial Control Systems.*

IEC 61158 Series, Parts 1-6 - *Digital Data Communications for Measurement and Control – Fieldbus for Use in Industrial Control Systems.*

## About the Author

**Samuel M. "Sam" Herb**, PE, is currently with Invensys Process Systems. Herb has worked with Honeywell, Leeds & Northrup, Moore Process Automation, and Siemens during his career. He has worked in the utility and process controls industries for nearly five decades, with various roles in marketing, business development, marketing communications, control applications, product management, systems engineering, project management, product evaluation, technical publications, and education.

Herb holds a BS/EE from Drexel University, is a senior life member of ISA, and is the author of dozens of journal articles for a variety of technical publications, seminars, and courses in instrumentation. In addition, he consulted on several ISA videotape series, authored chapters in the Practical Guide Series on Continuous Control, authored the textbook *Understanding Distributed Processor Systems for Control,* and developed five online courses and CDs on control systems. Herb received the ISA Eagle and Distinguished Society service awards and the Donald P. Eckman award for education.

# Index